STATISTICS

Concepts and Controversies

Fifth Edition

David S. Moore
Purdue University

W. H. Freeman and Company
NEW YORK

Acquisitions editor: Patrick Farace

Assistant editor: Danielle Swearengin

Marketing Manager: Claire Pearson

Project editor: Diane Davis

Cover and text designer: Blake Logan

Cover and text illustrations: Mark Chickinelli

Illustration coordinator: Bill Page

Production coordinator: Julia DeRosa

Photo editors: Vikii Wong, Elyse Rieder

Media/Supplements editor: Mark Santee

Illustration and composition: Publication Services

Manufacturing: R R Donnelley & Sons Company

Library of Congress Cataloging-in-Publication Data
Moore, David S.
Statistics: concepts and controversies / David S. Moore.—5th ed.
p. cm.
Includes bibliographical references and index.
ISBN 0-7167-4008-7
1. Statistics. I. Title

QA276.12.M66 2000
519.5—dc21

Printed in the United States of America

Second printing 2001

W. H. Freeman and Company
41 Madison Avenue, New York, NY 10010
Houndmills, Basingstoke RG21 6XS, England

Brief Contents

Part III Chance 343

Part IV Inference 417

*This material is optional.

Contents

5 Experiments, Good and Bad 71

6 Experiments in the Real World 89

7 Data Ethics 108

8 Measuring 126

9 Do the Numbers Make Sense? 146

*This material is optional.

*This material is optional.

To the Teacher

Statistics as a Liberal Discipline

Statistics: Concepts and Controversies (*SCC*) is a book on statistics as a liberal discipline, that is, as part of the general education of nonmathematical students. The book grew out of my experience in developing and teaching a course for freshmen and sophomores from Purdue University's School of Liberal Arts. I'm pleased that other teachers have found *SCC* useful for an unusually diverse readership, extending as far as students of philosophy and medicine. This fifth edition is an extensive revision of the text, amounting to almost a new book. It retains, however, the goals of the original: to present statistics not as a technical tool but as part of the intellectual culture that educated people share.

Statistics among the liberal arts

Statistics has a widespread reputation as the least liberal of subjects. When statistics is praised, it is most often for its usefulness. Health professionals need statistics to read accounts of medical research; managers need statistics because efficient crunching of numbers will find its way to the bottom line; citizens need statistics to understand opinion polls and the Consumer Price Index. Because data and chance are omnipresent, our propaganda line goes, everyone will find statistics useful, and perhaps even profitable.

This is true. I would even argue that for most students the conceptual and verbal approach in *SCC* is better preparation for future encounters with statistical studies than the usual methods-oriented introduction. The joint curriculum committee of the American Statistical Association and the Mathematical Association of America recommends that any first course in statistics "emphasize the elements of statistical thinking" and feature "more data and concepts, fewer recipes and derivations." *SCC* does this, with the flavor appropriate to a liberal education: more concepts, more thinking, only simple data, fewer recipes, and no formal derivations.

There is, however, another justification for learning about statistical ideas: statistics belongs among the liberal arts. A liberal education emphasizes fundamental intellectual skills, that is, general methods of inquiry that apply in a wide variety of settings. The traditional liberal arts present such methods: literary and historical studies, the political and social analysis of human societies, the probing of nature by experimental science, the power of abstraction and deduction in mathematics. Statistics belongs among the liberal arts because reasoning from uncertain empirical data is a similarly general intellectual method. *Data* and *chance,* the topics of this book, are pervasive aspects of our experience. Though we employ the tools of mathematics to work with data and chance, the mathematics implements ideas that are not strictly mathematical. Indeed, psychologists argue convincingly that mastering formal mathematics does little to improve our ability to reason effectively about data and chance in everyday life.

SCC is shaped, as far as the limitations of the author and the intended readers allow, by the view that statistics is an independent and fundamental intellectual method. The focus is on statistical thinking, on what others might call *quantitative literacy* or *numeracy.*

The nature of this book

There are books on statistical theory and books on statistical methods. This is neither. *SCC* is a book on statistical ideas and statistical reasoning and on their relevance to public policy and to the human sciences from medicine to sociology. I have included many elementary graphical and numerical techniques to give flesh to the ideas and muscle to the reasoning. Students learn to think about data by working with data. I have not, however, allowed technique to dominate concepts. My intention is to teach verbally rather than algebraically, to invite discussion and even argument rather than mere computation, though some computation remains essential. The coverage is considerably broader than that of traditional statistics texts, as the table of contents reveals. In the spirit of general education, I have preferred breadth to detail.

Despite its informal nature, *SCC* is a textbook. It is organized for systematic study and has abundant exercises, many of which ask students to offer a discussion or make a judgment. Even those admirable students who seek pleasure in uncompelled reading should look at the exercises as well as the text. Teachers should be aware that the book is more serious than its low mathematical level suggests. The emphasis on ideas and reasoning asks more of the reader than many recipe-laden methods texts.

New in this edition

I hope that this largely new version of a classic text fits our current teaching environment while continuing to present statistics to nonmathematical readers as an aid to clear thinking in personal and professional life. Here is an overview of what is new.

Structure. Previous editions of *SCC* contained 8 chapters; this fifth edition contains **25 chapters** in 4 parts and **4 part review chapters.** Each chapter concentrates on a single coherent set of ideas and ends with a single set of exercises. The part review chapters are entirely new. They are deliberately straightforward in presenting an **outline of skills** students should have and **additional exercises** that emphasize basic ideas and skills. I hope that this new structure will guide students more clearly. The review chapters also suggest **projects,** often suitable for teams of students, that require gathering information or producing data and emphasize writing a clear report on the work done.

Content. The content of the new edition, unlike its structure, represents evolutionary rather than revolutionary change. For example:

- Sections on "How to live with nonsampling errors" (Chapter 4) and "How to live with observational studies" (Chapter 5) alert students to the ways good practice deals with these difficulties.
- Data ethics receives a unified treatment in Chapter 7.
- Chapter 17, "Thinking about Chance," says more about the nature of probability, including personal probabilities.
- Descriptive material on relations among categorical variables has been joined with a new introduction to the chi-square test in Chapter 24.
- The early chapters on producing data say a bit more about confidence statements and statistical significance, helpful for instructors who (like me) find that much of Part IV must be omitted. There are of course many more improvements in detail.

Pedagogy. *SCC* is, I hope, a serious book that is light in tone and interesting to read. Many changes in the new edition advance this goal.

- Two-thirds of the **examples** and three-fourths of the **exercises** are new, as is appropriate in a book that interacts with contemporary society.
- Each chapter opens with a short **case study** and ends with a short **Statistics in Summary** narrative of the big ideas of the chapter.
- New and more numerous **cartoons** enliven the pages.
- **Statistical Controversies** features in nine chapters introduce disputed issues to which the ideas in the adjoining text are relevant. More than 30 short marginal **think pieces** offer interesting and often amusing anecdotes about statistics and its impact.

- The numerous **source notes** at the end of the book are organized by page number rather than pointed to by footnotes.
- **Answers to odd-numbered exercises** appear for the first time in the back of the book. I hope that instructors will remind students that these short answers are often guides rather than full discussions of the exercises.

The **Exploring the Web** boxes near the end of most chapters deserve special comment. Even the least quantitative students are now familiar with Web browsers and accustomed to using them. I choose to avoid statistical software in teaching from *SCC*, though I require a "two-variable statistics" calculator to reduce the pain of finding means, standard deviations, and correlations. I do ask students to use the Internet as part of each assignment, and the Exploring the Web boxes point to some appropriate sites. I have tried to choose sites that are relatively permanent, though of course some will move or disappear. Instructors may base some student work on these sites or simply leave them for curious students to explore.

Supplements

The new edition of *SCC* is accompanied by an extended set of supplements to help teachers and students:

Instructor's Guide with Test Bank. (ISBN 0-7167-4235-7) This guide contains **full solutions** of all exercises, prepared by Darryl Nester of Bluffton College, who also prepared the briefer in-text answers. I have added **hints on teaching** this somewhat nonstandard material and a good supply of **additional examples** for use in teaching.

The test bank included in the *Instructor's Guide* contains more than **900 multiple-choice items** for the use of instructors who must give exams that are at least in part multiple-choice. Many of these items are taken from my own exams, which are in multiple-choice format because enrollments of 470 students per semester make machine scoring necessary. The test bank items are included in the *Instructor's Guide* and are also available on CD-ROM.

***SCC* Web site.** The Web site www.whfreeman.com/scc contains many resources for both instructors and students. Instructors who adopt *SCC* receive full access by registering. Students who purchase the *Media and Activities Supplement* described on the next page gain access to the student material. The **data sets** from the text are here in ASCII (plain text) format. The site also contains a version of **EESEE** (the Electronic Encyclopedia of Statistical Examples and Exercises), a rich source of additional case studies with data and exercises for students. Instructors will find images of all **statistical graphics** from *SCC* for use in PowerPoint or other presentations.

The Web site also contains a number of Java **applets,** interactive statistical demonstrations and calculators that can enhance instruction, particularly if instructors choose (as I do) to avoid specialized software. Four of these applets, chosen for their usefulness with *SCC*, are open to all so that students can freely use them. The Exploring the Web boxes point students to these applets.

Media and Activities Supplement. (ISBN 0-7167-4234-9) This student supplement, by Dennis Pearl of Ohio State University, links EESEE to *SCC*, provides laboratory activities, extra homework problems, and several applet exercises, and it offers additional project ideas. It also includes a **Web password** giving students access to EESEE and other student resources on the *SCC* Web site.

Student version of Minitab. (ISBN 0-7167-4331-0) Teachers who wish to use a statistical software package in the course can adopt *SCC* bundled with the student version of Minitab.

Acknowledgments

The staff of W. H. Freeman and Company, especially Patrick Farace, Diane Davis, and Blake Logan, have done their usual excellent job in editing, designing, and producing the book. I am grateful to many colleagues who commented on successive drafts of the manuscript:

Richard Anderson-Sprecher, University of Wyoming
Marigene Arnold, Kalamazoo College
John Ferdinands, Calvin College
William S. Griffith, University of Kentucky
Jane L. Harvill, Mississippi State University
Robert E. Johnson, Virginia Commonwealth University
Vivian Lew, UCLA
Dave Olwell, University of Minnesota
Dennis Pearl, The Ohio State University
William S. Rayens, University of Kentucky
Timothy J. Robinson, Virginia Tech
Joseph P. Sedlacek, Kirkwood Community College
Jim Stallard, University of Calgary
David S. Wallace, Ohio University
Larry Wasserman, Carnegie Mellon University
Sheila O'Leary Weaver, University of Vermont
R. Webster West, University of South Carolina
Tom Wonacott, University of Western Ontario
Tilaka N. Vijithakumara, Illinois State University
Jill C. Zimmerman, Manchester Community College

Prelude

Making Sense of Statistics

Statistics is about data. Data are numbers, but they are not "just numbers." *Data are numbers with a context.* The number 10.5, for example, carries no information by itself. But if we hear that a friend's new baby weighed 10.5 pounds at birth, we congratulate her on the healthy size of the child. The context engages our background knowledge and allows us to make judgments. We know that a baby weighing 10.5 pounds is quite large, and that a human baby is unlikely to weigh 10.5 ounces or 10.5 kilograms. The context makes the number informative.

Statistics uses data to gain insight and to draw conclusions. The tools are graphs and calculations, but the tools are guided by ways of thinking that amount to educated common sense. Let's begin our study of statistics with a rapid and informal guide to coping with data and statistical studies in the news media and in the heat of political and social controversy. We will examine the examples in this prelude in more detail later.

Data beat anecdotes

Belief is no substitute for arithmetic.
HENRY SPENCER

An anecdote is a striking story that sticks in our minds exactly because it is striking. Anecdotes humanize an issue, so news reports usually start (and often stop) with anecdotes. But anecdotes are weak ground for making up your mind—they are often misleading exactly because they are striking. Always ask if a claim is backed by data, not just by an appealing personal story.

Does living near power lines cause leukemia in children? The National Cancer Institute spent 5 years and $5 million gathering data on the question. Result: no connection between leukemia and exposure to magnetic fields of the kind produced by power lines. The editorial that accompanied the study report in the *New England Journal of Medicine* thundered, "It is time to stop wasting our research resources" on the question.

Now compare the impact of a television news report of a 5-year, $5 million investigation against a televised interview with an articulate mother whose child has leukemia and who happens to live near a power line. In the public mind, the anecdote wins every time. Be skeptical. Data are more reliable than anecdotes because they systematically describe an overall picture rather than focusing on a few incidents.

I'm tempted to add, "Data beat self-proclaimed experts." The idea of balance held by much of the news industry is to present a quick statement by an "expert" on either side. We never learn that one expert expresses the consensus of an entire field of science, while the other is a quack with a special-interest axe to grind. As a result of the media's taste for conflict, the public now thinks that for every expert there is an equal and opposite expert. If you really care about an issue, try to find out what the data say and how good the data are. Many issues do remain unsettled, but many others are unsettled only in the minds of people who don't care about evidence. You can start by looking at the credentials of the "experts" and at whether the studies they cite have appeared in journals that require careful outside review before they publish a claim.

Where the data come from is important

Figures won't lie but liars will figure.
CHARLES GROSVENOR

Data are numbers, and numbers always seem solid. Some are and some are not. Where the data come from is the single most important fact about any statistical study. When Ann Landers asks readers of her advice column whether they would have children again and 70% of those who reply shout "No," you should just amuse yourself with Ann's excerpts from tear-stained letters describing what beasts the writers' children are. Ann Landers is in the entertainment business. Her invitation attracts parents who regret having their children. Most parents don't regret having children. We know this because opinion polls have asked large numbers of parents, chosen at random to avoid attracting one opinion or another. Opinion polls have their problems, as we will see, but they beat just asking upset people to write in.

Even the most reputable publications have not been immune to bad data. The *Journal of the American Medical Association* once printed an article claiming that pumping refrigerated liquid through tubes in the stomach relieves ulcers. The patients did respond, but only because patients often respond to *any* treatment given with the authority of a trusted doctor. That is, placebos (dummy treatments) work. When a skeptic finally tried a properly controlled study in which some patients got the tube and some got a placebo, the placebo actually did a bit better. "No comparison, no conclusion" is a good starting point for judging medical studies. I would be skeptical about the current boom in "natural remedies," for example. Few of these have passed a comparative trial to show that they are more than just placebos sold in bottles bearing pretty pictures of plants.

Beware the lurking variable

I have enough money to last me the rest of my life,
unless I buy something.

JACKIE MASON

You read that crime is higher in counties with gambling casinos. A college teacher says that students who took a course online did better than the students in the classroom. Government reports emphasize that well-educated people earn a lot more than people with less education. Don't jump to conclusions. Ask first, "What is there that they didn't tell me that might explain this?"

Crime is higher in counties with casinos, but it is also higher in urban counties and in poor counties. What kind of counties are casinos in? Did these counties have high crime rates before the casino arrived? The online students did better, but they were older and better prepared than the in-class students. No wonder they did better. Well-educated people do earn a lot. But educated people have (on the average) parents with more education and more money than the parents of poorly educated people have. They grew up in nicer places and went to better schools. These advantages help them get more education and would help them earn more even without that education.

All these studies report a connection between two variables and invite us to conclude that one of these variables influences the other. "Casinos increase crime" and "Stay in school if you want to be rich" are the messages we hear. Perhaps these messages are true. But perhaps much of the connection is explained by other variables lurking in the background, such as the nature of counties that accept casinos and the advantages that highly educated people were born with. Good statistical studies look at lots of background variables. This is tricky, but you can at least find out if it was done.

Variation is everywhere

When the facts change, I change my mind. What do you do, sir?

JOHN MAYNARD KEYNES

If a thermometer under your tongue reads higher than 98.6° F, do you have a fever? Maybe not. People vary in their "normal" temperature. Your own temperature also varies—it is higher around 6 A.M. and lower around 6 P.M. The government announces that the unemployment rate rose a tenth of a percent last month and that new home starts fell by 3%. The stock market promptly jumps (or sinks). Stocks are jumpier than is sensible. The government data come from samples that give good estimates but not the exact truth. Another run of the same samples would give slightly different answers. And economic facts jump around anyway, because of weather, strikes, holidays, and all sorts of other reasons.

Many people join the stock market in overreacting to minor changes in data that are really nothing but background noise. Here is Arthur Nielsen, head of the country's largest market research firm, describing his experience:

> *Too many business people assign equal validity to all numbers printed on paper. They accept numbers as representing Truth and find it difficult to work with the concept of probability. They do not see a number as a kind of shorthand for a range that describes our actual knowledge of the underlying condition.*

Variation is everywhere. Individuals vary; repeated measurements on the same individual vary; almost everything varies over time. Ignore the pundits who try to explain the deep reasons behind each day's stock market moves, or who condemn a team's ability and character after a game decided by a last-second shot that did or didn't go in.

Conclusions are not certain

> *As far as the laws of mathematics refer to reality they are not certain, and as far as they are certain they do not refer to reality.*
>
> ALBERT EINSTEIN

Because variation is everywhere, statistical conclusions are not certain. Most women who reach middle age have regular mammograms to detect breast cancer. Do mammograms really reduce the risk of dying of breast cancer? Statistical studies of high quality find that mammograms reduce the risk of death in women aged 50 to 64 years by 26%. That's an average over all women in the age group. Because variation is everywhere, the results are different for different women. Some women who have mammograms every year die of breast cancer, and some who never have mammograms live to 100 and die when they crash their motorcycles.

What the summary study actually said was "mammography reduces the risk of dying of breast cancer by 26% (95% confidence interval, 17% to 34%)." That 26% is, in Arthur Nielsen's words, "shorthand for a range that describes our actual knowledge of the underlying condition." The range is 17% to 34%, and we are 95% confident that the truth lies in that range.

We're pretty sure, in other words, but not certain. Once you get beyond news reports, you can look for words like "95% confident" and "statistically significant" that tell us that a study did produce findings that, while not certain, are pretty sure.

Data reflect social values

It's easy to lie with statistics. But it is easier to lie without them.
FREDERICK MOSTELLER

Good data do beat anecdotes. Data are more objective than anecdotes or loud arguments about what might happen. Statistics certainly lies on the factual, scientific, rational side of public discourse. Statistical studies deserve more weight than most other evidence about controversial issues. There is, however, no such thing as perfect objectivity. Statistics shares a social context that influences what we decide to measure and how we measure it.

Suicide rates, for example, vary greatly among nations. It appears that much of the difference in the reported rates is ascribable to social attitudes rather than to actual differences in suicides. Counts of suicides come from death certificates. The officials who complete the certificates (details vary depending on the state or nation) can choose to look more or less closely at, for example, drownings and falls that lack witnesses. Where suicide is stigmatized, deaths are more often reported as accidents. Countries that are predominantly Catholic have lower reported suicide rates than others, for example. Japanese culture has a tradition of honorable suicide as a response to shame. This tradition leads to better reporting of suicide in Japan because it reduces the stigma attached to suicide. In other nations, changes in social values may lead to higher suicide counts. It is becoming more common to view depression as a medical problem rather than a weakness of character and suicide as a tragic end to the illness rather than a moral flaw. Families and doctors then become more willing to report suicide as the cause of death.

Social values influence data on matters less sensitive than suicide. The percent of people who are unemployed in the United States is measured each month by the Bureau of Labor Statistics, using a large and very professional sample of people across the country. But what does it mean to be "unemployed"? It means that you don't have a job even though you want a job and have *actively looked for work in the last 2 weeks.* If you went 2 weeks without seeking work, you are not unemployed; you are "out of the labor force." This definition of unemployment reflects the value we attach to working. A different definition might give a very different unemployment rate.

My point is not that you should mistrust the unemployment rate. The definition of "unemployment" has been stable over time, so that we can see trends. The definition is reasonably consistent across nations, so that we can make international comparisons. The data are produced by professionals free of political interference. The unemployment rate is important and useful information. My point is that not everything important can be reduced to numbers and that reducing things to numbers is done by people influenced by many pressures, conscious and unconscious.

Statistics and You

What Lies Ahead in This Book

This isn't a book about the tools of statistics. It is a book about statistical ideas and their impact on everyday life, public policy, and many different fields of study. You will learn some tools, of course. Life will be easier if you have in hand a *calculator with built-in statistical functions.* Specifically, you need a calculator that will find means, standard deviations, and correlations. Look for a calculator that claims to do "two-variable statistics" or mentions "correlation." On the other hand, you need little formal mathematics. If you can read and use simple equations, you are in good shape. Be warned, however, that you will be asked to think. Thinking exercises the mind more deeply than following mathematical recipes. *Statistics: Concepts and Controversies* presents statistical ideas in four parts:

I. **Data production** describes methods for producing data that can give clear answers to specific questions. Where the data come from really is important—basic concepts about how to select samples and design experiments are the most influential ideas in statistics.

II. **Data analysis** concerns methods and strategies for exploring, organizing, and describing data using graphs and numerical summaries. You can learn to look at data intelligently even with quite simple tools.

III. **Probability** is the language we use to describe chance, variation, and risk. Because variation is everywhere, probabilistic thinking helps separate reality from background noise.

IV. **Statistical inference** moves beyond the data in hand to draw conclusions about some wider universe, taking into account that variation is everywhere and that conclusions are uncertain.

Statistical ideas and tools emerged only slowly from the struggle to work with data. Two centuries ago, astronomers and surveyors faced the problem of combining many observations that, despite the greatest care, did not exactly match. Their efforts to deal with variation in their data produced some of the first statistical tools. As the social sciences emerged in the nineteenth century, old statistical ideas were transformed and new ones were invented to describe the variation in individuals and societies. The study of heredity and of variable populations in biology brought more advance. The first half of the twentieth century gave birth to statistical designs for producing data and to statistical inference based on probability. By midcentury it was clear that a new discipline had been born. As all fields of study place more emphasis on data and increasingly recognize that variability in data is unavoidable, statistics has become a central intellectual method. Every educated person should be acquainted with statistical reasoning. Reading this book will enable you to make that acquaintance.

STATISTICS

Concepts and Controversies

Part II

Calvin and Hobbes by Bill Watterson

Producing Data

You and your friends are not typical. Your taste in music, for example, is probably not my taste. Of course, I and my friends are also not typical. To get a true picture of the country as a whole (or even of college students), we must recognize that the picture may not resemble us or what we see around us. We need *data*. Data from retail stores show that the top-selling music types are rhythm and blues (175 million albums sold in 1999) and alternative music (121 million albums). If you like metal (30 million albums) and I like jazz (only 20 million), we may have no clue about the tastes of the music-buying public as a whole. If we are in the music business, or even if we are interested in pop culture, we must put our own tastes aside and look at the data.

You can find data in the library or on the Internet (that's where I found the music sales data). How can we know whether data can be trusted? Good data are as much a human product as wool sweaters and DVD players. Sloppily produced data will frustrate you as much as a sloppily made sweater. You examine a sweater before you buy, and you don't buy if it is not well made. Neither should you use data that are not well made. The first part of this book shows how to tell if data are well made.

Chapter 1 Where Do Data Come From?

You can't see just by watching

You can read the newspaper and watch TV news for months without seeing an algebraic formula. No wonder algebra seems unconnected to life. You can't go a day, however, without meeting data and statistical studies. You hear that last month's unemployment rate was 4.5%. A newspaper article says that only 21% of people aged 18 to 29 say that they always vote, as opposed to 59% of people 65 or older. A longer article says that low-income children who received high-quality day care did better on academic tests given years later and were more likely to go to college and hold good jobs than other similar children.

Where do these data come from? Why can we trust them? Or maybe we can't trust them. It may be, as Yogi Berra says, that "You can observe a lot by watching." But you can't see just by watching that 21% of young people vote or that good day care sends kids to college 15 years later. Good data are the fruit of intelligent human effort. Bad data result from laziness or lack of understanding, or even the desire to mislead others. "Where do the data come from?" is the first question you should ask when someone throws a number at you.

Talking about data: individuals and variables

Statistics is the science of data. I almost said "the art of data," because good judgment and even good taste along with good math make good statistics. A big part of good judgment lies in deciding what you want to measure in order to produce data that will shed light on your concerns. We need some vocabulary to describe the raw materials that go into data. Here it is:

Individuals and variables

Individuals are the objects described by a set of data. Individuals may be people, but they may also be animals or things.

A **variable** is any characteristic of an individual. A variable can take different values for different individuals.

For example, here are the first lines of a professor's data set at the end of a statistics course:

Name	Major	Points	Grade
Advani, Sura	COMM	397	B
Barton, David	HIST	323	C
Brown, Annette	LIT	446	A
Chiu, Sun	PSYC	405	B
Cortez, Maria	PSYC	461	A

The *individuals* are students enrolled in the course. In addition to each student's name, there are three *variables*. The first says what major a student has chosen. The second variable gives the student's total points out of 500 for the course, and the third records the grade received.

Statistics deals with numbers, but not all variables are numerical. Of the three variables in the professor's data set, only total points has numbers as its values. To do statistics with the other variables, we use *counts* or *percents*. We might give the percent of students who got an A, for example, or the percent who are psychology majors.

Bad judgment in choosing variables can lead to data that cost lots of time and money but don't shed light on the world. What constitutes good judgment can be controversial. Here are examples of the challenges in deciding what data to collect.

Example 1. Who recycles?

Who takes the trouble to recycle? Researchers spent lots of time and money weighing the stuff put out for recycling in two neighborhoods in a California city, call them Upper Crust and Lower Mid. The *individuals* here are households, because trash and recycling pickup are done for residences, not for people one at a time. The *variable* measured was the weight in pounds of the curbside recycling basket each week.

The Upper Crust households contributed more pounds per week on the average than did the folk in Lower Mid. Can we say that the rich are more serious about recycling? No. Someone noticed that Upper Crust recycling baskets contained lots of heavy glass wine bottles. In Lower Mid, they put out lots of light plastic soda bottles and light metal beer and soda cans. Weight tells us little about commitment to recycling.

Example 2. What's your race?

The U.S. census asks "What is this person's race" for every person in every household. "Race" is a *variable*, and the Census Bureau must say exactly how to measure it. The census form does this by giving a list of races. Years of political fighting lie behind this list.

How many races shall we list, and what names shall we use for them? Shall we have a category for people of mixed race? Asians wanted more national categories, such as Filipino and Vietnamese, for the growing Asian population. Pacific Islanders wanted to be separated from the larger Asian group. Black leaders did not want a mixed-race category, fearing that many blacks would choose it and so reduce the official count of the black population.

The 2000 census form ended up with six Asian groups (plus "Other Asian") and three Pacific Island groups (plus "Other Pacific Islander"). There is no "mixed-race" group, but you can mark more than one race. That is, people claiming mixed-race can count as both, so that the total of the racial group counts in 2000 is larger than the population count. Unable to decide what the proper term for blacks should be, the Census Bureau settled on "Black, African American, or Negro." What about Hispanics? That's a separate question, because Hispanics can be of any race. Again unable to choose a short name that would satisfy everyone, the Census Bureau asked if you are "Spanish/Hispanic/Latino."

The fight over "race" reminds us that data reflect society. Race is a social idea, not a biological fact. In the census, you say what race you consider yourself to be. Race is a sensitive issue in the United States, so the fight is no surprise and the Census Bureau's diplomacy seems a good compromise.

Observational studies

Sometimes all you can do is watch. To learn how chimpanzees in the wild behave, watch. To study how a teacher and young children interact in a schoolroom, watch. It helps if the watcher knows what to look for. The

United States
Census 2000

U.S. Department of Commerce • Bureau of the Census

This is the official form for all the people at this address. It is quick and easy, and your answers are protected by law. Complete the Census and help your community get what it needs — today and in the future!

Start Here **Please use a black or blue pen.**

1. **How many people were living or staying in this house, apartment, or mobile home on April 1, 2000?**

☐☐ Number of people

INCLUDE in this number:
- foster children, roomers, or housemates
- people staying here on April 1, 2000 who have no other permanent place to stay
- people living here most of the time while working, even if they have another place to live

DO NOT INCLUDE in this number:
- college students living away while attending college
- people in a correctional facility, nursing home, or mental hospital on April 1, 2000
- Armed Forces personnel living somewhere else
- people who live or stay at another place most of the time

2. **Is this house, apartment, or mobile home —** *Mark* ☒ *ONE box.*
   - ☐ Owned by you or someone in this household with a mortgage or loan?
   - ☐ Owned by you or someone in this household free and clear (without a mortgage or loan)?
   - ☐ Rented for cash rent?
   - ☐ Occupied without payment of cash rent?

3. **Please answer the following questions for each person living in this house, apartment, or mobile home. Start with the name of one of the people living here who owns, is buying, or rents this house, apartment, or mobile home. If there is no such person, start with any adult living or staying here. We will refer to this person as Person 1.**

What is this person's name? *Print name below.*

Last Name

First Name MI

4. **What is Person 1's telephone number?** *We may call this person if we don't understand an answer.*

Area Code + Number

5. **What is Person 1's sex?** *Mark* ☒ *ONE box.*
   ☐ Male ☐ Female

6. **What is Person 1's age and what is Person 1's date of birth?**

Age on April 1, 2000

Print numbers in boxes.

Month Day Year of birth

→ **NOTE: Please answer BOTH Questions 7 and 8.**

7. **Is Person 1 Spanish/Hispanic/Latino?** *Mark* ☒ *the* ***"No"*** *box if* ***not*** *Spanish/Hispanic/Latino.*
   - ☐ **No,** not Spanish/Hispanic/Latino
   - ☐ Yes, Mexican, Mexican Am., Chicano
   - ☐ Yes, Puerto Rican
   - ☐ Yes, Cuban
   - ☐ Yes, other Spanish/Hispanic/Latino — *Print group.*

8. **What is Person 1's race?** ***Mark*** ☒ ***one or more races*** *to indicate what this person considers himself/herself to be.*
   - ☐ White
   - ☐ Black, African Am., or Negro
   - ☐ American Indian or Alaska Native — *Print name of enrolled or principal tribe.*
   - ☐ Asian Indian ☐ Japanese ☐ Native Hawaiian
   - ☐ Chinese ☐ Korean ☐ Guamanian or Chamorro
   - ☐ Filipino ☐ Vietnamese ☐ Samoan
   - ☐ Other Asian — *Print race.* ☐ Other Pacific Islander — *Print race.*
   - ☐ Some other race — *Print race.*

→ **If more people live here, continue with Person 2.**

OMB No. 0607-0856: Approval Expires 12/31/2000

Form **D-1**

Figure 1.1 The first page of the 2000 census form, mailed to all households in the country. Courtesy of U.S. Department of Commerce/Bureau of the Census.

chimpanzee expert may be interested in how males and females interact, in whether some chimps in the troop are dominant, in whether the chimps hunt and eat meat. Indeed, chimps were thought to be vegetarians until Jane Goodall watched them carefully in Gombe National Park, Tanzania. Now it is clear that meat is a natural part of the chimpanzee diet.

At first, the observer may not know what to record. Eventually patterns seem to emerge and we can decide what variables we want to measure. How often do chimpanzees hunt? Alone or in groups? How large are hunting groups? Males alone, or both males and females? How much of the diet is meat? Observation that is organized and measures clearly defined variables is more convincing than just watching. Here is an example of highly organized (and expensive) observation.

Example 3. Do power lines cause leukemia in children?

Electric currents generate magnetic fields. So living with electricity exposes people to magnetic fields. Living near power lines increases exposure to these fields. Really strong fields can disturb living cells in laboratory studies. What about the weaker fields we experience if we live near power lines? Some data suggested that more children in these locations might develop leukemia, a cancer of the blood cells.

We can't do experiments that expose children to magnetic fields. It's hard to compare cancer rates among children who happen to live in more and less exposed locations, because leukemia is quite rare and locations vary a lot in many ways other than magnetic fields. It is easier to start with children who have leukemia and compare them with children who don't. We can look at lots of possible causes—diet, pesticides, drinking water, magnetic fields, and others—to see where children with leukemia differ from those without. Some of these broad studies suggested a closer look at magnetic fields.

A really careful look at magnetic fields took five years and cost $5 million. The researchers compared 638 children who had leukemia and 620 who did not. They went into the homes and actually measured the magnetic fields in the children's bedrooms, in other rooms, and at the front door. They recorded facts about nearby power lines for the family home and also for the mother's residence when she was pregnant. Result: no evidence of more than a chance connection between magnetic fields and childhood leukemia.

"No evidence" that magnetic fields are connected with childhood leukemia doesn't prove that there is no risk. It says only that a very careful study could not find any risk that stands out from the play of chance that distributes leukemia cases across the landscape. Critics continue to argue that the study failed to measure some important variables, or that the children studied don't fairly represent all children. Nonetheless, a carefully designed observational study is a great advance over haphazard and sometimes emotional counting of cancer cases.

Observational study

An **observational study** observes individuals and measures variables of interest but does not attempt to influence the responses. The purpose of an observational study is to describe some group or situation.

Sample surveys

"You don't have to eat the whole ox to know that the meat is tough." That is the idea of sampling: to gain information about the whole by examining only a part. **Sample surveys** are an important kind of observational study. They survey some group of individuals by studying only some of its members, selected not because they are of special interest but because they represent the larger group. Here is the vocabulary we use to discuss sampling.

Populations and samples

The **population** in a statistical study is the entire group of individuals about which we want information.

A **sample** is a part of the population from which we actually collect information, used to draw conclusions about the whole.

Notice that the *population* is the group we want to study. If we want information about all U.S. college students, that is our population even if students at only one college are available for sampling. To make sense of any sample result, you must know what population the sample represents. Did that preelection poll, for example, ask the opinions of all adults? Citizens only? Registered voters only? Democrats only? The *sample* consists of the people we actually have information about. If the poll can't contact some of the people it selected, those people aren't in the sample.

The distinction between population and sample is basic to statistics. The following examples illustrate this distinction and also introduce some major uses of sampling. These brief descriptions also indicate the variables measured for each individual in the sample.

Example 4. Public opinion polls

Polls such as those conducted by Gallup and many news organizations ask people's opinions on a variety of issues. The *variables* measured are responses to questions about public issues. Though most noticed at election time, these polls are conducted on a regular basis throughout the year. For a typical opinion poll:

Population: U.S. residents 18 years of age and over. Noncitizens and even illegal immigrants are included.

Sample: Between 1000 and 1500 people interviewed by telephone.

You just don't understand

A sample survey of journalists and scientists found quite a communications gap. Journalists think that scientists are arrogant, while scientists think that journalists are ignorant. We won't take sides, but here is one interesting result from the survey: 82% of the scientists agree that the "media do not understand statistics well enough to explain new findings" in medicine and other fields.

Example 5. The Current Population Survey

Government economic and social data come from large sample surveys of a nation's individuals, households, or businesses. The monthly Current Population Survey (CPS) is the most important government sample survey in the United States. Many of the *variables* recorded by the CPS concern the employment or unemployment of everyone over 16 years old in a household. The government's monthly unemployment rate comes from the CPS. The CPS also records many other economic and social variables. For the CPS:

Population: The more than 100 million U.S. households. Notice that the individuals are households rather than people or families. A household consists of all people who share the same living quarters, regardless of how they are related to each other.

Sample: About 50,000 households interviewed each month.

Example 6. TV ratings

Market research is designed to discover what consumers want and what products they use. One example of market research is the television-rating service of Nielsen Media Research. The Nielsen ratings influence how much advertisers will pay to sponsor a program and whether or not the program stays on the air. For the Nielsen national TV ratings:

Population: The 100 million U.S. households that have a television set.

Sample: About 5000 households that agree to use a "people meter" to record the TV viewing of all people in the household.

The *variables* recorded include the number of people in the household and their age and sex, whether the TV set is in use at each time period, and, if so, what program is being watched and who is watching it.

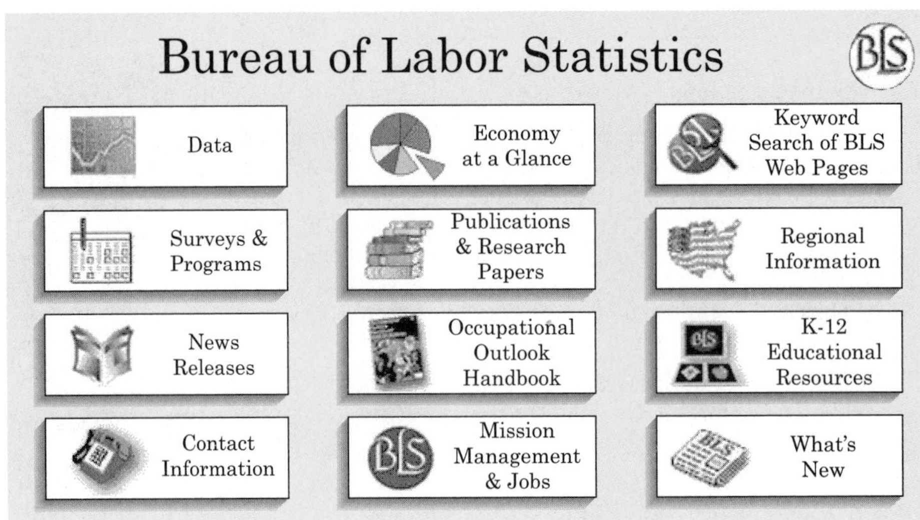

Figure 1.2 The Web site of the Bureau of Labor Statistics, stats.bls.gov. This is the home of the Current Population Survey and other important government sample surveys.

Example 7. The General Social Survey

Social science research makes heavy use of sampling. The General Social Survey (GSS), carried out every second year by the National Opinion Research Center at the University of Chicago, is the most important social science sample survey. The *variables* cover the subject's personal and family background, experiences and habits, and attitudes and opinions on subjects from abortion to war.

Population: Adults (age 18 and over) living in households in the United States. The population does not include adults in institutions such as prisons and college dormitories. It also does not include persons who cannot be interviewed in English.

Sample: About 3000 adults interviewed in person in their homes.

Most statistical studies use samples in the broad sense. For example, the 638 children with leukemia in Example 3 are supposed to represent all children with leukemia. We usually reserve the dignified term "sample survey" for studies that use an organized plan to choose a sample that represents some specific population. The children with leukemia were patients at centers that specialize in treating children's cancer. Expert judgment says they are typical of all leukemia patients, even though they come only from rather fancy hospitals. A sample survey doesn't rely on judgment: it starts with an entire population and chooses a sample to represent it. Chapters 2, 3, and 4 discuss the art and science of sample surveys.

Census

A sample survey looks at only a part of the population. Why not look at the entire population? A *census* tries to do this.

Census

A **census** is a sample survey that attempts to include the entire population in the sample.

The U.S. Constitution requires a census of the American population every 10 years. A census of so large a population is expensive and takes a long time. Even the federal government, which can afford a census, uses samples such as the Current Population Survey to produce timely data on employment and many other variables. If the government asked every adult in the country about his or her employment, this month's unemployment rate would be available next year rather than next month. Even the every-ten-years census includes a sample survey. A census "long form" that asks many more questions than the basic census form is sent to a sample of one-sixth of all households.

So time and money favor samples over a census. Samples can have other advantages as well. If you are testing fireworks or fuses, the individuals in the sample are destroyed. Moreover, a sample can produce more accurate data than a census. A careful sample of an inventory of spare parts will almost certainly give more accurate results than asking the clerks to count all 500,000 parts in the warehouse. Bored people do not count accurately.

The experience of the Census Bureau reminds us that a census can only *attempt* to sample the entire population. The bureau estimates that the 1990 census missed 1.8% of the American population. These missing persons included an estimated 4.4% of the black population, largely in inner cities. A census is not foolproof, even with the resources of the government behind it. Why take a census at all? The government needs block-by-block population figures to create election districts with equal population. The main function of the U.S. census is to provide this local information.

Is a census old-fashioned?

The United States has taken a census every 10 years since 1790. Technology marches on, however, and replacements for a national census look promising. Denmark has no census, and France plans to eliminate its census. Denmark has a national register of all its residents, who carry identification cards and change their register entry whenever they move. France will replace its census by a large sample survey that rotates among the nation's regions. The U.S. Census Bureau has a similar idea: the American Community Survey has already started and will eliminate the census "long form" after 2000.

Experiments

Our goal in choosing a sample is a picture of the population, disturbed as little as possible by the act of gathering information. All observational studies share the principle "observe but don't disturb." When Jane Goodall first began observing chimpanzees in Tanzania, she set up a feeding station where the chimps could eat bananas. She later said that was a mistake, because it might have changed the apes' behavior.

"Now eat that banana. The nice statistician is watching us."

In *experiments,* on the other hand, we want to change behavior. In doing an experiment, we don't just observe individuals or ask them questions. We actively impose some treatment in order to observe the response. Experiments can answer questions such as "Does aspirin reduce the chance of a heart attack?" and "Do a majority of college students prefer Pepsi to Coke when they taste both without knowing which they are drinking?"

Experiments

An **experiment** deliberately imposes some treatment on individuals in order to observe their responses. The purpose of an experiment is to study whether the treatment causes a change in the response.

Example 8. Helping welfare mothers find jobs

Most adult recipients of welfare are mothers of young children. Observational studies of welfare mothers show that many are able to increase their earnings and leave the welfare system. Some take advantage of voluntary job-training programs to improve their skills. Should participation in job-training and job-search programs be required of all able-bodied welfare mothers? Observational studies cannot tell us what the effects of such a policy would be. Even if the mothers studied are a properly chosen sample of all welfare recipients, those who seek out training and find jobs may differ in many ways from those who do not. They are observed to have more education, for example, but they may also differ in values and motivation, things that cannot be observed.

To see if a required jobs program will help mothers escape welfare, such a program must actually be tried. Choose two similar groups of mothers when they apply for welfare. Require one group to participate in a job-training program, but do not offer the program to the other group. This is an experiment. Comparing the income and work record of the two groups after several years will show whether requiring training has the desired effect.

The welfare example illustrates the big advantage of experiments over observational studies: *In principle, experiments can give good evidence for cause and effect.* If we design the experiment properly, we start with two very similar groups of welfare mothers. The *individual* women of course differ from each other in age, education, number of children, and other respects. But the two *groups* resemble each other when we look at the ages, years of education, and number of children for all women in each group. During the experiment, the women's lives differ, but there is only one systematic difference between the two groups: whether or not they are in the jobs program. All live through the same good or bad economic times, the same changes in public attitudes, and so on. If the training group does much better than the untrained group in holding jobs and earning money, we can say that the training program actually causes this happy outcome.

That experiments can give good evidence that a treatment causes a response is one of the big ideas of statistics. A big idea needs a big caution: statistical conclusions hold "on the average" for groups of individuals. They don't tell us much about one individual. *On the average,* the women in the training program earned more than those who were left out. That says that the program achieved its goal. It doesn't say that every woman will be helped. And a big idea may also raise big questions: if we hope the training will raise earnings, is it ethical to offer it to some women and not to others? Chapters 5 and 6 explain how to design good experiments, and Chapter 7 looks at ethical issues.

Exploring the Web

You can find the report of the National Institute of Environmental Health Sciences on the possible health effects of exposure to power lines at www.niehs.nih.gov/emfrapid.

All of the sample surveys mentioned in Examples 4 to 7 maintain Web sites:

- Gallup Poll (Example 4): www.gallup.com/poll
- Current Population Survey (Example 5): stats.bls.gov/cpshome.htm
- Nielsen Media Research (Example 6): www.nielsenmedia.com
- General Social Survey (Example 7): www.norc.uchicago.edu/gss/homepage.htm

I recommend the Gallup site for its current poll results and clear explanations of sample survey methods.

Statistics in Summary Any statistical study records data about some **individuals** (people, animals, or things) by giving the value of one or more **variables** for each individual. Some variables, such as age and income, take numerical values. Others, such as occupation and sex, do not. Be sure the variables in a study really do tell you what you want to know.

The most important fact about any statistical study is how the data were produced. **Observational studies** try to gather information without disturbing the scene they are observing. **Sample surveys** are an important kind of observational study. A sample survey chooses a **sample** from a specific **population** and uses the sample to get information about the entire population. A **census** attempts to measure every individual in a population. **Experiments** actually do something to individuals in order to see how they respond. The goal of an experiment is usually to learn whether some treatment actually causes a certain response.

CHAPTER 1 EXERCISES

1.1 Miles per gallon. Here is a small part of a data set that describes the fuel economy (in miles per gallon) of 2000 model motor vehicles:

Make and model	Vehicle type	Transmission type	Number of cylinders	City MPG	Highway MPG
⋮					
BMW 328CI	Subcompact car	Automatic	6	18	27
BMW 328CI	Subcompact car	Manual	6	20	29
Buick Regal	Midsize car	Automatic	6	20	29
Chevrolet Blazer	Sport utility vehicle (4WD)	Automatic	6	16	20
⋮					

(a) What are the individuals in this data set?

(b) For each individual, what variables are given? Which of these variables take numerical values?

1.2 Athletes' salaries. Here is a small part of a data set that describes major-league baseball players as of opening day of the 1999 season:

Player	Team	Position	Age	Salary
⋮				
Dunwoody, Todd	Marlins	Outfield	24	222
Osuna, Antonio	Dodgers	Pitcher	26	1050
Pettitte, Andy	Yankees	Pitcher	26	5950
Sosa, Sammy	Cubs	Outfield	30	9000
⋮				

(a) What individuals does this data set describe?

(b) In addition to the player's name, how many variables does the data set contain? Which of these variables take numerical values?

(c) What do you think are the *units* in which each of the numerical variables is expressed? For example, what does it mean to give Sammy Sosa's annual salary as 9000?

1.3 Who recycles? In Example 1, weight is not a good measure of the participation of households in different neighborhoods in a city recycling program. What variables would you measure in its place?

1.4 Sampling employed women. A sociologist wants to know the opinions of employed adult women about government funding for day care. She obtains a list of the 520 members of a local business and professional women's club and mails a questionnaire to 100 of these women selected at random. Only 48 questionnaires are returned. What is the population in this study? What is the sample from which information is actually obtained? What percent of the women whom the sociologist tried to contact responded?

1.5 Rating the president. A newspaper article about an opinion poll says that "43% of Americans approve of the president's overall job performance." Toward the end of the article, you read: "The poll is based on telephone interviews with 1210 adults from around the United States, excluding Alaska and Hawaii." What variable did this poll measure? What population do you think the newspaper wants information about? What was the sample?

1.6 The political gender gap. There may be a "gender gap" in political party preference in the United States, with women more likely than men to prefer Democratic candidates. A political scientist interviews a large sample of registered voters, both men and women. She asks each voter whether they voted for the Democratic or the Republican candidate in the last congressional election. Is this study an experiment? Why or why not? What variables does the study measure?

1.7 What is the population? For each of the following sampling situations, identify the population as exactly as possible. That is, say what kind of individuals the population consists of and say exactly which individuals fall in the population. If the information given is not sufficient, complete the description of the population in a reasonable way.

(a) An opinion poll contacts 1161 adults and asks them "Which political party do you think has better ideas for leading the country in the 21st century?"

(b) A furniture maker buys hardwood in large lots. The supplier is supposed to dry the wood before shipping—wood that is not dry won't hold its size and shape. The furniture maker chooses 5 pieces of wood from each lot and tests their moisture content. If any piece exceeds 12% moisture content, the entire lot is sent back.

(c) The American Community Survey will contact 3 million households, including some in every county in the United States. This new Census Bureau survey will ask each household questions about their housing, economic, and social status. The new survey will replace the census "long form."

1.8 What is the population? For each of the following sampling situations, identify the population as exactly as possible. That is, say what kind of individuals the population consists of and say exactly which individuals

fall in the population. If the information given is not sufficient, complete the description of the population in a reasonable way.
(a) A business school researcher wants to know what factors affect the survival and success of small businesses. She selects a sample of 150 eating-and-drinking establishments from those listed in the telephone directory Yellow Pages for a large city.

(b) Your local television station wonders if its viewers would rather watch your college basketball team play or an NBA game scheduled at the same time. It schedules the NBA game and receives 89 calls asking that it show the local game instead.

(c) An insurance company wants to monitor the quality of its procedures for handling loss claims from its auto insurance policyholders. Each month the company selects a sample from all auto insurance claims filed that month to examine them for accuracy and promptness.

1.9 Public housing. To study the effect of living in public housing on family stability in poverty-level households, researchers obtain a list of all applicants for public housing in Chicago last year. Some applicants were accepted, while others were turned down by the housing authority. The researchers interview both groups and compare them. Is this an experiment, a sample survey, or an observational study that is not a sample survey? Explain your answer.

1.10 Power lines and leukemia. The study of power lines and leukemia in Example 3 compared two groups of individuals and measured many variables that might influence differences between the groups. Explain carefully why this study is *not* an experiment.

1.11 Treating prostate disease. A large study used records from Canada's national health care system to compare the effectiveness of two ways to treat prostate disease. The two treatments are traditional surgery and a new method that does not require surgery. The records described many patients whose doctors had chosen each method. The study found that patients treated by the new method were more likely to die within 8 years.
(a) Explain why this is an observational study, not an experiment.

(b) Briefly describe the nature of an experiment to compare the two ways to treat prostate disease.

1.12 Exercise and heart attacks. Does regular exercise reduce the risk of a heart attack? Here are two ways to study this question.

1. A researcher finds 2000 men over 40 who exercise regularly and have not had heart attacks. She matches each with a similar man who does not exercise regularly, and she follows both groups for 5 years.

2. Another researcher finds 4000 men over 40 who have not had heart attacks and are willing to participate in a study. She assigns 2000 of the men to a regular program of supervised exercise. The other 2000 continue their usual habits. The researcher follows both groups for 5 years.

(a) Explain why the first is an observational study and the second is an experiment.

(b) Why does the experiment give more useful information about whether exercise reduces the risk of heart attacks?

1.13 Physical fitness and leadership. A study of the relationship between physical fitness and leadership uses as subjects middle-aged executives who have volunteered for an exercise program. The executives are divided into a low-fitness group and a high-fitness group on the basis of a physical examination. All subjects then take a psychological test designed to measure leadership, and the results for the two groups are compared.
(a) Is this an experiment? Explain your answer.

(b) We would prefer a sample survey to using men who volunteer for a fitness program. What population does it appear that the investigators were interested in? What variables did they measure?

1.14 Tom Clancy's words. Different types of writing can sometimes be distinguished by the lengths of the words used. A student interested in this fact wants to study the lengths of words used by Tom Clancy in his novels. She opens a Clancy novel at random and records the lengths of the first 250 words on the page. What is the population in this study? What is the sample? What variable does the student measure?

1.15 Choose your study type. What is the best way to answer each of the questions below: an experiment, a sample survey, or an observational study that is not a sample survey? Explain your choices.
(a) Are people generally satisfied with how things are going in the country right now?

(b) Do college students learn basic accounting better in a classroom or using an online course?

(c) How long do your teachers wait on the average after they ask their class a question?

1.16 Choose your study purpose. Give an example of a question about college students, their behavior, or their opinions that would best be answered by
(a) a sample survey.

(b) an observational study that is not a sample survey.

(c) an experiment.

Chapter 2
Samples, Good and Bad

The *Town Talk* takes an opinion poll

In Rapides Parish, Louisiana, only one company is allowed to provide ambulance service. The local paper, the *Town Talk,* asked readers to call in to offer their opinion on whether the company should keep its monopoly. Call-in polls are generally automated: call one telephone number to vote "Yes" and call another number to vote "No." Telephone companies often charge callers to these numbers.

The *Town Talk* got 3763 calls, which suggests unusual interest in ambulance service. Investigation showed that 638 calls came from the ambulance company office or from the homes of its executives. Many more no doubt came from lower-level employees. "We've got employees who are concerned about this situation, their job stability and their families and maybe called more than they should have," said a company vice-president. Other sources said employees were told to, as they say in Chicago, "vote early and often."

We will see that a sample size of 3763 is quite respectable—if the sample is properly designed. Inviting people to call (and call, and call . . .) isn't proper sample design. We are about to learn how the pros take a sample, and why it beats the *Town Talk* method.

How to sample badly

As the *Town Talk* learned, it is easier to sample badly than to sample well. The paper relied on *voluntary response,* allowing people to call in rather than actively selecting its own sample. The result was *biased*—the sample was overweighted with people favoring the ambulance monopoly. Voluntary response samples attract people who feel strongly about the issue in question. These people, like the employees of the ambulance company, may not fairly represent the opinions of the entire population.

There are other ways to sample badly. Suppose that I sell your company several crates of oranges each week. You examine a sample of oranges from each crate to determine the quality of my oranges. It is easy to inspect a few oranges from the top of each crate, but these oranges may not be representative of the entire crate. Those on the bottom are more often damaged in shipment. If I were less than honest, I might make sure that the rotten oranges are packed on the bottom with some good ones on top for you to inspect. If you sample from the top, your sample results are again *biased*—the sample oranges are systematically better than the population they are supposed to represent.

Biased sampling methods

The design of a statistical study is **biased** if it systematically favors certain outcomes.

Selection of whichever individuals are easiest to reach is called **convenience sampling**.

A **voluntary response sample** chooses itself by responding to a general appeal. Write-in or call-in opinion polls are examples of voluntary response samples.

Convenience samples and voluntary response samples are often biased.

Example 1. Interviewing at the mall

Squeezing the oranges on the top of the crate is one example of convenience sampling. Mall interviews are another. Manufacturers and advertising agencies often use interviews at shopping malls to gather information about the habits of consumers and the effectiveness of ads. A sample of mall shoppers is fast and cheap. But people contacted at shopping malls are not representative of the entire U.S. population. They are richer, for example, and more likely to be teenagers or retired. Moreover, the interviewers tend to select neat, safe-looking individuals from the stream of customers. Mall samples are biased: they systematically overrepresent some parts of the population (prosperous people, teenagers, and retired people) and underrepresent others. The opinions of such a convenience sample may be very different from those of the population as a whole.

Example 2. Write-in opinion polls

Ann Landers once asked the readers of her advice column, "If you had it to do over again, would you have children?" She received nearly 10,000 responses, almost 70% saying "NO!" Can it be true that 70% of parents regret having children? Not at all. This is a voluntary response sample. People who feel strongly about an issue, particularly people with strong negative feelings, are more likely to take the trouble to respond. Ann Landers's results are strongly biased—the percent of parents who would not have children again is much higher in her sample than in the population of all parents.

"Hey, Pops, what was that letter you sent off to Ann Landers yesterday?"

Write-in and call-in opinion polls are almost sure to lead to strong bias. In fact, only about 15% of the public have ever responded to a call-in poll, and these tend to be the same people who call radio talk shows. That's not a representative sample of the population as a whole.

Simple random samples

In a voluntary response sample, people choose whether to respond. In a convenience sample, the interviewer makes the choice. In both cases, personal choice produces bias. The statistician's remedy is to allow impersonal chance to choose the sample. A sample chosen by chance allows neither favoritism by the sampler nor self-selection by respondents. Choosing a sample by chance attacks bias by giving all individuals an equal chance to be chosen. Rich and poor, young and old, black and white, all have the same chance to be in the sample.

The simplest way to use chance to select a sample is to place names in a hat (the population) and draw out a handful (the sample). This is the idea of *simple random sampling.*

Simple random sample

A **simple random sample (SRS)** of size *n* consists of *n* individuals from the population chosen in such a way that every set of *n* individuals has an equal chance to be the sample actually selected.

An SRS not only gives each individual an equal chance to be chosen (thus avoiding bias in the choice) but also gives every possible sample an equal chance to be chosen. Drawing names from a hat does this. Write 100 names on identical slips of paper and mix them in a hat. This is a population. Now draw 10 slips, one after the other. This is an SRS, because any 10 slips have the same chance as any other 10.

Drawing names from a hat makes clear what it means to give each individual and each possible set of *n* individuals the same chance to be chosen. That's the idea of an SRS. Of course, drawing tags from a hat would be a bit awkward for a sample of the country's 100 million households. In practice, we use computer-generated *random digits* to choose samples. If you don't use software directly, you can use a *table of random digits* to choose small samples by hand.

Random digits

A **table of random digits** is a long string of the digits 0, 1, 2, 3, 4, 5, 6, 7, 8, 9 with these two properties:

1. Each entry in the table is equally likely to be any of the 10 digits 0 through 9.

2. The entries are independent of each other. That is, knowledge of one part of the table gives no information about any other part.

Table A at the back of the book is a table of random digits. You can think of Table A as the result of asking an assistant (or a computer) to mix the digits 0 to 9 in a hat, draw one, then replace the digit drawn, mix again, draw a second digit, and so on. The assistant's mixing and drawing save us the work of mixing and drawing when we need to randomize. Table A begins with the digits 19223950340575628713. To make the table easier to read, the digits appear in groups of five and in numbered rows. The groups and rows have no meaning—the table is just a long list of randomly chosen digits. Here's how to use the table to choose an SRS.

Example 3. How to choose an SRS

Joan's small accounting firm serves 30 business clients. Joan wants to interview a sample of 5 clients to find ways to improve client satisfaction. To avoid bias, she chooses an SRS of size 5.

Step 1: Label. Give each client a numerical label, using as few digits as possible. Two digits are needed to label 30 clients, so we use labels

01, 02, 03, . . . , 28, 29, 30

It is also correct to use labels 00 to 29 or even another choice of 30 two-digit labels. Here is the list of clients, with labels attached:

01	A-1 Plumbing	16	JL Records
02	Accent Printing	17	Johnson Commodities
03	Action Sport Shop	18	Keiser Construction
04	Anderson Construction	19	Liu's Chinese Restaurant
05	Bailey Trucking	20	MagicTan
06	Balloons, Inc.	21	Peerless Machine
07	Bennett Hardware	22	Photo Arts
08	Best's Camera Shop	23	River City Books
09	Blue Print Specialties	24	Riverside Tavern
10	Central Tree Service	25	Rustic Boutique
11	Classic Flowers	26	Satellite Services
12	Computer Answers	27	Scotch Wash
13	Darlene's Dolls	28	Sewer's Center
14	Fleisch Realty	29	Tire Specialties
15	Hernandez Electronics	30	Von's Video Store

Are these random digits really random?

Not a chance. The random digits in Table A were produced by a computer program. Computer programs do exactly what you tell them to do. Give the program the same input and it will produce exactly the same "random" digits. Of course, clever people have devised computer programs that produce output that *looks* like random digits. These are called "pseudo-random numbers," and that's what Table A contains. Pseudo-random numbers work fine for statistical randomizing, but they have hidden nonrandom patterns that can mess up more refined uses.

Step 2: Table. Enter Table A anywhere and read two-digit groups. Suppose we enter at line 130, which is

69051 64817 87174 09517 84534 06489 87201 97245

The first 10 two-digit groups in this line are

69 05 16 48 17 87 17 40 95 17

Each two-digit group in Table A is equally likely to be any of the 100 possible groups, 00, 01, 02, . . . , 99. So two-digit groups choose two-digit labels at random. That's just what we want.

Joan used only labels 01 to 30, so we ignore all other two-digit groups. The first 5 labels between 01 and 30 that we encounter in the table choose our sample. Of the first 10 labels in line 130, we ignore 5 because they are too high (over 30). The others are

05, 16, 17, 17, and 17. The clients labeled 05, 16, and 17 go into the sample. Ignore the second and third 17s because that client is already in the sample. Now run your finger across line 130 (and continue to line 131 if needed) until 5 clients are chosen.

The sample is the clients labeled 05, 16, 17, 20, 19. These are Bailey Trucking, JL Records, Johnson Commodities, MagicTan, and Liu's Chinese Restaurant.

Using the table of random digits is much quicker than drawing names from a hat. As Example 3 shows, choosing an SRS has two steps.

Choose an SRS in two steps

Step 1: Label. Assign a numerical label to every individual in the population. Be sure that all labels have the same number of digits.

Step 2: Table. Use random digits to select labels at random.

You can assign labels in any convenient manner, such as alphabetical order for names of people. As long as all labels have the same number of digits, all individuals will have the same chance to be chosen. Use the shortest possible labels: one digit for a population of up to 10 members, two digits for 11 to 100 members, three digits for 101 to 1000 members, and so on. As standard practice, I recommend that you begin with label 1 (or 01 or 001, as needed). You can read digits from Table A in any order—across a row, down a column, and so on—because the table has no order. As standard practice, I recommend reading across rows.

Sample surveys use computer software to choose an SRS, but the software just automates the steps in Example 3. The computer doesn't look in a table of random digits because it can generate them on the spot.

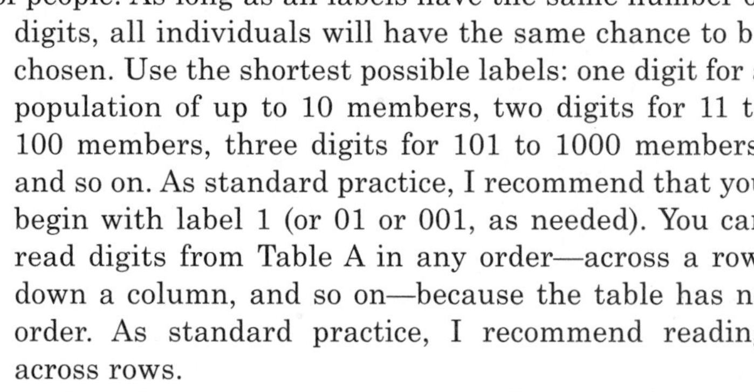

Golfing at random

Random drawings give all the same chance to be chosen, so they offer a fair way to decide who gets a scarce good—like a round of golf. Lots of golfers want to play the famous Old Course at St. Andrews, Scotland. A few can reserve in advance. Most must hope that chance favors them in the daily random drawing for tee times. At the height of the summer season, only 1 in 6 wins the right to pay $120 for a round.

Can you trust a sample?

The *Town Talk,* Ann Landers, and mall interviews produce samples. We can't trust results from these samples, because they are chosen in ways that invite bias. We have more confidence in results from an SRS, because it uses impersonal chance to avoid bias. The first question to ask of any sample is whether it was chosen at random. Opinion polls and other sample surveys carried out by people who know what they are doing use random sampling.

Example 4. A Gallup poll

A Gallup poll on smoking began with the question "Have you, yourself, smoked any cigarettes in the past week?" The press release reported that "just 23% of Americans smoke." Can we trust this fact? Ask first how Gallup selected its sample. Later in the press release we read this: "The results are based on telephone interviews with a randomly selected national sample of 1,039 adults, 18 years and older, conducted September 23–26, 1999."

This is a good start toward gaining our confidence. Gallup tells us what population it has in mind (people at least 18 years old living anywhere in the United States). We know that the sample from this population was of size 1039 and, most important, that it was chosen at random. There is more to say, and we will soon say it, but we have at least heard the comforting words "randomly selected."

Exploring the Web

The Internet brings voluntary response polls to the computer nearest you. Visit www.misterpoll.com to become part of the sample in any of dozens of online polls. As the site says, "None of these polls are 'scientific,' but do represent the collective opinion of everyone who participates."

To see how software speeds up choosing an SRS, go to the Research Randomizer at www.randomizer.org. Click on "Randomizer" and fill in the boxes. You can even ask the Randomizer to arrange your sample in order.

Statistics in Summary We select a **sample** in order to get information about some **population.** How can we choose a sample that fairly represents the population? **Convenience samples** and **voluntary response samples** are common but do not produce trustworthy data. These sampling methods are usually **biased.** That is, they systematically favor some parts of the population over others in choosing the sample.

The deliberate use of chance in producing data is one of the big ideas of statistics. Random samples use chance to choose a sample, thus avoiding bias due to personal choice. The basic type of random sample is the **simple random sample,** which gives all samples of the same size the same chance to be the sample we actually choose. To choose an SRS by hand, use a **table of random digits** such as Table A in the back of the book.

CHAPTER 2 EXERCISES

2.1 Letters to Congress. You are on the staff of a member of Congress who is considering a bill that would provide government-sponsored insurance for nursing home care. You report that 1128 letters have been received on the issue, of which 871 oppose the legislation. "I'm surprised that most voters in my district oppose the bill. I thought it would be quite popular," says the congresswoman. Are you convinced that a majority of the voters oppose the bill? How would you explain the statistical issue to the congresswoman?

2.2 Instant opinion. The Harris/Excite instant poll can be found online at news.excite.com/news/poll. The question appears on the screen, and you simply click buttons to vote "Yes," "No," or "Don't know." On January 25, 2000, the question was "Should female athletes be paid the same as men for the work they do?" In all, 13,147 (44%) said "Yes," another 15,182 (50%) said "No," and the remaining 1448 said "Don't know."

(a) What is the sample size for this poll?

(b) That's a much larger sample than standard sample surveys. In spite of this, we can't trust the result to give good information about any clearly defined population. Why?

(c) It is still true that more men than women use the Web. How might this fact affect the poll results?

2.3 Ann Landers takes a sample. Advice columnist Ann Landers once asked her female readers whether they would be content with affectionate treatment by men, with no sex ever. Over 90,000 women wrote in, with 72% answering "Yes." Many of the letters described unfeeling treatment by men. Explain why this sample is certainly biased. What is the likely direction of the bias? That is, is that 72% probably higher or lower than the truth about the population of all adult women?

2.4 We don't like one-way streets. Highway planners decided to make a main street in West Lafayette, Indiana, a one-way street. The *Lafayette Journal and Courier* took a one-day poll by inviting readers to call a telephone number to record their comments. The next day, the paper reported:

> Journal and Courier *readers overwhelmingly prefer two-way traffic flow in West Lafayette's Village area to one-way streets. By nearly a 7-1 margin, callers to the newspaper's Express Yourself opinion line on Wednesday complained about the one-way streets that have been in place since May. Of the 98 comments received, all but 14 said no to one-way.*

(a) What population do you think the newspaper wants information about?

(b) Is the proportion of this population who favor one-way streets almost certainly larger or smaller than the proportion 14/98 in the sample? Why?

2.5 Design your own bad sample. Your college wants to gather student opinion about parking for students on campus. It isn't practical to contact all students.

(a) Give an example of a way to choose a sample of students that is poor practice because it depends on voluntary response.

(b) Give another example of a bad way to choose a sample that doesn't use voluntary response.

2.6 A call-in opinion poll. Should the United Nations continue to have its headquarters in the United States? A television program asked its viewers to call in with their opinions on that question. There were 186,000 callers, 67% of whom said "No." A nationwide random sample of 500 adults found that 72% answered "Yes" to the same question. Explain to someone who knows no statistics why the opinions of only 500 randomly chosen respondents are a better guide to what all Americans think than the opinions of 186,000 callers.

2.7 Choose an SRS. A firm wants to understand the attitudes of its minority managers toward its system for assessing management performance. Below is a list of all the firm's managers who are members of minority groups. Use Table A at line 139 to choose 6 to be interviewed in detail about the performance appraisal system.

Agarwal	Dewald	Huang	Puri
Alfonseca	Fleming	Kim	Richards
Baxter	Fonseca	Liao	Rodriguez
Bowman	Gates	Mourning	Santiago
Brown	Goel	Nunez	Shen
Cortez	Gomez	Peters	Vega
Cross	Hernandez	Pliego	Watanabe

2.8 Choose an SRS. Your class in ancient Ugaritic religion is poorly taught and has decided to complain to the dean. The class decides to choose four of its members at random to carry the complaint. The class list appears below. Choose an SRS of 4 using the table of random digits, beginning at line 145.

Anderson	Gutierrez	Patnaik
Aspin	Gwynn	Pirelli
Benitez	Harter	Rao
Bock	Henderson	Rider
Breiman	Hughes	Robertson
Castillo	Johnson	Rodriguez
Dixon	Kempthorn	Siegel
Edwards	Liang	Tompkins
Fernande	Montoya	Vandegraff
Gupta	Olds	Wang

2.9 An election day sample. You want to choose an SRS of 25 of a city's 440 voting precincts for special voting-fraud surveillance on election day.
(a) Explain clearly how you would label the 440 precincts. How many digits make up each of your labels? What is the greatest number of precincts you could label using this number of digits?

(b) Use Table A to choose the SRS, and list the labels of the precincts you selected. Enter Table A at line 117.

2.10 Is this an SRS? A university has 1000 male and 500 female faculty members. A survey of faculty opinion selects 100 of the 1000 men at random and then separately selects 50 of the 500 women at random. The 150 faculty members chosen make up the sample.
(a) Explain why this sampling method gives each member of the faculty an equal chance to be chosen.

(b) Nonetheless, this is not an SRS. Why not?

2.11 How much do students earn? A university's financial aid office wants to know how much it can expect students to earn from summer employment. This information will be used in setting the level of financial aid. The population contains 3478 students who have completed at least one year of study but have not yet graduated. The university will send a questionnaire to an SRS of 100 of these students, drawn from an alphabetized list.
(a) Describe how you will label the students in order to select the sample.

(b) Use Table A, beginning at line 105, to select the first 5 students in the sample.

2.12 Apartment living. You are planning a report on apartment living in a college town. You decide to select three apartment complexes at random for in-depth interviews with residents. Use Table A, starting at line 117, to select a simple random sample of three of the following apartment complexes.

Ashley Oaks	Country View	Mayfair Village
Bay Pointe	Country Villa	Nobb Hill
Beau Jardin	Crestview	Pemberly Courts
Bluffs	Del-Lynn	Peppermill
Brandon Place	Fairington	Pheasant Run
Briarwood	Fairway Knolls	Richfield
Brownstone	Fowler	Sagamore Ridge
Burberry	Franklin Park	Salem Courthouse
Cambridge	Georgetown	Village Manor
Chauncey Village	Greenacres	Waterford Court
Country Squire	Lahr House	Williamsburg

2.13 How do random digits behave? Which of the following statements are true of a table of random digits, and which are false? Explain your answers.

(a) There are exactly four 0s in each row of 40 digits.

(b) Each pair of digits has chance 1/100 of being 00.

(c) The digits 0000 can never appear as a group, because this pattern is not random.

2.14 You call the shots. A newspaper advertisement for *USA Today: The Television Show* once said:

> *Should handgun control be tougher? You call the shots in a special call-in poll tonight. If yes, call 1-900-720-6181. If no, call 1-900-720-6182. Charge is 50 cents for the first minute.*

Explain why this opinion poll is almost certainly biased.

2.15 More randomization. Most sample surveys call residential telephone numbers at random. They do not, however, always ask their questions of the person who picks up the phone. Instead, they ask about the adults who live in the residence and choose one at random to be in the sample. Why is this a good idea?

2.16 Rating the police. The Miami Police Department wants to know how black residents of Miami feel about police service. A sociologist prepares several questions about the police. The police department chooses an SRS of 300 mailing addresses in predominantly black neighborhoods and sends a uniformed black police officer to each address to ask the questions of an adult living there.

(a) What are the population and the sample?

(b) Why are the results likely to be biased even though the sample is an SRS?

2.17 Random selection? Choosing at random is a "fair" way to decide who gets some scarce good, in the sense that everyone has the same chance to win. Random choice isn't always a good idea—sometimes we don't want to treat everyone the same, because some people have a better claim. In each of the following situations, would you support choosing at random? Give your reasons in each case.

(a) The basketball arena has 4000 student seats, and 7000 students want tickets. Shall we choose 4000 of the 7000 at random?

(b) The list of people waiting for liver transplants is much larger than the number of available livers. Shall we let impersonal chance decide who gets a transplant?

(c) During the Vietnam War, young men were chosen for army service at random, by a "draft lottery." Is this the best way to decide who goes and who stays home?

Chapter 3
What Do Samples Tell Us?

Do you lotto?

You know that lotteries are popular. How popular? Here's what the Gallup Poll says: "They can offer massive jackpots—and a ticket just costs a dollar at your neighborhood store. For many Americans, picking up a lottery ticket has become routine, despite the massive odds against striking it rich. A new Gallup Poll Social Audit on gambling shows that 57% of Americans have bought a lottery ticket in the last 12 months, making lotteries by far the favorite choice of gamblers."

Reading further, we find that Gallup talked with 1523 randomly selected adults to reach these conclusions. We're happy Gallup chooses at random—we wouldn't get unbiased information about gambling by asking people standing in line to buy lotto tickets. However, the Census Bureau says that there are about 200 million adults in the United States. How can 1523 people, even a random sample of 1523 people, tell us about the habits of 200 million people?

A sample won't tell us the exact truth about the population. So Gallup gives us a margin of error. Here's Gallup's claim: "For results based on a sample of this size, one can say with 95% confidence that the error attributable to sampling and other random effects could be plus or minus 3 percentage points for adults." In the news, this gets reduced to "The margin of error for the poll was plus or minus 3 percentage points." Either way, it isn't exactly obvious what the statement means. To find out, read on.

From sample to population

Gallup's finding that "57% of Americans have bought a lottery ticket in the last 12 months" makes a claim about the population of 200 million adults. But Gallup doesn't know the truth about this population. The poll contacted 1523 people and found that 57% of them said they had bought a lottery ticket in the past year. Because the sample of 1523 people was chosen at random, it's reasonable to think that they represent the entire population pretty well. So Gallup turns the *fact* that 57% of the *sample* bought lottery tickets into an *estimate* that about 57% of *all adults* bought tickets. That's a basic move in statistics: use a fact about a sample to estimate the truth about the whole population. To think about such moves, we must keep straight whether a number describes a sample or a population. Here is the vocabulary we use.

Parameters and statistics

A **parameter** is a number that describes the **population.** A parameter is a fixed number, but in practice we don't know its value.

A **statistic** is a number that describes a **sample.** The value of a statistic is known when we have taken a sample, but it can change from sample to sample. We often use a statistic to estimate an unknown parameter.

So parameter is to population as statistic is to sample. Want to estimate an unknown parameter? Choose a sample from the population and use a sample statistic as your estimate. That's what Gallup did.

Example 1. Do you lotto?

The proportion of all adults who bought a lottery ticket in the past 12 months is a *parameter* describing the population of 200 million adults. Call it p, for "proportion." Alas, we do not know the numerical value of p. To estimate p, Gallup took a sample of 1523 adults. The proportion of the sample who bought lottery tickets is a *statistic*. Call it $\hat{p}$ read as "p-hat." It happens that 868 of this sample of size 1523 bought tickets, so for this sample,

$$\hat{p} = \frac{868}{1523} = 0.57 \qquad \text{(that is, 57\%)}$$

Because all adults had the same chance to be among the chosen 1523, it seems reasonable to use the statistic $\hat{p} = 0.57$ as an estimate of the unknown parameter p. It's a *fact* that 57% of the sample bought lottery tickets—we know because we asked them. We don't know what percent of all adults bought tickets, but we *estimate* that about 57% did.

Sampling variability

If Gallup took a second random sample of 1523 adults, the new sample would have different people in it. It is almost certain that there would not be exactly 868 positive responses. That is, the value of the statistic $\hat{p}$ will *vary* from sample to sample. Could it happen that one random sample finds that 57% of adults recently bought a lottery ticket and a second random sample finds that only 37% had done so? Random samples eliminate *bias* from the act of choosing a sample, but they can still be wrong because of the *variability* that results when we choose at random. If the variation when we take repeat samples from the same population is too great, we can't trust the results of any one sample.

We are saved by the second great advantage of random samples. The first advantage is that choosing at random eliminates favoritism. That is, random sampling attacks bias. The second advantage is that if we took lots of random samples of the same size from the same population, the variation from sample to sample would follow a predictable pattern. This predictable pattern shows that results of bigger samples are less variable than the results of smaller samples.

Example 2. Lots and lots of samples

Here's another big idea of statistics: to see how trustworthy one sample is likely to be, ask what would happen if we took many samples from the same population. Let's try it and see. Suppose that in fact (unknown to Gallup), exactly 60% of all adults have bought a lottery ticket in the past 12 months. That is, the truth about the population is that $p = 0.6$. What if Gallup used the sample proportion $\hat{p}$ from an SRS of size 100 to estimate the unknown value of the population proportion p?

Figure 3.1 illustrates the process of choosing many samples and finding $\hat{p}$ for each one. In the first sample, 56 of the 100 people had bought lottery tickets, so $\hat{p} = 56/100 = 0.56$. Only 46 in the next sample had bought tickets, so for that sample $\hat{p} = 0.46$. Choose 1000 samples and make a histogram of the 1000 values of $\hat{p}$. That's the graph at the right of Figure 3.1. The different values of $\hat{p}$ run along the horizontal axis. The heights of the bars show how many of our 1000 samples gave each group of values.

Of course, Gallup interviewed 1523 people, not just 100. Figure 3.2 shows the results of 1000 SRSs, each of size 1523, drawn from a population in which the true sample proportion is $p = 0.6$. Figures 3.1 and 3.2 are drawn on the same scale. Comparing them shows what happens when we increase the size of our samples from 100 to 1523.

Look carefully at Figures 3.1 and 3.2. We flow from the population, to many samples from the population, to the many values of $\hat{p}$ from these many samples. Gather these values together and study the histograms that display them.

- In both cases, the values of the sample proportion $\hat{p}$ vary from sample to sample, but the values are centered at 0.6. Recall that $p = 0.6$ is

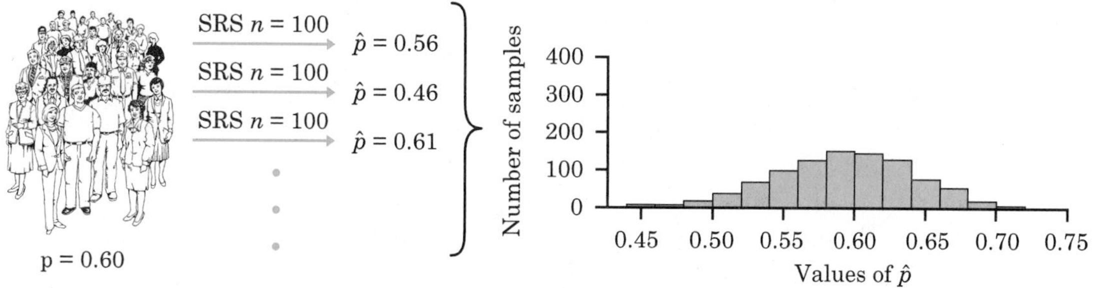

Figure 3.1 The results of many SRSs have a regular pattern. Here, we draw 1000 SRSs of size 100 from the same population. The population proportion is $p = 0.60$. The sample proportions vary from sample to sample, but their values center at the truth about the population.

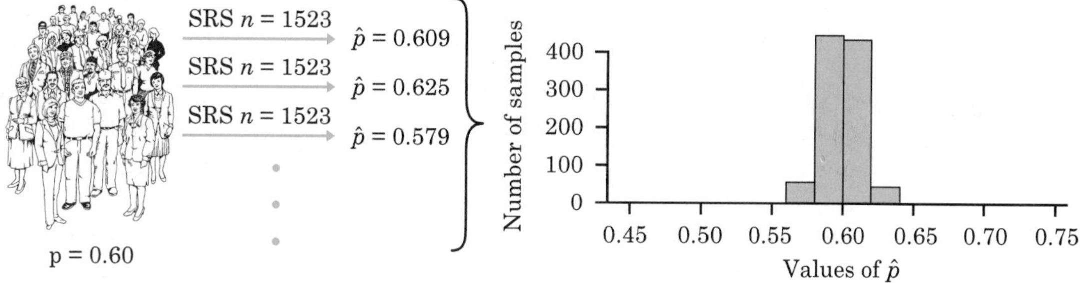

Figure 3.2 Draw 1000 SRSs of size 1523 from the same population as in Figure 3.1. The 1000 values of the sample proportion are much less spread out than was the case for smaller samples.

the true population parameter. Some samples have a $\hat{p}$ less than 0.6 and some greater, but there is no tendency to be always low or always high. That is, $\hat{p}$ has no **bias** as an estimator of p. This is true for both large and small samples.

- The values of $\hat{p}$ from samples of size 100 are much more spread out than the values from samples of size 1523. In fact, 95% of our 1000 samples of size 1523 have a $\hat{p}$ lying between 0.576 and 0.624. That's within 0.024 on either side of the population truth 0.6. Our samples of size 100, on the other hand, spread the middle 95% of their values between 0.50 and 0.69. That goes out 0.1 from the truth, about four times as far as the larger samples. So larger random samples have less **variability** than smaller samples.

The upshot is that we can rely on a sample of size 1523 to almost always give an estimate $\hat{p}$ that is close to the truth about the population. Figure 3.2 illustrates this fact for just one value of the population proportion, but it is true for any population. Samples of size 100, on the other hand, might give an estimate of 50% or 70% when the truth is 60%.

Thinking about Figures 3.1 and 3.2 helps us restate the idea of bias when we use a statistic like $\hat{p}$ to estimate a parameter like p. It also reminds us that variability matters as much as bias.

Two types of error in estimation

Bias is consistent, repeated deviation of the sample statistic from the population parameter in the same direction when we take many samples.

Variability describes how spread out the values of the sample statistic are when we take many samples. Large variability means that the result of sampling is not repeatable.

A good sampling method has both small bias and small variability.

We can think of the true value of the population parameter as the bull's-eye on a target, and of the sample statistic as an arrow fired at the bull's-eye. Bias and variability describe what happens when an archer fires many arrows at the target. *Bias* means that the aim is off, and the arrows land consistently off the bull's-eye in the same direction. The sample values do not center about the population value. Large *variability* means that repeated shots are widely scattered on the target. Repeated samples do not give similar results but differ widely among themselves. Figure 3.3 shows this target illustration of the two types of error.

Notice that small variability (repeated shots are close together) can accompany large bias (the arrows are consistently away from the bull's-eye in one direction). And small bias (the arrows center on the bull's-eye) can accompany large variability (repeated shots are widely scattered). A good sampling scheme, like a good archer, must have both small bias and small variability. Here's how we do this:

Managing bias and variability

To reduce bias, use random sampling. When we start with a list of the entire population, simple random sampling produces *unbiased* estimates—the values of a statistic computed from an SRS neither consistently overestimate nor consistently underestimate the value of the population parameter.

To reduce the variability of an SRS, use a larger sample. You can make the variability as small as you want by taking a large enough sample.

In practice, Gallup only takes one sample. We don't know how close to the truth an estimate from this one sample is, because we don't know what

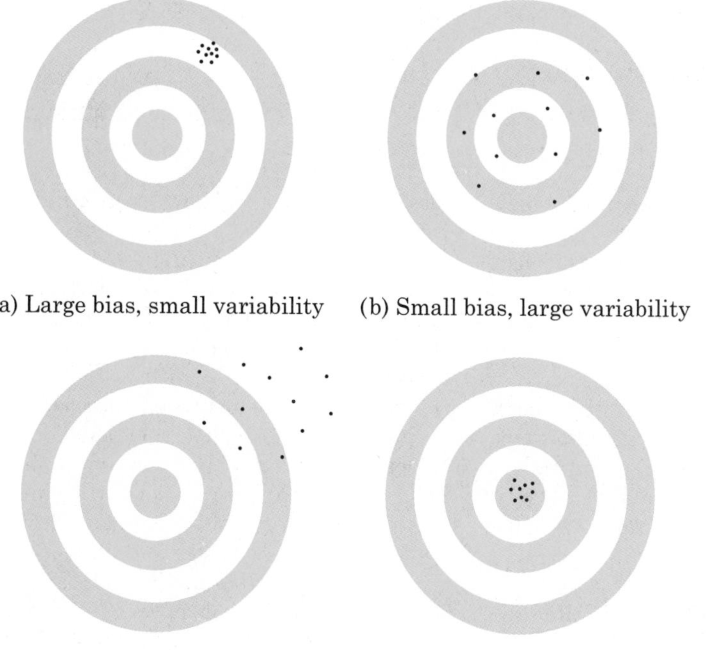

(a) Large bias, small variability (b) Small bias, large variability

(c) Large bias, large variability (d) Small bias, small variability

Figure 3.3 Bias and variability in shooting arrows at a target. Bias means the archer systematically misses in the same direction. Variability means that the arrows are scattered.

the truth about the population is. But *large random samples almost always give an estimate that is close to the truth.* Looking at the pattern of many samples shows that we can trust the result of one sample.

Margin of error and all that

The "margin of error" that sample surveys announce translates sampling variability of the kind pictured in Figures 3.1 and 3.2 into a statement of how much confidence we can have in the results of a survey. Let's start with the kind of language we hear so often in the news.

What margin of error means

"Margin of error plus or minus three percentage points" is shorthand for this statement:

> If we took many samples using the same method we used to get this one sample, 95% of the samples would give a result within plus or minus 3 percentage points of the truth about the population.

Statistical Controversies

Should Election Polls Be Banned?

Preelection polls tell us that Senator So-and-So is the choice of 58% of Ohio voters. The media love these polls. Statisticians don't love them, because elections often don't go as forecast even when the polls use all the right statistical methods. Many people who respond to the polls change their minds before the election. Others say they are undecided. Still others say which candidate they favor but won't bother to vote when the election arrives. Election forecasting is one of the less satisfactory uses of sample surveys because we must ask people now how they will vote in the future.

Exit polls, which interview voters as they leave the voting place, don't share these problems. The people in the sample have just voted. A good exit poll, based on a national sample of election precincts, can often call a presidential election correctly long before the polls close. That fact sharpens the debate over the political effects of election forecasts.

Arguments *against* public preelection polls charge that they influence voter behavior. Voters may decide to stay home if the polls predict a landslide—why bother to vote if the result is a foregone conclusion? Exit polls are particularly worrisome, because they in effect report actual election results before the election is complete. The U.S. television networks agree not to release the results of their exit surveys in any state until the polls close in that state. If a presidential election is not close, the networks may know the winner by midafternoon, but they only forecast the vote one state at a time as the polls close across the country. Even so, a presidential election result may be known before voting ends in the West. Some countries have laws restricting election forecasts. In France, no poll results can be published in the week before a presidential election. Canada forbids poll results in the 72 hours before federal elections. In all, some 30 countries restrict publication of election surveys.

The argument *for* preelection polls is simple: democracies should not forbid publication of information. Voters can decide for themselves how to use the information. After all, supporters of a candidate who is far behind know that even without polls. Restricting publication of polls just invites abuses. In France, candidates continue to take private polls (less reliable than the public polls) in the week before the election. They then leak the results to reporters in the hope of influencing press reports.

Daemmirch/Stock, Boston.

Take this step by step. A sample chosen at random will usually not estimate the truth about the population exactly. We need a margin of error to tell us how close our estimate comes to the truth. But we can't be *certain* that the truth differs from the estimate by no more than the margin of error. Ninety-five percent of all samples come this close to the truth, but 5% miss by more than the margin of error. We don't know the truth about the population, so we don't know if our sample is one of the 95% that hit or one of the 5% that miss. We say we are **95% confident** that the truth lies within the margin of error.

Example 3. Understanding the news

Here's what the TV news announcer says: "A new Gallup poll finds that 57% of American adults bought a lottery ticket in the last 12 months. The margin of error for the poll was 3 percentage points." Plus or minus 3 percent starting at 57% is 54% to 60%. Most people think Gallup claims that the truth about the entire population lies in that range.

This is what Gallup actually said: "For results based on a sample of this size, one can say with 95% confidence that the error attributable to sampling and other random effects could be plus or minus 3 percentage points for adults." That is, Gallup tells us that the margin of error only works for 95% of all its samples. "95% confidence" is shorthand for that. The news report left out the "95% confidence."

Finding the margin of error exactly is a job for statisticians. You can, however, use a simple formula to get a rough idea of the size of a sample survey's margin of error.

A quick method for the margin of error

Use the sample proportion $\hat{p}$ from a simple random sample of size n to estimate an unknown population proportion p. The margin of error for 95% confidence is roughly equal to $1/\sqrt{n}$.

Example 4. What is the margin of error?

The Gallup Poll in Example 1 interviewed 1523 people. The margin of error for 95% confidence will be about

$$\frac{1}{\sqrt{1523}} = \frac{1}{39.03} = 0.026 \qquad \text{(that is, 2.6\%)}$$

Gallup actually announced a margin of error of 3%. Our result differs a bit from Gallup's for two reasons. First, polls usually round their announced margin of error to the nearest whole percent to keep their press releases simple. Second, our rough formula works for an SRS. We will see in the next chapter that most national samples are more complicated than an SRS in ways that tend to slightly increase the margin of error. Nonetheless, our quick method comes pretty close.

"Seventy-three percent are in favor of one through five, forty-one percent find six unfair, thirteen percent are opposed to seven, sixty-two percent applauded eight, thirty-seven percent..."

Our quick method also reveals an important fact about how margins of error behave. Because the sample size n appears in the denominator of the fraction, larger samples have smaller margins of error. We knew that. Because the formula uses the square root of the sample size, however, *to cut the margin of error in half, we must use a sample four times as large.*

Example 5.

In Example 2 we compared the results of taking many SRSs of size $n = 100$ and many SRSs of size $n = 1523$ from the same population. We found that the spread of the middle 95% of the sample results was about four times as large for the smaller samples.

Our quick formula estimates the margin of error for SRSs of size 1523 to be about 2.6%. The margin of error for SRSs of size 100 is about

$$\frac{1}{\sqrt{100}} = \frac{1}{10} = 0.1 \qquad \text{(that is, 10\%)}$$

Because 1523 is roughly 16 times 100 and the square root of 16 is 4, the margin of error is about four times larger for samples of 100 people than for samples of 1523 people.

Confidence statements

Here is Gallup's conclusion about buying lottery tickets in short form: "The poll found that 57% of adults bought a lottery ticket in the past 12 months. We are 95% confident that the truth about all adults is within plus or minus

3 percentage points of this sample result." Here is an even shorter form: "We are 95% confident that between 54% and 60% of all adults bought a lottery ticket in the last 12 months." These are *confidence statements.*

Confidence statements

A **confidence statement** has two parts: a **margin of error** and a **level of confidence.** The margin of error says how close the sample statistic lies to the population parameter. The level of confidence says what percent of all possible samples satisfy the margin of error.

A confidence statement is a fact about what happens in all possible samples, used to say how much we can trust the result of one sample. "95% confidence" means "We used a sampling method that gives a result this close to the truth 95% of the time." Here are some hints for interpreting confidence statements:

- *The conclusion of a confidence statement always applies to the population, not to the sample.* We know exactly how the 1523 people in the sample acted, because Gallup interviewed them. The confidence statement uses the sample result to say something about the population of all adults.
- *Our conclusion about the population is never completely certain.* Gallup's sample *might* be one of the 5% that miss by more than 3 percentage points.
- *A sample survey can choose to use a confidence level other than 95%.* We pay for higher confidence with a larger margin of error. For the same sample, a 99% confidence statement requires a larger margin of error than 95% confidence. If you are content with 90% confidence, you get in return a smaller margin of error. Remember that our quick method only gives the margin of error for 95% confidence.
- *It is usual to report the margin of error for 95% confidence.* If a news report gives a margin of error but leaves out the confidence level, it's pretty safe to assume 95% confidence.
- *Want a smaller margin of error with the same confidence? Take a larger sample.* Remember that

The telemarketer's pause

People who do sample surveys hate telemarketing. We all get so many unwanted sales pitches by phone that many people hang up before learning that the caller is conducting a survey rather than selling vinyl siding. Here's a tip. Both sample surveys and telemarketers dial telephone numbers at random. Telemarketers automatically dial many numbers, and their sellers only come on the line after you pick up the phone. Once you know this, the telltale "telemarketer's pause" gives you a chance to hang up before the seller arrives. Sample surveys have a live interviewer on the line when you answer.

larger samples have less variability. You can get as small a margin of error as you want and still have high confidence by paying for a large enough sample.

Example 6. Do you approve of gambling?

Gallup began its poll on gambling with this question: "First, generally speaking, do you approve or disapprove of legal gambling or betting?" In addition to the 1523 adults (ages 18 and older) whose lottery habits we have explored, Gallup took a random sample of 501 teenagers (ages 13 to 17). The sample results were

Adults:	959 out of 1523 approve	$\hat{p} = 959/1523 = 0.63$
Teens:	261 out of 501 approve	$\hat{p} = 261/501 = 0.52$

After reporting these and other results, Gallup says: "For results based on a sample of this size, one can say with 95 percent confidence that the error attributable to sampling and other random effects could be plus or minus 3 percentage points for adults (18+), and plus or minus 5 percentage points for teens (13–17 year olds)."

There you have it: the sample of teens is smaller, so the margin of error for conclusions about teens is wider. We are 95% confident that between 47% (that's 52% minus 5%) and 57% (that's 52% plus 5%) of all teenagers approve of legal gambling.

Sampling from large populations

Gallup's sample of 1523 adults is only 1 out of every 130,000 adults in the United States. Does it matter whether 1523 is 1-in-100 individuals in the population or 1-in-130,000?

Population size doesn't matter

The variability of a statistic from a random sample does not depend on the size of the population, as long as the population is at least 100 times larger than the sample.

Why does the size of the population have little influence on the behavior of statistics from random samples? Imagine sampling harvested corn by thrusting a scoop into a lot of corn kernels. The scoop doesn't know whether it is surrounded by a bag of corn or by an entire truckload. As long as the corn is well mixed (so that the scoop selects a random sample), the variability of the result depends only on the size of the scoop.

This is good news for national sample surveys like the Gallup Poll. A random sample of size 1000 or 1500 has small variability because the sample size is large. But remember that even a very large voluntary response sample or convenience sample is worthless because of bias. Taking a larger sample doesn't fix bias.

However, the fact that the variability of a sample statistic depends on the size of the sample and not on the size of the population is bad news for anyone planning a sample survey in a university or a small city. For example, it takes just as large an SRS to estimate the proportion of Ohio State University students who call themselves political conservatives as to estimate with the same margin of error the proportion of all adult U.S. residents who are conservatives. We can't use a smaller SRS at Ohio State just because there are 48,000 Ohio State students and 200 million adults in the United States.

Exploring the Web

You can read the Gallup Organization's own explanation of why surprisingly small samples can give trustworthy results about large populations at www.gallup.com/poll/faq.asp.

Statistics in Summary The purpose of sampling is to use a sample to gain information about a population. We often use a sample **statistic** to estimate the value of a population **parameter.** This chapter has one big idea: to describe how trustworthy a sample is, ask "What would happen if we took a large number of samples from the same population?" If almost all samples would give a result close to the truth, we can trust our one sample even though we can't be certain that it is close to the truth.

In planning a sample survey, first aim for small **bias** by using random sampling and avoiding bad sampling methods such as voluntary response. Next, choose a large enough random sample to reduce the **variability** of the result. Using a large random sample guarantees that almost all samples will give accurate results. To say how accurate our conclusions about the population are, make a **confidence statement.** News reports often mention only the **margin of error.** Most often this margin of error is for **95% confidence.** That is, if we chose many samples, the truth about the population would be within the margin of error 95% of the time. We can estimate the margin of error for 95% confidence based on a simple random sample of size n by the formula $1/\sqrt{n}$. As this formula suggests, only the size of the sample, not the size of the population, matters. This is true as long as the population is much larger than the sample.

CHAPTER 3 EXERCISES

Each boldface number in Exercises 3.1 to 3.4 is the value of either a **parameter** or a **statistic.** In each case, state which it is.

3.1 The Bureau of Labor Statistics announces that last month it interviewed all members of the labor force in a sample of 50,000 households; **4.5%** of the people interviewed were unemployed.

3.2 A carload lot of ball bearings has an average diameter of **2.503** centimeters (cm). This is within the specifications for acceptance of the lot by the purchaser. The inspector happens to inspect 100 bearings from the lot with an average diameter of **2.515** cm. This is outside the specified limits, so the lot is mistakenly rejected.

3.3 A telemarketer in Los Angeles uses a random digit dialing device to dial residential phone numbers in that city at random. Of the first 100 numbers dialed, **43** are unlisted numbers. This is not surprising, because **52%** of all Los Angeles residential phones are unlisted.

3.4 Voter registration records show that **68%** of all voters in Indianapolis are registered as Republicans. To test a random digit dialing device, you use the device to call 150 randomly chosen residential telephones in Indianapolis. Of the registered voters contacted, **73%** are registered Republicans.

3.5 **A sampling experiment.** Figures 3.1 and 3.2 show how the sample proportion $\hat{p}$ behaves when we take many samples from the same population. You can follow the steps in this process on a small scale.

Figure 3.4 is a small population. Each circle represents an adult. The colored circles are people who disapprove of legal gambling, and the white circles are people who approve. You can check that 60 of the 100 circles are white, so in this population the proportion who approve of gambling is $p = 60/100 = 0.6$.

(a) The circles are labeled 00, 01, . . . , 99. Use line 101 of Table A to draw an SRS of size 5. What is the proportion $\hat{p}$ of the people in your sample who approve of gambling?

(b) Take 9 more SRSs of size 5 (10 in all), using lines 102 to 110 of Table A, a different line for each sample. You now have 10 values of the sample proportion $\hat{p}$.

(c) Because your samples have only 5 people, the only values $\hat{p}$ can take are 0/5, 1/5, 2/5, 3/5, 4/5, and 5/5. That is, $\hat{p}$ is always 0, 0.2, 0.4, 0.6, 0.8, or 1. Mark these numbers on a line and make a histogram of your 10 results by putting a bar above each number to show how many samples had that outcome.

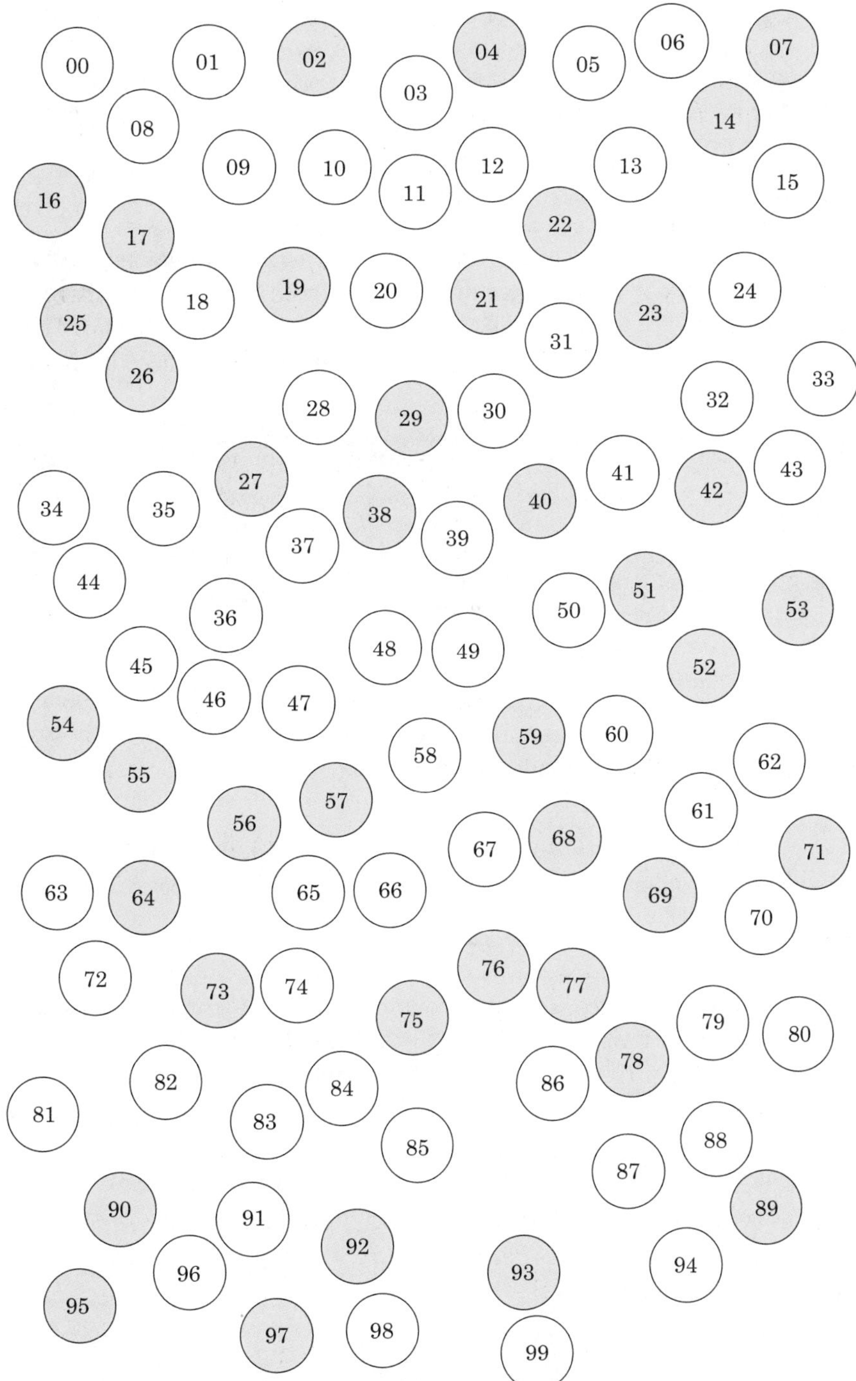

Figure 3.4 A population of 100 individuals for Exercise 3.5. Some individuals (white circles) approve of legal gambling, and the others disapprove.

(d) Taking samples of size 5 from a population of size 100 is not a practical setting, but let's look at your results anyway. How many of your 10 samples estimated the population proportion $p = 0.6$ exactly correctly? Is the true value 0.6 roughly in the center of your sample values? Explain why 0.6 would be in the center of the sample values if you took a large number of samples.

3.6 A sampling experiment. Let us illustrate sampling variability in a small sample from a small population. Ten of the 25 club members listed below are female. Their names are marked with asterisks in the list. The club chooses 5 members at random to receive free trips to the national convention.

Alonso	Darwin	Herrnstein	Myrdal	Vogt*
Binet*	Epstein	Jimenez*	Perez*	Went
Blumenbach	Ferri	Luo	Spencer*	Wilson
Chase*	Gonzales	Moll*	Thomson	Yerkes
Chen*	Gupta*	Morales*	Toulmin	Zimmer

(a) Draw 20 SRSs of size 5, using a different part of Table A each time. Record the number of females in each of your samples. Make a histogram like that in Figure 3.1 to display your results. What is the average number of females in your 20 samples?

(b) Do you think the club members should suspect discrimination if none of the 5 tickets go to women?

3.7 Canada's national health care. The Ministry of Health in the Canadian Province of Ontario wants to know whether the national health care system is achieving its goals in the province. Much information about health care comes from patient records, but that source doesn't allow us to compare people who use health services with those who don't. So the Ministry of Health conducted the Ontario Health Survey, which interviewed a random sample of 61,239 people who live in the Province of Ontario.

(a) What is the population for this sample survey? What is the sample?

(b) The survey found that 76% of males and 86% of females in the sample had visited a general practitioner at least once in the past year. Do you think these estimates are close to the truth about the entire population? Why?

3.8 Bigger samples, please. Explain in your own words the advantages of bigger random samples in a sample survey.

3.9 Sampling variability. In thinking about Gallup's sample of size 1523, we asked, "Could it happen that one random sample finds that 57% of adults recently bought a lottery ticket and a second random sample finds that only 37% had done so?" Look at Figure 3.2, which shows the results of 1000 samples of this size when the population truth is $p = 0.6$, or 60%. Would you be surprised if a sample from this population gave 57%? Would you be surprised if a sample gave 37%?

3.10 Balancing the budget. A poll of 1190 adults finds that 702 prefer balancing the budget over cutting taxes. The announced margin of error for this result is plus or minus 4 percentage points. The news report does not give the confidence level. You can be quite sure that the confidence level is 95%.
(a) What is the value of the sample proportion $\hat{p}$ who prefer balancing the budget? Explain in words what the population parameter p is in this setting.
(b) Make a confidence statement about the parameter p.

3.11 Bias and variability. Figure 3.5 shows the behavior of a sample statistic in many samples in four situations. These graphs are like Figures 3.1 and 3.2. That is, the heights of the bars show how often the sample statistic took various values in many samples from the same population. The true value of the population parameter is marked on each graph. Label each of the graphs in Figure 3.5 as showing high or low bias and as showing high or low variability.

3.12 Predict the election. Just before a presidential election, a national opinion poll increases the size of its weekly sample from the usual 1500

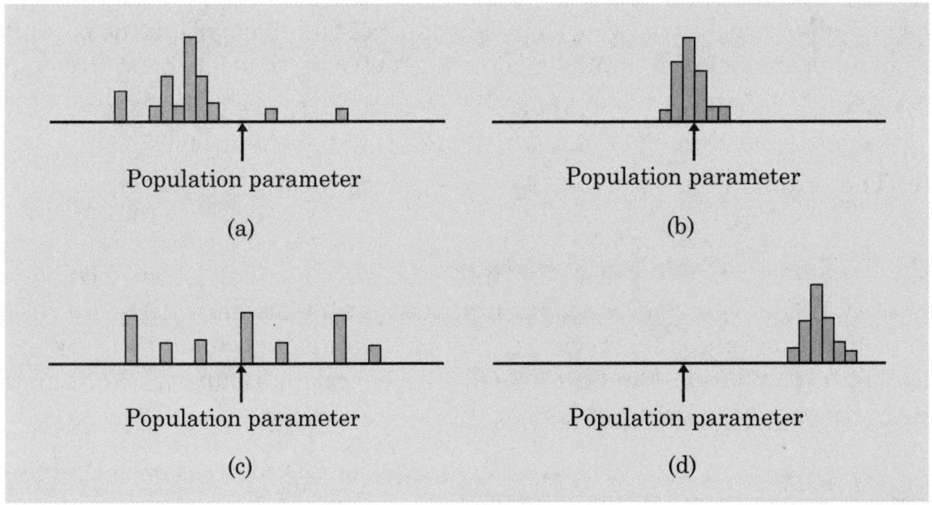

Figure 3.5 Take many samples from the same population and make a histogram of the values taken by a sample statistic. Here are the results for four different sampling methods, for Exercise 3.11.

people to 4000 people. Does the larger random sample reduce the bias of the poll result? Does it reduce the variability of the result?

3.13 Take a bigger sample. A management student is planning a project on student attitudes toward part-time work while attending college. She develops a questionnaire and plans to ask 25 randomly selected students to fill it out. Her faculty advisor approves the questionnaire but suggests that the sample size be increased to at least 100 students. Why is the larger sample helpful? Back up your statement by using the quick method to estimate the margin of error for samples of size 25 and for samples of size 100.

3.14 Sampling in the states. An agency of the federal government plans to take an SRS of residents in each state to estimate the proportion of owners of real estate in each state's population. The population of the states ranges from about 525,000 people in Wyoming to about 33 million in California.
(a) Will the variability of the sample proportion change from state to state if an SRS of size 2000 is taken in each state? Explain your answer.

(b) Will the variability of the sample proportion change from state to state if an SRS of 1/10 of 1% (0.001) of the state's population is taken in each state? Explain your answer.

3.15 Polling women. A *New York Times* poll on women's issues interviewed 1025 women randomly selected from the United States, excluding Alaska and Hawaii. The poll found that 47% of the women said they do not get enough time for themselves.
(a) The poll announced a margin of error of ±3 percentage points for 95% confidence in its conclusions. Make a 95% confidence statement about the percent of all adult women who think they do not get enough time for themselves.

(b) Explain to someone who knows no statistics why we can't just say that 47% of all adult women do not get enough time for themselves.

(c) Then explain clearly what "95% confidence" means.

3.16 The Harris Poll. Here is the language used by the Harris Poll to explain the accuracy of its results: "In theory, with a sample of this size, one can say with 95 percent certainty that the results have a statistical precision of plus or minus 3 percentage points of what they would be if the entire adult population had been polled with complete accuracy." What does Harris mean by "95 percent certainty"?

3.17 Polling men and women. The sample survey described in Exercise 3.15 interviewed 472 randomly selected men as well as 1025 women. The poll announced a margin of error of ±3 percentage points for 95% confidence in conclusions about women. The margin of error for results concerning men was ±5 percentage points. Why is this larger than the margin of error for women?

3.18 Explaining confidence. A student reads that we are 95% confident that the average score of young men on the quantitative part of the National Assessment of Educational Progress is 267.8 to 276.2. Asked to explain the meaning of this statement, the student says, "95% of all young men have scores between 267.8 and 276.2." Is the student right? Explain your answer.

3.19 Gun violence. The Harris Poll asked a sample of 1009 adults which causes of death they thought would become more common in the future. Topping the list was gun violence—70% of the sample thought deaths from guns would increase.
(a) How many of the 1009 people interviewed thought deaths from gun violence would increase?

(b) Harris says that the margin of error for this poll is plus or minus 3 percentage points. Explain to someone who knows no statistics what "margin of error plus or minus 3 percentage points" means.

3.20 Find the margin of error. Example 6 tells us that Gallup asked 501 teenagers whether they approved of legal gambling; 52% said they did. Use the quick method to estimate the margin of error for conclusions about all teenagers. How does your result compare with Gallup's margin of error quoted in Example 6?

3.21 Find the margin of error. Exercise 3.19 concerns a Harris Poll sample of 1009 people. Use the quick method to estimate the margin of error for statements about the population of all adults. Is your result close to the 3% margin of error announced by Harris?

3.22 Find the margin of error. Exercise 3.7 describes a sample survey of 61,239 adults living in Ontario. About what is the margin of error for conclusions having 95% confidence about the entire adult population of Ontario?

3.23 Is there a hell? A news article reports that in a recent opinion poll, 78% of a sample of 1108 adults said they believe there is a heaven. Only 60% said they believe there is a hell.
(a) Use the quick method to estimate the margin of error for a sample of this size.

(b) Make a confidence statement about the percent of all adults who believe there is a hell.

3.24 Fear of crime. The Gallup Poll asked a random sample of 1493 adults, "Are you afraid to go out at night within a mile of your home because of crime?" Of the sample, 672 said "Yes." Make a confidence statement about the percent of all adults who fear to go out at night because of crime. (Use the quick method to find the margin of error.)

3.25 Smaller margin of error. Exercise 3.21 describes an opinion poll that interviewed 1009 people. Suppose you want a margin of error half as large as the one you found in that exercise. How many people must you plan to interview?

3.26 Satisfying Congress. Exercise 3.10 describes a sample survey of 1190 adults, with margin of error ±4% for 95% confidence.
(a) A member of Congress thinks that 95% confidence is not enough. He wants to be 99% confident. How would the margin of error for 99% confidence based on the same sample compare with the margin of error for 95% confidence?

(b) Another member of Congress is satisfied with 95% confidence, but she wants a smaller margin of error than ±4 percentage points. How can we get a smaller margin of error, still with 95% confidence?

3.27 The Current Population Survey. Though opinion polls usually make 95% confidence statements, some sample surveys use other confidence levels. The monthly unemployment rate, for example, is based on the Current Population Survey of about 50,000 households. The margin of error in the unemployment rate is announced as about two-tenths of one percentage point with 90% confidence. Would the margin of error for 95% confidence be smaller or larger? Why?

3.28 The millennium begins with optimism. In January of the year 2000, the Gallup Poll asked a random sample of 1633 adults, "In general, are you satisfied or dissatisfied with the way things are going in the United States at this time?" It found that 1127 said that they were satisfied. Write a short report of this finding, as if you were writing for a newspaper. Be sure to include a margin of error.

3.29 Simulation. Random digits can be used to *simulate* the results of random sampling. Suppose that you are drawing simple random samples of size 25 from a large number of college students and that 20% of the students are unemployed during the summer. To simulate this SRS, let 25 consecutive digits in Table A stand for the 25 students in your sample. The digits 0 and 1 stand for unemployed students, and other digits stand for employed students. This is an accurate imitation of the SRS because 0 and 1 make up 20% of the 10 equally likely digits.

Simulate the results of 50 samples by counting the number of 0s and 1s in the first 25 entries in each of the 50 rows of Table A. Make a histogram like that in Figure 3.1 to display the results of your 50 samples. Is the truth about the population (20% unemployed, or 5 in a sample of 25) near the center of your graph? What are the smallest and largest counts of unemployed students you obtained in your 50 samples? What percent of your samples had either 4, 5, or 6 unemployed?

Chapter 4
Sample Surveys in the Real World

The whole truth about opinion polls

An opinion poll talks to 1000 people chosen at random, announces its results, and announces a margin of error. Should we be happy? Perhaps not. Many polls don't tell the whole truth about their samples. The Pew Research Center for the People and the Press imitated the methods of the better opinion polls and did tell the whole truth. Here it is.

Most polls are taken by telephone, dialing numbers at random to get a random sample of households. After eliminating fax and business numbers, Pew had to call 2879 residential numbers to get their sample of 1000 people. Here's the breakdown:

Never answered phone	938
Answered but refused	678
Not eligible: no person age 18, or language barrier	221
Incomplete interview	42
Complete interview	1000
Total called	2879

Out of 2879 good residential phone numbers, 33% never answered. Of those who answered, 35% refused to talk. The overall rate of nonresponse (people who never answered, refused, or would not complete the interview) was 1658 out of 2879, or 58%. Pew called every number five times over a five-day period, at different times of day and on different days of the week. Many polls call only once, and it is usual to find that over half those who answer refuse to talk. In the real world, a simple random sample isn't simple and may not be random. That's our issue for this chapter.

How sample surveys go wrong

Random sampling eliminates bias in choosing a sample and allows control of variability. So once we see the magic words "randomly selected" and "margin of error," do we know we have trustworthy information before us? It certainly beats voluntary response, but not always by as much as we might hope. Sampling in the real world is more complex and less reliable than choosing an SRS from a list of names in a textbook exercise. Confidence statements do not reflect all of the sources of error that are present in practical sampling.

Errors in sampling

Sampling errors are errors caused by the act of taking a sample. They cause sample results to be different from the results of a census.

Random sampling error is the deviation between the sample statistic and the population parameter caused by chance in selecting a random sample. The margin of error in a confidence statement includes *only* random sampling error.

Nonsampling errors are errors not related to the act of selecting a sample from the population. They can be present even in a census.

Most sample surveys are afflicted by errors other than random sampling errors. These errors can introduce bias that makes a confidence statement meaningless. Good sampling technique includes the art of reducing all sources of error. Part of this art is the science of statistics, with its random samples and confidence statements. In practice, however, good statistics isn't all there is to good sampling. Let's look at sources of errors in sample surveys and at how samplers combat them.

Sampling errors

Random sampling error is one kind of sampling error. The margin of error tells us how serious random sampling error is, and we can control it by choosing the size of our random sample. Another source of sampling error is the use of *bad sampling methods,* such as voluntary response. We can avoid bad methods. Other sampling errors are not so easy to handle. Sampling begins with a list of individuals from which we will draw our sample. This list is called the **sampling frame.** Ideally, the sampling frame should list every individual in the population. Because a list of the entire population is rarely available, most samples suffer from some degree of *undercoverage.*

Undercoverage

Undercoverage occurs when some groups in the population are left out of the process of choosing the sample.

If the sampling frame leaves out certain classes of people, even random samples from that frame will be biased. Using telephone directories as the frame for a telephone survey, for example, would miss everyone with an unlisted telephone number. More than half the households in many large cities have unlisted numbers, so massive undercoverage and bias against urban areas would result. In fact, telephone surveys use random digit dialing equipment that dials telephone numbers in selected regions at random. In effect, the sampling frame contains all residential telephone numbers.

Example 1. We undercover

Most opinion polls can't afford to even attempt full coverage of the population of all adult residents of the United States. The interviews are done by telephone, thus missing the 6% of households without phones. Only households are contacted, so that students in dormitories, prison inmates, and most members of the armed forces are left out. So are the homeless and people staying in shelters. Because calls to Alaska and Hawaii are expensive, most polls restrict their samples to the continental states. Many polls interview only in English, which leaves some immigrant households out of their samples.

Opting out of modern society was tiring at times, but Ted felt that increasing the undercoverage of opinion polls made it all worthwhile.

The kinds of undercoverage found in most sample surveys are most likely to leave out people who are young or poor or who move often. Nonetheless, random digit dialing comes close to producing a random sample of households with phones outside Alaska and Hawaii. Sampling errors in careful sample surveys are usually quite small. The real problems start when someone picks up (or doesn't pick up) the phone. Now nonsampling errors take over.

Nonsampling errors

Nonsampling errors are those that can plague even a census. They include **processing errors,** mistakes in mechanical tasks such as doing arithmetic or entering responses into a computer. The spread of computers has made processing errors less common than in the past.

Example 2. Computer-assisted interviewing

The days of the interviewer with a clipboard are past. Contemporary interviewers carry a laptop computer for face-to-face interviews or watch a computer screen as they carry out a telephone interview. Computer software manages the interview. The interviewer reads questions from the computer screen and uses the keyboard to enter the responses. The computer skips irrelevant items—once a respondent says that she has no children, further questions about her children never appear. The computer can check that answers to related questions are consistent with each other. It can even present questions in random order to avoid any bias due to always asking questions in the same order.

Computer software also manages the sampling process. It keeps records of who has responded and prepares a file of data from the responses. The tedious process of transferring responses from paper to computer, once a source of processing errors, has disappeared. The computer even schedules the calls in telephone surveys, taking account of the respondent's time zone and honoring appointments made by people who were willing to respond but did not have time when first called.

Another type of nonsampling error is **response error,** which occurs when a subject gives an incorrect response. A subject may lie about her age or income, or about whether she has used illegal drugs. She may remember incorrectly when asked how many packs of cigarettes she smoked last week. A subject who does not understand a question may guess at an answer rather than appear ignorant. Questions that ask subjects about their behavior during a fixed time period are notoriously prone to response errors due to faulty memory. For example, the National Health Survey asks people how many times they have visited a doctor in the past year. Checking their responses against health records found that they failed to remember 60% of their visits to a doctor. A survey that asks about sensitive issues can also expect response errors, as the next example illustrates.

Example 3. The effect of race

In 1989, New York City elected its first black mayor and the state of Virginia elected its first black governor. In both cases, samples of voters interviewed as they left their polling places predicted larger margins of victory than the official vote counts. The polling organizations were certain that some voters lied when interviewed because they felt uncomfortable admitting that they had voted against the black candidate.

Technology and attention to detail can minimize processing errors. Skilled interviewers greatly reduce response errors, especially in face-to-face interviews. There is no simple cure, however, for the most serious kind of nonsampling error, *nonresponse.*

Nonresponse

Nonresponse is the failure to obtain data from an individual selected for a sample. Most nonresponse happens because some subjects can't be contacted or because some subjects who are contacted refuse to cooperate.

"You can call, you can send email, you can stand at the door all day. The answer is still NO!"

Nonresponse is the most serious problem facing sample surveys. People are increasingly reluctant to answer questions, particularly over the phone. The rise of telemarketing, answering machines, and caller ID drives down response to telephone surveys. Gated communities and buildings guarded by doormen prevent face-to-face interviews. Nonresponse can bias sample survey results, because different groups have different rates of nonresponse. Refusals are higher in large cities and among the elderly, for example. Bias due to nonresponse can easily overwhelm the random sampling error described by a survey's margin of error.

Example 4. How bad is nonresponse?

The Current Population Survey has the best response rate of any poll I know: only about 6% or 7% of the households in the CPS sample don't respond. People are more likely to respond to a government survey such as the CPS, and the CPS contacts its sample in person before doing later interviews by phone.

The General Social Survey (Example 7 in Chapter 1) also contacts its sample in person, and it is run by a university. Despite these advantages, its recent surveys have a 24% rate of nonresponse.

What about polls done by the media and by market research and opinion polling firms? We don't know their rates of nonresponse, because they won't say. That itself is a bad sign. The Pew study we looked at suggests how bad things are. Pew got 1221 responses (of whom 1000 were in the population they targeted) and 1658 who were never at home, refused, or would not finish the interview. That's a nonresponse rate of 1658 out of 2879, or 58%. The Pew researchers were more thorough than many polls. Insiders say that nonresponse often reaches 75% or 80% of an opinion poll's original sample.

Sample surveyors know some tricks to reduce nonresponse. Carefully trained interviewers can keep people on the line if they answer at all. Calling back over longer time periods helps. So do letters sent in advance. Letters and many callbacks slow down the survey, so opinion polls that want fast answers to satisfy the media don't use them. Even the most careful surveys find that nonresponse is a problem that no amount of expertise can fully overcome. That makes this reminder even more important.

What the margin of error doesn't say

The announced margin of error for a sample survey covers only random sampling error. Undercoverage, nonresponse, and other practical difficulties can cause large bias that is not covered by the margin of error.

Careful sample surveys tell us this. Gallup, for example, says, "In addition to sampling error, question wording and practical difficulties in conducting surveys can introduce error or bias into the findings of public opinion polls." How true it is.

Does nonresponse make many sample surveys useless? Maybe not. We began this chapter with an account of a "standard" telephone survey done by the Pew Center. The Pew researchers also carried out a "rigorous" survey, with letters sent in advance, unlimited calls over eight weeks, letters by priority mail to people who refused, and so on. All this drove the rate of nonresponse down to 30%, compared with 58% for the standard survey. Pew then compared the answers to the same questions from the two surveys. The two samples were quite similar in age, sex, and race, though the rigorous sample was a bit more prosperous. The two samples also held similar opinions on all issues except one: race. People who at first refused to respond were less sympathetic to blacks. Overall, it appears that standard polls give reasonably accurate results. But, as in Example 3, race is again an exception.

Wording questions

A final influence on the results of a sample survey is the exact wording of questions. It is surprisingly difficult to word questions that are completely clear. A survey that asked about "ownership of stock" found that most Texas ranchers owned stock, though probably not the kind traded on the New York Stock Exchange.

Example 5. A few words make a big difference

How do Americans feel about government help for the poor? Only 13% think we are spending too much on "assistance to the poor," but 44% think we are spending too much on "welfare." How do the Scots feel about the movement to become independent from England? Well, 51% would vote for "independence for Scotland," but only 34% support "an independent Scotland separate from the United Kingdom."

It seems that "assistance to the poor" and "independence" are nice, hopeful words. "Welfare" and "separate" are negative words. Small changes in how a question is worded can make a big difference in the response.

The wording of questions always influences the answers. If the questions are slanted to favor one response over others, we have another source of nonsampling error. A favorite trick is to ask if the subject favors some policy as a means to a desirable end: "Do you favor banning private ownership of handguns in order to reduce the rate of violent crime?" and "Do you favor imposing the death penalty in order to reduce the rate of violent crime?" are loaded questions that draw positive responses from people who are worried about crime. Here is an example of the influence of a slanted question.

Example 6. Campaign finance

The financing of political campaigns is a perennial issue. Here are two opinion poll questions on this topic:

> *Should laws be passed to eliminate all possibilities of special interests giving huge sums of money to candidates?*

> *Should laws be passed to prohibit interest groups from contributing to campaigns, or do groups have a right to contribute to the candidate they support?*

The first question was posed by Ross Perot, the third-party candidate in the 1992 presidential election. It drew a 99% "Yes" response in a mail-in poll. We know that voluntary response polls are worthless, so the Yankelovich survey firm asked the same question of a nationwide random sample. The result: 80% "Yes." Perot's question almost demands a "Yes" answer, so Yankelovich wrote the second question as a more neutral way of presenting the issue. Only 40% of a nationwide random sample wanted to prohibit contributions when asked this question.

How to live with nonsampling errors

Nonsampling errors, especially nonresponse, are always with us. What should a careful sample survey do about this? First, **substitute other households** for the nonresponders. Because nonresponse is higher in cities, replacing nonresponders with other households in the same neighborhood may reduce bias. Once the data are in, all professional surveys use statistical methods to **weight the responses** in an attempt to correct sources of bias. If many urban households did not respond, the survey gives more weight to those that did respond. If too many women are in the sample, the survey gives more weight to the men. Here, for example, is part of a statement in the *New York Times* describing one of its sample surveys:

> *The results have been weighted to take account of household size and number of telephone lines into the residence and to adjust for variations in the sample relating to geographic region, sex, race, age and education.*

Doonesbury BY GARRY TRUDEAU

The goal is to get results "as if" the sample matched the population in age, gender, place of residence, and other variables.

The practice of weighting creates job opportunities for statisticians. It also means that the results announced by a sample survey are rarely as simple as they seem to be. Gallup announces that it interviewed 1523 adults and found that 57% of them bought a lottery ticket in the last 12 months. It would seem that because 57% of 1523 is 868, Gallup found that 868 people in its sample had played the lottery. Not so. Gallup no doubt used some quite fancy statistics to weight the actual responses: 57% is Gallup's best estimate of what it would have found in the absence of nonresponse. Weighting does help correct bias. It usually also increases variability. The announced margin of error must take this into account, creating more work for statisticians.

He started it!

A study of deaths in bar fights showed that in 90% of the cases, the person who died started the fight. You shouldn't believe this. If you killed someone in a fight, what would you say when the police ask you who started the fight? After all, dead men tell no tales. Now that's nonresponse.

Sample design in the real world

The basic idea of sampling is straightforward: take an SRS from the population and use a statistic from your sample to estimate a parameter of the population. We now know that the sample statistic is fiddled with behind the scenes to partly correct for nonresponse. The statisticians also have their hands on our beloved SRS. In the real world, most sample surveys use more complex designs.

Example 7. The Current Population Survey

The CPS population consists of all households in the United States (including Alaska and Hawaii). The sample is **chosen in stages.** The Census Bureau divides the nation into 2007 geographic areas called Primary Sampling Units (PSUs). These are generally groups of neighboring counties. At the first stage, 792 PSUs are chosen. This isn't an SRS. If all PSUs had the same chance to be chosen, the sample might miss Chicago and Los Angeles. So 432 highly populated PSUs are automatically in the sample. The other 1575 are grouped into 360 **strata** by combining PSUs that are similar in various ways. One PSU is chosen at random to represent each stratum.

Each of the 792 PSUs in the first-stage sample is divided into census blocks (smaller geographic areas). The blocks are also grouped into strata, based on such things as housing types and minority population. The households in each block are arranged in order of their location and grouped into **clusters** of about four households each. The final sample is a sample of clusters (not of individual households) from each stratum of blocks. Interviewers go to all households in the chosen clusters. The samples of clusters within each stratum of blocks are also not SRSs. To be sure that the clusters spread out geographically, the sample starts at a random cluster and then takes (say) every 10th cluster in the list.

The design of the CPS illustrates several ideas that are common in real-world samples that use face-to-face interviews. Taking the sample **in several stages** with **clusters** at the final stage saves travel time for interviewers by grouping the sample households first in PSUs and then in clusters. The most important refinement mentioned in Example 7 is *stratified sampling.*

Stratified sample

To choose a **stratified random sample:**

Step 1. Divide the sampling frame into distinct groups of individuals, called **strata.** Choose the strata because you have a special interest in these groups within the population or because the individuals in each stratum resemble each other.

Step 2. Take a separate SRS in each stratum and combine these to make up the complete sample.

We must of course choose the strata using facts about the population that are known before we take the sample. You might group a university's students into undergraduate and graduate students or into those who live on campus and those who commute. Stratified samples have some advantages over an SRS. First, by taking a separate SRS in each stratum, we can set sample sizes to allow separate conclusions about each stratum. Second, a stratified sample usually has a smaller margin of error than an SRS of the same size. The reason is that the individuals in each stratum are more alike than the population as a whole, so working stratum-by-stratum eliminates some variability in the sample.

It may surprise you that stratified samples can violate one of the most appealing properties of the SRS—stratified samples need not give all individuals in the population the same chance to be chosen. Some strata may be deliberately overrepresented in the sample.

Example 8. Stratifying a sample of students

A large university has 30,000 students, of whom 3000 are graduate students. An SRS of 500 students gives every student the same chance to be in the sample. That chance is

$$\frac{500}{30{,}000} = \frac{1}{60}$$

We expect an SRS to contain only about 50 grad students—because grad students make up 10% of the population, we expect them to make up about 10% of an SRS. A sample of size 50 isn't large enough to estimate grad student opinion with reasonable accuracy. We might prefer a stratified random sample of 200 grad students and 300 undergraduates.

You know how to select such a stratified sample. Label the graduate students 0001 to 3000 and use Table A to select an SRS of 200. Then label the undergraduates 00001 to 27000 and use Table A a second time to select an SRS of 300 of them. These two SRSs together form the stratified sample.

In the stratified sample, each grad student has chance

$$\frac{200}{3000} = \frac{1}{15}$$

to be chosen. Each of the undergraduates has a smaller chance,

$$\frac{300}{27{,}000} = \frac{1}{90}$$

Because we have two SRSs, it is easy to estimate opinions in the two groups separately. The quick method (page 37) tells us that the margin of error for a sample proportion will be about

$$\frac{1}{\sqrt{200}} = 0.07 \quad \text{(that is, 7\%)}$$

for grad students and about

$$\frac{1}{\sqrt{300}} = 0.058 \quad \text{(that is, 5.8\%)}$$

for undergraduates.

New York, New York

New York City, they say, is bigger, richer, faster, ruder. Maybe there's something to that. The sample survey firm Zogby International says that as a national average it takes 5 telephone calls to reach a live person. When calling to New York, it takes 12 calls. Survey firms assign their best interviewers to make calls to New York and often pay them bonuses to cope with the stress.

Because the sample in Example 8 deliberately overrepresents graduate students, the final analysis must adjust for this to get unbiased estimates of overall student opinion. Remember that our quick method only works for an SRS. In fact, a professional analysis would also take account of the fact that the population contains "only" 30,000 individuals—more job opportunities for statisticians.

Example 9. The woes of telephone samples

In principle, it would seem that telephone surveys based on dialing numbers at random can really use an SRS. Telephone surveys have little need for clustering. Stratifying can still reduce variability, so telephone surveys often take samples in two stages: a stratified sample of telephone number prefixes (area code plus first three digits) followed by individual numbers (last four digits) dialed at random in each prefix.

The real problem with an SRS of telephone numbers is that too few numbers lead to households. Blame technology. Fax machines, modems, and cell phones demand new phone numbers. Between 1988 and 1999, the number of households in the United States grew by 11%, but the number of possible residential phone numbers grew by 90%. Many new prefixes have as yet no household telephones behind them. Telephone surveys now use "list-assisted samples" that check electronic telephone directories to eliminate prefixes that have no listed numbers before random sampling begins. Fewer calls are wasted, but anyone living where all numbers are unlisted is missed. Prefixes with no listed numbers are therefore separately sampled (stratification again) to fill the gap.

It's clear from Examples 7, 8, and 9 that designing samples is a business for experts. Even most statisticians don't qualify. We won't worry about such details. The big idea is that good sample designs use chance to select individuals from the population. That is, all good samples are *probability samples.*

Probability sample

A **probability sample** is a sample chosen by chance. We must know what samples are possible and what chance, or probability, each possible sample has. Some probability samples, such as stratified samples, don't allow all possible samples from the population and may not give an equal chance to all the samples they do allow.

A stratified sample of 300 undergraduate students and 200 grad students, for example, only allows samples with exactly that makeup. An SRS would allow any 500 students. Both are probability samples. We need only know that estimates from any probability sample share the nice properties of estimates from an SRS. Confidence statements can be made without bias and have smaller margins of error as the size of the sample increases. Nonprobability samples such as voluntary response samples do not share these advantages and cannot give trustworthy information about a population. Now that we know that most nationwide samples are more complicated than an SRS, we will usually go back to acting as if good samples were SRSs. That keeps the big idea and hides the messy details.

Questions to ask before you believe a poll

Opinion polls and other sample surveys can produce accurate and useful information if the pollster uses good statistical techniques and also works hard at preparing a sampling frame, wording questions, and reducing nonresponse. Many surveys, however, especially those designed to influence public opinion rather than just record it, do not produce accurate or useful information. Here are some questions to ask before you pay much attention to poll results.

- **Who carried out the survey?** Even a political party should hire a professional sample survey firm whose reputation demands that they follow good survey practices.
- **What was the population?** That is, whose opinions were being sought?
- **How was the sample selected?** Look for mention of random sampling.
- **How large was the sample?** Even better, find out both the sample size and the **margin of error** within which the results of 95% of all samples drawn as this one was would fall.
- **What was the response rate?** That is, what percent of the original subjects actually provided information?
- **How were the subjects contacted?** By telephone? Mail? Face-to-face interview?
- **When was the survey conducted?** Was it just after some event that might have influenced opinion?
- **What were the exact questions asked?**

Academic survey centers and government statistical offices answer these questions when they announce the results of a sample survey. National opinion polls usually don't announce their response rate (which is often low) but do give us the other information. Editors and newscasters have the bad habit of cutting out these dull facts and reporting only the sample results. Many sample surveys by interest groups and local newspapers and TV stations don't answer these questions because their polling methods are in fact unreliable. If a politician, an advertiser, or your local TV station announces the results of a poll without complete information, be skeptical.

Exploring the Web

This chapter began by pointing to a study by the Pew Research Center for the People and the Press. The Center's Web site is www.people-press.org. The site contains reports on studies such as the one we have looked at and the results of the Center's own opinion polls. You might look at the "Poll Analysis" section for careful comments on what current opinion polls, both the Center's and others such as Gallup Polls, tell us about public views on important issues.

Statistics in Summary Even professional sample surveys don't give exactly correct information about the population. There are many potential sources of error in sampling. The margin of error announced by a sample survey covers only **random sampling error,** the variation due to chance in choosing a random sample. Other types of error are in addition to the margin of error and can't be directly measured. **Sampling errors** come from the act of choosing a sample. Random sampling error and **undercoverage** are common types of sampling error. Undercoverage occurs when some members of the population are left out of the **sampling frame,** the list from which the sample is actually chosen.

The most serious errors in most careful surveys, however, are **nonsampling errors.** These have nothing to do with choosing a sample—they are present even in a census. The single biggest problem for sample surveys is **nonresponse:** subjects can't be contacted or refuse to answer. Mistakes in handling the data (**processing errors**) and incorrect answers by respondents (**response errors**) are other examples of nonsampling errors. Finally, the exact **wording of questions** has a big influence on the answers. People who design sample surveys use statistical techniques that help correct nonsampling errors, and they also use **probability samples** more complex than simple random samples, such as **stratified samples.**

You can assess the quality of a sample survey quite well by just looking at the basics: use of random samples, sample size and margin of error, the rate of nonresponse, and the wording of the questions.

CHAPTER 4 EXERCISES

4.1 Not in the margin of error. A recent Gallup Poll found that 68% of adult Americans favor teaching creationism along with evolution in public schools. The Gallup press release says:

> *For results based on samples of this size, one can say with 95 percent confidence that the maximum error attributable to sampling and other random effects is plus or minus 3 percentage points.*

Give one example of a source of error in the poll result that is *not* included in this margin of error.

4.2 What kind of error? Which of the following are sources of *sampling error* and which are sources of *nonsampling error?* Explain your answers.

(a) The subject lies about past drug use.

(b) A typing error is made in recording the data.

(c) Data are gathered by asking people to mail in a coupon printed in a newspaper.

4.3 What kind of error? Each of the following is a source of error in a sample survey. Label each as *sampling error* or *nonsampling error*, and explain your answers.

(a) The telephone directory is used as a sampling frame.

(b) The subject cannot be contacted in five calls.

(c) Interviewers choose people on the street to interview.

4.4 Internet users. A survey of users of the Internet found that males outnumbered females by nearly 2 to 1. This was a surprise, because earlier surveys had put the ratio of men to women closer to 9 to 1. Later in the article we find this information:

> *Detailed surveys were sent to more than 13,000 organizations on the Internet; 1,468 usable responses were received. According to Mr. Quarterman, the margin of error is 2.8 percent, with a confidence level of 95 percent.*

(a) What was the *response rate* for this survey? (The response rate is the percent of the planned sample that responded.)

(b) Use the quick method (page 37) to estimate the margin of error of this survey. Is your result close to the 2.8% claimed?

(c) Do you think that the small margin of error is a good measure of the accuracy of the survey's results? Explain your answer.

4.5 Polling students. A university chooses an SRS of 100 students from the registrar's list of all undergraduates to interview about student life. If it selected two SRSs of 100 students at the same time, the two samples would give somewhat different results. Is this variation a source of sampling error or of nonsampling error? Does the survey's announced margin of error take this source of error into account?

4.6 Ring-no-answer. A common form of nonresponse in telephone surveys is "ring-no-answer." That is, a call is made to an active number but no one answers. The Italian National Statistical Institute looked at nonresponse to a government survey of households in Italy during the periods January 1 to Easter and July 1 to August 31. All calls were made between 7 and 10 P.M., but 21.4% gave "ring-no-answer" in one period versus 41.5% "ring-no-answer" in the other period. Which period do you think had the higher rate of no answers? Why? Explain why a high rate of nonresponse makes sample results less reliable.

4.7 Should he go or stay? In December 1998, the House of Representatives impeached President Clinton, the first step in removing him from office. The Senate then conducted a trial and found the president not guilty. Here are two opinion poll questions asked after the House had acted:

> *What do you think President Clinton should do: fight the charges in the Senate, or resign from office?*
>
> *What do you think President Clinton should do: continue to serve and stand trial in the Senate, or resign from office?*

In response to the first question, 58% thought the president should resign. But only 43% of those asked the second question thought he should resign. Why do you think the first wording encouraged more people to favor resignation?

4.8 Amending the Constitution. You are writing an opinion poll question about a proposed amendment to the Constitution. You can ask if people are in favor of "changing the Constitution" or "adding to the Constitution" by approving the amendment. One of these choices of wording will produce a much higher percent in favor. Which one? Why?

4.9 Do the people want a tax cut? Before the 2000 presidential election, the candidates debated what to do with the large government surplus. The Pew Center asked two questions of random samples of adults. Both said that Social Security would be "fixed." Here are the uses suggested for the remaining surplus:

> *Should the money be used for a tax cut, or should it be used to fund new government programs?*
>
> *Should the money be used for a tax cut, or should it be spent on programs for education, the environment, health care, crime-fighting and military defense?*

One of these questions drew 60% favoring a tax cut; the other, only 22%. Which wording pulls respondents toward a tax cut? Why?

4.10 Wording survey questions. Comment on each of the following as a potential sample survey question. Is the question clear? Is it slanted toward a desired response?

(a) Which of the following best represents your opinion on gun control?

1. The government should take away our guns.

2. We have the right to keep and bear arms.

(b) The United States should build a system to defend against intercontinental missiles because it would prevent terrorists from threatening us and would keep us ahead of the rest of the world. Do you agree or disagree?

(c) In view of escalating environmental degradation and incipient resource depletion, would you favor economic incentives for recycling of resource-intensive consumer goods?

4.11 Bad survey questions. Write your own examples of bad sample survey questions.

(a) Write a biased question designed to get one answer rather than another.

(b) Write a question that is confusing, so that it is hard to answer.

4.12 Appraising a poll. The *Wall Street Journal* published an article on attitudes toward the Social Security system based on a sample survey. It found, for example, that 36% of people aged 18 to 34 years expected Social Security to pay nothing at all when they retire. News articles tend to be brief in describing sample surveys. Here is part of the *Wall Street Journal*'s description of this poll:

> *The Wall Street Journal/NBC News poll was based on nationwide telephone interviews of 2,012 adults, conducted Thursday through Sunday by the polling organizations of Peter Hart and Robert Teeter.*
>
> *The sample was drawn from 520 randomly selected geographic points in the continental U.S. Each region was represented in proportion to its population. Households were selected by a method that gave all telephone numbers, listed and unlisted, an equal chance of being included.*

Page 61 lists several "questions to ask" about an opinion poll. What answers does the *Wall Street Journal* give to each of these questions?

4.13 Appraising a poll. A *New York Times* article on attitudes toward the political parties discussed the results of a sample survey that found, for example, that 44% of adults think the Democrats "have better ideas for leading the country into the 21st century." Another 37% chose the Republicans;

the others had no opinion. Here is part of the *Times*'s statement on "How the Poll Was Conducted":

> *The latest New York Times/CBS News poll is based on telephone interviews conducted Nov. 4 through 7 with 1,162 adults throughout the United States. . . .*
>
> *The sample of telephone exchanges called was randomly selected by a computer from a complete list of more than 42,000 active residential exchanges across the country.*
>
> *Within each exchange, random digits were added to form a complete telephone number, thus permitting access to both listed and unlisted numbers. Within each household, one adult was designated by a random procedure to be the respondent for the survey.*

Page 61 lists several "questions to ask" about an opinion poll. What answers does the *Times* give to each of these questions?

4.14 Closed versus open questions. Two basic types of questions are closed questions and open questions. A closed question asks the subject for one or more of a fixed set of responses. An open question allows the subject to answer in his or her own words. The interviewer writes down the responses and sorts them later. An example of an open question is

> *How do you feel about broccoli?*

An example of a closed question is

> *What is your opinion about broccoli? Do you*
> *a. Like it very much?*
> *b. Like it somewhat?*
> *c. Neither like nor dislike it?*
> *d. Dislike it somewhat?*
> *e. Dislike it very much?*

What are the advantages and disadvantages of open and closed questions?

4.15 Telling the truth? Many subjects don't give honest answers to questions about activities that are illegal or sensitive in some other way. One study divided a large group of white adults into thirds at random. All were asked if they had ever used cocaine. The first group was interviewed by telephone: 21% said "Yes." In the group visited at home by an interviewer, 25% said "Yes." The final group was interviewed at home but answered the question on an anonymous form that they sealed in an envelope. Of this group, 28% said they had used cocaine.

(a) Which result do you think is closest to the truth? Why?

(b) Give two other examples of behavior you think would be underreported in a telephone survey.

4.16 Did you vote? When the Current Population Survey asked the adults in its sample of 50,000 households if they voted in the 1996 presidential election, 54% said they had. In fact, only 49% of the adult population voted in that election. Why do you think the CPS result missed by much more than the margin of error?

4.17 A party poll. At a party there are 30 students over age 21 and 20 students under age 21. You choose at random 3 of those over 21 and separately choose at random 2 of those under 21 to interview about attitudes toward alcohol. You have given every student at the party the same chance to be interviewed: what is that chance? Why is your sample not an SRS?

4.18 A stratified sample. A club has 30 student members and 10 faculty members. The students are

Abel	Fisher	Huber	Lopez	Reinmann
Carson	Ghosh	Jimenez	Miranda	Santos
Chen	Griswold	Jones	Neyman	Shaw
David	Hein	Kim	O'Brien	Thompson
Deming	Hernandez	Klotz	Pearl	Utts
Elashoff	Holland	Liu	Potter	Varga

The faculty members are

Aliaga	Fernandez	Kim	Moore	West
Besicovitch	Gupta	Lightman	Vicario	Yang

The club can send 4 students and 2 faculty members to a convention. It decides to choose those who will go by random selection.

(a) Use Table A to choose a stratified random sample of four students and two faculty members.

(b) What is the chance that the student named Santos is chosen? What is the chance that faculty member Kim is chosen?

4.19 A stratified sample. A university has 2000 male and 500 female faculty members. The equal opportunity employment officer wants to poll the opinions of a random sample of faculty members. In order to give adequate attention to female faculty opinion, he decides to choose a stratified random sample of 200 males and 200 females. He has alphabetized lists of female and male faculty members.

(a) Explain how you would assign labels and use random digits to choose the desired sample. Enter Table A at line 122 and give the first 5 females and the first 5 males in your sample.

(b) What is the chance that any one of the 2000 males will be in your sample? What is the chance that any one of the 500 females will be in your sample?

4.20 Sampling by accountants. Accountants use stratified samples during audits to verify a company's records of such things as accounts receivable. The stratification is based on the dollar amount of the item and often includes 100% sampling of the largest items. One company reports 5000 accounts receivable. Of these, 100 are in amounts over $50,000; 500 are in amounts between $1000 and $50,000; and the remaining 4400 are in amounts under $1000. Using these groups as strata, you decide to verify all of the largest accounts and to sample 5% of the midsize accounts and 1% of the small accounts. How would you label the two strata from which you will sample? Use Table A, starting at line 115, to select *only the first 5* accounts from each of these strata.

4.21 A sampling paradox? Example 8 compares two SRSs, of a university's undergraduate and graduate students. The sample of undergraduates contains a smaller fraction of the population, 1 out of 90, versus 1 out of 15 for graduate students. Yet sampling 1 out of 90 undergraduates gives a smaller margin of error than sampling 1 out of 15 graduate students. Explain to someone who knows no statistics why this happens.

4.22 Exercise 4.12 gives part of the description of a sample survey from the *Wall Street Journal.* It appears that the sample was taken in several stages. Why can we say this? The first stage no doubt used a stratified sample, though the *Journal* does not say this. Explain why it would be bad practice to use an SRS from a large number of "geographic points" across the country rather than a stratified sample of such points.

4.23 An article in the journal *Science* looks at differences in attitudes toward genetically modified foods between Europe and the United States. This calls for sample surveys. The European survey chose a sample of 1000 adults in each of 17 European countries. Here's part of the description: "The Eurobarometer survey is a multistage, random-probability face-to-face sample survey."

(a) What does "multistage" mean?

(b) You can see that the first stage was stratified. What were the strata?

(c) What does "random-probability sample" mean?

4.24 What do schoolkids want? What are the most important goals of schoolchildren? Do girls and boys have different goals? Are goals different in urban, suburban, and rural areas? To find out, researchers wanted to ask children in the fourth, fifth, and sixth grades this question:

What would you most like to do at school?
A. Make good grades.
B. Be good at sports.
C. Be popular.

Because most children live in heavily populated urban and suburban areas, an SRS might contain few rural children. Moreover, it is too expensive to choose children at random from a large region—we must start by choosing schools rather than children. Describe a suitable sample design for this study and explain the reasoning behind your choice of design.

4.25 Systematic random samples. The last stage of the Current Population Survey (Example 7) uses a **systematic random sample.** An example will illustrate the idea of a systematic sample. Suppose that we must choose 4 rooms out of the 100 rooms in a dormitory. Because 100/4 = 25, we can think of the list of 100 rooms as four lists of 25 rooms each. Choose 1 of the first 25 rooms at random, using Table A. The sample will contain this room and the rooms 25, 50, and 75 places down the list from it. If 13 is chosen, for example, then the systematic random sample consists of the rooms numbered 13, 38, 63, and 88. Use Table A to choose a systematic random sample of 5 rooms from a list of 200. Enter the table at line 120.

4.26 Systematic isn't simple. Exercise 4.25 describes a systematic random sample. Like an SRS, a systematic sample gives all individuals the same chance to be chosen. Explain why this is true, then explain carefully why a systematic sample is nonetheless *not* an SRS.

4.27 Planning a survey of students. The student government plans to ask a random sample of students about their priorities for changing campus life. The college provides a list of the 3500 enrolled students to serve as a sampling frame.

(a) How would you choose an SRS of 250 students?

(b) How would you choose a systematic sample of 250 students? (See Exercise 4.25 to learn about systematic samples.)

(c) The list shows whether students live on campus (2400 students) or off campus (1100 students). How would you choose a stratified sample of 200 on-campus students and 50 off-campus students?

4.28 Sampling students. You want to investigate the attitudes of students at your school about the school's policy on sexual harassment. You have a grant that will pay the costs of contacting about 500 students.
(a) Specify the exact population for your study. For example, will you include part-time students?

(b) Describe your sample design. For example, will you use a stratified sample with men and women as strata?

(c) Briefly discuss the practical difficulties that you anticipate. For example, how will you contact the students in your sample?

4.29 Mall interviews. Example 1 in Chapter 2 (page 20) describes mall interviewing. This is an example of a convenience sample. Why do mall interviews not produce probability samples?

4.30 Homosexuals in the military? Here are three opinion poll questions on the same issue, with the poll results:

> *Do you approve or disapprove of allowing openly homosexual men and women to serve in the armed forces of the United States?* Result: 47% strongly or somewhat disapproved; 45% strongly or somewhat approved.
>
> *Do you think that gays and lesbians should be banned from the military or not?* Result: 37% said they should be banned; 57% said not.
>
> *Should President Clinton change military policy to allow gays in the military?* Result: 53% said "No"; 35% said "Yes."

Using this example, discuss the difficulty of using responses to opinion polls to understand public opinion.

Chapter 5
Experiments, Good and Bad

A tale of three studies

An optimistic account of learning online reports a study at Nova Southeastern University, Fort Lauderdale, Florida. The authors of the study claim that students taking undergraduate courses online were "equal in learning" to students taking the same courses in class. Replacing college classes with Web sites saves colleges money, so this study suggests we should all move online.

Ulcers seem to accompany modern life. "Gastric freezing" is a clever treatment for stomach ulcers. The patient swallows a deflated balloon with tubes attached; then a refrigerated solution is pumped through the balloon for an hour. The idea is that cooling the stomach will reduce its production of acid and so relieve ulcers. An experiment reported in the *Journal of the American Medical Association* claimed that gastric freezing did relieve ulcer pain.

Should the government provide day care for low-income children? If day care helps these children stay in school and hold good jobs later in life, the government would save money by paying less welfare and collecting more taxes, so even those who only count dollars might support day care programs. The Carolina Abecedarian Project (the name suggests learning the ABCs) has followed a group of children since 1972. The results show that good day care makes a big difference in later school and work.

Three big issues, three studies that claim to shed light on the issues. In fact, the first of these is an observational study that can't be trusted to give good evidence about learning in online college courses. The gastric freezing experiment now appears to be plainly misleading. Most ulcers are caused by bacteria, and cooling the stomach is not an effective treatment. The results of the Abecedarian Project, on the other hand, are about as convincing as evidence about the long-term effects of day care can be. What makes some studies—especially some experiments—convincing? Why should we ignore others? This chapter shows what to look for.

Talking about experiments

Observational studies are passive data collection. We observe, record, or measure, but we don't interfere. Experiments are active data production. Experimenters actively intervene by imposing some treatment in order to see what happens. All experiments and many observational studies are interested in the effect one variable has on another variable. Here is the vocabulary we use to distinguish the variable that acts from the variable that is acted upon.

The vocabulary of experiments

A **response variable** is a variable that measures an outcome or result of a study.

An **explanatory variable** is a variable that we think explains or causes changes in the response variables.

The individuals studied in an experiment are often called **subjects.**

A **treatment** is any specific experimental condition applied to the subjects. If an experiment has several explanatory variables, a treatment is a combination of specific values of these variables.

Example 1. Learning on the Web and the effects of day care

College students are the *subjects* in the Nova Southeastern study. The *explanatory variable* is the setting for learning (in class or online). The *response variable* is a student's score on a test at the end of the course.

The Abecedarian Project is an experiment in which the *subjects* are 111 people who in 1972 were healthy but low-income black infants in Chapel Hill, North Carolina. All the infants received nutritional supplements and help from social workers. Half, chosen at random, were also placed in an intensive preschool program. The experiment compares these two treatments. The *explanatory variable* is just "preschool, yes or no." There are many *response variables,* recorded over more than 20 years, including academic test scores, college attendance, and employment.

You will often see explanatory variables called *independent variables* and response variables called *dependent variables.* The idea is that the response variables depend on the explanatory variables. I don't care for these older terms, partly because "independent" has other and very different meanings in statistics.

How to experiment badly

Do students who take a course via the Web learn as well as those who take the same course in a traditional classroom? The best way to find out is to assign some students to the classroom and others to the Web. That's an experiment. The Nova Southeastern study was not an experiment, because it imposed no treatment on the student subjects. Students chose for themselves whether to enroll in a classroom or in an online version of a course. The study simply measured their learning. The students who chose the online course were very different from the classroom students. For example, their average score on tests on the course material given before the courses started was 40.70, against only 27.64 for the classroom students. It's hard to compare in-class versus online learning when the online students have a big head start. The effect of online versus in-class instruction is hopelessly mixed up with influences lurking in the background. Figure 5.1 shows the mixed-up influences in picture form.

Confounding

A **lurking variable** is a variable that has an important effect on the relationship among the variables in a study but is not one of the explanatory variables studied.

Two variables are **confounded** when their effects on a response variable cannot be distinguished from each other. The confounded variables may be either explanatory variables or lurking variables.

In the Nova Southeastern study, student preparation (a lurking variable) is confounded with the explanatory variable. The study report claims that the two groups did equally well on the final test. We can't say how much of the online group's performance is due to their head start. That a group that started with a big advantage did no better than the more poorly prepared classroom students is not very impressive evidence of the wonders of Web-based instruction. Here is another example, one in which a second experiment untangled the confounding.

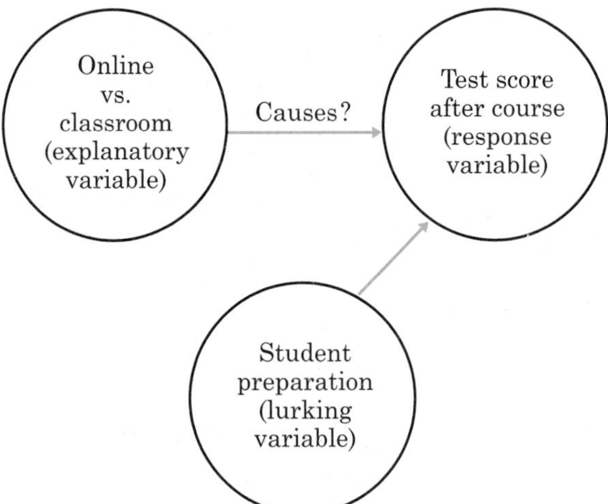

Figure 5.1 Confounding in the Nova Southeastern University study. The influence of course setting (the explanatory variable) can't be distinguished from the influence of student preparation (a lurking variable).

Example 2. Gastric freezing flunks

Experiments that study the effectiveness of medical treatments on actual patients are called **clinical trials.** The clinical trial that made gastric freezing a popular treatment for stomach ulcers had this "one-track" design:

Impose treatment	⟶	Measure response
Gastric freezing	⟶	Reduced pain?

The patients did report reduced pain, but we can't say that gastric freezing caused the reduced pain. It might be just the **placebo effect.** A **placebo** is a dummy treatment with no active ingredients. Many patients respond favorably to *any* treatment, even a placebo. This response to a dummy treatment is the placebo effect. Perhaps the placebo effect is in our minds, based on trust in the doctor and expectations of a cure. Perhaps it is just a name for the fact that many patients improve for no visible reason. The one-track design of the experiment meant that the placebo effect was confounded with any effect gastric freezing might have.

A second clinical trial, done several years later, divided ulcer patients into two groups. One group was treated by gastric freezing as before. The other group received a placebo treatment in which the solution in the balloon was at body temperature rather than freezing. The results: 34% of the 82 patients in the treatment group improved, but so did 38% of the 78 patients in the placebo group. This and other properly designed experiments showed that gastric freezing was no better than a placebo, and doctors abandoned it.

Both observation and one-track experiments often yield useless data because of confounding with lurking variables. It is hard to avoid confounding when only observation is possible. Experiments offer better possibilities, as the second gastric freezing experiment shows. This experiment included a group of subjects who received only a placebo. This allows us to see whether the treatment being tested does better than a placebo and so has more than the placebo effect going for it. Effective medical treatments pass the placebo test. Gastric freezing flunks the test.

"I want to make one thing perfectly clear, Mr. Smith. The medication I prescribe *will* cure that run-down feeling."

Randomized comparative experiments

The first goal in designing an experiment is to ensure that it will show us the effect of the explanatory variables on the response variables. Confounding often prevents one-track experiments from doing this. The remedy is to *compare* two or more treatments. Here is an example of a new medical treatment that passes the placebo test in a direct comparison.

Example 3. Sickle cell anemia

Sickle cell anemia is an inherited disorder of the red blood cells that in the United States affects mostly blacks. It can cause severe pain and many complications. The National Institutes of Health carried out a clinical trial of the drug hydroxyurea for treatment of sickle cell anemia. The subjects were 299 adult patients who had had at least three episodes of pain from sickle cell anemia in the previous year.

Simply giving hydroxyurea to all 299 subjects would confound the effect of the medication with the placebo effect and other lurking variables such as the effect of knowing that you are a subject in an experiment. Instead, half of the subjects received hydroxyurea, and the other half received a placebo that looked and tasted the same. All subjects were treated exactly the same (same schedule of medical checkups, for example) except for the content of the medicine they took. Lurking variables therefore affected both groups equally and should not cause any differences between their average responses.

The two groups of subjects must be similar in all respects before they start taking the medication. Just as in sampling, the best way to avoid bias in choosing which subjects get hydroxyurea is to allow impersonal chance to make the choice. An SRS of 152 of the subjects formed the hydroxyurea group; the remaining 147 subjects made up the placebo group. Figure 5.2 outlines the experimental design.

The experiment was stopped ahead of schedule because the hydroxyurea group had many fewer pain episodes than the placebo group. This was compelling evidence that hydroxyurea is an effective treatment for sickle cell anemia, good news for those who suffer from this serious illness.

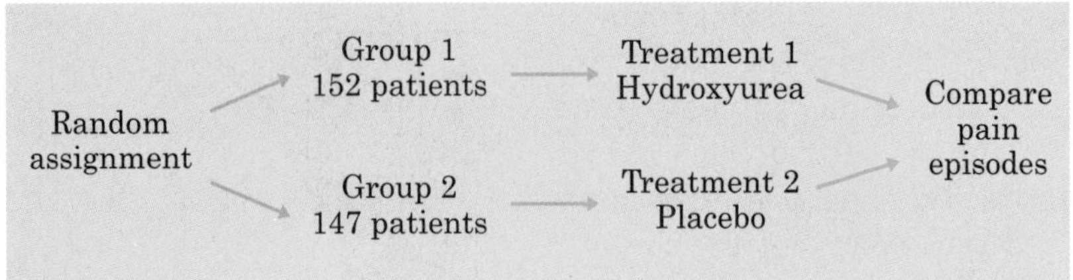

Figure 5.2 The design of a randomized comparative experiment to compare hydroxyurea with a placebo for treating sickle cell anemia.

Figure 5.2 illustrates the simplest **randomized comparative experiment,** one that compares just two treatments. The diagram outlines the essential information about the design: random assignment; one group for each treatment; the number of subjects in each group (it is generally best to keep the groups similar in size); what treatment each group gets; and the response variable we compare. You know how to carry out the random assignment of subjects to groups. Label the 299 subjects 001 to 299, then read three-digit groups from the table of random digits (Table A) until you have chosen the 152 subjects for Group 1. The remaining 147 subjects form Group 2.

The placebo group in Example 3 is called a **control group** because comparing the treatment and control groups allows us to control the effects of lurking variables. A control group need not receive a dummy treatment such as a placebo. Clinical trials often compare a new treatment for a medical condition, not with a placebo, but with a treatment that is already on the market. Patients who are randomly assigned to the existing treatment form the control group. To compare more than two treatments, we can randomly assign the available experimental subjects to as many groups as there are treatments. Here is an example with three groups.

Example 4. Conserving energy

Many utility companies have introduced programs to encourage energy conservation among their customers. An electric company considers placing electronic meters in households to show what the cost would be if the electricity use at that moment continued for a month. Will meters reduce electricity use? Would cheaper methods work almost as well? The company decides to design an experiment.

One cheaper approach is to give customers a chart and information about monitoring their electricity use. The experiment compares these two approaches (meter, chart) and also a control. The control group of customers receives information about energy conservation but no help in monitoring electricity use. The response variable is

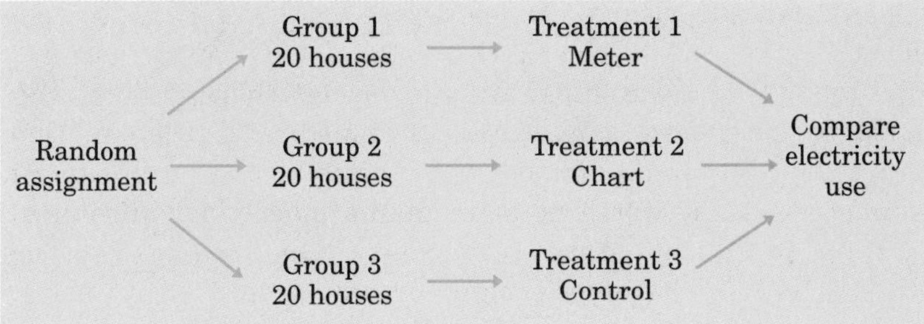

Figure 5.3 The design of a randomized comparative experiment to compare three programs to reduce electricity use by households.

total electricity used in a year. The company finds 60 single-family residences in the same city willing to participate, so it assigns 20 residences at random to each of the 3 treatments. Figure 5.3 outlines the design.

To carry out the random assignment, label the 60 households 01 to 60. Enter Table A to select an SRS of 20 to receive the meters. Continue in Table A, selecting 20 more to receive charts. The remaining 20 form the control group.

The logic of experimental design

The randomized comparative experiment is one of the most important ideas in statistics. It is designed to allow us to draw cause-and-effect conclusions. Be sure you understand the logic:

- Randomization produces groups of subjects that should be similar in all respects before we apply the treatments.
- Comparative design ensures that influences other than the experimental treatments operate equally on all groups.
- Therefore, differences in the response variable must be due to the effects of the treatments.

We use chance to choose the groups in order to eliminate any systematic bias in assigning the subjects to groups. In the sickle cell study, for example, a doctor might subconsciously assign the most seriously ill patients to the hydroxyurea group, hoping that the untested drug will help them. That would bias the experiment against hydroxyurea. Choosing an SRS of the subjects to be Group 1 gives everyone the same chance to be in either group. We expect the two groups to be similar in all respects—age, seriousness of illness, smoker or not, and so on. Chance tends to assign equal numbers of smokers to both groups, for example, even if we don't know which subjects are smokers.

It should not surprise you to learn that medical researchers adopted randomized comparative experiments only slowly—many doctors think they can tell "just by watching" whether a new therapy helps their patients. Not so. There are many examples of medical treatments that (like gastric freezing) became popular on the basis of one-track experiments and were shown to be worth no more than a placebo when some skeptic tried a randomized comparative experiment. One search of the medical literature looked for therapies studied both by proper comparative trials and by trials with "historical controls." A study with historical controls compares the results of a new treatment, not with a control group, but with how well similar patients had done in the past. Of the 56 therapies studied, 44 came out winners with historical controls. Only 10 passed the placebo test in proper randomized comparative experiments. Expert judgment is too optimistic even when aided by comparison with past patients. At present, the law requires that new drugs be shown to be both safe and effective by randomized comparative trials. There is no such requirement for other medical treatments, such as surgery. You can expect new drugs to beat a placebo. New surgical ideas, like gastric freezing in the past, may not.

There is one important caution about randomized experiments. Like random samples, they are subject to the laws of chance. Just as an SRS of voters might by bad luck choose nearly all Republicans, a random assignment of subjects might by bad luck put nearly all the smokers in one group. We know that if we choose *large* random samples, it is very likely that the sample will match the population well. In the same way, if we use *many* experimental subjects, it is very likely that random assignment will produce groups that match closely. More subjects means that there is less chance variation among the treatment groups and less chance variation in the outcomes of the experiment. "Use enough subjects" joins "compare two or more treatments" and "randomize" as a basic principle of statistical design of experiments.

Principles of experimental design

The basic principles of statistical design of experiments are:

1. **Control** the effects of lurking variables on the response, most simply by comparing two or more treatments.

2. **Randomize**—use impersonal chance to assign subjects to treatments.

3. **Use enough subjects** in each group to reduce chance variation in the results.

Statistical significance

The presence of chance variation requires us to look more closely at the logic of randomized comparative experiments. We cannot say that *any* difference in the average number of pain episodes between the hydroxyurea and control groups must be due to the effect of the drug. There will be some differences even if both treatments are the same because there will always be some differences among the individuals who are our subjects. Even though randomization eliminates systematic differences between the groups, there will still be chance differences. We should insist on a difference in the responses so large that it is unlikely to happen just because of chance variation.

Statistical significance

An observed effect so large that it would rarely occur by chance is called **statistically significant.**

The difference between the average number of pain episodes for subjects in the hydroxyurea group and the average for the control group was "highly statistically significant." That means that a difference this large would almost never happen just by chance. We do indeed have strong evidence that hydroxyurea beats a placebo in helping sickle cell disease sufferers. You will often see the phrase "statistically significant" in reports of investigations in many fields of study. It tells you that the investigators found good evidence for the effect they were seeking.

Of course, the actual results of an experiment are more important than the seal of approval given by statistical significance. The treatment group in the sickle cell experiment had an average of 2.5 pain episodes per year, against 4.5 per year in the control group. That's a big enough difference to be important to people with the disease. A difference of 2.5 versus 2.8 would be much less interesting even if it were statistically significant.

How to live with observational studies

Does regular church attendance lengthen people's lives? Do doctors discriminate against women in treating heart disease? Does talking on a cell phone while driving increase the risk of having an accident? These are cause-and-effect questions, so we reach for our favorite tool, the randomized comparative experiment. Sorry. We can't randomly assign people to attend church or not, because going to religious services is an expression of beliefs or their absence. We can't use random digits to assign heart disease patients to be men or women. We are reluctant to require drivers to use cell phones in traffic, because talking while driving may be risky.

The best data we have about these and many other cause-and-effect questions come from observational studies. We know that observation is a weak second best to experiment, but good observational studies are far from worthless. What makes a good observational study?

First, good studies are **comparative** even when they are not experiments. We compare random samples of people who do and who don't attend religious services regularly. We compare how doctors treat men and women patients. We might compare drivers talking on cell phones with the *same* drivers when they are not on the phone. We can often combine comparison with **matching** in creating a control group. To see the effects of taking a painkiller during pregnancy, we compare women who did so with women who did not. From a large pool of women who did not take the drug, we select individuals who match the drug group in age, education, number of children, and other lurking variables. We now have two groups that are similar in all these ways, so that these lurking variables should not affect our comparison of the groups.

Comparison does not eliminate confounding. People who attend church or synagogue or mosque take better care of themselves than nonattenders. They are less likely to smoke, more likely to exercise, and less likely to be overweight. Matching can reduce some but not all of these differences. A direct comparison of the ages at death of attenders and nonattenders would confound any effect of religion with the effects of healthy living. So a good comparative study **measures and adjusts for confounding variables.** If we measure weight, smoking, and exercise, there are statistical techniques that reduce the effects of these variables on length of life so that (we hope) only the effect of religion itself remains.

Example 5. Living longer through religion

One of the better studies of the effect of regular attendance at religious services gathered data from a random sample of 3617 adults. Random sampling is a good start. The researchers then measured lots of variables, not just the explanatory variable (religious activities) and the response variable (length of life). A news article said:

> *Churchgoers were more likely to be nonsmokers, physically active, and at their right weight. But even after health behaviors were taken into account, those not attending religious services regularly still were about 25% more likely to have died.*

That "taken into account" means that the final results were adjusted for differences between the two groups. Adjustment reduced the advantage of religion but still left a large benefit.

Example 6. Sex bias in treating heart disease?

Doctors are less likely to give aggressive treatment to women with symptoms of heart disease than to men with similar symptoms. Is this because doctors are sexist? Not necessarily. Women tend to develop heart problems much later than men, so that female heart patients are older and often have other health problems. That might explain why doctors proceed more cautiously in treating them.

This is a case for a comparative study with statistical adjustments for the effects of confounding variables. There have been several such studies, and they produce conflicting results. Some show, in the words of one doctor, "When men and women are otherwise the same and the only difference is gender, you find that treatments are very similar." Other studies find that women are undertreated even after adjusting for differences between the female and male subjects.

As Example 6 suggests, statistical adjustment is tricky. Randomization creates groups that are similar in *all* variables known and unknown. Matching and adjustment, on the other hand, can't work with variables the study didn't think to measure. Even if you believe the researchers thought of everything, you should be a bit skeptical about statistical adjustment. There's lots of room for cheating in deciding which variables to adjust for. And the "adjusted" conclusion is really something like this:

> *If female heart disease patients were younger and healthier than they really are, and if male patients were older and less healthy than they really are, then the two groups would get the same medical care.*

"Statistics say that religious people live longer, so I practice a different religion every day of the week to be sure I'm covered."

This may be the best we can get, and we should thank statistics for making such wisdom possible. But we end up longing for the clarity of a good experiment.

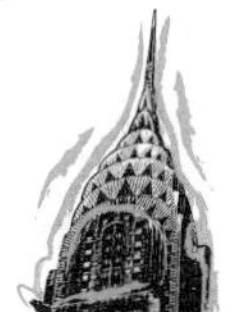

Exploring the Web

You can find the latest medical research in the *Journal of the American Medical Association* (jama.ama-assn.org) and the *New England Journal of Medicine* (www.nejm.org). Many of the articles describe randomized comparative experiments, and even more use the language of statistical significance.

Statistics in Summary

Statistical studies often try to show that changing one variable (the **explanatory variable**) causes changes in another variable (the **response variable**). In an **experiment,** we actually set the explanatory variables ourselves rather than just observing them. Observational studies and one-track experiments that simply apply a single treatment often fail to produce useful data because **confounding** with **lurking variables** makes it impossible to say what the effect of the treatment was. The remedy is to use a **randomized comparative experiment**. Compare two or more treatments, use chance to decide which subjects get each treatment, and use enough subjects so that the effects of chance are small. Comparing two or more treatments **controls** lurking variables such as the **placebo effect** because they act on all the treatment groups.

Differences among the effects of the treatments so large that they would rarely happen just by chance are called **statistically significant**. Statistically significant results from randomized comparative experiments are the best available evidence that changing the explanatory variable really **causes** changes in the response. Observational studies of cause-and-effect questions are more impressive if they **compare matched groups** and measure as many lurking variables as possible to allow **statistical adjustment.** Observational studies remain a weak second best to experiments for answering questions about causation.

CHAPTER 5 EXERCISES

5.1 Treating breast cancer. What is the preferred treatment for breast cancer that is detected in its early stages? The most common treatment was once removal of the breast. It is now usual to remove only the tumor and nearby lymph nodes, followed by radiation. To study whether these treatments differ in their effectiveness, a medical team examines the records of 25 large hospitals and compares the survival times after surgery of all women who have had either treatment.

(a) What are the explanatory and response variables?

(b) Explain carefully why this study is not an experiment.

(c) Explain why confounding will prevent this study from discovering which treatment is more effective. (The current treatment was in fact recommended after a large randomized comparative experiment.)

5.2 Teaching reading. An educator wants to compare the effectiveness of computer software that teaches reading with that of a standard reading curriculum. He tests the reading ability of each student in a class of fourth graders, then divides them into two groups. One group uses the computer regularly, while the other studies a standard curriculum. At the end of the year, he retests all the students and compares the increase in reading ability in the two groups. Is this an experiment? Why or why not? What are the explanatory and response variables?

5.3 Nursing your baby. An article in a women's magazine says that women who nurse their babies feel warmer and more receptive toward the infants than mothers who bottle-feed. The author concludes that nursing has desirable effects on the mother's attitude toward the child. But women choose whether to nurse or bottle-feed. Explain why this fact makes any conclusion about cause and effect untrustworthy. Use the language of lurking variables and confounding in your explanation, and draw a picture like Figure 5.1 to illustrate it.

5.4 Does job training work? A state institutes a job-training program for manufacturing workers who lose their jobs. After five years, the state reviews how well the program works. Critics claim that because the state's unemployment rate for manufacturing workers was 6% when the program began and 10% five years later, the program is ineffective.

Explain why higher unemployment does not necessarily mean that the training program failed. In particular, identify some lurking variables whose effect on unemployment may be confounded with the effect of the training program. Draw a picture like Figure 5.1 to illustrate your explanation.

5.5 Aspirin and heart attacks. Can aspirin help prevent heart attacks? The Physicians' Health Study, a large medical experiment involving 22,000 male physicians, attempted to answer this question. One group of about 11,000 physicians took an aspirin every second day, while the rest took a placebo. After several years the study found that subjects in the aspirin group had significantly fewer heart attacks than subjects in the placebo group.

(a) Identify the experimental subjects, the explanatory variable and the values it can take, and the response variable.

(b) Use a diagram to outline the design of the Physicians' Health Study. (When you outline the design of an experiment, be sure to indicate the size of the treatment groups and the response variable. The diagrams in Figures 5.2 and 5.3 are models.)

5.6 Learning about markets. Your economics professor wonders if playing market games online will help students understand how markets set prices. You suggest an experiment: have some students use the online games, while others discuss markets in recitation sections. The course has two lectures, at 8:30 A.M. and 2:30 P.M. There are 10 recitation sections attached to each lecture. The students are already assigned to recitations. For practical reasons, all students in each recitation must follow the same program. The professor says, "Let's just have the 8:30 group do online work in recitation and the 2:30 group do discussion." Why is this a bad idea?

5.7 The effects of propaganda. In 1940, a psychologist conducted an experiment to study the effect of propaganda on attitude toward a foreign government. He administered a test of attitude toward the German government to a group of American students. After the students read German propaganda for several months, he tested them again to see if their attitudes had changed.

Unfortunately, Germany attacked and conquered France while the experiment was in progress. Explain clearly why confounding makes it impossible to determine the effect of reading the propaganda. Use a picture like Figure 5.1 to help your explanation.

5.8 Learning about markets, continued.

(a) Outline a better design than that of Exercise 5.6 for an experiment to compare the two methods of learning about economic markets. Use the 20 recitation sections as individuals. What do you suggest as a response variable? (When you outline the design of an experiment, be sure to indicate the size of the treatment groups and the response variable. The diagrams in Figures 5.2 and 5.3 are models.)

(b) Use Table A, starting at line 116, to do the randomization your design requires.

5.9 Learning on the Web. The discussion following Example 1 notes that the Nova Southeastern study doesn't tell us much about Web versus classroom learning because the students who chose the Web version were much better prepared. Describe the design of an experiment to get better information.

5.10 Do antioxidants prevent cancer? People who eat lots of fruits and vegetables have lower rates of colon cancer than those who eat little of these foods. Fruits and vegetables are rich in "antioxidants" such as vitamins A, C, and E. Will taking antioxidants help prevent colon cancer? A clinical trial studied this question with 864 people who were at risk for colon cancer. The subjects were divided into four groups: daily beta carotene, daily vitamins C and E, all three vitamins every day, and daily placebo. After four years, the researchers were surprised to find no significant difference in colon cancer among the groups.

(a) What are the explanatory and response variables in this experiment?

(b) Outline the design of the experiment. (The diagrams in Figures 5.2 and 5.3 are models.)

(c) Assign labels to the 864 subjects and use Table A, starting at line 118, to choose the *first 5* subjects for the beta carotene group.

(d) What does "no significant difference" mean in describing the outcome of the study?

(e) Suggest some lurking variables that could explain why people who eat lots of fruits and vegetables have lower rates of colon cancer. The experiment suggests that these variables, rather than the antioxidants, may be responsible for the observed benefits of fruits and vegetables.

5.11 Conserving energy. Example 4 describes an experiment to learn whether providing households with electronic meters or charts will reduce their electricity consumption. An executive of the electric company objects to including a control group. He says, "It would be cheaper to just compare electricity use last year [before the meter or chart was provided] with consumption in the same period this year. If households use less electricity this year, the meter or chart must be working." Explain clearly why this design is inferior to that in Example 4.

5.12 Improving Chicago's schools. The National Science Foundation (NSF) paid for "systemic initiatives" to help cities reform their public education systems in ways that should help students learn better. Does this program work? The initiative in Chicago focused on improving the teaching of mathematics in high schools. The average scores of students on a standard test of math skills were higher after two years of the program in 51 out of 60 high schools in the city. Leaders of NSF said this was evidence that the Chicago program was succeeding. Critics said this doesn't say anything about the effect of the systemic initiative. Explain why the critics are right. What are some lurking variables that might help explain an increase in test scores over a two-year period?

5.13 Sealing food packages. A manufacturer of food products uses package liners that are sealed at the top by applying heated jaws after the package is filled. The customer peels the sealed pieces apart to open the package. What effect does the temperature of the jaws have on the force needed to peel the liner? To answer this question, engineers obtain 20 pairs of pieces of package liner. They seal five pairs each at of 250°, 275°, 300°, and 325° F. Then they measure the force needed to peel each seal.

(a) The individuals studied in this experiment are not people. What are they?

(b) What is the explanatory variable, and what values does it take?

(c) What is the response variable?

5.14 Reducing health care spending. Will people spend less on health care if their health insurance requires them to pay some part of the cost themselves? An experiment on this issue asked if the percent of medical costs that are paid by health insurance has an effect either on the amount of medical care that people use or on their health. The treatments were four insurance plans. Each plan paid all medical costs above a ceiling. Below the ceiling, the plans paid 100%, 75%, 50%, or 0% of costs incurred.

(a) Outline the design of a randomized comparative experiment suitable for this study.

(b) Briefly describe the practical and ethical difficulties that might arise in such an experiment.

5.15 Sealing food packages. Use a diagram to describe a randomized comparative experimental design for the package liner experiment of Exercise 5.13. Use Table A, starting at line 120, to do the randomization required by your design.

5.16 Treating drunk drivers. Once a person has been convicted of drunk driving, one purpose of court-mandated treatment or punishment is to prevent future offenses of the same kind. Suggest three different treatments that a court might require. Then outline the design of an experiment to compare their effectiveness. Be sure to specify the response variables you will measure.

5.17 Statistical significance. A randomized comparative experiment examines whether a calcium supplement in the diet reduces the blood pressure of healthy men. The subjects receive either a calcium supplement or a placebo for 12 weeks. The researchers conclude that "the blood pressure of the calcium group was significantly lower than that of the placebo group." "Significant" in this conclusion means statistically significant. Explain what statistically significant means in the context of this experiment, as if you were speaking to a doctor who knows no statistics.

5.18 Statistical significance. The financial aid office of a university asks a sample of students about their employment and earnings. The report says that "for academic year earnings, a significant difference was found between the sexes, with men earning more on the average. No significant difference was found between the earnings of black and white students." Explain the meaning of "a significant difference" and "no significant difference" in plain language.

5.19 Calcium and blood pressure. Some medical researchers suspect that added calcium in the diet reduces blood pressure. You have available 40 men with high blood pressure who are willing to serve as subjects.

(a) Outline an appropriate design for the experiment, taking the placebo effect into account.

(b) The names of the subjects appear below. Use Table A, beginning at line 119, to do the randomization required by your design. List the subjects to whom you will give the drug.

Alomar	Denman	Han	Liang	Rosen
Asihiro	Durr	Howard	Maldonado	Solomon
Bennett	Edwards	Hruska	Marsden	Tompkins
Bikalis	Farouk	Imrani	Mondesi	Townsend
Chen	Fratianna	James	O'Brian	Tullock
Clemente	George	Kaplan	Ogle	Underwood
Cranston	Green	Krushchev	Orosco	Veras
Curtis	Guillen	Lawless	Rodriguez	Zhang

5.20 Treating prostate disease. A large study used records from Canada's national health care system to compare the effectiveness of two ways to treat prostate disease. The two treatments are traditional surgery and a new method that does not require surgery. The records described many patients whose doctors had chosen each method. The study found that patients treated by the new method were significantly more likely to die within 8 years.

(a) Further study of the data showed that this conclusion was wrong. The extra deaths among patients treated with the new method could be explained by lurking variables. What lurking variables might be confounded with a doctor's choice of surgical or nonsurgical treatment?

(b) You have 300 prostate patients who are willing to serve as subjects in an experiment to compare the two methods. Use a diagram to outline the design of a randomized comparative experiment.

5.21 Prayer and meditation. You read in a magazine that "nonphysical treatments such as meditation and prayer have been shown to be effective in controlled scientific studies for such ailments as high blood pressure, insomnia, ulcers, and asthma." Explain in simple language what the article means by "controlled scientific studies" and why such studies might show that meditation and prayer are effective treatments for some medical problems.

5.22 Exercise and heart attacks. Does regular exercise reduce the risk of a heart attack? Here are two ways to study this question. Explain clearly why the second design will produce more trustworthy data.

1. A researcher finds 2000 men over 40 who exercise regularly and have not had heart attacks. She matches each with a similar man who does not exercise regularly, and she follows both groups for 5 years.
2. Another researcher finds 4000 men over 40 who have not had heart attacks and are willing to participate in a study. She assigns 2000 of the men to a regular program of supervised exercise. The other 2000 continue their usual habits. The researcher follows both groups for 5 years.

5.23 Safety of anesthetics. The death rates of surgical patients are different for operations in which different anesthetics are used. An observational study found these death rates for four anesthetics:

Anesthetic	Halothane	Pentothal	Cyclopropane	Ether
Death rate	1.7%	1.7%	3.4%	1.9%

This is *not* good evidence that cyclopropane is more dangerous than the other anesthetics. Suggest some lurking variables that may be confounded with the choice of anesthetic in surgery and that could explain the different death rates.

5.24 Randomization at work. To demonstrate how randomization reduces confounding, consider the following situation. A nutrition experimenter intends to compare the weight gain of newly weaned male rats fed Diet A with that of rats fed Diet B. To do this, she will feed each diet to 10 rats. She has available 10 rats of genetic strain 1 and 10 of strain 2. Strain 1 is more vigorous, so if the 10 rats of strain 1 were fed Diet A, the effects of strain and diet would be confounded, and the experiment would be biased in favor of Diet A.

(a) Label the rats 00, 01, . . . , 19. Use Table A to assign 10 rats to Diet A. Do this four times, using different parts of the table, and write down the four groups assigned to Diet A.

(b) Unknown to the experimenter, the rats labeled 00, 02, 04, 06, 08, 10, 12, 14, 16, and 18 are the 10 strain 1 rats. How many of these rats were in each of the four Diet A groups that you generated? What was the average number of strain 1 rats assigned to Diet A?

Chapter 6 Experiments in the Real World

The case of the fickle mice

The randomized comparative experiment may be the most important idea in statistics. Experiments are the gold standard for evidence that changing one variable really causes changes in another variable. In the real world, however, experiments don't always go smoothly. Even if they do go smoothly, we can't always take a firm stand on the findings. Consider the case of the fickle mice.

That our behavior is (in part) coded into our genes is one of the big scientific issues of the day. No human experiments allowed, so mice serve in our place. Researchers "knock out" a gene in one group of mice and compare their behavior with a control group of normal mice. All the mice have the same genetic makeup before treatment and are assigned at random to the two groups. This is an iron-clad randomized comparative experiment: if the behaviors differ, the gene that was knocked out must influence the behavior that changes. Alas, as a writer in the journal *Science* says, "No sooner has one group of researchers tied a gene to a behavior when along comes the next study, proving that the link is spurious or even that the gene in question has exactly the opposite effect."

This is frustrating for the scientists, and also for news reporters who want to say "science shows" that this or that behavior is all in our genes. What goes wrong?

Some frustrated scientists tried to find out. They did the same experiments with the same genetic strain of mice in three different labs. One researcher said they "went nuts" trying to make the conditions exactly alike in Oregon, Alberta (Canada), and New York. The results were often very different. It appears that very small differences in the lab environments have big effects on the behavior of the mice. Remember this the next time you read that our genes control our behavior.

Equal treatment for all

Probability samples are a big idea, but sampling in practice has difficulties that just using random samples doesn't solve. Randomized comparative experiments are also a big idea, but they don't solve all the difficulties of experimenting. A sampler must know exactly what information she wants and must compose questions that extract that information from her sample. An experimenter must know exactly what treatments and responses he wants information about, and he must construct the apparatus needed to apply the treatments and measure the responses. This is what psychologists or medical researchers or engineers mean when they talk about "designing an experiment." We are concerned with the *statistical* side of designing experiments, ideas that apply to experiments in psychology, medicine, engineering, and other areas as well. Even at this general level, you should understand the practical problems that can prevent an experiment from producing useful data.

The logic of a randomized comparative experiment assumes that all the subjects are treated alike except for the treatments that the experiment is designed to compare. Any other unequal treatment can cause bias. Treating subjects exactly alike is hard to do.

Example 1. Mice, rats, and rabbits

Mice, rats, and rabbits, specially bred to be uniform in their inherited characteristics, are the subjects in many experiments. As the case of the fickle mice illustrates, animals, like people, can be quite sensitive to how they are treated. Here are two amusing examples of how unequal treatment can create bias.

Does a new breakfast cereal provide good nutrition? To find out, compare the weight gains of young rats fed the new product and rats fed a standard diet. The rats are randomly assigned to diets and are housed in large racks of cages. It turns out that rats in upper cages grow a bit faster than rats in bottom cages. If the experimenters put rats fed the new product at the top and those fed the standard diet below, the experiment is biased in favor of the new product. Solution: Assign the rats to cages at random.

Another study looked at the effects of human affection on the cholesterol level of rabbits. All of the rabbit subjects ate the same diet. Some (chosen at random) were regularly removed from their cages to have their furry heads scratched by friendly people. The rabbits who received affection had lower cholesterol. So affection for some but not other rabbits could bias an experiment in which the rabbits' cholesterol level is a response variable.

Double-blind experiments

Placebos work. That bare fact means that medical studies must take special care to show that a new treatment is not just a placebo. Part of equal treatment for all is to be sure that the placebo effect operates on all subjects.

Example 2. The powerful placebo

Want to help balding men keep their hair? Give them a placebo—one study found that 42% of balding men maintained or increased the amount of hair on their heads when they took a placebo. Another study told 13 people who were very sensitive to poison ivy that the stuff being rubbed on one arm was poison ivy. It was a placebo, but all 13 broke out in a rash. The stuff rubbed on the other arm really was poison ivy, but the subjects were told it was harmless—and only 2 of the 13 developed a rash.

When the ailment is vague and psychological, like depression, some experts think that about three-quarters of the effect of the most widely used drugs is just the placebo effect. Others disagree. The strength of the placebo effect in medical treatments is hard to pin down because it depends on the exact environment, much as the behavior of the fickle mice does. How enthusiastic the doctor is seems to matter a lot. But "placebos work" is a good place to start when you think about planning medical experiments.

The strength of the placebo effect is a strong argument for randomized comparative experiments. In the baldness study, 42% of the placebo group kept or increased their hair, but 86% of the men getting a new drug to fight baldness did so. The drug beats the placebo, so it has something besides the placebo effect going for it. Of course, the placebo effect is still part of the reason this and other treatments work.

Because the placebo effect is so strong, it would be foolish to tell subjects in a medical experiment whether they are receiving a new drug or a placebo. Knowing that they are getting "just a placebo" might weaken the placebo effect and bias the experiment in favor of the other treatments. It is also foolish to tell doctors and other medical personnel what treatment each subject received. If they know that a subject is getting "just a placebo," they may expect less than if they know the subject is receiving a promising experimental drug. Doctors' expectations change

Statistical Controversies

Is It or Isn't It a Placebo?

Natural remedies are big business: elk antler velvet to enhance athletic performance; valerian root for stress, headaches, and menstrual cramps; yohimbe bark to help your sex life. Store shelves and Web sites are filled with exotic substances claiming to improve your health.

A therapy that has not been compared with a placebo in a randomized experiment may itself be just a placebo. In the United States, the law requires that new prescription drugs and new medical devices show their safety and effectiveness in randomized trials. You can be confident that a drug your doctor prescribes is more than a placebo.

What about those "natural remedies"? The law allows makers of herbs, vitamins, and dietary supplements to claim without any evidence that they are safe and will help "natural conditions." They can't claim to treat "diseases." Of course, the boundary between natural conditions and diseases is vague. I can claim that Dr. Moore's Old Indiana Extract promotes healthy hearts without any evidence whatsoever. I can't claim that it reduces the risk of heart disease without clinical trials and an OK by the Food and Drug Administration (FDA). No doubt lots of folks will think that "promotes healthy hearts" means the same thing as "reduces the risk of heart disease" when they see my advertisements. I also don't have to worry about what dose of Old Indiana Extract my pills contain, or about what dose might actually be toxic.

Should the FDA require natural remedies to meet the same standards as prescription drugs? That's hard to do in practice, because natural substances can't be patented. Drug companies spend millions of dollars on clinical trials because they can patent the drugs that prove effective. Nobody can patent an herb, so nobody has a financial incentive to pay for a clinical trial. Don't look for big changes in the regulations.

Meanwhile, it's easy to find claims that ginkgo biloba is good for (as one Web site says) "hearing and vision problems as well as impotence, edema, varicose veins, leg ulcers, and strokes." Common sense says you should be suspicious of claims that a substance is good for lots of unrelated conditions. Statistical sense says you should be suspicious of claims not backed by comparative experiments. Many untested remedies are no doubt just placebos. Yet they may have real effects in many people—the placebo effect is strong. Just remember that the safety of these concoctions is also untested.

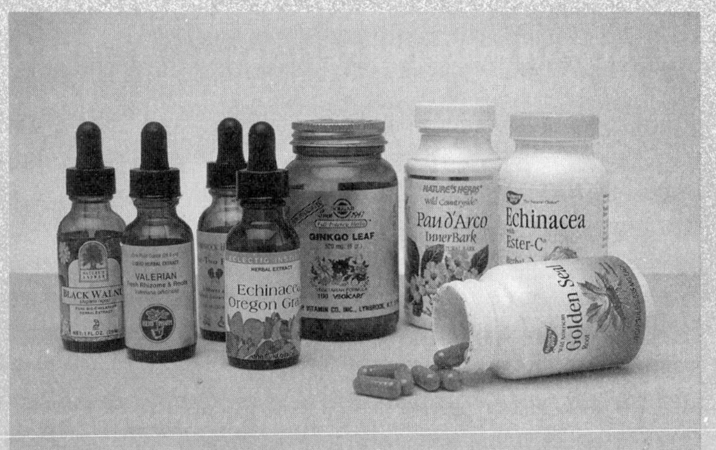

Ogust/The Image Works

"Dr. Burns, are you sure this is what the statisticians call a double-blind experiment?"

how they interact with patients and even the way they diagnose a patient's condition. Whenever possible, experiments with human subjects should be *double-blind.*

Double-blind experiments

In a **double-blind experiment**, neither the subjects nor the people who work with them know which treatment each subject is receiving.

Until the study ends and the results are in, only the study's statistician knows for sure. Reports in medical journals regularly begin with words like these, from a study of a flu vaccine given as a nose spray: "This study was a randomized, double-blind, placebo-controlled trial. Participants were enrolled from 13 sites across the continental United States between mid-September and mid-November 1997." Doctors are supposed to know what this means. Now you also know.

Refusals, nonadherers, and dropouts

Sample surveys suffer from nonresponse due to failure to contact some people selected for the sample and the refusal of others to participate. Experiments with human subjects suffer from similar problems.

Example 3. Minorities in clinical trials

Refusal to participate is a serious problem for medical experiments on treatments for serious diseases such as cancer. As in the case of samples, bias can result if those who refuse are systematically different from those who cooperate.

Minorities, women, the poor, and the elderly have long been underrepresented in clinical trials. In many cases, they weren't asked. The law now requires representation of women and minorities, and data show that most clinical trials now have fair representation. But refusals remain a problem. Minorities, especially blacks, are more likely to refuse to participate. The government's Office of Minority Health says, "Though recent studies have shown that African Americans have increasingly positive attitudes toward cancer medical research, several studies corroborate that they are still cynical about clinical trials. A major impediment for lack of participation is a lack of trust in the medical establishment." Some remedies for lack of trust are complete and clear information about the experiment, insurance coverage for experimental treatments, participation of black researchers, and cooperation with doctors and health organizations in black communities.

Subjects who participate but don't follow the experimental treatment, called **nonadherers**, can also cause bias. AIDS patients who participate in trials of a new drug sometimes take other treatments on their own, for example. What is more, some AIDS subjects have their medication tested and drop out or add other medications if they were not assigned to the new drug. This may bias the trial against the new drug.

Experiments that continue over an extended period of time also suffer **dropouts**, subjects who begin the experiment but do not complete it. If the reasons for dropping out are unrelated to the experimental treatments, no harm is done other than reducing the number of subjects. If subjects drop out because of their reaction to one of the treatments, bias can result.

Example 4. Dropouts in a medical study

Orlistat is a new drug that may help reduce obesity by preventing absorption of fat from the foods we eat. As usual, the drug was compared with a placebo in a double-blind randomized trial. Here's what happened.

Start with 1187 obese subjects. First give a placebo for four weeks and drop the subjects who won't take a pill regularly. This attacks the problem of nonadherers. There are 892 subjects left. Randomly assign these subjects to orlistat or a placebo, along with a weight-loss diet. After a year devoted to losing weight, 576 subjects are still participating. On the average, the orlistat group lost 3.15 kilograms (about 7 pounds) more than the placebo group. Keep going for another year, now emphasizing maintaining the weight loss from the first year. At the end of the second year, 403 subjects are left. That's only 45% of the 892 who were randomized. Orlistat again beat the placebo, reducing the weight regained by an average of 2.25 kilograms (about 5 pounds).

Can we trust the results when so many subjects dropped out? The overall dropout rates were similar in the two groups: 54% of the subjects taking orlistat and 57% of those in the placebo group dropped out. Were dropouts related to the treatments? Placebo subjects in weight-loss experiments often drop out because they aren't losing weight. This would bias the study against orlistat because the subjects in the placebo group at the end may be those who could lose weight just by following a diet. The researchers looked carefully at the data available for subjects who dropped out. Dropouts from both groups had lost less weight than those who stayed, but careful statistical study suggested that there was little bias. Perhaps so, but the results aren't as clean as our first look at experiments promised.

The League to Mess Up Experiments meets...

"Agent B, you will scratch the heads of the lab rabbits. Agent Q, you will join a clinical trial and not take your pills. Agent K, you will sign up for an experiment, then drop out just before the end."

Can we generalize?

A well-designed experiment tells us that changes in the explanatory variable cause changes in the response variable. More exactly, it tells us that this happened for specific subjects in the specific environment of this specific experiment. No doubt we had grander things in mind. We want to proclaim that our new method of teaching math does better for high school students in general, or that our new drug beats a placebo for some broad class of patients. Can we generalize our conclusions from our little group of subjects to a wider population?

The first step is to be sure that our findings are *statistically significant*, that they are too strong to often occur just by chance. That's important, but it's a technical detail that the study's statistician can reassure us about. The serious threat is that the treatments, the subjects, or the environment of our experiment may not be realistic. Let's look at some examples.

Example 5. Studying frustration

A psychologist wants to study the effects of failure and frustration on the relationships among members of a work team. She forms a team of students, brings them to the psychology laboratory, and has them play a game that requires teamwork. The game is rigged so that they lose regularly. The psychologist observes the students through a one-way window and notes the changes in their behavior during an evening of game-playing.

Playing a game in a laboratory for small stakes, knowing that the session will soon be over, is a long way from working for months developing a new product that never works right and is finally abandoned by your company. Does the behavior of the students in the lab tell us much about the behavior of the team whose product failed?

In Example 5, the subjects (students who know they are subjects in an experiment), the treatment (a rigged game), and the environment (the psychology lab) are all unrealistic if the psychologist's goal is to reach conclusions about the effects of frustration on teamwork in the workplace. Psychologists do their best to devise realistic experiments for studying human behavior, but lack of realism limits the usefulness of experiments in this area.

Example 6. Center brake lights

Cars sold in the United States since 1986 have been required to have high center brake lights in addition to the usual two brake lights at the rear of the vehicle. This safety requirement was justified by randomized comparative experiments with fleets of rental and business cars. The experiments showed that the third brake light reduced rear-end collisions by as much as 50%.

After almost a decade in actual use, the Insurance Institute found only a 5% reduction in rear-end collisions, helpful but much less than the experiments predicted. What happened? Most cars did not have the extra brake light when the experiments were carried out, so it caught the eye of following drivers. Now that almost all cars have the third light, it no longer captures attention. The experimental conclusions did not generalize as well as safety experts hoped because the environment changed.

Example 7. Are subjects treated too well?

Surely medical experiments are realistic? After all, the subjects are real patients in real hospitals really being treated for real illnesses.

Even here, there are some questions. Patients participating in medical trials get better medical care than most other patients, even if they are in the placebo group. Their doctors are specialists doing research on their specific ailment. They are watched more carefully than other patients. They are more likely to take their pills regularly because they are constantly reminded to do so. Providing "equal treatment for all" except for the experimental and control therapies translates into "provide the best possible medical care for all." The result: ordinary patients may not do as well as the clinical trial subjects when the new therapy comes into general use. It's likely that a therapy that beats a placebo in a clinical trial will beat it in ordinary medical care, but "cure rates" or other measures of success from the trial may be optimistic.

The Carolina Abecedarian Project (Chapter 5) faces the same "too good to be realistic" question. That long and expensive experiment does show that intensive day care has substantial benefits in later life. The day care in the study was intensive indeed—lots of highly qualified staff, lots of parent participation, and detailed activities starting at a very young age, all costing about $11,000 per year for each child. It's unlikely that society will decide to offer such care to all low-income children. The unanswered question is a big one: how good must day care be to really help children succeed in life?

When experiments are not fully realistic, statistical analysis of the experimental data cannot tell us how far the results will generalize. Experimenters generalizing from students in a lab to workers in the real world must argue based on their understanding of how people function, not based just on the data. It is even harder to generalize from rats in a lab to people in the real world. This is one reason why a single experiment is rarely completely convincing, despite the compelling logic of experimental design. The true scope of a new finding must usually be explored by a number of experiments in various settings.

A convincing case that an experiment is sufficiently realistic to produce useful information is based not on statistics but on the experimenter's knowledge of the subject matter of the experiment. The attention to detail required to avoid hidden bias also rests on subject matter knowledge. Good experiments combine statistical principles with understanding of a specific field of study.

Meta-analysis

A single study of an important issue is rarely decisive. We often find several studies in different settings, with different designs, and of different quality. Can we combine their results to get an overall conclusion? That is the idea of "meta-analysis." Of course, differences among the studies prevent us from just lumping them together. Statisticians have more sophisticated ways of combining the results. Meta-analysis has been applied to issues from the effect of secondhand smoke to whether coaching improves SAT scores.

Experimental design in the real world

The experimental designs we have met all have the same pattern: divide the subjects at random into as many groups as there are treatments, then apply each treatment to one of the groups. These are *completely randomized* designs.

Completely randomized design

In a **completely randomized** experimental design, all the experimental subjects are allocated at random among all the treatments.

What is more, our examples to this point have had only a single explanatory variable (drug versus placebo, classroom versus Web instruction). A completely randomized design can have any number of explanatory variables. Here is an example with two.

		Variable B Repetitions		
		1 time	3 times	5 times
Variable A Length	30 seconds	Treatment 1	Treatment 2	Treatment 3
	90 seconds	Treatment 4	Treatment 5	Treatment 6

Figure 6.1 The treatments in the experiment of Example 8. Combinations of two explanatory variables form 6 treatments.

Example 8. Effects of TV advertising

What are the effects of repeated exposure to an advertising message? The answer may depend both on the length of the ad and on how often it is repeated. An experiment investigated this question using undergraduate students as subjects. All subjects viewed a 40-minute television program that included ads for a digital camera. Some subjects saw a 30-second commercial; others, a 90-second version. The same commercial was repeated either 1, 3, or 5 times during the program. After viewing, all of the subjects answered questions about their recall of the ad, their attitude toward the camera, and their intention to purchase it. These are the response variables.

This experiment has two explanatory variables: length of the commercial, with 2 levels, and repetitions, with 3 levels. The 6 combinations of one level of each variable form 6 treatments. Figure 6.1 shows the layout of the treatments.

Experimenters often want to study the combined effects of several variables simultaneously. The interaction of several factors can produce effects that could not be predicted from looking at the effect of each factor alone. Perhaps longer commercials increase interest in a product, and more commercials also increase interest, but if we both make a commercial longer and show it more often, viewers get annoyed and their interest in the product drops. The experiment in Example 8 will help us find out.

Matched pairs and block designs

Completely randomized designs are the simplest statistical designs for experiments. They illustrate clearly the principles of control and randomization. However, completely randomized designs are often inferior to more elaborate statistical designs. In particular, matching the subjects in various ways can produce more precise results than simple randomization.

One common design that combines matching with randomization is the **matched pairs design.** A matched pairs design compares just two treatments. Choose pairs of subjects that are as closely matched as possible. Assign one of the treatments to each subject in a pair by tossing a coin or reading odd and even digits from Table A. Sometimes each "pair" in a matched pairs design consists of just one subject, who gets both treatments one after the other. Each subject serves as his or her own control. The *order* of the treatments can influence the subject's response, so we randomize the order for each subject, again by a coin toss.

Example 9. Coke versus Pepsi

Pepsi wanted to demonstrate that Coke drinkers prefer Pepsi when they taste both colas blind. The subjects, all people who said they were Coke drinkers, tasted both colas from glasses without brand markings and said which they liked better. This is a matched pairs design in which each subject compares the two colas. Because responses may depend on which cola is tasted first, the order of tasting should be chosen at random for each subject.

When more than half the Coke drinkers chose Pepsi, Coke claimed that the experiment was biased. The Pepsi glasses were marked *M* and Coke glasses were marked *Q*. Aha, said Coke, the results could just mean that people like the letter *M* better than the letter *Q*. The matched pairs design is OK, but a more careful experiment would avoid any distinction other than Coke vs. Pepsi.

Matched pairs designs use the principles of comparison of treatments and randomization. However, the randomization is not complete—we do not randomly assign all the subjects at once to the two treatments. Instead, we only randomize within each matched pair. This allows matching to reduce the effect of variation among the subjects. Matched pairs are an example of *block designs.*

Block design

A **block** is a group of experimental subjects that are known before the experiment to be similar in some way that is expected to affect the response to the treatments. In a **block design,** the random assignment of subjects to treatments is carried out separately within each block.

A block design combines the idea of creating equivalent treatment groups by matching with the principle of forming treatment groups at random. Blocks are another form of *control.* They control the effects of some outside variables by bringing those variables into the experiment to form the blocks. Here are some typical examples of block designs.

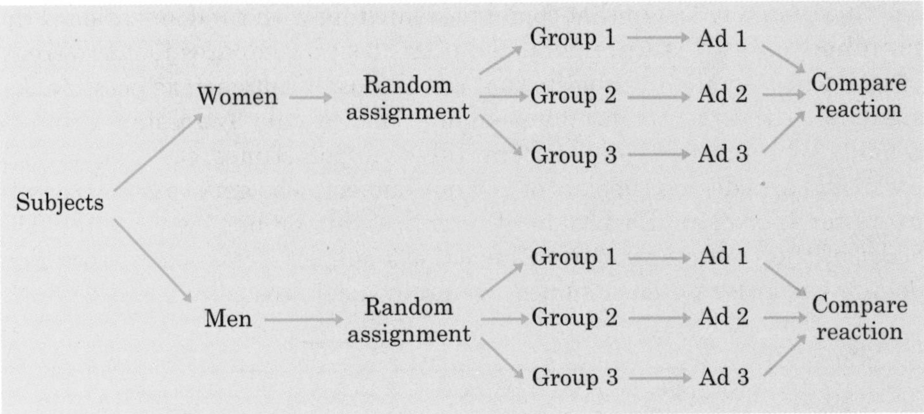

Figure 6.2 A block design to compare the effectiveness of three TV advertisements. Female and male subjects form two blocks.

Example 10. Men, women, and advertising

Women and men respond differently to advertising. An experiment to compare the effectiveness of three television commercials for the same product will want to look separately at the reactions of men and women, as well as assess the overall response to the ads.

A completely randomized design considers all subjects, both men and women, as a single pool. The randomization assigns subjects to three treatment groups without regard to their sex. This ignores the differences between men and women. A better design considers women and men separately. Randomly assign the women to three groups, one to view each commercial. Then separately assign the men at random to three groups. Figure 6.2 outlines this improved design.

Example 11. Comparing welfare policies

A social policy experiment will assess the effect on family income of several proposed new welfare systems and compare them with the present welfare system. Because the future income of a family is strongly related to its present income, the families who agree to participate are divided into blocks of similar income levels. The families in each block are then allocated at random among the welfare systems.

A block is a group of subjects formed before an experiment starts. We reserve the word "treatment" for a condition that we impose on the subjects. We don't speak of 6 treatments in Example 10 even though we can compare the responses of 6 groups of subjects formed by the 2 blocks (men, women) and the 3 commercials. Block designs are similar to stratified samples. Blocks and strata both group similar individuals together.

We use two different names only because the idea developed separately for sampling and experiments. The advantages of block designs are the same as the advantages of stratified samples. Blocks allow us to draw separate conclusions about each block—for example, about men and women in the advertising study in Example 10. Blocking also allows more precise overall conclusions, because the systematic differences between men and women can be removed when we study the overall effects of the three commercials. The idea of blocking is an important additional principle of statistical design of experiments. A wise experimenter will form blocks based on the most important unavoidable sources of variability among the experimental subjects. Randomization will then average out the effects of the remaining variation and allow an unbiased comparison of the treatments.

Like the design of samples, the design of complex experiments is a job for experts. Now that we have seen a bit of what is involved, we will usually just act as if most experiments were completely randomized.

Exploring the Web

The Office of Minority Health newsletter *Closing the Gap* has a special issue on clinical trials. Available at www.omhrc.gov/ctg/ctg-ct.htm, the issue includes several articles on minority participation in medical experiments.

You can find a (very) skeptical look at health claims not backed by proper clinical trials at the "Quackwatch" site, www.quackwatch.com.

Statistics in Summary As with samples, experiments require a combination of good statistical design and steps to deal with practical problems. Because the **placebo effect** is strong, **clinical trials** and other experiments with human subjects should be **double-blind** whenever this is possible. The double-blind method helps achieve a basic requirement of comparative experiments: **equal treatment for all subjects** except for the actual treatments the experiment is comparing.

Just as samples suffer from nonresponse, experiments suffer from uncooperative subjects. Some subjects refuse to participate; others drop out before the experiment is complete; others don't follow instructions, as when some subjects in a drug trial don't take their pills. The most common weakness in experiments is that we can't **generalize** the conclusions widely. Some experiments apply unrealistic treatments, some use subjects from some special group such as college students, and all take place at some specific place and time. We want to see similar experiments at other places and times confirm important findings.

Many experiments use designs that are more complex than the basic **completely randomized design** that divides all the subjects among all the treatments in one randomization. **Matched pairs designs** compare two treatments by giving one to each of a pair of similar subjects or by giving both to the same subject in random order. **Block designs** form blocks of similar subjects and assign treatments at random separately in each block. The big ideas of randomization, control, and adequate numbers of subjects remain the keys to convincing experiments.

CHAPTER 6 EXERCISES

6.1 Medical news. When it was found that hydroxyurea reduced the symptoms of sickle cell anemia, the National Institutes of Health released a medical bulletin. The bulletin said, "These findings are the results of data analyzed from the Multicenter Study of Hydroxyurea in Sickle Cell Anemia (MSH), which was a double-blind, placebo-controlled trial in which half of the patients received hydroxyurea and half received a placebo capsule." Explain to someone who knows no statistics what the terms "placebo-controlled" and "double-blind" mean here.

6.2 Does meditation reduce anxiety? An experiment that claimed to show that meditation reduces anxiety proceeded as follows. The experimenter interviewed the subjects and rated their level of anxiety. Then the subjects were randomly assigned to two groups. The experimenter taught one group how to meditate and they meditated daily for a month. The other group was simply told to relax more. At the end of the month, the experimenter interviewed all the subjects again and rated their anxiety level. The meditation group now had less anxiety. Psychologists said that the results were suspect because the ratings were not blind. Explain what this means and how lack of blindness could bias the reported results.

6.3 Emergency room care. An article in a medical journal reports an experiment to see if injecting an oxygen-carrying fluid in addition to performing standard emergency room procedures would help patients in shock from loss of blood. The article describes the experiment as a "randomized, controlled, single-blinded efficacy trial conducted between February 1997 and January 1998 at 18 US trauma centers." What do you think "single-blinded" means here? Why isn't a double-blind experiment possible?

6.4 Daytime running lights. Canada requires that cars be equipped with "daytime running lights," headlights that automatically come on at a low level when the car is started. Some manufacturers are now equipping cars sold in the United States with running lights. Will running lights reduce accidents by making cars more visible?

(a) Briefly discuss the design of an experiment to help answer this question. In particular, what response variables will you examine?

(b) Example 6 discusses center brake lights. What cautions do you draw from that example that apply to an experiment on the effects of running lights?

6.5 Diet and cancer. Substances that cause cancer should not appear in our food. We don't want to experiment on people to learn what substances cause cancer, so we experiment on rats instead. The rats are specially bred to have more tumors than humans do. They are fed large doses of the test chemical for most of their natural life, about two years. Briefly discuss the questions that arise in using these experiments to decide what is safe in human diets.

6.6 Dealing with cholesterol. Clinical trials have shown that reducing blood cholesterol using either drugs or diet reduces heart attacks. The first researchers followed their subjects for five to seven years. In order to see results in this relatively short time, the subjects were chosen from the group at greatest risk, middle-aged men with high cholesterol levels or existing heart disease. The experiments generally showed that reducing blood cholesterol does decrease the risk of a heart attack. Some doctors questioned whether these experimental results applied to many of their patients. Why?

6.7 Testing a natural remedy. Although the law doesn't require it, I decide to subject Dr. Moore's Old Indiana Extract to a clinical trial. I hope to show that the extract reduces pain from arthritis. Sixty patients suffering from arthritis and needing pain relief are available. I will give a pill to each patient and ask them an hour later, "About what percentage of pain relief did you experience?"

(a) Why should I not simply give the extract to all 60 patients and record the responses?

(b) Outline the design of an experiment to compare the extract's effectiveness with that of aspirin and of a placebo.

(c) Should patients be told which remedy they are receiving? How might this knowledge affect their reactions?

(d) If patients are not told which treatment they are receiving, the experiment is single-blind. Should this experiment be double-blind also? Explain.

6.8 Testing a natural remedy. The National Institutes of Health is at last sponsoring proper clinical trials of some natural remedies. In one study at Duke University, 330 patients with mild depression are enrolled in a trial to compare St. John's wort with a placebo and with Zoloft, a common

prescription drug for depression. The Beck Depression Inventory is a common instrument that rates the severity of depression on a 0 to 3 scale.
(a) What would you use as the response variable to measure change in depression after treatment?

(b) Outline the design of a completely randomized clinical trial for this study.

(c) What other precautions would you take in this trial?

6.9 The placebo effect. A survey of physicians found that some doctors give a placebo to a patient who complains of pain for which the physician can find no cause. If the patient's pain improves, these doctors conclude that it had no physical basis. The medical school researchers who conducted the survey claimed that these doctors do not understand the placebo effect. Why?

6.10 Repairing knees in comfort. Injured knees are repaired by arthroscopic surgery that does not require opening up the knee. Can we reduce patient discomfort by giving them a nonsteroidal anti-inflammatory drug (NSAID)? Eighty-three patients were placed in three groups. Group A received the NSAID both before and after surgery. Group B was given a placebo before and the NSAID after. Group C received a placebo both before and after surgery. The patients recorded a pain score by answering questions one day after the surgery.
(a) Outline the design of this experiment. You do not need to do the randomization that your design requires.

(b) You read that "the patients, physicians, and physical therapists were blinded" during the study. What does this mean?

(c) You also read that "the pain scores for Group A were significantly lower than Group C but not significantly lower than Group B." What does this mean? What does this finding lead you to conclude about the use of NSAIDs?

6.11 Comparing corn varieties. New varieties of corn with altered amino acid content may have higher nutritional value than standard corn, which is low in the amino acid lysine. An experiment compares two new varieties, called opaque-2 and floury-2, with normal corn. The researchers mix corn-soybean meal diets using each type of corn at each of three protein levels: 12% protein, 16% protein, and 20% protein. They feed each diet to 10 one-day-old male chicks and record their weight gains after 21 days. The weight gain of the chicks is a measure of the nutritional value of their diet.
(a) What are the individuals and the response variable in this experiment?

(b) How many explanatory variables are there? How many treatments? Use a diagram like Figure 6.1 to describe the treatments. How many experimental individuals does the experiment require?

(c) Use a diagram to describe a completely randomized design for this experiment. (Don't actually do the randomization.)

6.12 Baking cakes. A food company is preparing to market a new cake mix. It is important that the taste of the cake not be changed by small variations in baking time or temperature. In an experiment, cakes made from the mix are baked at 300°, 320°, and 340° F, and for 1 hour and 1 hour and 15 minutes. Ten cakes are baked at each combination of temperature and time. A panel of tasters scores each cake for texture and taste.

(a) What are the explanatory variables and the response variables for this experiment?

(b) Make a diagram like Figure 6.1 to describe the treatments. How many treatments are there? How many cakes are needed?

(c) Explain why it is a bad idea to bake all 10 cakes for one treatment at once, then bake the 10 cakes for the second treatment, and so on. Instead, the experimenters will bake the cakes in a random order determined by the randomization in your design.

6.13 Comparing hand strength. Is the right hand generally stronger than the left in right-handed people? You can crudely measure hand strength by placing a bathroom scale on a shelf with the end protruding, then squeezing the scale between the thumb below and the four fingers above it. The reading of the scale shows the force exerted. Describe the design of a matched pairs experiment to compare the strength of the right and left hands, using 10 right-handed people as subjects. (You need not actually do the randomization.)

6.14 Does charting help investors? Some investors believe that charts of past trends in the prices of securities can help predict future prices. Most economists disagree. In an experiment to examine the effects of using charts, business students trade (hypothetically) a foreign currency at computer screens. There are 20 student subjects available, named for convenience A, B, C, . . . , T. Their goal is to make as much money as possible, and the best performances are rewarded with small prizes. The student traders have the price history of the foreign currency in dollars in their computers. They may or may not also have software that highlights trends. Describe two designs for this experiment, a completely randomized design and a matched pairs design in which each student serves as his or her own control. In both cases, carry out the randomization required by the design.

6.15 Temperature and work performance. An expert on worker performance is interested in the effect of room temperature on the performance of tasks requiring manual dexterity. She chooses temperatures of 70° and 90° F as treatments. The response variable is the number of correct insertions, during a 30-minute period, in a peg-and-hole apparatus that requires the use of both hands simultaneously. Each subject is trained on the

apparatus and then asked to make as many insertions as possible in 30 minutes of continuous effort.
(a) Outline a completely randomized design to compare dexterity at 70° and 90°. Twenty subjects are available.

(b) Because individuals differ greatly in dexterity, the wide variation in individual scores may hide the systematic effect of temperature unless there are many subjects in each group. Describe in detail the design of a matched pairs experiment in which each subject serves as his or her own control.

6.16 Comparing cancer treatments. The progress of a type of cancer differs in women and men. A clinical experiment to compare three therapies for this cancer therefore treats sex as a blocking variable.
(a) You have 500 male and 300 female patients who are willing to serve as subjects. Use a diagram to outline a block design for this experiment. Figure 6.2 is a model.

(b) What are the advantages of a block design over a completely randomized design using these 800 subjects? What are the advantages of a block design over a completely randomized design using 800 male subjects?

6.17 Comparing weight-loss treatments. Twenty overweight females have agreed to participate in a study of the effectiveness of 4 weight-loss treatments: A, B, C, and D. The researcher first calculates how overweight each subject is by comparing the subject's actual weight with her "ideal" weight. The subjects and their excess weights in pounds are

Birnbaum	35	Hernandez	25	Moses	25	Smith	29
Brown	34	Jackson	33	Nevesky	39	Stall	33
Brunk	30	Kendall	28	Obrach	30	Tran	35
Cruz	34	Loren	32	Rodriguez	30	Wilansky	42
Deng	24	Mann	28	Santiago	27	Williams	22

The response variable is the weight lost after 8 weeks of treatment. Because a subject's excess weight will influence the response, a block design is appropriate.
(a) Arrange the subjects in order of increasing excess weight. Form 5 blocks of 4 subjects each by grouping the 4 least overweight, then the next 4, and so on.

(b) Use Table A to randomly assign the 4 subjects in each block to the 4 weight-loss treatments. Be sure to explain exactly how you used the table.

6.18 In the corn field. An agronomist wants to compare the yield of 5 corn varieties. The field in which the experiment will be carried out increases in fertility from north to south. The agronomist therefore divides

the field into 30 plots of equal size, arranged in 6 east-west rows of 5 plots each, and employs a block design with the rows of plots as the blocks.
(a) Draw a sketch of the field, divided into 30 plots. Label the rows Block 1 to Block 6.

(b) Do the randomization required by the block design. That is, randomly assign the 5 corn varieties A, B, C, D, and E to the 5 plots in each block. Mark on your sketch which variety is planted in each plot.

6.19 Speeding the mail? Is the number of days a letter takes to reach another city affected by the time of day it is mailed and whether or not the zip code is used? Describe briefly the design of an experiment with two explanatory variables to investigate this question. Be sure to specify the treatments exactly and to tell how you will handle lurking variables such as the day of the week on which the letter is mailed.

6.20 McDonald's versus Wendy's. Do consumers prefer the taste of a cheeseburger from McDonald's or from Wendy's in a blind test in which neither burger is identified? Describe briefly the design of a matched pairs experiment to investigate this question.

6.21 What do you want to know? The previous two exercises illustrate the use of statistically designed experiments to answer questions that arise in everyday life. Select a question of interest to you that an experiment might answer and briefly discuss the design of an appropriate experiment.

6.22 Doctors and nurses. Nurse practitioners are nurses with advanced qualifications who often act much like primary care physicians. An experiment assigned 1316 patients who had no regular source of medical care to either a doctor (510 patients) or a nurse practitioner (806 patients). All of the patients had been diagnosed with either asthma, diabetes, or high blood pressure before being assigned. The response variables included measures of the patients' health and of their satisfaction with their medical care after 6 months.
(a) Is the diagnosis (asthma, etc.) a treatment variable or a block? Why?

(b) Is the type of care (nurse or doctor) a treatment variable or a block? Why?

Chapter 7 Data Ethics

Are sham surgeries ethical?

"Randomized, double-blind, placebo-controlled trials are the gold standard for evaluating new interventions and are routinely used to assess new medical therapies." So says an article in the *New England Journal of Medicine* that discusses the treatment of Parkinson's disease. The article isn't about the new treatment, which offers hope of reducing the tremors and lack of control brought on by the disease, but about the ethics of studying the treatment.

The law requires well-designed experiments to show that new drugs work and are safe. Not so with surgery—only about 7% of studies of surgery use randomized comparisons. Surgeons think their operations succeed, but innovators always think their innovations work. Even if the patients are helped, the placebo effect may deserve most of the credit. So we don't really know whether many common surgeries are worth the risk they carry. To find out, do a proper experiment. That includes a "sham surgery" to serve as a placebo. In the case of Parkinson's disease, the promising treatment involves surgery to implant new cells. The placebo subjects get the same surgery, but the cells are not implanted.

Is this ethical? One side says: Only a randomized comparative experiment can tell us if the new treatment works. If it doesn't work, thousands of patients will be spared operations that will do no good. If it doeswork, we can do those thousands of operations knowing that we offer more than a placebo. The other side says: Any surgery carries risks.

Sham operations place patients at risk without the hope that the operation will help them. Doctors should not risk harm to some patients today just because other patients may benefit tomorrow.

Reasonable people disagree about sham surgery as part of a test of real surgery for Parkinson's disease. But there is wide agreement on some basic ethical principles for statistical studies. To think about the hard cases, we should first think about these principles.

First principles

The production and use of data, like all human endeavors, raise ethical questions. We won't discuss the telemarketer who begins a telephone sales pitch with "I'm conducting a survey." Such deception is clearly unethical. It enrages legitimate survey organizations, which find the public less willing to talk with them. Neither will we discuss those few researchers who, in the pursuit of professional advancement, publish fake data. There is no ethical question here—faking data to advance your career is just wrong. It will end your career when uncovered. But just how honest must researchers be about real, unfaked data? Here is an example that suggests the answer is "More honest than they often are."

Example 1. Missing details

Papers reporting scientific research are supposed to be short, with no extra baggage. Brevity can allow the researchers to avoid complete honesty about their data. Did they choose their subjects in a biased way? Did they report data on only some of their subjects? Did they try several statistical analyses and only report the ones that looked best? The statistician John Bailar screened more than 4000 medical papers in more than a decade as consultant to the *New England Journal of Medicine*. He says, "When it came to the statistical review, it was often clear that critical information was lacking, and the gaps nearly always had the practical effect of making the authors' conclusions look stronger than they should have." The situation is no doubt worse in fields that screen published work less carefully.

The most complex issues of data ethics arise when we collect data from people. The ethical difficulties are more severe for experiments that impose some treatment on people than for sample surveys that simply gather information. Trials of new medical treatments, for example, can do harm as well as good to their subjects. Here are some basic standards of data ethics that must be obeyed by any study that gathers data from human subjects, whether sample survey or experiment.

Basic data ethics

The organization that carries out the study must have an **institutional review board** that reviews all planned studies in advance in order to protect the subjects from possible harm.

All individuals who are subjects in a study must give their **informed consent** before data are collected.

All individual data must be kept **confidential.** Only statistical summaries for groups of subjects may be made public.

The law requires that studies funded by the federal government obey these principles. But neither the law nor the consensus of experts is completely clear about the details of their application.

"That's the gist of what I want to say. Now get me some statistics to base it on."

Institutional review boards

The purpose of an institutional review board is not to decide whether a proposed study will produce valuable information or whether it is statistically sound. The board's purpose is, in the words of one university's board, "to protect the rights and welfare of human subjects (including patients) recruited to participate in research activities." The board reviews the plan of the study and can require changes. It reviews the consent form to be sure that subjects are informed about the nature of the study and about any potential risks. Once research begins, the board monitors its progress at least once a year.

The most pressing issue concerning institutional review boards is whether their workload has become so large that their effectiveness in protecting subjects drops. When the government temporarily stopped human subject research at Duke University Medical Center in 1999 due to inadequate protection of subjects, more than 2000 studies were going on. That's a lot of review work. There are shorter review procedures for projects that involve only minimal risks to subjects, such as most sample surveys. When a board is overloaded, there is a temptation to put more proposals in the minimal risk category to speed the work.

Informed consent

Both words in the phrase "informed consent" are important, and both can be controversial. Subjects must be *informed* in advance about the nature of a study and any risk of harm it may bring. In the case of a sample survey, physical harm is not possible. The subjects should be told what kinds of questions the survey will ask and about how much of their time it will take. Experimenters must tell subjects the nature and purpose of the study and outline possible risks. Subjects must then *consent* in writing.

Example 2. Who can consent?

Are there some subjects who can't give informed consent? It was once common, for example, to test new vaccines on prison inmates who gave their consent in return for good-behavior credit. Now we worry that prisoners are not really free to refuse, and the law forbids medical experiments in prisons.

Children can't give fully informed consent, so the usual procedure is to ask their parents. A study of new ways to teach reading is about to start at a local elementary school, so the study team sends consent forms home to parents. Many parents don't return the forms. Can their children take part in the study because the parents did not say "No," or should we allow only children whose parents returned the form and said "Yes"?

What about research into new medical treatments for people with mental disorders? What about studies of new ways to help emergency room patients who may be unconscious or have suffered a stroke? In most cases, there is not time even to get the consent of the family. Does the principle of informed consent bar realistic trials of new treatments for unconscious patients?

These are questions without clear answers. Reasonable people differ strongly on all of them. There is nothing simple about informed consent.

The difficulties of informed consent do not vanish even for capable subjects. Some researchers, especially in medical trials, regard consent as a barrier to getting patients to participate in research. They may not explain all possible risks; they may not point out that there are other therapies that might be better than those being studied; they may be too optimistic in talking with patients even when the consent form has all the right details. On the other hand, mentioning every possible risk leads to very long consent forms that really are barriers. "They are like rental car contracts," one lawyer said. Some subjects don't read forms that run five or six printed pages. Others are frightened by the large number of possible (but unlikely) disasters that might happen and so refuse to participate. Of course, unlikely disasters sometimes happen. When they do, lawsuits follow and the consent forms become yet longer and more detailed.

Confidentiality

Ethical problems do not disappear once a study has been cleared by the review board, has obtained consent from its subjects, and has actually collected data about the subjects. It is important to protect the subjects' privacy by keeping all data about individuals confidential. The report of an opinion poll may say what percent of the 1500 respondents felt that legal immigration should be reduced. It may not report what *you* said about this or any other issue.

Confidentiality is not the same as **anonymity**. Anonymity means that subjects are anonymous—their names are not known even to the director of the study. Anonymity is rare in statistical studies. Even where anonymity is possible (mainly in surveys conducted by mail), it prevents any follow-up to improve nonresponse or inform subjects of results.

Any breach of confidentiality is a serious violation of data ethics. The best practice is to separate the identity of the subjects from the rest of the data at once. Sample surveys, for example, use the identification only to check on who did or did not respond. In an era of advanced technology, however, it is no longer enough to be sure that each individual set of data protects people's privacy. The government, for example, maintains a vast amount of informa-

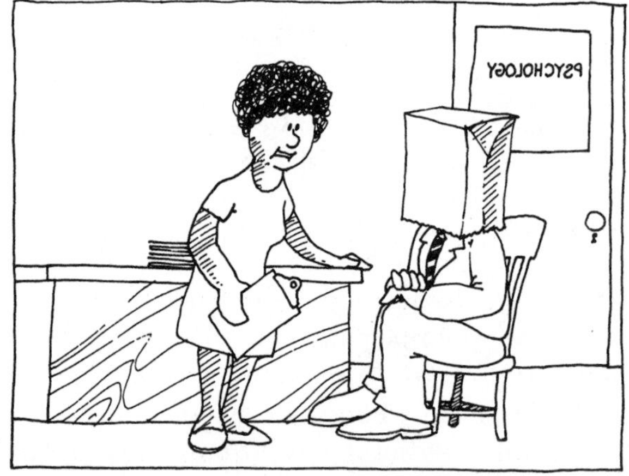

"I realize the participants in this study are to be anonymous, but you're going to have to expose your eyes."

tion about citizens in many separate data bases—census responses, tax returns, Social Security information, data from surveys such as the Current Population Survey, and so on. Many of these data bases can be searched by computers for statistical studies. A clever computer search of several data bases might be able, by combining information, to identify you and learn a great deal about you even if your name and other identification have been removed from the data available for search. Privacy and confidentiality of data are hot issues among statisticians in the computer age.

Statisticians honest and dishonest

Developed nations rely on government statisticians to produce honest data. We trust the monthly unemployment rate, for example, to guide both public and private decisions. Honesty can't be taken for granted everywhere, however. In 1998, the Russian government arrested the top statisticians in the State Committee for Statistics. They were accused of taking bribes to fudge data to help companies avoid taxes. "It means that we know nothing about the performance of Russian companies," said one newspaper editor.

Example 3. Use of government data bases

Citizens are required to give information to the government. Think of tax returns and Social Security contributions. The government needs these data for administrative purposes—to see if I paid the right amount of tax and how large a Social Security benefit I am owed when I retire. Some people feel that individuals should be able to forbid any other use of their data, even with all identification removed. This would prevent using government records to study, say, the ages, incomes, and household sizes of Social Security recipients. Such a study could well be vital to debates on reforming Social Security.

Clinical trials

Clinical trials are experiments that study the effectiveness of medical treatments on actual patients. Medical treatments can harm as well as heal, so clinical trials spotlight the ethical problems of experiments with human subjects. Here are the starting points for a discussion:

- Randomized comparative experiments are the only way to see the true effects of new treatments. Without them, risky treatments that are no better than placebos will become common.
- Clinical trials produce great benefits, but most of these benefits go to future patients. The trials also pose risks, and these risks are borne by the subjects of the trial. So we must balance future benefits against present risks.
- Both medical ethics and international human rights standards say that "the interests of the subject must always prevail over the interests of science and society."

The quoted words are from the 1964 Helsinki Declaration of the World Medical Association, the most respected international standard. The most outrageous examples of unethical experiments are those that ignore the interests of the subjects.

Example 4. The Tuskegee syphilis study

In the 1930s, syphilis was common among black men in the rural South, a group that had almost no access to medical care. The Public Health Service recruited 399 poor black sharecroppers with syphilis and 201 others without the disease in order to observe how syphilis progressed when no treatment was given. Beginning in 1943, penicillin became available to treat syphilis. The study subjects were not treated. In fact, the Public Health Service prevented any treatment until word leaked out and forced an end to the study in the 1970s.

The Tuskegee study is an extreme example of investigators following their own interests and ignoring the well-being of their subjects. A 1996 review said, "It has come to symbolize racism in medicine, ethical misconduct in human research, paternalism by physicians, and government abuse of vulnerable people." In 1997, President Clinton formally apologized to the surviving participants in a White House ceremony.

The Tuskegee study helps explain the lack of trust that lies behind the reluctance of many blacks to take part in clinical trials.

Because "the interests of the subject must always prevail," medical treatments can be tested in clinical trials only when there is reason to hope that they will help the patients who are subjects in the trials. Future benefits aren't enough to justify experiments with human subjects. Of course, if there is already strong evidence that a treatment works and is safe, it is unethical *not* to give it. Here are the words of Dr. Charles Hennekens of the Harvard Medical School, who directed the large clinical trial that showed that aspirin reduces the risk of heart attacks:

> *There's a delicate balance between when to do or not do a randomized trial. On the one hand, there must be sufficient belief in the agent's potential to justify exposing half the subjects to it. On the other hand, there must be sufficient doubt about its efficacy to justify withholding it from the other half of subjects who might be assigned to placebos.*

Why is it ethical to give a control group of patients a placebo? Well, we know that placebos often work. What is more, placebos have no harmful side effects. So in the state of balanced doubt described by Dr. Hennekens, the placebo group may be getting a better treatment than the drug group. If we *knew* which treatment was better, we would give it to everyone. When we don't know, it is ethical to try both and compare them. Here are some harder questions about placebos, with arguments on both sides.

Statistical Controversies

Hope for Sale?

We have pointed to the ethical problems of experiments with human subjects, clinical trials in particular. *Not* doing proper experiments can also pose problems. Here is an example. Women with advanced breast cancer will eventually die. A promising but untried treatment appears. Should we wait for controlled clinical trials to show that it works, or should we make it available right now?

The promising treatment is "bone marrow transplant" (BMT for short). The idea of BMT is to harvest a patient's bone marrow cells, blast the cancer with very high doses of drugs, then return the harvested cells to keep the drugs from killing the patient. BMT has become popular. It is painful, expensive, and dangerous.

New anticancer drugs are first available through clinical trials, but there is no constraint on therapies such as BMT. When small, uncontrolled trials seemed to show success, BMT became widely available. The economics of medicine had a lot to do with this. The early leaders in offering BMT were for-profit hospitals that advertise heavily to attract patients. Others soon jumped in. The *New York Times* reported: "Every entity offering the experimental procedure tried a different sales pitch. Some promoted the prestige of their institutions, others the convenience of their locations, others their caring attitudes and patient support. . . ." The profits for hospitals and doctors are high.

But does BMT keep patients alive longer than standard treatments? We don't know, but the answer appears to be "probably not." The patients naturally want to try anything that might keep them alive. Some doctors are willing to offer hope not backed by good evidence. Patients would not join controlled trials that might assign them to standard treatments rather than to BMT. Results from such trials were delayed for years by the difficulty in recruiting subjects. Of the first five trials reported, four found no significant difference between BMT and standard treatments. The fifth favored BMT—but the researcher soon admitted "a serious breach of scientific honesty and integrity." The *New York Times* put it more bluntly: "he falsified data."

Compassion seems to support making untested treatments available to dying patients. Reason responds that this opens the door to sellers of hope and delays development of treatments that really work. Compare children's cancer, where doctors agree not to offer experimental treatments outside controlled trials. Result: 60% of all children with cancer are in clinical trials, and progress in saving lives has been much faster for children than for adults. BMT for a rare cancer in children was tested immediately and found to be effective. In contrast, one of the pioneers in BMT for breast cancer, in the light of better evidence, now says, "We deceived ourselves and we deceived our patients."

Example 5. Placebo controls?

You are testing a new drug. Is it ethical to give a placebo to a control group if an effective drug already exists?

Yes: The placebo gives a true baseline for the effectiveness of the new drug. There are three groups: new drug, best existing drug, and placebo. Every clinical trial is a bit different, and not even genuinely effective treatments work in every setting. The placebo control helps us see if the study is flawed so that even the best existing drug does not beat the placebo. Sometimes the placebo wins, so the doubt needed to justify its use is present. Placebo controls are ethical except for life-threatening conditions.

No: It isn't ethical to deliberately give patients an inferior treatment. We don't know whether the new drug is better than the existing drug, so it is ethical to give both in order to find out. If past trials showed that the existing drug is better than a placebo, it is no longer right to give patients a placebo. After all, the existing drug includes the placebo effect. A placebo group is ethical only if the existing drug is an older one that did not undergo proper clinical trials or doesn't work well or is dangerous.

Example 6. Sham surgery

Placebos work. As more doctors recognize this fact, more begin to ask, "If we accept a placebo in drug trials, why don't we accept it in surgery trials?" As the Parkinson's disease case that opened this chapter illustrates, this is a very controversial question.

Yes: Most surgeries have not been tested in comparative experiments, and some are no doubt just placebos. Unlike placebo pills, these surgeries carry risks. Comparing real surgeries to placebo surgeries can eliminate thousands of unnecessary operations and save many lives. The placebo surgery can be made quite safe. For example, placebo patients can be given a safe drug that removes their memory of the operation rather than a more risky anesthetic. Subjects are told that they are in a placebo-controlled trial and they agree to take part. Placebo-controlled trials of surgery are ethical (except for life-threatening conditions) if the risk to the placebo group is small and there is informed consent.

No: Placebo surgery, unlike placebo drugs, always carries some risk. Remember that "the interests of the subject must always prevail." Even great future benefits can't justify risks to subjects today unless those subjects receive some benefit. We might give a patient a placebo drug as a medical therapy, because placebos work and are not risky. No doctor would do a sham surgery as ordinary therapy, because there is some risk. If we would not use it in medical practice, it isn't ethical to use it in a clinical trial.

Behavioral and social science experiments

When we move from medicine to the behavioral and social sciences, the direct risks to experimental subjects are less acute, but so are the possible benefits to the subjects. Consider, for example, the experiments conducted by psychologists in their study of human behavior.

Example 7. Keep out of my space

Psychologists observe that people have a "personal space" and get annoyed if others come too close to them. We don't like strangers to sit at our table in a coffee shop if other tables are available, and we see people move apart in elevators if there is room to do so. Americans tend to require more personal space than people in most other cultures. Can violations of personal space have physical as well as emotional effects?

Investigators set up shop in a men's public rest room. They blocked off urinals to force men walking in to use either a urinal next to an experimenter (treatment group) or a urinal separated from the experimenter (control group). Another experimenter, using a periscope from a toilet stall, measured how long the subject took to start urinating and how long he kept at it.

This personal space experiment illustrates the difficulties facing those who plan and review behavioral studies.

- There is no risk of harm to the subjects, although they would certainly object to being watched through a periscope. What should we protect subjects from when physical harm is unlikely? Possible emotional harm? Undignified situations? Invasion of privacy?
- What about informed consent? The subjects in Example 7 did not even know they were participating in an experiment. Many behavioral experiments rely on hiding the true purpose of the study. The subjects would change their behavior if told in advance what the investigators were looking for. Subjects are asked to consent on the basis of vague information. They receive full information only after the experiment.

"I'm doing a little study on the effects of emotional stress. Now, just take the axe from my assistant."

The "Ethical Principles" of the American Psychological Association require consent unless a study merely observes behavior in a public place. They allow deception only when it is necessary to the study, does not hide information that might influence a subject's willingness to participate, and is explained to subjects as soon as possible. The personal space study (from the 1970s) does not meet current ethical standards.

We see that the basic requirement for informed consent is understood differently in medicine and psychology. Here is an example of another setting with yet another interpretation of what is ethical. The subjects get no information and give no consent. They don't even know that an experiment may be sending them to jail for the night.

Example 8. Domestic violence

How should police respond to domestic violence calls? In the past, the usual practice was to remove the offender and order him to stay out of the household overnight. Police were reluctant to make arrests because the victims rarely pressed charges. Women's groups argued that arresting offenders would help prevent future violence even if no charges were filed. Is there evidence that arrest will reduce future offenses? That's a question that experiments have tried to answer.

A typical domestic violence experiment compares two treatments: arrest the suspect and hold him overnight, or warn the suspect and release him. When police officers reach the scene of a domestic violence call, they calm the participants and investigate. Weapons or death threats require an arrest. If the facts permit an arrest but do not require it, an officer radios headquarters for instructions. The person on duty opens the next envelope in a file prepared in advance by a statistician. The envelopes contain the treatments in random order. The police either arrest the suspect or warn and release him, depending on the contents of the envelope. The researchers then watch police records and visit the victim to see if the domestic violence reoccurs.

The first such experiment appeared to show that arresting domestic violence suspects does reduce their future violent behavior. As a result of this evidence, arrest has become the common police response to domestic violence.

The domestic violence experiments shed light on an important issue of public policy. Because there is no informed consent, the ethical rules that govern clinical trials and most social science studies would forbid these experiments. They were cleared by review boards because, in the words of one domestic violence researcher, "These people became subjects by committing acts that allow the police to arrest them. You don't need consent to arrest someone."

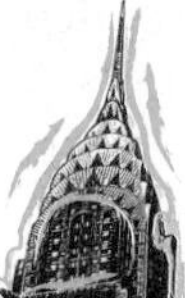

Exploring the Web

For a glimpse at the work of an institutional review board, visit the site of the University of Pittsburgh's board, www.ofres-hs.upmc.edu/irb. Look at the *Reference Manual for the Use of Human Subjects in Research* (the first item listed on this page) to see how elaborate the review process can be.

The official ethical codes of the American Statistical Association (www.amstat.org/profession/ethicalstatistics.html) and the American Psychological Association (www.apa.org/ethics/code.html) are long documents that address many issues in addition to those discussed in this chapter.

Statistics in Summary Ordinary honesty says you shouldn't make up data or claim to be taking a survey when you are really selling roofing. Data ethics begin with some principles that go beyond just being honest. Studies with human subjects must be screened in advance by an **institutional review board.** All subjects must give their **informed consent** before taking part. All information about individual subjects must be kept **confidential**. These principles are a good start, but many ethical debates remain, especially in the area of experiments with human subjects. Many of the debates concern the right balance between the welfare of the subjects and the future benefits of the experiment. Remember that randomized comparative experiments can answer questions that can't be answered without them. Also remember that "the interests of the subject must always prevail over the interests of science and society."

CHAPTER 7 EXERCISES

Most of the exercises in this chapter pose issues for discussion. There are no right or wrong answers, but there are more and less thoughtful answers.

7.1 Minimal risk? You are a member of your college's institutional review board. You must decide whether several research proposals qualify for lighter review because they involve only minimal risk to subjects. Federal regulations say that "minimal risk" means the risks are no greater than "those ordinarily encountered in daily life or during the performance of routine physical or psychological examinations or tests." That's vague. Which of these do you think qualifies as "minimal risk"?
(a) Draw a drop of blood by pricking a finger in order to measure blood sugar.

(b) Draw blood from the arm for a full set of blood tests.

(c) Insert a tube that remains in the arm, so that blood can be drawn regularly.

7.2 Who serves on the review board? Government regulations require that institutional review boards consist of at least five people, including at least one scientist, one nonscientist, and one person from outside the institution. Most boards are larger, but many contain just one outsider.
(a) Why should review boards contain people who are not scientists?

(b) Do you think that one outside member is enough? How would you choose that member? (For example, would you prefer a medical doctor? A member of the clergy? An activist for patients' rights?)

7.3 Institutional review boards. Your college or university has an institutional review board that screens all studies that use human subjects. Get a copy of the document that describes this board (you can probably find it online).

(a) According to this document, what are the duties of the board?

(b) How are members of the board chosen? How many members are not scientists? How many members are not employees of the college? Do these members have some special expertise, or are they simply members of the "general public"?

7.4 Informed consent. A researcher suspects that traditional religious beliefs tend to be associated with an authoritarian personality. She prepares a questionnaire that measures authoritarian tendencies and also asks many religious questions. Write a description of the purpose of this research to be read by subjects in order to obtain their informed consent. You must balance the conflicting goals of not deceiving the subjects as to what the questionnaire will tell about them and of not biasing the sample by scaring off religious people.

7.5 Is consent needed? In which of the circumstances below would you allow collecting personal information without the subjects' consent?

(a) A government agency takes a random sample of income tax returns to obtain information on the average income of people in different occupations. Only the incomes and occupations are recorded from the returns, not the names.

(b) A social psychologist attends public meetings of a religious group to study the behavior patterns of members.

(c) The social psychologist pretends to be converted to membership in a religious group and attends private meetings to study the behavior patterns of members.

7.6 Students as subjects. Students taking Psychology 001 are required to serve as experimental subjects. Students in Psychology 002 are not required to serve, but they are given extra credit if they do so. Students in Psychology 003 are required either to sign up as subjects or to write a term paper. Serving as an experimental subject may be educational, but current ethical standards frown on using "dependent subjects" such as prisoners or charity medical patients. Students are certainly somewhat dependent on their teachers. Do you object to any of these course policies? If so, which ones, and why?

7.7 How common is HIV infection? Researchers from Yale, working with medical teams in Tanzania, wanted to know how common infection with the AIDS virus is among pregnant women in that African country. To do this, they planned to test blood samples drawn from pregnant women.

Yale's institutional review board insisted that the researchers get the informed consent of each woman and tell her the results of the test. This is the usual procedure in developed nations. The Tanzanian government did not want to tell the women why blood was drawn or tell them the test results. The government feared panic if many people turned out to have an incurable disease for which the country's medical system could not provide care. The study was canceled. Do you think that Yale was right to apply its usual standards for protecting subjects?

7.8 Anonymous or confidential? One of the most important non-government surveys in the United States is the General Social Survey (Example 7 in Chapter 1). The GSS regularly monitors public opinion on a wide variety of political and social issues. Interviews are conducted in person in the subject's home. Are a subject's responses to GSS questions anonymous, confidential, or both? Explain your answer.

7.9 Anonymous or confidential? Texas A&M, like many universities, offers free screening for HIV, the virus that causes AIDS. The announcement says, "Persons who sign up for the HIV Screening will be assigned a number so that they do not have to give their name." They can learn the results of the test by telephone, still without giving their name. Does this practice offer *anonymity* or just *confidentiality?*

7.10 Not really anonymous. Some common practices may appear to offer anonymity while actually delivering only confidentiality. Market researchers often use mail surveys that do not ask the respondent's identity but contain hidden codes on the questionnaire that identify the respondent. A false claim of anonymity is clearly unethical. If only confidentiality is promised, is it also unethical to say nothing about the identifying code, perhaps causing respondents to believe their replies are anonymous?

7.11 Human biological materials. Long ago, doctors drew a blood specimen from you as part of treating minor anemia. Unknown to you, the sample was stored. Now researchers plan to use stored samples from you and many other people to look for genetic factors that may influence anemia. It is no longer possible to ask your consent. Modern technology can read your entire genetic makeup from the blood sample.

(a) Do you think it violates the principle of informed consent to use your blood sample if your name is on it but you were not told that it might be saved and studied later?

(b) Suppose that your identity is not attached. The blood sample is known only to come from (say) "a 20-year-old white female being treated for anemia." Is it now OK to use the sample for research?

(c) Perhaps we should use biological materials such as blood samples only from patients who have agreed to allow the material to be stored for later use in research. It isn't possible to say in advance what kind of research, so this falls short of the usual standard for informed consent. Is it nonetheless acceptable, given complete confidentiality and the fact that using the sample can't physically harm the patient?

7.12 Equal treatment. Researchers on aging proposed to investigate the effect of supplemental health services on the quality of life of older people. Eligible patients on the rolls of a large medical clinic were to be randomly assigned to treatment and control groups. The treatment group would be offered hearing aids, dentures, transportation, and other services not available without charge to the control group. The review board felt that providing these services to some but not other persons in the same institution raised ethical questions. Do you agree?

7.13 Sham surgery? Clinical trials like the Parkinson's disease study cited at the beginning of this chapter are becoming more common. One medical researcher says, "This is just the beginning. Tomorrow, if you have a new procedure, you will have to do a double-blind placebo trial." Example 6 outlines the arguments for and against testing surgery just as drugs are tested. When would you allow sham surgery in a clinical trial of a new surgery?

7.14 AIDS clinical trials. Now that effective treatments for AIDS are at last available, is it ethical to test treatments that may be less effective? Combinations of several powerful drugs reduce the level of the HIV virus in the blood and at least delay illness and death from AIDS. But effectiveness depends on how damaged the patient's immune system is and what drugs he has previously taken. There are strong side effects, and patients must be able to take more than a dozen pills on time every day. Because AIDS is often fatal and the combination therapy works, we might argue that it isn't ethical to test any treatment for AIDS that doesn't give the full combination. That might prevent discovery of better treatments. This is a strong example of the conflict between doing the best we know for patients now and finding better treatments for other patients in the future. How can we ethically test new drugs for AIDS?

7.15 AIDS trials in Africa. Effective drugs for treating AIDS are very expensive, so most African nations cannot afford to give them to large numbers of people. Yet AIDS is more common in parts of Africa than anywhere else. Several clinical trials are looking at ways to prevent pregnant mothers infected with HIV from passing the infection to their unborn children, a major source of HIV infections in Africa. Some people say these trials are unethical because they do not give effective AIDS drugs to their subjects, as would be required in rich nations. Others reply that the trials are looking for treatments that can work in the real world in Africa and that they promise benefits at least to the children of their subjects. What do you think?

7.16 AIDS trials in Africa. One of the most important goals of AIDS research is to find a vaccine that will protect against the HIV virus. Because AIDS is so common in parts of Africa, that is the easiest place to test a vaccine. It is likely, however, that a vaccine would be so expensive that it could not (at least at first) be widely used in Africa. Is it ethical to test in Africa if the benefits go mainly to rich countries? The treatment group of subjects would get the vaccine and the placebo group would later be given the vaccine if it proved effective. So the actual subjects would benefit—it is the future benefits that would go elsewhere. What do you think?

7.17 Opinion polls. The presidential election campaign is in full swing, and the candidates have hired polling organizations to take regular polls to find out what the voters think about the issues. What information should the pollsters be required to give out?

(a) What does the standard of informed consent require the pollsters to tell potential respondents?

(b) The standards accepted by polling organizations also require giving respondents the name and address of the organization that carries out the poll. Why do you think this is required?

(c) The polling organization usually has a professional name such as "Samples Incorporated," so respondents don't know that the poll is being paid for by a political party or candidate. Would revealing the sponsor to respondents bias the poll? Should the sponsor always be announced whenever poll results are made public?

7.18 A right to know? Some people think that the law should require that all political poll results be made public. Otherwise, the possessors of poll results can use the information to their own advantage. They can act on the information, release only selected parts of it, or time the release for best effect. A candidate's organization replies that they are paying for the poll in order to gain information for their own use, not to amuse the public. Do you favor requiring complete disclosure of political poll results? What about other private surveys, such as market research surveys of consumer tastes?

7.19 Telling the government. The 2000 census long form asked 53 detailed questions, for example:

> *Do you have COMPLETE plumbing facilities in this house, apartment, or mobile home; that is, 1) hot and cold piped water, 2) a flush toilet, and 3) a bathtub or shower?*

The form also asked your income in dollars, broken down by source, and whether any "physical, mental, or emotional condition" causes you difficulty in "learning, remembering, or concentrating." Some members of Congress objected to these questions, even though Congress had approved them.

Give brief arguments on both sides of the debate over the long form: the government has legitimate uses for such information, but the questions seem to invade people's privacy.

7.20 Charging for data? Data produced by the government are often available free or at low cost to private users. For example, satellite weather data produced by the U.S. National Weather Service are available free to TV stations for their weather reports and to anyone on the Web. European governments, on the other hand, charge TV stations for weather data. *Opinion 1:* Government data should be available to everyone at minimal cost. *Opinion 2:* The satellites are expensive, and the TV stations are making a profit from their weather services, so they should share the cost.

Which opinion do you support, and why?

7.21 Surveys of youth. The Centers for Disease Control and Prevention, in a survey of teenagers, asked the subjects if they were sexually active. Those who said "Yes" were then asked,

> *How old were you when you had sexual intercourse for the first time?*

Should consent of parents be required to ask minors about sex, drugs, and other such issues, or is consent of the minors themselves enough? Give reasons for your opinion.

7.22 Deceiving subjects. Students sign up to be subjects in a psychology experiment. When they arrive, they are told that interviews are running late and are taken to a waiting room. The experimenters then stage a theft of a valuable object left in the waiting room. Some subjects are alone with the thief, and others are in pairs—these are the treatments being compared. Will the subject report the theft?

The students had agreed to take part in an unspecified study, and the true nature of the experiment is explained to them afterward. Do you think this study is ethically OK?

7.23 Tempting subjects. A psychologist conducts the following experiment: she measures the attitude of subjects toward cheating, then has them play a game rigged so that winning without cheating is impossible. The computer that organizes the game also records—unknown to the subjects—whether or not they cheat. Then attitude toward cheating is retested.

Subjects who cheat tend to change their attitudes to find cheating more acceptable. Those who resist the temptation to cheat tend to condemn cheating more strongly on the second test of attitude. These results confirm the psychologist's theory.

This experiment tempts subjects to cheat. The subjects are led to believe that they can cheat secretly when in fact they are observed. Is this experiment ethically objectionable? Explain your position.

7.24 Decency and public money. Congress has often objected to spending public money on projects that seem "indecent" or "in bad taste." The arts are most often affected, but science can also be a target. Congress once refused to fund an experiment to study the effect of marijuana on sexual response. The journal *Science* reported:

> *Dr. Harris B. Rubin and his colleagues at the Southern Illinois Medical School proposed to exhibit pornographic films to people who had smoked marihuana and to measure the response with sensors attached to the penis.*
>
> *Marihuana, sex, pornographic films—all in one package, priced at $120,000. The senators smothered the hot potato with a ketchup of colorful oratory and mixed metaphors.*
>
> *"I am firmly convinced we can do without this combination of red ink, 'blue' movies, and Acapulco 'gold,'" Senator John McClellan of Arkansas opined in a persiflage of purple prose. . . .*

(a) The subjects were volunteers who gave properly informed consent. If you were a member of the review board, would you veto this experiment on grounds of "decency" or "good taste"?

(b) Suppose we concede that a free society should permit any legal experiment with volunteer subjects. It is a further step to say that any such experiment is entitled to government funding if the usual review procedure finds it scientifically worthwhile. If you were a member of Congress, would you ever refuse to pay for an experiment on grounds of "decency" or "good taste"?

Chapter 8
Measuring

Got some spare time?

Do people have more or less leisure time now than they did a generation ago? A book titled *The Overworked Americans* says we are working more than ever. Another book, titled *Time for Life,* says that we have more free time than ever. Is somebody lying with statistics?

To see if leisure time is increasing, we must **measure** "leisure time." That is, we must reduce the vague idea to a number that can go up or down. The first step is to say what we mean by leisure time. Something like "time when you're not working, not traveling to or from work, not doing household chores, not . . ." You see that we can get different numbers by changing our list of "nots."

Once we decide what leisure time is, we must actually produce the numbers. We might ask a sample of people how they spent their time yesterday. Don't trust their memory? Afraid they will exaggerate how long they worked? Then we might ask people to keep a diary of how they spend their time today. Of course, the really busy people may forget to record all their work time. Not only is it hard to say exactly what "leisure time" is, it's hard to attach a number to measure whatever we say it is.

Random samples are wonderful. Randomized comparative experiments are even more wonderful. But at the end, we need to turn ideas like "leisure time" or "pain relief" or "income" into numbers. Don't trust the numbers until you know how this was done.

Measurement basics

Statistics deals with numbers. Planning the production of data through a sample or an experiment does not by itself produce numbers. Once we have our sample respondents or our experimental subjects, we must still *measure* whatever characteristics interest us. First, think broadly: are we trying to measure the right things? Are we overlooking some outcomes that are important even though they may be hard to measure?

Example 1. But what about the patients?

Clinical trials tend to measure things that are easy to measure: blood pressure, tumor size, virus concentration in the blood. They often don't directly measure what matters most to patients—does the treatment really improve their lives? One study found that only 5% of trials published between 1980 and 1997 measured the effect of treatments on patients' emotional well-being or their ability to function in social settings.

Once we have decided what properties we want to measure, we can think about how to do the measurements.

Measurement

We **measure** a property of a person or thing when we assign a number to represent the property.

We often use an **instrument** to make a measurement. We may have a choice of the **units** we use to record the measurements.

The result of measurement is a numerical **variable** that takes different values for people or things that differ in whatever we are measuring.

Example 2. Length, college readiness, highway safety

To measure the length of my bed, I use a tape measure as the *instrument*. I can choose either inches or centimeters as the *unit of measurement*. If I choose centimeters, my *variable* is the length of the bed in centimeters.

To measure a student's readiness for college, I might ask the student to take the SAT reasoning exam. The exam form is the *instrument*. The *variable* is the student's score in points, somewhere between 400 and 1600 if I combine the verbal and mathematics sections of the SAT.

How can I measure the safety of traveling on the highway? I might decide to count the number of people who die in motor vehicle accidents in a year. The government's Fatal Accident Reporting System collects data on all fatal traffic crashes. I can use the government count of traffic deaths as a *variable* to measure highway safety.

What are your units?

Not paying attention to units of measurement can get you into trouble. In 1999, the Mars Climate Orbiter spacecraft burned up in the Martian atmosphere. It was supposed to be 93 miles (150 kilometers) above the planet, but was in fact only 35 miles (57 kilometers) up. It seems that Lockheed Martin, which built the Orbiter, specified important measurements in English units (pounds, miles). The National Aeronautics and Space Administration team who flew the spacecraft thought the numbers were in metric system units (kilograms, kilometers). There went $125 million.

Here are some questions you should ask about the variables in any statistical study:

1. Exactly how is the variable defined?
2. Is the variable a valid way to describe the property it claims to measure?
3. How accurate are the measurements?

We don't often design our own measuring devices—we use the results of the SAT or the Fatal Accident Reporting System—so we won't go deeply into these questions. Any consumer of numbers, however, should know a bit about them.

Know your variables

Measurement is the process of turning concepts like length or employment status into precisely defined variables. Using a tape measure to turn the idea of "length" into a number is straightforward because we know exactly what we mean by length. Measuring college readiness is controversial because it isn't clear exactly what makes a student ready for college work. Using SAT scores at least says exactly how we will get numbers. Measuring leisure time requires that we first say what time counts as leisure. Even counting highway deaths requires us to say exactly what counts as a highway death: Pedestrians hit by cars? People in cars hit by a train at a crossing? People who die from injuries 6 months after an accident? We can simply accept the government's counts, but someone had to answer those and other questions in order to know what to count. For example, a person must die within 30 days of an accident to count as a traffic death. These details are a nuisance, but they can make a difference.

Example 3. Measuring unemployment

Each month the Bureau of Labor Statistics (BLS) announces the *unemployment rate* for the previous month. People who are not available for work (retired people, for example, or students who do not want to work while in school) should not be counted as unemployed just because they don't have a job. To be unemployed, a person must first be in the labor force. That is, she must be available for work and looking for work. The unemployment rate is

$$\text{unemployment rate} = \frac{\text{number of people unemployed}}{\text{number of people in the labor force}}$$

To complete the exact definition of the unemployment rate, the BLS has very detailed descriptions of what it means to be "in the labor force" and what it means to be "employed." For example, if you are on strike but expect to return to the same job, you count as employed. If you are not working and did not look for work in the last two weeks, you are not in the labor force. So people who say they want to work but are too discouraged to keep looking for a job don't count as unemployed. The details matter. The official unemployment rate would be different if the government used a different definition of unemployment.

"This is our new college admissions sizer. I got the idea from the carry-on luggage sizers at the airport."

The BLS estimates the unemployment rate based on interviews with the sample in the monthly Current Population Survey. The interviewer can't simply ask "Are you in the labor force?" and "Are you employed?" Many questions are needed to classify a person as employed, unemployed, or not in the labor force. Changing the questions can change the unemployment rate. At the beginning of 1994, after several years of planning, the BLS introduced computer-assisted interviewing and improved its questions. Figure 8.1 is a graph of the unemployment rate that appeared on the front page of the BLS monthly news

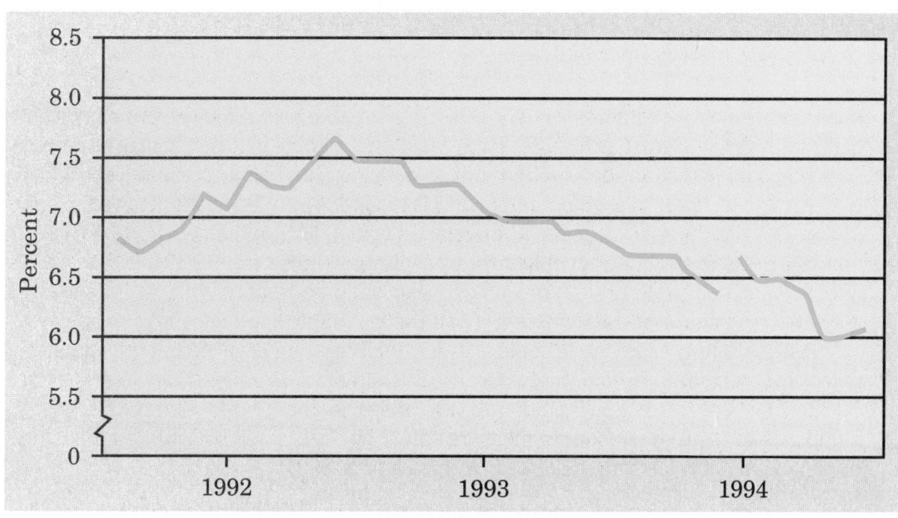

Figure 8.1 The unemployment rate from August 1991 to July 1994. The gap shows the effect of a change in how the government measures unemployment.

release on the employment situation. There is a gap in the graph in January 1994 because of the change in the interviewing process. The unemployment rate would have been 6.3% under the old system. It was 6.7% under the new system. That's a big enough change to make politicians unhappy.

Measurements valid and invalid

No one would object to using a tape measure reading in centimeters to measure the length of my bed. Many people object to using SAT scores to measure readiness for college. Let's shortcut that debate: just measure the height in inches of all applicants and accept the tallest. Bad idea, you say. Why? Because height has nothing to do with being prepared for college. In more formal language, height is not a *valid* measure of a student's academic background.

Valid measurement

A variable is a **valid** measure of a property if it is relevant or appropriate as a representation of that property.

It is valid to measure length with a tape measure. It isn't valid to measure a student's readiness for college by recording her height. The BLS unemployment rate is a valid measure, even though changes in the official definitions would give a somewhat different measure. Let's think about measures valid and invalid in some other settings.

Example 4. Measuring highway safety

Roads get better. Speed limits increase. Big SUVs replace cars. Enforcement campaigns reduce drunk driving. How has highway safety changed over time in this changing environment?

We could just count deaths from motor vehicles. The Fatal Accident Reporting System says there were 45,582 deaths in 1989 and 41,471 deaths 10 years later in 1998. But the number of licensed drivers rose from 166 million in 1989 to 185 million in 1998. The number of miles that people drove rose from 2096 billion to 2619 billion. If more people drive more miles, there may be more deaths even if the roads are safer. The count of deaths is not a valid measure of highway safety.

Rather than a *count*, we should use a *rate*. The number of deaths per mile driven takes into account the fact that more people drive more miles than in the past. In 1998, vehicles drove 2,619,000,000,000 miles in the United States. Because this number is so large, it is usual to measure safety by deaths per 100 million miles driven rather than deaths per mile. For 1998, this death rate is

$$\frac{\text{motor vehicle deaths}}{\text{100s of millions of miles driven}} = \frac{41{,}471}{26{,}190} = 1.6$$

The death rate fell from 2.2 deaths per 100 million miles in 1989 to 1.6 in 1998. That's a big change—there were 27% fewer deaths per mile driven in 1998 than a decade earlier. Driving has been getting safer.

Rates and counts

Often a **rate** (a fraction, proportion, or percent) at which something occurs is a more valid measure than a simple **count** of occurrences.

Using height to measure readiness for college, and even using counts when rates are needed, are examples of clearly invalid measures. The tougher questions concern measures that are neither clearly invalid nor obviously valid.

Example 5. Achievement tests

When you take a statistics exam, you hope that it will ask you about the main points of the course syllabus. If it does, the exam is a valid measure of how much you know about the course material. The College Board, which administers the SAT, also offers achievement tests in a variety of disciplines (including an advanced placement exam in statistics). These achievement tests are not very controversial. Experts can judge validity by comparing the test questions with the syllabus of material they are supposed to cover.

Example 6. IQ tests

Psychologists would like to measure aspects of the human personality that can't be observed directly, such as "intelligence" or "authoritarian personality." Does an IQ test measure intelligence? Some psychologists say "Yes" rather loudly. There is such a thing as general intelligence, they argue, and the various standard IQ tests do measure it, though not perfectly. Other experts say "No" equally loudly. There is no single intelligence, just a variety of mental abilities that no one instrument can measure.

The disagreement over the validity of IQ tests is rooted in disagreement about the nature of intelligence. If we can't agree on exactly what intelligence is, we can't agree on how to measure it.

Statistics is little help in these examples. They start with an idea like "knowledge of statistics" or "intelligence." If the idea is vague, validity becomes a matter of opinion. However, statistics can help a lot if we refine the idea of validity a bit.

What can't be measured matters

One member of the young Edmonton Oilers hockey team of 1981 finished last in almost everything one can measure: strength, speed, reflexes, eyesight. That was Wayne Gretzky, soon to be known as "the great one." He broke the National Hockey League scoring record that year, then scored yet more points in seven different seasons. Somehow the physical measurements didn't catch what made Gretzky the best hockey player ever. Not everything that matters can be measured.

Example 7. The SAT again

"SAT bias will illegally cheat thousands of young women out of college admissions and scholarship aid they have earned by superior classroom performance." That's what the organization FairTest said when the 1999 SAT scores were released. The gender gap was larger on the math part of the test, where women averaged 495 and men averaged 531. The federal Office of Civil Rights says that tests on which women and minorities score lower are discriminatory.

The College Board, which administers the SAT, replies that there are many reasons some groups have lower average scores than others. For example, more women than men from families with low incomes and little education sign up for the SAT. Students whose parents have low incomes and little education have on the average fewer advantages at home and in school than richer students. They have lower SAT scores because their backgrounds have not prepared them as well for college. The mere fact of lower scores doesn't imply that the test is not valid.

Is the SAT a valid measure of readiness for college? "Readiness for college academic work" is a vague concept that probably combines inborn intelligence (whatever we decide that is), learned knowledge, study and test-taking skills, and motivation to work at academic subjects. Opinions will always differ about whether SAT scores (or any other measure) accurately reflect this vague concept.

Instead, we ask a simpler and more easily answered question: Do SAT scores help predict students' success in college? Success in college is a clear concept, measured by whether students graduate and by their college grades. Students with high SAT scores are more likely to graduate and earn (on the average) higher grades than students with low SAT scores. We say that SAT scores have *predictive validity* as measures of readiness for college. This is the only kind of validity that data can assess directly.

Predictive validity

A measurement of a property has **predictive validity** if it can be used to predict success on tasks that are related to the property measured.

Predictive validity is the clearest and most useful form of validity from the statistical viewpoint. "Do SAT scores help predict college grades?" is a much clearer question than "Do IQ test scores measure intelligence?" However, predictive validity is not a yes-or-no idea. We must ask *how accurately* SAT scores predict college grades. Moreover, we must ask *for what groups* the SAT has predictive validity. It is possible, for example, that the SAT predicts college grades well for men but not for women. There are statistical ways to describe "how accurately." The *Statistical Controversies*

Statistical Controversies

SAT Exams in College Admissions

Colleges use a variety of measures to make admissions decisions. The student's record in school is the most important, but SAT scores do matter, especially at selective colleges. The SAT has the advantage of being a national test. An A in algebra means different things in different high schools, but a math SAT score of 625 means the same thing everywhere. The SAT can't measure willingness to work hard or creativity, so it won't predict college performance exactly, but most colleges have long found it helpful.

How well do SAT scores predict first-year college grades? The table at the bottom of this page gives some results from a sample of 48,039 students. The numbers in the table say what percent of the variation among students in college grades can be predicted by SAT scores (math and verbal combined), by high school grades, and by SAT and high school grades together. An entry of 0% would mean no predictive validity, and 100% would mean predictions were always exactly correct.

We see that *SAT scores predict college grades about as well as high school grades do.* Combining SAT scores and high school grades does a better job than either by itself. The predictions are actually a bit better for women than for men. College grades are predicted slightly less well for blacks, but the SAT again does as well as high school grades. We also see that *neither SAT scores nor high school grades predict college grades very well.* Students with the same grades and SAT scores often perform quite differently in college. Motivation and study habits matter a lot.

Selective colleges are justified in paying some attention to SAT scores, but they are also justified in looking beyond SAT scores for the motivation that can bring success to students with weaker academic preparation. The SAT debate is not really about the numbers. It is about how colleges should use all the information they have in deciding whom to admit, and also about the goals colleges should have in forming their entering classes.

Mark Richards/ Photo Edit

	All students	Men only	Women only	Black students
SAT	27%	26%	31%	25%
School grades	29%	28%	28%	24%
Both together	37%	36%	38%	34%

discussion on the previous page uses one of these descriptions to give the big picture. It appears that SAT scores do have moderate predictive validity, and that they are about equally valid for different groups of students. Differences among groups in SAT scores generally reflect unequal environments that bring about unequal preparation for college.

Measurements accurate and inaccurate

Using a bathroom scale to measure your weight is valid. If your scale is like mine, however, the measurement may not be very accurate. Think about my bathroom scale. It measures my weight, but it may not give my true weight. My scale always reads 3 pounds too high, so

$$\text{measured weight} = \text{true weight} + 3 \text{ pounds}$$

If that is the whole story, the scale will always give the same reading for the same true weight. Most scales vary a bit—they don't always give the same reading when you step off and step right back on. My scale is somewhat old and rusty. It always reads 3 pounds too high because its aim is off, but it is also erratic. This morning it sticks a bit and reads one-half pound too low for that reason. So the reading is

$$\text{measured weight} = \text{true weight} + 3 \text{ pounds} - 0.5 \text{ pound}$$

When I step off and step right back on, the scale sticks in a different spot that makes it read one-quarter pound too high. The reading I get is now

$$\text{measured weight} = \text{true weight} + 3 \text{ pounds} + 0.25 \text{ pounds}$$

If I have nothing better to do than keep stepping on and off the scale, I will keep getting different readings. They center on a reading 3 pounds too high, but they vary about that center.

My scale has two kinds of errors. If it didn't stick, the scale would always read 3 pounds high. That is true every time anyone steps on the scale. We call this systematic error that occurs every time we make a measurement *bias*. My scale does stick—but how much this changes the reading is different every time someone steps on the scale. Sometimes stickiness pushes the scale reading up; sometimes it pulls it down. The result is that the scale weighs 3 pounds too high on the average, but its reading varies when we weigh the same thing repeatedly. We can't predict the error due to stickiness, so we call it *random error*.

A scale that always reads the same when it weighs the same item is perfectly reliable even if it is biased. Reliability only says that the result is repeatable. Bias and lack of reliability are different kinds of error. And don't confuse reliability with validity just because both sound like good qualities. Using a scale to measure weight is valid even if the scale is not reliable. Here's an example of a measurement that is reliable but not valid.

Errors in measurement

We can think about errors in measurements this way:

$$\text{measured value} = \text{true value} + \text{bias} + \text{random error}$$

A measurement process has **bias** if it systematically overstates or understates the true value of the property it measures.

A measurement process has **random error** if repeated measurements on the same individual give different results. If the random error is small, we say the measurement is **reliable.**

Example 8. Do big skulls house smart brains?

In the mid-nineteenth century, it was thought that measuring the volume of a human skull would measure the intelligence of the skull's owner. It was difficult to measure a skull's volume reliably, even after it was no longer attached to its owner. Paul Broca, a professor of surgery, showed that filling a skull with small lead shot, then pouring out the shot and weighing it, gave quite reliable measurements of the skull's volume. These accurate measurements do not, however, give a valid measure of intelligence. Skull volume turned out to have no relation to intelligence or achievement.

By permission of Johnny Hart and Creators Syndicate, Inc.

Improving reliability, reducing bias

What time is it? Much modern technology, such as the Global Positioning System, which uses satellite signals to tell you where you are, requires very exact measurements of time. Time starts with the earth's path around the sun, which lasts one year. But the earth is much too erratic. Since 1967, time starts with the standard second, and the second is defined to be the time required for 9,192,631,770 vibrations of a cesium atom. Physical clocks are bothered by changes in temperature, humidity, and air pressure. The cesium atom doesn't care. People who need really accurate time can buy atomic clocks. The National Institute of Standards and Technology (NIST) keeps an even more accurate atomic clock and broadcasts the results (with some loss in transmission) by radio, telephone, and Internet.

Example 9. Really accurate time

NIST's atomic clock is very accurate, but not perfectly accurate. The world standard is Universal Coordinated Time, compiled by the International Bureau of Weights and Measures (BIPM) in Sèvres, France. BIPM doesn't have a better clock than NIST. It calculates the time by averaging the results of more than 200 atomic clocks around the world. NIST tells us (after the fact) how much it misses the correct time by. Here are the last 10 errors as I write, in seconds:

0.000000007	0.000000000
0.000000005	−0.000000003
0.000000006	−0.000000005
0.000000000	−0.000000001
0.000000002	−0.000000001

In the long run, NIST's measurements of time are not biased. The NIST second is sometimes shorter than the BIPM second and sometimes longer, not always off in the same direction. NIST's measurements are very reliable, but the numbers above do show some variation. There is no such thing as a perfectly reliable measurement. The average (mean) of several measurements is more reliable than a single measurement. That's one reason BIPM combines the time measurements of many atomic clocks.

Scientists everywhere repeat their measurements and use the average to get more reliable results. Even students in a chemistry lab often do this. Just as larger samples reduce variation in a sample statistic, averaging over more measurements reduces variation in the final result.

Use averages to improve reliability

No measuring process is perfectly reliable. The average of several repeated measurements of the same individual is more reliable (less variable) than a single measurement.

Figure 8.2 This atomic clock at the National Institute of Standards and Technology is accurate to 1 second in 6 million years. (Photo courtesy of John Wessels, NIST Time and Frequency Division.)

Unfortunately, there is no similarly straightforward way to reduce the bias of measurements. Bias depends on how good the measuring instrument is. To reduce the bias, you need a better instrument. The atomic clock at NIST (Figure 8.2) is accurate to 1 second in 6 million years and is a bit large to put beside your bed.

Example 10. Measuring unemployment again

Measuring unemployment is also "measurement." The concepts of bias and reliability apply here just as they do to measuring length or time.

The Bureau of Labor Statistics checks the *reliability* of its measurements of unemployment by having supervisors reinterview about 5% of the sample. This is repeated measurement on the same individual, just as a student in a chemistry lab measures a weight several times.

The BLS attacks *bias* by improving its instrument. That's what happened in 1994, when the Current Population Survey was given its biggest overhaul in more than 50 years. The old system for measuring unemployment, for example, underestimated unemployment among women because the detailed procedures had not kept up with changing patterns of women's work. The new measurement system corrected that bias—and raised the reported rate of unemployment.

Pity the poor psychologist

Statisticians are creatures of habit: they think about measurement much the same way they think about sampling. In both settings, the big idea is to ask, "What would happen if we did this many times?" In sampling we want to estimate a population parameter, and we worry that our estimate may be biased or vary too much from sample to sample. Now we want to measure the true value of some property, and we worry that our measurement may be biased or vary too much when we repeat the measurement on the same individual. Bias is systematic error that happens every time; high variability (low reliability) means our result can't be trusted because it isn't repeatable.

Thinking of measurement this way is pretty straightforward when we are measuring my weight. To start with, we have a clear idea of what my "true weight" is. We know that there are really good scales around: start at the doctor's office, go to the physics lab, end up at NIST. We can measure my weight as accurately as we wish. This makes it easy to see that my bathroom scale always reads 3 pounds too high. Reliability is also easy to describe—step on and off the scale many times and see how much its readings vary.

Asking "what would happen if we did this many times" is a lot harder to put into practice when we want to measure "intelligence" or "readiness for college." Consider as an example the poor psychologist who wants to measure "authoritarian personality."

Example 11. Authoritarian personality?

Do some people have a personality type that disposes them to rigid thinking and to following strong leaders? Psychologists looking back on the Nazis after World War II thought so. In 1950, a group of psychologists developed the "F-scale" as an instrument to measure "authoritarian personality." The F-scale asks how strongly you agree or disagree with statements such as

- *Obedience and respect for authority are the most important virtues children should learn.*
- *Science has its place, but there are many important things that can never be understood by the human mind.*

Strong agreement with such statements marks you as authoritarian. The F-scale and the idea of the authoritarian personality continue to be prominent in psychology, especially in studies of prejudice and right-wing extremist movements.

Here are some questions we might ask about using the F-scale to measure "authoritarian personality." The same questions come to mind when we think about IQ tests or the SAT exam.

1. Just what is an "authoritarian personality"? We understand this much less well than we understand my weight. The answer in practice seems to

be "whatever the F-scale measures." Any claim for validity must rest on what kinds of behavior high F-scale scores go along with. That is, we fall back on predictive validity.

2. The name sounds unpleasant, and F stands for Fascist. As the second question in Example 11 suggests, people who hold traditional religious beliefs are likely to get higher F-scale scores than similar people who don't hold those beliefs. Does the instrument reflect the beliefs of those who developed it, so that people with different beliefs might come up with a quite different instrument?

3. We think we understand what my true weight is. What is the true value of my F-scale score? NIST can help us find a true weight, but not a true authoritarianism score. If we suspect that the instrument is biased as a measure of "authoritarian personality" because it penalizes religious beliefs, how can we check that?

4. We can weigh me many times to learn the reliability of my bathroom scale. If I take the F-scale test many times, I remember what answers I gave the first time. That is, repeats of the same psychological measurement are not really repeats. So reliability is hard to check in practice. Psychologists may develop several forms of the same instrument in order to repeat their measurements. How do we know these forms are really equivalent?

The point is not that psychologists lack answers to these questions. The first two are controversial because not all psychologists think about human personality in the same way. The second two questions have at least partial answers, but not simple answers. The point is that "measurement," which seems so straightforward when we measure weight, is complicated indeed when we try to measure human personality.

There is a larger lesson here. Be wary of statistical "facts" about squishy topics like authoritarian personality, intelligence, and even readiness for college. The numbers look solid, as numbers always do. But data are a human product and reflect human desires, prejudices, and weaknesses. If we don't understand and agree on what we are measuring, the numbers may produce more heat than light.

Exploring the Web

You can get the time direct from the atomic clock at NIST at www.time.gov. There is some error due to Internet delays, but the display even tells you roughly how accurate the time on the screen is.

For a glimpse of the elaborate system that lies behind everyday measurements of length, weight, and time, visit the home of the International Bureau of Weights and Measures, www.bipm.fr.

Statistics in Summary To **measure** something means to assign a number to some property of an individual. When we measure many individuals, we have values of a **variable** that describes them. When you work with data or read about a statistical study, ask exactly how the variables are defined and whether they leave out some things you want to know. Ask if the variables are **valid** as numerical measures of the concepts the study discusses. Validity is simple for measurements of physical properties such as length, weight, and time. When we want to measure human personality and other vague properties, **predictive validity** is the most useful way to say whether our measures are valid.

Also ask if there are **errors in measurements** that reduce the value of the data. You can think about errors in measurement like this:

measured value = true value + bias + random error

Some ways of measuring are **biased,** or systematically wrong in the same direction. To reduce bias, you must use a better **instrument** to make the measurements. Other measuring processes lack **reliability,** so that measuring the same individuals again would give quite different results due to **random error.** You can improve the reliability of a measurement by repeating it several times and using the average result.

CHAPTER 8 EXERCISES

8.1 Counting the unemployed? We could measure the extent of unemployment by a count (the number of people who are unemployed) or by giving a rate (the percent of the labor force that is unemployed). The number of people in the labor force grew from 107 million in 1980, to 126 million in 1990, and to 140 million at the beginning of 2000. Use these facts to explain why the count of unemployed people is not a valid measure of the extent of unemployment.

8.2 Measuring physical fitness. You want to measure the "physical fitness" of college students. Give an example of a clearly invalid way to measure fitness. Then briefly describe a measurement process that you think is valid.

8.3 School bus safety. The National Highway Traffic Safety Administration says that an average of 11 children die each year in school bus accidents, and an average of 600 school-age children die each year in auto accidents during school hours. These numbers suggest that riding the bus is safer than driving to school with a parent. The *counts* aren't fully convincing, however. What *rates* would you like to know to compare the safety of bus and private auto?

8.4 Rates versus counts. Customers returned 36 coats to Sears this holiday season, and only 12 to La Boutique Classique next door. Sears sold 1100 coats this season, while La Boutique sold 200.

(a) Sears had a greater number of coats returned. Why does this not show that Sears's coat customers were less satisfied than those of La Boutique?

(b) What is the rate of returns (percent of coats returned) at each of the stores?

8.5 Capital punishment. Between 1977 and 1998, 432 convicted criminals were put to death in the United States. Here are data on the number of executions in several states during those years, as well as the 1990 population of these states:

State	Population (thousands)	Executions
Alabama	4,040	16
Arkansas	2,351	16
Florida	12,938	39
Missouri	5,117	29
Nevada	1,202	6
Texas	16,986	144
Virginia	6,189	46

Texas and Florida are among the leaders in executions. Because these are large states, we might expect them to have many executions. Find the *rate* of executions for each of the states listed above, in executions per million population. Because population is given in thousands, you can find the rate per million as

$$\text{rate per million} = \frac{\text{executions}}{\text{population in thousands}} \times 1000$$

Arrange the states in order of the number of executions relative to population. Are Florida and Texas still high by this measure?

8.6 Measuring intelligence. "Intelligence" means something like "general problem-solving ability." Explain why it is *not* valid to measure intelligence by a test that asks questions such as

Who wrote The Star-Spangled Banner?

Who won the last soccer World Cup?

8.7 Measuring life's quality. Is life in Britain getting better or worse? The usual government data don't say. So the British government announced that it wants to add measures of such things as housing, traffic, and air pollution. "The quality of life is not simply economic," said a deputy prime minister. Help them out: how would you measure "traffic" and its impact on the quality of life?

8.8 Measuring pain. The Department of Veterans Affairs offers medical care to 3.4 million patients. It wants doctors and nurses to treat pain as a "fifth vital sign," to be recorded along with blood pressure, pulse, temperature, and breathing rate. Help out the VA: how would you measure a patient's pain?

8.9 Fighting cancer. Congress wants the medical establishment to show that progress is being made in fighting cancer. Some variables that might be used are:

(a) Total deaths from cancer. These have risen sharply over time, from 331,000 in 1970, to 505,000 in 1990, and to 539,000 in 1998.

(b) The percent of all Americans who die from cancer. The percent of deaths due to cancer rose steadily, from 17.2% in 1970 to 23.5% in 1990, then leveled off to 23.0% in 1998.

(c) The percent of cancer patients who survive for 5 years from the time the disease was discovered. These rates are rising slowly. For whites, the 5-year survival rate was 50.3% in the 1974 to 1976 period and 60.9% from 1989 to 1995.

None of these variables is fully valid as a measure of the effectiveness of cancer treatment. Explain why both (a) and (b) could increase even if treatment is getting more effective, and why (c) could increase even if treatment is getting less effective.

8.10 Testing job applicants. The law requires that tests given to job applicants must be shown to be directly job-related. The Department of Labor believes that an employment test called the General Aptitude Test Battery (GATB) is valid for a broad range of jobs. As in the case of the SATs, blacks and Hispanics get lower average scores on the GATB than do whites. Describe briefly what must be done to establish that the GATB has predictive validity as a measure of future performance on the job.

8.11 Validity, bias, reliability. Give your own example of a measurement process that is valid but has large bias. Then give your own example of a measurement process that is invalid but highly reliable.

8.12 An activity on bias. Let's study bias in an intuitive measurement. Figure 8.3 is a drawing of a tilted glass. Reproduce this drawing on 10 sheets of paper. Choose 10 people: 5 men and 5 women. Explain that the drawing represents a tilted glass of water. Ask each subject to draw the water level when the glass is full of water.

Figure 8.3 A tilted glass, for Exercise 8.12. Can you draw the level of water in the glass when it is full?

The correct level is horizontal (straight back from the lower lip of the glass). Many people make large errors in estimating the level. Use a protractor to measure the angle of each subject's error. Were your subjects systematically wrong in the same direction? How large was the average error? Was there a clear difference between the average errors made by men and women?

8.13 An activity on bias and reliability. Cut 5 pieces of string having these lengths in inches:

2.9	9.5	5.7	4.2	7.6

(a) Show the pieces to another student one at a time, asking the subject to estimate the length to the nearest tenth of an inch by eye. The error your subject makes is measured value minus true value and can be either positive or negative. What is the average of the 5 errors? Explain why this average would be close to 0 if there were no bias and we used many pieces of string rather than just 5.

(b) The following day, ask the subject to again estimate the length of each piece of string. (Present them in a different order the second day.) Explain why the five differences between the first and second guesses would all be 0 if your subject were a perfectly reliable measurer of length. The bigger the differences, the less reliable your subject is. What is the average difference (ignoring whether they are positive or negative) for your subject?

8.14 More on bias and reliability. The previous exercise gives 5 true values for lengths. A subject measures each length twice by eye. Make up a set of results from this activity that matches each of the descriptions below. For simplicity, assume that bias means the same fixed error every time, rather than an "on the average" error in many measurements.

(a) The subject has a bias of 0.5 inch too long and is perfectly reliable.

(b) The subject has no bias but is not perfectly reliable, so that the average difference in repeated measurements is 0.5 inch.

8.15 Does job training work? To measure the effectiveness of government training programs, it is usual to compare workers' pay before and after training. But many workers sign up for training when their pay drops or they are laid off. So the "before" pay is unusually low and the pay gain looks large.

(a) Is this bias or random error in measuring the effect of training on pay? Why?

(b) How would you measure the success of training programs?

8.16 A recipe for poor reliability. Every month, the government releases data on "personal savings." This number tells us how many dollars individuals saved the previous month. Savings are calculated by subtracting personal spending (an enormously large number) from personal income (another enormous number). The result is one of the government's least reliable statistics.

Give a numerical example to show that small percentage changes in two very large numbers can produce a big percentage change in the difference between those numbers. A variable that is the difference between two big numbers is usually not very reliable.

8.17 Measuring crime. Crime data make headlines. We measure the amount of crime by the number of crimes committed or (better) by crime rates (crimes per 100,000 population). The FBI publishes data on crime in the United States by compiling crimes reported to police departments. The National Crime Victimization Survey publishes data based on a national probability sample of more than 43,000 households. The victim survey shows about two and a half times as many crimes as the FBI report. Explain why the FBI reports have a large downward bias for many types of crime. (Here is a case in which bias in producing data leads to bias in measurement.)

8.18 Measuring crime. Each year, the National Crime Victimization Survey asks a random sample of more than 43,000 households whether they have been victims of crime and, if so, the details. In all, more than 80,000 people answer these questions. If other people in a household are in the room while one person is answering questions, the measurement of, for example, rape and other sexual assaults could be seriously biased. Why? Would the presence of other people lead to overreporting or underreporting of sexual assaults?

8.19 Measuring pulse rate. You want to measure your resting pulse rate. You might count the number of beats in 6 seconds and multiply by 10 to get beats per minute. Why is this method less reliable than actually measuring the number of beats in a minute?

8.20 How good is your college? Each year, *U.S. News and World Report* ranks the nation's colleges and universities. You can find the current rankings on the magazine's Web site, www.usnews.com. Colleges often dispute the validity of the rankings, and some even say that *U.S. News* changes its measurement process a bit each year to produce different "winners" and thus generate news. The magazine describes its methods on the Web site. Give three variables that you would use as indicators of a college's "academic excellence." You may or may not decide to use some of *U.S. News*'s variables.

8.21 Tough course? A friend tells you, "In American History, 20 students failed. Only 11 students failed Russian History. That American History prof is a tougher grader than the Russian History teacher." Explain why the conclusion may not be true. What additional information would you need to compare the courses?

8.22 Testing job applicants. A company used to give IQ tests to all job applicants. This is now illegal, because IQ is not related to the performance of workers in all the company's jobs. Does the reason for the policy change involve the *reliability,* the *bias,* or the *validity* of IQ tests as a measure of future job performance? Explain your answer.

8.23 Better dishwashers. You are writing an article for a consumer magazine based on a survey of the magazine's readers that asked about the reliability of their household appliances. Of 13,376 readers who reported owning Brand A dishwashers, 2942 required a service call during the past year. Only 192 service calls were reported by the 480 readers who owned Brand B dishwashers. Describe an appropriate variable to measure the reliability of a make of dishwasher, and compute the values of this variable for Brand A and for Brand B. Which brand is more reliable?

8.24 Where to live? Each year, *Money* magazine ranks the 300 largest metropolitan areas in the United States in an article on the best places to live. First place in 1997 went to Nashua, New Hampshire. Nashua was ranked 42nd in 1996 and 19th in 1995. Monmouth and Ocean Counties in New Jersey were ranked 167th in 1995, 38th in 1996, and 3rd in 1997. Are these facts evidence that *Money*'s ratings are invalid, biased, or unreliable? Explain your choice.

Chapter 9
Do the Numbers Make Sense?

The case of the missing vans

Auto manufacturers lend their dealers money to help them keep vehicles on their lots. The loans are repaid when the vehicles are sold. A Long Island auto dealer named John McNamara borrowed over $6 billion from General Motors between 1985 and 1991. In December 1990 alone, Mr. McNamara borrowed $425 million to buy 17,000 GM vans customized by an Indiana company, allegedly for sale overseas. GM happily lent McNamara the money because he always repaid the loans.

Let's pause to consider the numbers, as GM should have done but didn't. The entire van-customizing industry produces only about 17,000 customized vans a month. So McNamara was claiming to buy an entire month's production. These large, luxurious, and gas-guzzling vehicles are designed for U.S. interstate highways. The recreational vehicle trade association says that only 1.35% were exported in 1990. It's not plausible to claim that 17,000 vans in a single month are being bought for export. McNamara's claimed purchases were large even when compared with total production of vans. Chevrolet, for example, produced 100,067 full-sized vans in all of 1990.

Having looked at the numbers, you can guess the rest. McNamara admitted in federal court in 1992 that he was defrauding GM on a massive scale. The Indiana company was a shell set up by McNamara, its invoices were phony, and the vans didn't exist. McNamara borrowed vastly from GM, used most of each loan to pay off the previous loan (thus establishing a record as a good credit risk), and skimmed off a bit for himself. The bit he skimmed amounted to over $400 million. GM set aside $275 million to cover its losses. Two executives, who should have looked at the numbers relevant to their business, were fired.

Business data, advertising claims, debate on public issues—we are assailed daily by numbers intended to prove a point, buttress an argument, or assure us that all is well. Some, like John McNamara, feed us fake data. Others use data to argue a cause and care more for the cause than for the accuracy of the data. Others simply lack the skills needed to employ numbers carefully. We know that we should always ask

- How were the data produced?
- What exactly was measured?

We also know quite a bit about what good answers to these questions sound like. That's wonderful, but it isn't enough. It may be that the General Motors executives taken in by John McNamara's scheme knew something about random samples and reliable measurements. What they lacked was "number sense," the habit of asking if numbers make sense. To help develop number sense, we will look at how bad data, or good data wrongly used, can trick the unwary.

What didn't they tell us?

The most common way to mislead with data is to cite correct numbers that don't quite mean what they appear to say because we aren't told the full story. The numbers are not made up, so the fact that the information is a bit incomplete may be an innocent oversight. Here are some examples. You decide how innocent they are.

Example 1. Snow! Snow! Snow!

Crested Butte attracts skiers by advertising that it has the highest average snowfall of any ski town in Colorado. That's true. But skiers want snow on the ski slopes, not in the town—and many other Colorado resorts get more snow on the slopes.

Example 2. Yet more snow

News reports of snowstorms say things like "A winter storm spread snow across the area, causing 28 minor traffic accidents." Eric Meyer, a reporter in Milwaukee, Wisconsin, says he often called the sheriff to gather such numbers. One day he decided to ask the sheriff how many minor accidents are typical in good weather: about 48, said the sheriff. Perhaps, says Meyer, the news should say, "Today's winter storm prevented 20 minor traffic accidents."

"Sure your patients have 50% fewer cavities. That's because they have 50% fewer teeth!"

Example 3. We attract really good students

Colleges know that many prospective students look at popular guidebooks to decide where to apply for admission. The guidebooks print information supplied by the colleges themselves. Surely no college would simply lie about, say, the average SAT score of its entering students. But we do want our scores to look good. How about leaving out the scores of our international and remedial students? Northeastern University did this, making the average SAT score of its freshman class 50 points higher than if all students were included. If we admit economically disadvantaged students under a special program sponsored by the state, surely no one will complain if we leave their SAT scores out of our average? New York University did this.

The point of these examples is that numbers have a context. If you don't know the context, the lonely, isolated, naked number doesn't tell you much.

Are the numbers consistent with each other?

John McNamara fooled General Motors because GM didn't compare his numbers with others. No one asked how a dealer could buy 17,000 vans in a single month for export when the entire custom van industry produces just 17,000 vans a month and only a bit over 1% are exported. Speaking of GM, here's another example in which the numbers don't line up with each other.

Example 4. We won!

GM's Cadillac brand was the best-selling luxury car in the United States for 57 years in a row. In 1998, Ford's Lincoln brand seemed to be winning until the last moment. Said the *New York Times,* "After reporting almost unbelievable sales results in December, Cadillac eked out a come-from-behind victory by just 222 cars." The final count was 187,343 for Cadillac, 187,121 for Lincoln. Then GM reported that Cadillac sales dropped 38% in January. How could sales be so different in December and January? Could it be that some January sales were counted in the previous year's total? Just enough, say, to win by 222 cars? Yes, indeed. In May, GM confessed that it sold 4773 fewer Cadillacs in December than it had claimed.

In the General Motors examples, we suspect something is wrong because numbers don't agree as we think they should. Here's an example where we *know* something is wrong because the numbers don't agree. This is part of an article on a cancer researcher at the Sloan-Kettering Institute who was accused of committing the ultimate scientific sin, falsifying data.

Example 5. Fake data

"One thing he did manage to finish was a summary paper dealing with the Minnesota mouse experiments. . . . That paper, cleared at SKI and accepted by the *Journal of Experimental Medicine,* contains a statistical table that is erroneous in such an elementary way that a bright grammar school pupil could catch the flaw. It lists 6 sets of 20 animals each, with the percentages of successful takes. Although any percentage of 20 has to be a multiple of 5, the percentages that Summerlin recorded were 53, 58, 63, 46, 48, and 67."

Are the numbers plausible?

As the General Motors examples illustrate, you can often detect dubious numbers simply because they don't seem plausible. Sometimes you can check an implausible number against data in reliable sources such as the annual *Statistical Abstract of the United States.* Sometimes, as the next example illustrates, you can do a calculation to show that a number can't be right.

Example 6. The abundant melon field

The very respectable journal *Science,* in an article on insects that attack plants, mentioned a California field that produces 750,000 melons per acre. A reader responded, "I learned as a farm boy that an acre covers 43,560 square feet, so this remarkable field produces about 17 melons per square foot. If these are cantaloupes, with each fruit covering about 1 square foot, I guess they must grow in a stack 17 deep." Here is the calculation the reader did:

$$\text{melons per square foot} = \frac{\text{melons per acre}}{\text{square feet per acre}} = \frac{750{,}000}{43{,}560} = 17.2$$

The editor, a bit embarrassed, replied that the correct figure was about 11,000 melons per acre.

Are the numbers too good to be true?

In Example 5, lack of consistency led to the suspicion that the data were phony. *Too much precision or regularity* can lead to the same suspicion, as when a student's lab report contains data that are exactly as the theory predicts. The laboratory instructor knows that the accuracy of the equipment and the student's laboratory technique are not good enough to give such perfect results. He suspects that the student made them up. Here is an example drawn from an article in *Science* about fraud in medical research.

Example 7. More fake data

"Lasker had been asked to write a letter of support. But in reading two of Slutsky's papers side by side, he suspected that the same 'control' animals had been used in both without mention of the fact in either. Identical data points appeared in both articles, but . . . the actual number of animals cited in each case was different. This suggested at best a sloppy approach to the facts. Almost immediately after being asked about the statistical discrepancies, Slutsky resigned and left San Diego."

In this case, suspicious regularity (identical data points) combined with inconsistency (different numbers of animals) led a careful reader to suspect fraud.

Is the arithmetic right?

Conclusions that are wrong or just incomprehensible are often the result of plain old-fashioned blunders. Rates and percentages cause particular trouble.

Example 8. Oh, those percents

Here are some examples from Australia. The *Canberra Times* reported, "Of those aged more than 60 living alone, 34% are women and only 15% are men." That's 49% of those living alone. I suppose the other 51% are neither women nor men.

Even smart people have problems with percents. A newsletter for female university teachers asked, "Does it matter that women are 550% (five and a half times) less likely than men to be appointed to a professional grade?" Now 100% of something is all there is. If you take away 100%, there is nothing left. I have no idea what "550% less likely" might mean.

It seems that few people do arithmetic once they leave school. Those who do are less likely to be taken in by meaningless numbers. A little thought and a calculator go a long way.

Just a little arithmetic mistake

In 1994, an investment club of grandmotherly women wrote a best-seller, *The Beardstown Ladies' Common-Sense Investment Guide: How We Beat the Stock Market—and How You Can, Too*. On the book cover and in their many TV appearances, the down-home authors claimed a 23.4% annual return, beating the market and most professionals. Four years later, a skeptic discovered that the club treasurer had entered data incorrectly. The Beardstown ladies' true return was only 9.1%, far short of the overall stock market return of 14.9% in the same period. We all make mistakes, but most of them don't earn as much money as this one did.

Example 9. Summertime is burglary time

An advertisement for a home security system says, "When you go on vacation, burglars go to work. According to FBI statistics, over 26% of home burglaries take place between Memorial Day and Labor Day."

This is supposed to convince us that burglars are more active in the summer vacation period. Look at your calendar. There are 14 weeks between Memorial Day and Labor Day. As a percent of the 52 weeks in the year, this is

$$\frac{14}{52} = 0.269 \qquad \text{(that is, 26.9\%)}$$

So the ad claims that 26% of burglaries occur in 27% of the year. You should not be impressed.

Example 10. The old folks are coming

A writer in *Science* claimed in 1976 that "people over 65, now numbering 10 million, will number 30 million by the year 2000, and will constitute an unprecedented 25 percent of the population." Sound the alarm: the elderly were going to triple in a quarter century to become a fourth of the population.

Let's check the arithmetic. Thirty million is 25% of 120 million, because

$$\frac{30}{120} = 0.25$$

So the writer's numbers only make sense if the population in 2000 is 120 million. The U.S. population in 1975 was already 216 million. Something is wrong.

Thus alerted, we can check the *Statistical Abstract* to learn the truth. In 1975, there were 22.4 million people over age 65, not 10 million. That's more than 10% of the total population. The estimate of 30 million by the year 2000 was only about 12% of the projected population for that year. Looking back from the year 2000, we now know that people at least 65 years old are 13% of the total U.S. population. As people live longer, the numbers of the elderly are growing. But growth from 10% to 13% over 25 years is far slower than the *Science* writer claimed.

Calculating the percent increase or decrease in some quantity seems particularly prone to mistakes. The percent change in a quantity is found by

$$\text{percent change} = \frac{\text{amount of change}}{\text{starting value}} \times 100$$

Example 11. Stocks go up, stocks go down

In 1999, the NASDAQ composite index of stock prices rose from 2192.69 to 4069.31. What percentage increase was this?

$$\begin{aligned}\text{percent change} &= \frac{\text{amount of change}}{\text{starting value}} \times 100 \\ &= \frac{4069.31 - 2192.69}{2192.69} \times 100 \\ &= \frac{1876.62}{2192.69} \times 100 = 0.856 \times 100 = 85.6\%\end{aligned}$$

That's pretty impressive. Of course, stock prices go down as well as up. In the week ending April 14, 2000, the NASDAQ index dropped from 4446.17 to 3321.29. That's a percentage decrease of

$$\frac{\text{amount of change}}{\text{starting value}} \times 100 = \frac{-1124.88}{4446.17} \times 100 = -25.3\%$$

Remember to always use the *starting* value, not the smaller value, in the denominator of your fraction.

A quantity can increase by any amount—a 100% increase just means it has doubled. But nothing can go down more than 100%—it has then lost 100% of its value, and 100% is all there is.

Is there a hidden agenda?

Lots of people feel strongly about various issues, so strongly that they would like the numbers to support their feelings. Often they can find support in numbers by choosing carefully which numbers to report or by working hard to squeeze the numbers into the shape they prefer. Here are two examples.

Example 12. Heart disease in women

A highway billboard says simply, "Half of all heart disease victims are women." What might be the agenda behind this true statement? Perhaps the billboard sponsors just want to make women aware that they do face risks from heart disease. (Surveys show that many women underestimate the risk of heart disease.)

On the other hand, perhaps the sponsors want to fight what some people see as an overemphasis on male heart disease. In that case, we might want to know that although half of heart disease victims are women, they are on the average much older than male victims. Roughly 36,000 women under age 65 and 85,000 men under age 65 die from heart disease each year. The American Heart Association says, "Risk of death due to coronary heart disease in women is roughly similar to that of men 10 years younger."

Example 13. Income inequality

During the economic boom of the 1980s and 1990s in the United States, the gap between the highest and lowest earners widened. In 1980, the bottom fifth of households received 4.3% of all income, and the top fifth received 43.7%. By 1998, the share of the bottom fifth had fallen to 3.6% of all income, and the share of the top fifth of households had risen to 49.2%. That is, the top fifth's share was almost 14 times the bottom fifth's share.

Can we massage the numbers to reduce the income gap? An article in *Forbes* (a magazine read mainly by rich folk) tried. First, more people live in the average rich household than in poor households, so let's change to income per person. The rich pay more taxes, so look at income after taxes. The poor receive food stamps and other assistance, so let's count that. Finally, high earners work more hours than low earners, so we should adjust for hours worked. After all this, the share of the top fifth is only 3 times that of the bottom fifth. Of course, hours worked are reduced by illness, disability, care of children and aged parents, and so on. If *Forbes*'s hidden agenda is to show that income inequality isn't important, we may not agree.

Yet other adjustments are possible. Income, in these Census Bureau figures, does not include capital gains from, for example, selling stocks that have gone up. Almost all capital gains go to the rich, so including them would widen the income gap. *Forbes* didn't make this adjustment. Making every imaginable adjustment in the meaning of "income," says the Census Bureau, gives the bottom fifth of households 4.7% of total income in 1998 and the top fifth 45.8%.

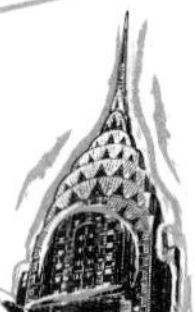

Exploring the Web

The *Statistical Abstract of the United States* is an essential compilation of data. You can find it online at the Census Bureau Web site. The 1999 edition was at www.census.gov/statab/www/. To find what you want, look in the index first.

The CHANCE Web site at Dartmouth College contains lots of interesting stuff (at least if you are interested in statistics). In particular, the *Chance News* section, www.dartmouth.edu/~chance/chance_news/news.html, offers a monthly newsletter that keeps track of statistics in the press, including dubious statistics.

Statistics in Summary The aim of statistics is to provide insight by means of numbers. Numbers are most likely to yield their insights to those who examine them closely. Pay attention to voluntary response samples and to confounding. Ask exactly what a number measures and decide if it is a valid measure. Look for the context of the numbers and ask if there is important **missing information.** Look for **inconsistencies,** numbers that don't agree as they should, and check for **incorrect arithmetic.** Compare numbers that are **implausible**—surprisingly large or small—with numbers you know are right. Be suspicious when numbers are **too regular or agree too well** with what their author would like to see. Look with special care if you suspect the numbers are put forward in support of some **hidden agenda.** If you form the habit of looking at numbers closely, your friends will soon think that you are brilliant. They might even be right.

CHAPTER 9 EXERCISES

9.1 Drunk driving. A newspaper article on drunk driving cited data on traffic deaths in Rhode Island: "Forty-two percent of all fatalities occurred on Friday, Saturday, and Sunday, apparently because of increased drinking on the weekends." What percent of the week do Friday, Saturday, and Sunday make up? Are you surprised that 42% of fatalities occur on those days?

9.2 Advertising painkillers. An advertisement for the pain reliever Tylenol was headlined "Why Doctors Recommend Tylenol More Than All Leading Aspirin Brands Combined." The makers of Bayer Aspirin, in a reply headlined "Makers of Tylenol, Shame on You!" accused Tylenol of misleading by giving the truth but not the whole truth. You be the detective. How is Tylenol's claim misleading even if true?

9.3 Advertising painkillers. Anacin was long advertised as containing "more of the ingredient doctors recommend most." Another over-the-counter pain reliever claimed that "doctors specify Bufferin most" over other "leading brands." Both advertising claims were literally true; the Federal Trade Commission found them both misleading. Explain why. (Hint: What is the active pain reliever in both Anacin and Bufferin?)

9.4 Deer in the suburbs. Westchester County is a suburban area covering 438 square miles immediately north of New York City. A garden magazine claimed that the county is home to 800,000 deer. Do a calculation that shows this claim to be implausible.

9.5 Suicides among Vietnam veterans. Did the horrors of fighting in Vietnam drive many veterans of that war to suicide? A figure of 150,000 suicides among Vietnam veterans in the 20 years following the end of the war has been widely quoted. Explain why this number is not plausible. To help you, here are some facts: about 25,000 American men commit suicide each year; about 3 million men served in Southeast Asia during the Vietnam War; there are roughly 93 million adult men in the United States.

9.6 Trash at sea? A report on the problem of vacation cruise ships polluting the sea by dumping garbage overboard said:

> *On a seven-day cruise, a medium-size ship (about 1,000 passengers) might accumulate 222,000 coffee cups, 72,000 soda cans, 40,000 beer cans and bottles, and 11,000 wine bottles.*

Are these numbers plausible? Do some arithmetic to back up your conclusion. Suppose, for example, that the crew is as large as the passenger list. How many cups of coffee must each person drink every day?

9.7 Funny numbers. Here's a quotation from a book review in a scientific journal:

> *. . . a set of 20 studies with 57 percent reporting significant results, of which 42 percent agree on one conclusion while the remaining 15 percent favor another conclusion, often the opposite one.*

Do the numbers given in this quotation make sense? Can you decide how many of the 20 studies agreed on "one conclusion," how many favored another conclusion, and how many did not report significant results?

9.8 Airport delays. An article in a midwestern newspaper about flight delays at major airports said:

> *According to a Gannett News Service study of U.S. airlines' performance during the past five months, Chicago's O'Hare Field scheduled 114,370 flights. Nearly 10 percent, 1,136, were canceled.*

Check the newspaper's arithmetic. What percent of scheduled flights from O'Hare were actually canceled?

9.9 Battered women? A letter to the editor of the *New York Times* complained about a *Times* editorial that said "an American woman is beaten by her husband or boyfriend every 15 seconds." The writer of the letter claimed that "at that rate, 21 million women would be beaten by their husbands or boyfriends every year. That is simply not the case." He cited the National Crime Victimization Survey, which estimated 56,000 cases of violence against women by their husbands and 198,000 by boyfriends or former boyfriends. The survey showed 2.2 million assaults against women in all, most by strangers or someone the woman knew who was not her past or present husband or boyfriend.

(a) First do the arithmetic. Every 15 seconds is 4 per minute. At that rate, how many beatings would take place in an hour? In a day? In a year? Is the letter-writer's arithmetic correct?

(b) Is the letter-writer correct to claim that the *Times* overstated the number of cases of domestic violence against women?

9.10 We can read, but can we count? The Census Bureau once gave a simple test of literacy in English to a random sample of 3400 people. The *New York Times* printed some of the questions under the headline "113% of Adults in U.S. Failed This Test." Why is the percent in the headline clearly wrong?

9.11 Stocks go down. On August 4, 1998, the Dow Jones Industrial Average dropped 299.43 points from its opening level of 8780.94. Some newspapers headlined this as "the third-biggest point drop ever." By what percent did the Dow drop that day? Because the index had gone up so much in the previous years, this wasn't even in the top 20 of percentage drops. It's another example of rates being clearer than counts.

9.12 Poverty. The number of Americans living below the official poverty line increased from 26,072,000 to 34,476,000 in the 20 years between 1979 and 1998. What percent increase was this? You should not conclude from this that poverty grew more common in these years, however. Why not?

9.13 We don't lose your baggage. Continental Airlines once advertised that it had "decreased lost baggage by 100% in the past six months." Do you believe this claim?

9.14 Rinse your mouth. A new mouth rinse claimed to "reduce plaque on teeth by 300%." Explain carefully why it is impossible to reduce anything by 300%.

9.15 Greeks on campus. The question-and-answer column of a campus newspaper was asked what percent of the campus was "Greek" (that is, members of fraternities or sororities). The answer given was that "the figures for the fall semester are approximately 13 percent for the girls and 15–18 percent for the guys, which produces a 'Greek' figure of approximately 28–31 percent of the undergraduates." Discuss the campus newspaper's arithmetic.

9.16 Don't dare to drive? A university sends a monthly newsletter on health to its employees. A recent issue included a column called "What Is the Chance" that said:

> *Chance that you'll die in a car accident this year: 1 in 75.*

There are about 275 million people in the United States. About 40,000 people die each year from motor vehicle accidents. What is the chance a typical person will die in a motor vehicle accident this year?

9.17 How many miles of highways? *Organic Gardening* magazine once said that "the U.S. Interstate Highway System spans 3.9 million miles and is wearing out 50% faster than it can be fixed. Continuous road deterioration adds $7 billion yearly in fuel costs to motorists." The distance from the east coast to the west coast of the United States is about 3000 miles. How many separate highways across the continent would be needed to account for 3.9 million miles of roads? What do you conclude about the number of miles in the interstate system?

9.18 In the garden. *Organic Gardening* magazine, describing how to improve your garden's soil, said, "Since a 6-inch layer of soil in a 100-square-foot plot weighs about 45,000 pounds, adding 230 pounds of compost will give you an instant 5% organic matter."

(a) What percent of 45,000 is 230?

(b) Water weighs about 62 pounds per cubic foot. There are 50 cubic feet in a garden layer 100 square feet in area and 6 inches deep. What would 50 cubic feet of water weigh? Is it plausible that 50 cubic feet of soil weighs 45,000 pounds?

(c) It appears from (b) that the 45,000 pounds isn't right. In fact, soil weighs about 75 pounds per cubic foot. If we use the correct weight, is the "5% organic matter" conclusion roughly correct?

9.19 No eligible men? A news report quotes a sociologist as saying that for every 233 unmarried women in their 40s in the United States, there are only 100 unmarried men in their 40s. These numbers point to an unpleasant social situation for women of that age. Are the numbers plausible? (Optional: The *Statistical Abstract* has a table titled "Marital status of the population by age and sex" that gives the actual counts.)

9.20 Too good to be true? The late English psychologist Cyril Burt was known for his studies of the IQ scores of identical twins who were raised apart. The high correlation between the IQs of separated twins in Burt's studies pointed to heredity as a major factor in IQ. ("Correlation" measures how closely two variables are connected. We will meet correlation in Chapter 14.) Burt wrote several accounts of his work, adding more pairs of twins over time. Here are his reported correlations as he published them:

Publication date	Twins reared apart	Twins reared together
1955	0.771 (21 pairs)	0.944 (83 pairs)
1966	0.771 (53 pairs)	0.944 (95 pairs)

What is suspicious here?

9.21 Where you start matters. When comparing numbers over time, you can slant the comparison by choosing your starting point. Say the Chicago Cubs lose 5 games, then win 4, then lose 1. You can truthfully say that the Cubs have lost 6 of their last 10 games (sounds bad) or that they have won 4 of their last 5 (sounds good).

The median income of American families (in dollars of constant buying power) was $44,284 in 1989, $41,051 in 1993, and $44,568 in 1997. By what percent did family income increase between 1989 and 1997? Between 1993 and 1997? You see that you can make the income trend sound bad or good by choosing your starting point.

9.22 Being on top also matters. The previous exercise noted that median family income hardly changed between 1989 and 1997. The top 5% of households earned $120,607 or more in 1989 and $128,521 or more in 1997. (These amounts are in dollars of constant buying power.) By what percent did the income of top earners increase between 1989 and 1997?

9.23 Boating safety. Data on accidents in recreational boating in the *Statistical Abstract* show that the number of deaths has dropped from 1360 in 1980 to 865 in 1990 and 819 in 1997. However, the number of injuries reported grew from 2650 in 1980 to 3822 in 1990 and 4555 in 1997. Why are there so few injuries in these government data relative to the number of deaths? Which count (deaths or injuries) is probably more accurate? Why might the injury count rise when deaths did not?

9.24 Time off. The pointy-haired boss in the *Dilbert* cartoon once noted that 40% of all sick days taken by the staff are Fridays and Mondays. Is this evidence that the staff are trying to take long weekends?

9.25 Find an example of one of the following. Explain in detail the statistical shortcomings of your example.

Leaving out essential information

Lack of consistency

Implausible numbers

Faulty arithmetic

One place to look is in the online *Chance News* mentioned in the "Exploring the Web" box.

Part I Review

The first and most important question to ask about any statistical study is "Where did the data come from?" Chapter 1 addressed this question. The distinction between observational and experimental data is a key part of the answer. Good statistics starts with good designs for producing data. Chapters 2, 3, and 4 discussed sampling, the art of choosing part of a population to represent the whole. Figure I.1 summarizes the big idea of a simple random sample. Chapters 5 and 6 dealt with the statistical aspects of designing experiments, studies that impose some treatment in order to learn about the response. The big idea is the randomized comparative experiment. Figure I.2 outlines the simplest design.

Random sampling and randomized comparative experiments are perhaps the most important statistical inventions of the 20th century. Both were slow to gain acceptance, and you will still see many voluntary response samples and uncontrolled experiments. Both random samples and randomized experiments involve the deliberate use of

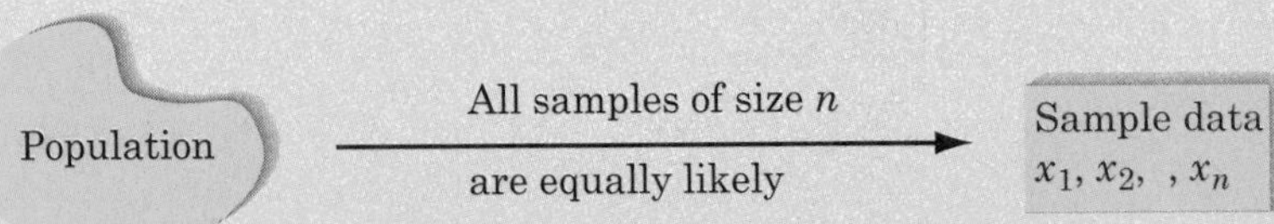

Figure I.1 The idea of a simple random sample.

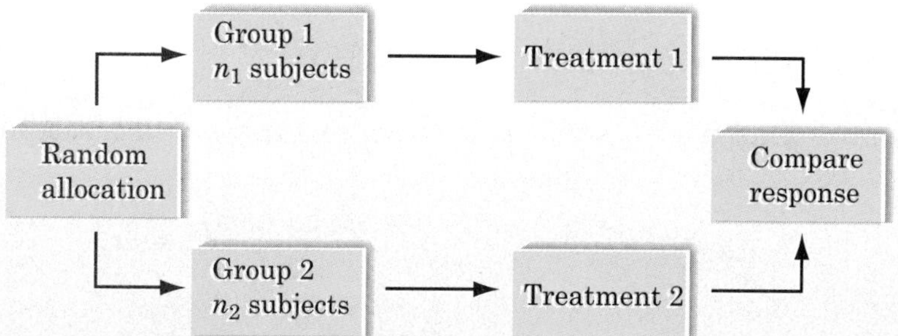

Figure I.2 The idea of a randomized comparative experiment.

chance to eliminate bias and produce a regular pattern of outcomes. The regular pattern allows us to give margins of error, make confidence statements, and assess the statistical significance of conclusions based on samples or experiments.

When we collect data about people, ethical issues can be important. Chapter 7 discussed these issues and introduced three principles that apply to any study with human subjects. The last step in producing data is to measure the characteristics of interest to produce numbers we can work with. Measurement was the subject of Chapter 8. "Where do the data come from?" is the first question we should ask about a study, and "Do the numbers make sense?" is the second. Chapter 9 encouraged the valuable habit of looking skeptically at numbers before accepting what they seem to say.

PART I SUMMARY

Here are the most important skills you should have after reading Chapters 1 to 9.

A. DATA

1. Recognize the individuals and variables in a statistical study.

2. Distinguish observational from experimental studies.

3. Identify sample surveys, censuses, and experiments.

B. SAMPLING

1. Identify the population in a sampling situation.

2. Recognize bias due to voluntary response samples and other inferior sampling methods.

3. Use Table A of random digits to select a simple random sample (SRS) from a population.

4. Explain how sample surveys deal with bias and variability in their conclusions. Explain in simple language what the margin of error for a sample survey result tells us and what "95% confidence" means.

5. Use the quick method to get an approximate margin of error for 95% confidence.

6. Understand the distinction between sampling errors and nonsampling errors. Recognize the presence of undercoverage and nonresponse as sources of error in a sample survey. Recognize the effect of the wording of questions on the responses.

7. Use random digits to select a stratified random sample from a population when the strata are identified.

C. EXPERIMENTS

1. Identify the explanatory variables, treatments, response variables, and subjects in an experiment.

2. Recognize bias due to confounding of explanatory variables with lurking variables in either an observational study or an experiment.

3. Outline the design of a completely randomized experiment using a diagram like that in Figure I.2. Such a diagram should show the sizes of the groups, the specific treatments, and the response variable.

4. Use Table A of random digits to carry out the random assignment of subjects to groups in a completely randomized experiment.

5. Make use of matched pairs or block designs when appropriate.

6. Recognize the placebo effect. Recognize when the double-blind technique should be used. Be aware of weaknesses in an experiment, especially in the ability to generalize its conclusions.

7. Explain why a randomized comparative experiment can give good evidence for cause-and-effect relationships.

8. Explain the meaning of statistical significance.

D. OTHER TOPICS

1. Explain the three first principles of data ethics. Discuss how they might apply in specific settings.

2. Explain how measuring leads to clearly defined variables in specific settings.

3. Evaluate the validity of a variable as a measure of a given characteristic, including predictive validity.

4. Explain how to reduce bias and improve reliability in measurement.

5. Recognize inconsistent numbers, implausible numbers, numbers so good they are suspicious, and arithmetic mistakes.

6. Calculate percent increase or decrease correctly.

PART I REVIEW EXERCISES

Review exercises are short and straightforward exercises that help you solidify the basic ideas and skills in each part of this book.

I.1 Know these terms. A friend who knows no statistics has encountered some statistical terms in reading for her psychology course. Explain each of the following terms in one or two simple sentences.
(a) Simple random sample.

(b) 95% confidence.

(c) Nonsampling error.

(d) Informed consent.

I.2 Know these terms. A friend who knows no statistics has encountered some statistical terms in her biology course. Explain each of the following terms in one or two simple sentences.
(a) Observational study.

(b) Placebo effect.

(c) Statistically significant.

(d) Institutional review board.

I.3 A biased sample. You see a woman student standing in front of the student center, now and then stopping other students to ask them questions. She says that she is collecting student opinions for a class assignment. Explain why this sampling method is almost certainly biased.

I.4 Select an SRS. A student at a large university wants to study the responses that students receive when calling an academic department for information. She selects an SRS of 6 departments from the following list for her study. Use Table A at line 116 to do this.

Agronomy	Education	Nursing
Art and Design	Electrical Eng	Pharmacology
Audiology	English	Philosophy
Biochemistry	Foreign Languages	Physics
Biology	History	Political Science
Chemistry	Horticulture	Psychology
Communication	Industrial Eng	Sociology
Computer Science	Management	Statistics
Consumer Sciences	Mathematics	Veterinary Anatomy

I.5 Select an SRS. The faculty grievance system at a university specifies that a 5-member hearing panel shall be drawn at random from the 30-member grievance committee. Use Table A at line 107 to draw an SRS of size 5 from the following committee members.

Abel	Foley	Lowe	Santogrossi	Walther
Adler	Gonzalez	Martinez	Schrag	Weinstein
Calderon	Hay	Miller	Staaks	Winer
DeVito	Ismail	Mork	Steer	Wu
Eisinger	Kirby	Perrucci	Targ	Young
Ewbank	Levy	Rothenberg	Toal	Zhang

I.6 Errors in surveys. Give an example of a source of nonsampling error in a sample survey. Then give an example of a source of sampling error.

I.7 Errors in surveys. An overnight opinion poll calls randomly selected telephone numbers. This polling method misses all people without a phone. Is this a source of nonsampling error or of sampling error? Does the poll's announced margin of error take this source of error into account?

I.8 Errors in surveys. A college chooses an SRS of 100 students from the registrar's list of all undergraduates to interview about student life. If it selected two SRSs of 100 students at the same time, the two samples would give somewhat different results. Is this variation a source of sampling error or of nonsampling error? Does the survey's announced margin of error take this source of error into account?

I.9 Errors in surveys. Exercises I.7 and I.8 each mention a source of error in a sample survey. Would each kind of error be reduced by doubling the size of the sample with no other changes in the survey procedures? Explain your answers.

I.10 Errors in surveys. A Gallup Poll found that 68% of adult Americans favor teaching creationism along with evolution in public schools. The Gallup press release says:

> *For results based on samples of this size, one can say with 95 percent confidence that the maximum error attributable to sampling and other random effects is plus or minus 3 percentage points.*

Give one example of a source of error in the poll result that is *not* included in this margin of error.

I.11 Find the margin of error. A poll of 586 adults who had used the Internet in the past week asked whether "the Internet has made your life much better, somewhat better, somewhat worse, much worse, or has it not affected your life either way." In all, 152 of the 586 subjects said "much better."

(a) What is the population for this sample survey?

(b) Use the quick method to find a margin of error. Then give a complete confidence statement for a conclusion about the population.

I.12 Find the margin of error. Should the government spend money to help low-income families who want to send their children to private or religious schools? A poll of 1006 adults found that 362 said "Yes."

(a) What is the population for this sample survey?

(b) Use the quick method to find a margin of error. Make a confidence statement about the opinion of the population.

I.13 What kind of sample? At a party there are 30 students over age 21 and 20 students under age 21. You choose at random 3 of those over 21 and separately choose at random 2 of those under 21 to interview about attitudes toward alcohol. You have given every student at the party the same chance to be interviewed: What is that chance? Why is your sample not an SRS? What is this kind of sample called?

I.14 Design an experiment. A university's department of statistics wants to attract more majors. It prepares two advertising brochures. Brochure A stresses the intellectual excitement of statistics. Brochure B stresses how much money statisticians make. Which will be more attractive to first-year students? You have a questionnaire to measure interest in majoring in stat, and you have 50 first-year students to work with. Outline the design of an experiment to decide which brochure works better.

I.15 Design an experiment. Gary thinks that the way to get a date is to tell a woman about yourself. Greg thinks that getting her to talk about herself works better. You recruit 20 guys who are willing to try either method in conversation, then call the woman a day later to ask for a date. Outline the design of an experiment to decide which method succeeds more often.

Exercises I.16 to I.19 are based on an article in the *Journal of the American Medical Association* that asks if flu vaccine works. The article reports a study of the effectiveness of a nasal spray vaccine called "trivalent LAIV." Here is part of the article's summary:

> **Design** Randomized, double-blind, placebo-controlled trial conducted from September 1997 through March 1998.
>
> **Participants** A total of 4561 healthy, working adults aged 18 to 64 years recruited through health insurance plans, at work sites, and from the general population.
>
> **Intervention** Participants were randomized 2:1 to receive intranasally administered trivalent LAIV vaccine (n = 3041) or placebo (n = 1520) in the fall of 1997.
>
> **Results** Vaccination also led to . . . fewer days of work lost (17.9% reduction for severe febrile illnesses; 28.4% reduction for febrile upper respiratory tract illnesses), and fewer days with health care provider visits (24.8% reduction for severe febrile illnesses; 40.9% reduction for febrile upper respiratory tract illnesses).

I.16 Know these terms. Explain in one sentence each what "randomized," "double-blind," and "placebo-controlled" mean in the description of the design of the study.

I.17 Experiment basics. Identify the subjects, the explanatory variable, and several response variables for this study.

I.18 Design an experiment. Use a diagram to outline the design of the experiment in this medical study.

I.19 Ethics. What are the three first principles of data ethics? Explain briefly what the flu vaccine study must do to apply each of these principles.

I.20 Measuring. Joni wants to measure the degree to which male college students belong to the political left. She decides simply to measure the length of their hair—longer hair will mean more left-wing.

(a) Is this method likely to be reliable? Why?

(b) This measurement appears to be invalid. Why?

(c) Nevertheless, it is possible that measuring politics by hair length might have some predictive validity. Explain how this could happen.

I.21 Reliability. You are laboring through a chemistry laboratory assignment in which you measure the conductivity of a solution. What does it mean for your measurement to be reliable? How can you improve the reliability of your final result?

I.22 Observation or experiment? The Nurses Health Study has queried a sample of over 100,000 female registered nurses every two years since 1976. Beginning in 1980, the study asked questions about diet, including alcohol consumption. The researchers concluded that "light-to-moderate drinkers had a significantly lower risk of death" than either nondrinkers or heavy drinkers.

(a) Is the Nurses Health Study an observational study or an experiment? Why?

(b) What does "significant" mean in a statistical report?

(c) Suggest some lurking variables that might explain why moderate drinkers have lower death rates than nondrinkers. (The study did adjust for these variables.)

I.23 Observation or experiment? In a study of the relationship between physical fitness and personality, middle-aged college faculty who have volunteered for an exercise program are divided into low-fitness and high-fitness groups on the basis of a physical examination. All subjects then take a personality test. The high-fitness group has a higher average score for "self-confidence."

(a) Is this an observational study or an experiment? Why?

(b) We cannot conclude that higher fitness causes higher self-confidence. Suggest other relationships among these variables and perhaps lurking variables that might explain the higher self-confidence of the high-fitness group.

I.24 Percents up and down. Between January 1999 and March 2000 the average price of regular gasoline in California increased from $1.20 per gallon to $1.80 per gallon.

(a) Verify that this is a 50% increase in price.

(b) If the price of gasoline decreases by 50% from its March level of $1.80 per gallon, what would be the new price? Notice that a 50% increase followed by a 50% decrease does not take us back to the starting point.

I.25 Percent decrease. On Tuesday, April 4, 2000, the NASDAQ stock index began the day at 4283 and by shortly after noon dropped to 3649 before recovering. By what percent did the index drop?

I.26 An implausible number? *Newsweek* once said in a story that a woman who is not currently married at age 40 has a better chance of being killed by a terrorist than of getting married. Do you think this is plausible? What kind of data would help you check this claim?

PART I PROJECTS

Projects are longer exercises that require gathering information or producing data and emphasize writing a short essay to describe your work. Many are suitable for teams of students.

Project 1. Design your own sample survey. Choose an issue of current interest to students at your school. Prepare a short (no more than five questions) questionnaire to determine opinions on this issue. Choose a sample of about 25 students, administer your questionnaire, and write a brief description of your findings. Also write a short discussion of your experiences in designing and carrying out the survey.

(Although 25 students are too few for you to be statistically confident of your results, this project centers on the practical work of a survey. You must first identify a population; if it is not possible to reach a wider student population, use students enrolled in this course. Did the subjects find your questions clear? Did you write the questions so that it was easy to tabulate the responses? At the end, did you wish you had asked different questions?)

Project 2. Measuring. Exercise 6.13 (page 105) asks you to outline the design of a matched pairs experiment to see if the right hand is stronger than the left hand in right-handed people. You can measure hand strength by asking a subject to squeeze a bathroom scale placed on a shelf with its end protruding. Using several subjects, try to determine whether this method of measuring hand strength is reliable. Write an account of your findings. For example, did you find that subjects used different grips, so that careful instructions were needed to get a consistent way of measuring? Prepare written instructions for subjects.

Project 3. Experimenting. After you or other members of your team have refined the measurement of hand strength in the previous project, carry out the matched pairs experiment of Exercise 6.13 with at least 10 subjects. Write a report that describes the randomization, gives the data, reports the differences in strength (right hand minus left hand), and says whether your small experiment seems to show that the right hand is stronger on the average.

Project 4. Describe a medical study. Go to the Web site of the *Journal of the American Medical Association* (jama.ama-assn.org). Unlike the *New England Journal of Medicine* (www.nejm.org), *JAMA* makes the full text of its articles freely available online. Select an article from the current issue or from a past issue that describes a study whose topic interests you. Write a newspaper article that summarizes the design and the findings of the study. (Be sure to include statistical aspects, such as observational study versus experiment and any randomization used. News accounts often neglect these facts.)

Project 5. Cars domestic and foreign. Are students at your college more or less likely than staff to drive foreign cars? Design and carry out a study to find out, and write a report describing your design and your findings. You must first be clear about what "domestic" and "foreign" mean so that each car is clearly in one class. Then you must locate a suitable sample of cars—perhaps from a student parking area and a staff parking area. If the areas are large, you will want to sample the cars rather than look at all of them. Consider using a systematic sample (Exercise 4.25, page 69).

Project 6. Data ethics. Locate a news discussion of an ethical issue that concerns statistical studies. Write your own brief summary of the debate and any conclusions you feel you can reach.

Here is an example of one way to approach this project. As I write, a debate continues as to whether it is ethical to study AIDS in African countries without offering all subjects the expensive treatments available in rich nations. Searching the archives at the Web site of the *New York Times* (www.nytimes.com) for "AIDS and ethics" finds many articles, including a promising one in the issue of March 30, 2000. To read the article, you must either pay $2.50 or go to the library.

Project 7. Measuring income. What is the "income" of a household? Household income may determine eligibility for government programs that assist "low-income" people. Income statistics also have political effects. Political conservatives often argue that government data overstate how many people are poor because the data include only money income, leaving out the value of food stamps and subsidized housing. Political liberals reply that the government should measure money income so it can see how many people need help.

You are on the staff of a member of Congress who is considering new welfare legislation. Write an exact definition of "income" for the purpose of determining which households are eligible for welfare. A short essay will be needed. Will you include nonmoney income such as the value of food stamps or subsidized housing? Will you allow deductions for the cost of child care needed to permit the parent to work? What about assets that are worth a lot but do not produce income, such as a house?

Part III

"Tonight we're going to let the statistics speak for themselves."

Organizing Data

Words alone don't tell a story. A writer organizes words into sentences and organizes the sentences into a story line. If the words are badly organized, the story isn't clear. Data also need organizing if they are to tell a clear story. Too many words obscure a subject rather than illuminate it. Vast amounts of data are even harder to digest—we often need a brief summary to highlight essential facts. How to organize, summarize, and present data are our topics in the second part of this book.

Organizing and summarizing a large body of facts opens the door to distortions, both unintentional and deliberate. This is no less (but also no more) the case when the facts take the form of numbers rather than words. I will point out some of the traps that data presentations can set for the unwary. Those who picture statistics as primarily a piece of the liar's art concentrate on the part of statistics that deals with summarizing and presenting data. I claim that misleading summaries and selective presentations go back to that after-the-apple conversation among Adam, Eve, and God. Don't blame statistics. Do remember the saying "Figures won't lie, but liars will figure," and beware.

Chapter 10
Graphs, Good and Bad

Getting rich

The end of the twentieth century saw a great bull market in U.S. common stocks. How great? Pictures tell the tale more clearly than words.

Look first at Figure 10.1. This shows the percent increase or decrease in stocks (measured by the Standard & Poor's 500 Index) in each year from 1971 to 1999. Until 1982, stock prices bounce up and down. Sometimes they go down a lot—stocks lost 15% of their value in 1973 and another 27% in 1974. But starting in 1982, stocks go up in 17 of the next 18 years, often by a lot.

Figure 10.2 shows how to get rich. If you had invested $1000 in stocks at the end of 1970, the graph shows how much money you had at the end of each following year. After 1974, your $1000 was down to $854, and at the end of 1981 it had grown to only $2121. That's only 7% a year. Your money would have grown faster in a bank during these years. Then the great bull market begins its work. By the end of 1999, it has turned your $1000 into $44,875.

The lesson is supposed to be that if you can wait long enough, you should put your money in stocks. Perhaps the real lesson will turn out to be that you should have put your money in stocks in 1980 and taken it out of stocks in 2000. Our statistical lesson, however, appeals to the mind rather than the pocketbook. It is the power of graphs to make clear what data say. Figure 10.1 shows how irregular the year-to-year behavior of stocks has been. Figure 10.2 shows the long-term growth in wealth that has rewarded patient investors. It points in particular to the effect of the 5 years of boom starting in 1995.

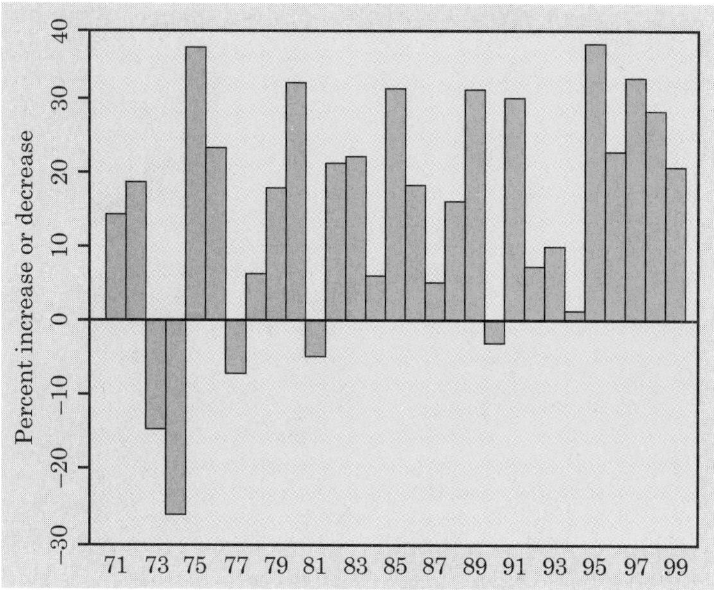

Figure 10.1 Percent increase or decrease in the S&P 500 Index of common stock prices, 1971 to 1999.

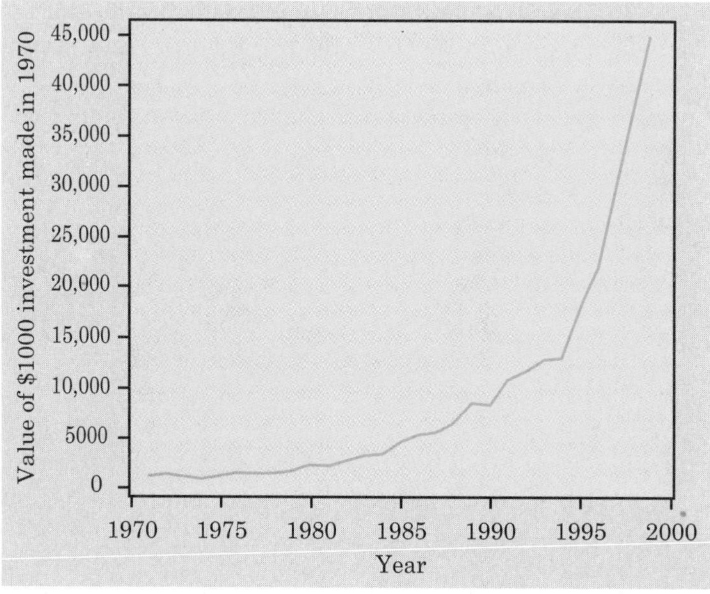

Figure 10.2 Value at the end of each year, 1971 to 1999, of $1000 invested in the S&P 500 Index at the end of 1970.

What mental picture does the word *statistics* call to mind? Very likely, images of tables crowded with numbers and graphs zigging up or zagging down. That's not a bad picture: statistics deals with data, and we use tables and graphs to present data. Random sampling, randomized comparative experiments, and valid measurement produce good data. Now we must see what the data say.

Data tables

Take a look at the *Statistical Abstract of the United States,* an annual volume packed with every variety of numerical information. Has the number of private elementary and secondary schools grown over time? What about minority enrollments in these schools? How many college degrees were given in each of the past several years, and how were these degrees divided among fields of study and by the age, race, and sex of the students? You can find all this and more in the education section of the *Statistical Abstract.* The tables *summarize* data. We don't want to see information on every college degree individually, only the counts in categories of interest to us.

Example 1. What makes a clear table?

How well educated are 30-something young adults? Table 10.1 presents the data for people aged 25 to 34 years. This table illustrates some good practices for data tables. It is clearly *labeled* so that we can see the subject of the data at once. The main heading describes the general subject of the data and gives the date because these data will change over time. Labels within the table identify the variables and state the *units* in which they are measured. Notice, for example, that the counts are in thousands. The *source* of the data appears at the foot of the table. This Census Bureau publication in fact presents data from our old friend, the Current Population Survey.

Table 10.1 starts with the *counts* of young adults with each level of education. *Rates* (percents or proportions) are often clearer than counts—it is more helpful to hear that 12.1% of young adults did not finish high school than to hear that there are 4,754,000 such people. The percents also appear in Table 10.1. The two columns of the table present the *distribution* of the variable "level of education" in two alternate forms. Each column gives information about what values the variable takes and how often it takes each value.

Distribution of a variable

The **distribution** of a variable tells us what values it takes and how often it takes these values.

Table 10.1 Education of people 25 to 34 years old, 1998

	Number of persons (thousands)	Percent
Less than high school	4,754	12.1
High school graduate	12,568	31.9
Some college	11,220	28.5
Bachelor's degree	8,367	21.3
Advanced degree	2,444	6.2
Total	39,354	100.0

SOURCE: Census Bureau, *Educational Attainment in the United States: March 1998.*

Example 2. Roundoff errors

Did you check Table 10.1 for consistency? The total number of people should be

$$4.754 + 12.568 + 11.220 + 8.367 + 2.444 = 39.353 \text{ (thousands)}$$

The table gives the total as 39,354. What happened? Each entry is rounded to the nearest thousand. The rounded entries don't quite add to the total, which is rounded separately. Such **roundoff errors** will be with us from now on as we do more arithmetic.

Pie charts and bar graphs

The distribution in Table 10.1 is quite simple because "level of education" has only 5 possible values. To picture this distribution in a graph, we might use a **pie chart.** Figure 10.3 is a pie chart of the level of education of young adults. Pie charts show how a whole is divided into parts. To make a pie chart, first draw a circle. The circle represents the whole, in this case all people aged 25 to 34 years. Wedges within the circle represent the parts, with the angle spanned by each wedge in proportion to the size of that part. For example, 21.3% of young adults have a bachelor's degree but not an advanced degree. Because there are 360 degrees in a circle, the "bachelor's degree" wedge spans an angle of

$$0.213 \times 360 = 77 \text{ degrees}$$

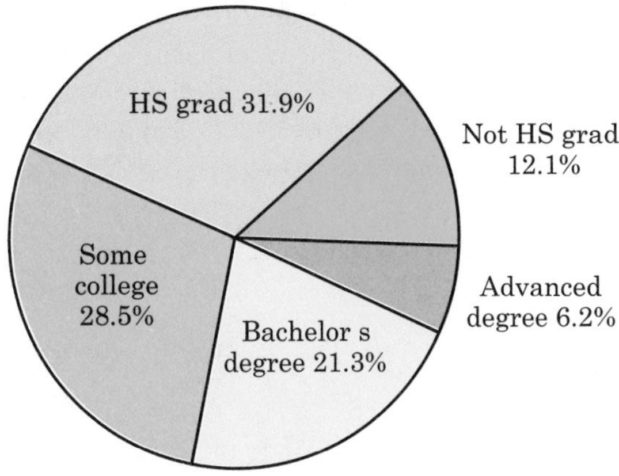

Figure 10.3 Pie chart of the distribution of level of education among persons aged 25 to 34 in 1998.

Pie charts force us to see that the parts do make a whole. But because angles are harder to compare than lengths, a pie chart is not a good way to compare the sizes of the various parts of the whole.

Figure 10.4 is a **bar graph** of the same data. The height of each bar shows the percent of young adults with the level of education marked at the bar's base. The bar graph makes it clear that there are more high school graduates than people who also have some college—the "HS grad" bar is taller. Because it is hard to see this from the wedges in the pie chart, I added labels to the wedges that include the actual percents. The bar graph is also easier to draw than the pie chart unless a computer is doing the drawing for you.

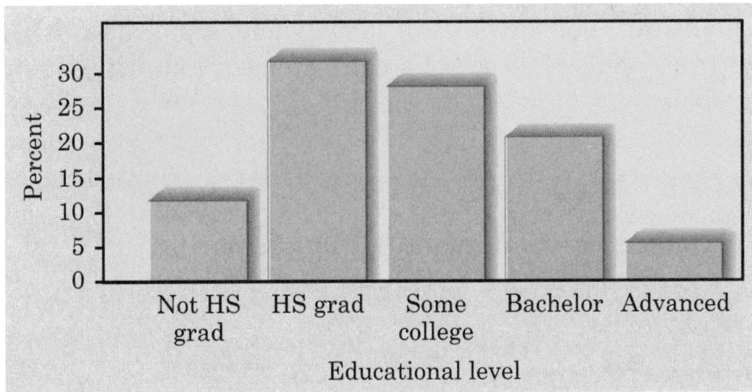

Figure 10.4 Bar graph of the distribution of level of education among persons aged 25 to 34 in 1998.

When we think about graphs, it is helpful to distinguish between variables whose values have a meaningful numerical scale (height in centimeters, SAT scores) and variables such as sex, occupation, or level of education that just place individuals into categories. Pie charts and bar graphs are most useful for the second kind of variable.

Categorical and quantitative variables

A **categorical variable** places an individual into one of several groups or categories.

A **quantitative variable** takes numerical values for which arithmetic operations such as adding and averaging make sense.

To display the distribution of a categorical variable, use a pie chart or a bar graph.

Although both pie charts and bar graphs can show the distribution (either counts or percents) of a categorical variable such as level of education, bar graphs have wider uses as well.

Example 3. High taxes?

Figure 10.5 compares the level of taxation in eight democratic nations. The height of the bars shows the percent of each nation's gross domestic product (GDP, the total value of all goods and services produced) that is taken in taxes. Americans accustomed to complaining about high taxes may be surprised to see that the United States, at 28.5% of GDP, is near the bottom of the group.

We cannot replace Figure 10.5 by a pie chart, because it compares 8 separate quantities, not the parts of some whole. A pie chart can only compare parts of a whole. Bar graphs can compare quantities that are not parts of a whole.

Beware the pictogram

Bar graphs compare several quantities by comparing the heights of bars that represent the quantities. Our eyes, however, react to the *area* of the bars as well as to their height. When all bars have the same width, the area (width $\times$ height) varies in proportion to the height and our eyes receive the right impression. When you draw a bar graph, make the bars equally wide. Artistically speaking, bar graphs are a bit dull. It is tempting to replace the bars with pictures for greater eye appeal.

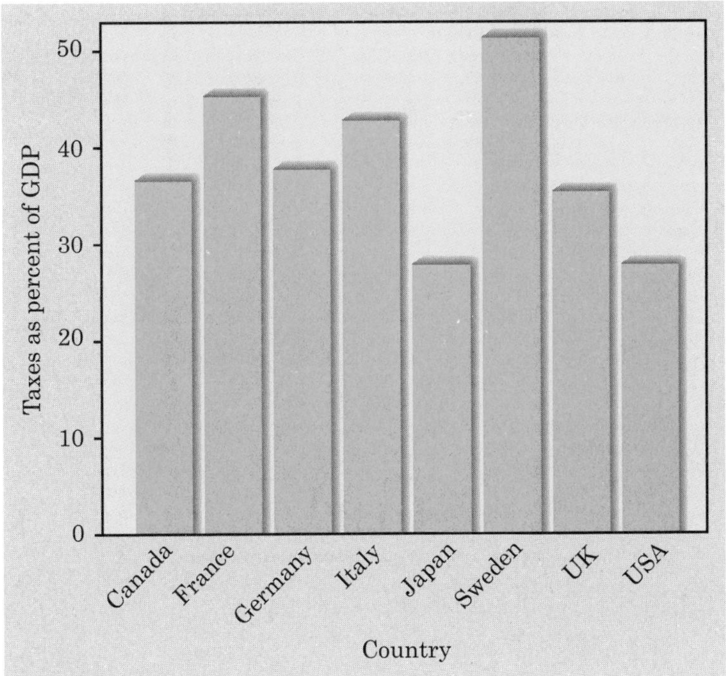

Figure 10.5 Total taxes as a percent of gross domestic product in eight countries in 1996. (From Office of Economic Cooperation and Development, *OECD in Figures, 1999.*)

Example 4. A misleading graph

Figure 10.6 is a pictogram. It is a bar graph in which pictures replace the bars. The graph is aimed at advertisers deciding where to spend their budgets. It shows that *Time* magazine attracts the lion's share of the advertising spending. Or does it? The numbers above the pens show that advertising spending in *Time* is 1.64 times as great as in *Newsweek*. Why does the graph suggest that *Time* is much farther ahead?

To magnify a picture, the artist must increase *both* height and width to avoid distortion. If both the height and width of *Time*'s pen are 1.64 times larger as *Newsweek*'s, the area is 1.64×1.64, or 2.7 times as large. Our eyes, responding to the area of the pens, see *Time* as the big winner.

Change over time: line graphs

Many quantitative variables are measured at intervals over time. We might, for example, measure the height of a growing child or the price of a stock at the end of each month. In these examples, our main interest is change over time. To display change over time, make a *line graph*.

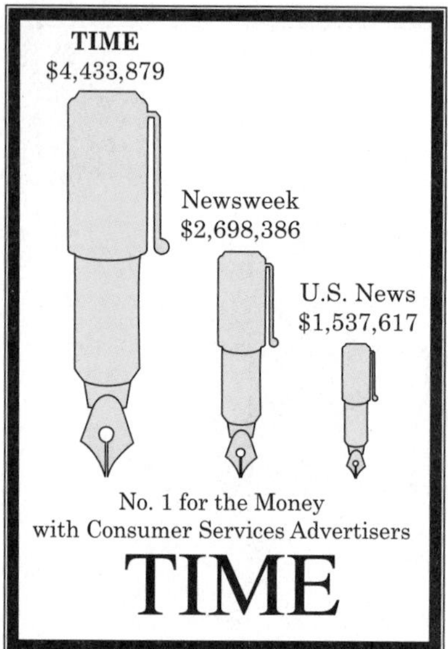

Figure 10.6 A pictogram. This variation of a bar graph is attractive but misleading.

Line graph

A **line graph** of a variable plots each observation against the time at which it was measured. Always put time on the horizontal scale of your plot and the variable you are measuring on the vertical scale. Connect the data points by lines to display the change over time.

Example 5. The price of gasoline

How has the price of gasoline at the pump changed over time? Figure 10.7 is a line graph of the average price of regular unleaded gasoline each month during the 1990s. For example, the January 1990 price was 105.1 cents per gallon. The first point in the figure is above the beginning of 1990 (January) and at 105.1 on the vertical scale.

It would be difficult to see patterns in a long table of monthly prices. Figure 10.7 makes the patterns clearer. What should we look for?

- First, look for an **overall pattern.** For example, a **trend** is a long-term upward or downward movement over time. There was no consistent overall up or down trend in gasoline prices in the decade of the 1990s.

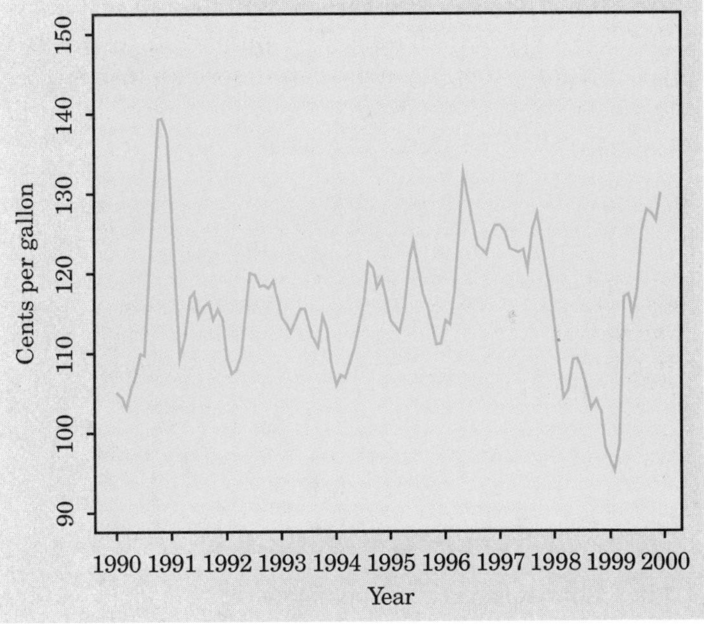

Figure 10.7 A line graph of the average cost of regular unleaded gasoline each month between January 1990 and December 1999. (Data from the Bureau of Labor Statistics.)

- Next, look for striking **deviations** from the overall pattern. The sharp rise in prices in 1990 and the plunging prices in late 1997 and all of 1998 stand out from the overall lack of any trend. The 1990 price spike was due to the invasion of Kuwait by Iraq. The sharp fall in 1997 and 1998 reflects an Asian economic crisis (lower demand for oil products) and an oversupply of oil until the OPEC nations drove the price up in 1999 by reducing their oil production.
- Change over time often has a regular pattern of **seasonal variation** that repeats each year. Gasoline prices are usually highest during the summer driving season and lowest in the winter, when there is less demand for gasoline. You can see this up-in-summer then down-in-the-fall pattern for many years in Figure 10.7.

Because seasonal variation is common, many government statistics are adjusted to remove seasonal effects. For example, the unemployment rate rises every year in January as holiday sales jobs end and outdoor work slows in the north due to winter weather. It would cause confusion (and perhaps political trouble) if the government's official unemployment rate jumped every January. The Bureau of Labor Statistics knows about how much it expects unemployment to rise in January, so it adjusts the published data for this expected change. The published unemployment rate goes up only if actual unemployment rises more than expected. We can then see the underlying changes in the employment situation without being confused by regular seasonal changes.

Seasonal variation, seasonal adjustment

A pattern that repeats itself at known regular intervals of time is called **seasonal variation**. Many series of regular measurements over time are **seasonally adjusted.** That is, the expected seasonal variation is removed before the data are published.

The Vietnam effect

Folklore says that during the Vietnam war many men went to college to avoid being drafted. Statistician Howard Wainer looked for traces of this "Vietnam effect" by plotting data against time. He found that scores on the Armed Forces Qualifying Test (an IQ test given to recruits) dipped sharply during the war, then rebounded. Scores on the SAT exams, taken by students applying to college, also dropped at the beginning of the war. It appears that the men who chose college over the army lowered the average test scores in both places.

Watch those scales!

Because graphs speak so strongly, they can mislead the unwary. The careful reader of a line graph looks closely at the *scales* marked off on the axes.

Example 6. Living together

The number of unmarried couples living together has increased in recent years, to the point that some people say that cohabitation is delaying or even replacing marriage. Figure 10.8 presents two line graphs of the number of unmarried-couple households in the United States. The data once again come from the Current Population Survey. The graph on the left suggests a steady but moderate increase. The right-hand graph says that cohabitation is thundering upward.

The secret is in the scales. You can transform the left-hand graph into the right-hand graph by stretching the vertical scale, squeezing the horizontal scale, and cutting off the vertical scale just above and below the values to be plotted. Now you know how to either exaggerate or play down a trend in a line graph.

Which graph is correct? Both are accurate graphs of the data, but both have scales chosen to create a specific effect. Because there is no one "right" scale for a line graph, correct graphs can give different impressions by their choices of scale. Watch those scales!

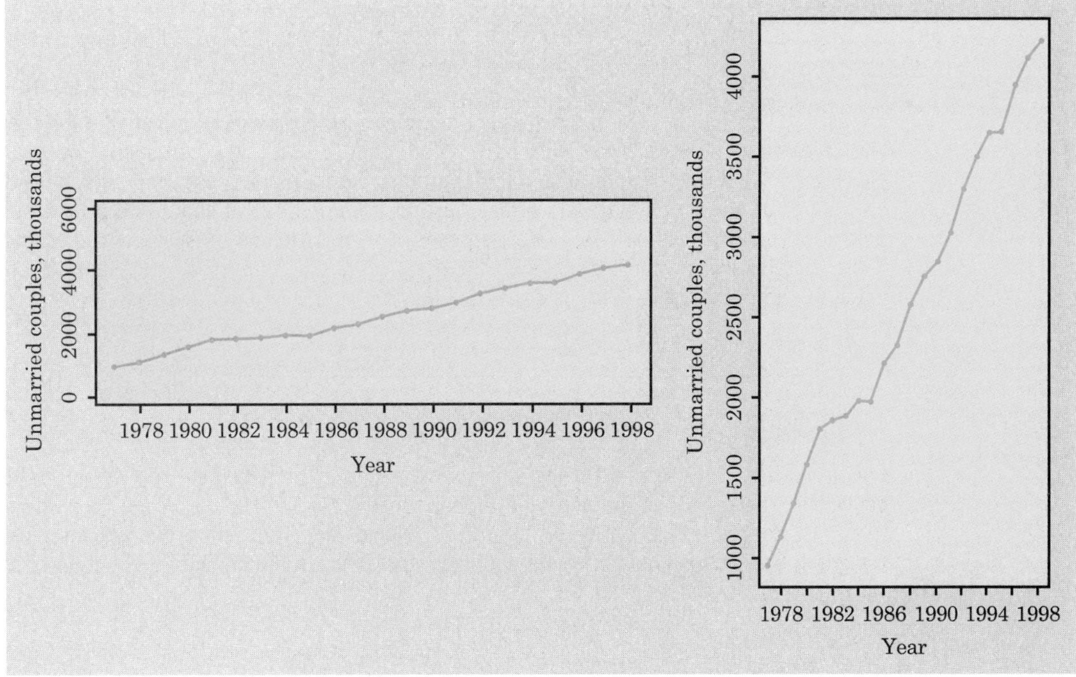

Figure 10.8 The effect of changing the scales in a line graph. Both graphs plot the same data, but the right-hand graph makes the increase appear much more rapid.

Making good graphs

Graphs are the most effective way to communicate using data. A good graph frequently reveals facts about the data that would be difficult or impossible to detect from a table. What is more, the immediate visual impression of a graph is much stronger than the impression made by data in numerical form. Here are some principles for making good graphs:

- Make sure **labels and legends** tell what variables are plotted, what their units are, and what is the source of the data.
- **Make the data stand out.** Be sure that the actual data, not labels, grids, or background art, catch the viewer's attention. You are drawing a graph, not a piece of creative art.
- **Pay attention to what the eye sees.** Avoid pictograms and be careful choosing scales. Avoid fancy "three-dimensional" effects that confuse the eye without adding information. Ask if a simple change in a graph would make the message clearer.

Example 7. The rise in college education

Figure 10.9 shows the rise in the percent of women 25 years old and over who have at least a bachelor's degree. There are only 5 data points, so a line graph should be simple. Figure 10.9 isn't simple. The artist couldn't resist a nice background sketch, and also cluttered the graph with grid lines. We find it harder than it should be to see the data. Grid lines on a graph serve no purpose—if your audience must know the actual numbers, give them a table along with the graph. A good graph uses no more ink than is needed to present the data clearly.

Example 8. High taxes, reconsidered

Figure 10.5 is a respectable bar graph comparing taxes as a percent of GDP in eight countries. The countries are arranged alphabetically. Wouldn't it be clearer to arrange them in order of their tax burdens? Figure 10.10 does this. This simple change improves the graph by making it clearer where each country stands in the group of eight countries.

Statistics in Summary To see what data say, start with graphs. The choice of graph depends on the type of data. Do you have a **categorical variable,** such as level of education or occupation, which puts individuals into categories? Or do you have a **quantitative variable** measured in meaningful numerical units? To display the **distribution** of a categorical variable, use a **pie chart** or a **bar graph.** Pie charts always show the parts of some whole, but bar graphs can compare any set of numbers measured in the same units. To show how a quantitative variable changes over time, use a **line graph** that plots values of the variable (vertical scale) against time (horizontal scale).

Graphs can mislead the eye. Avoid **pictograms** that replace the bars of a bar graph by pictures whose height and width both change. Look at the scales of a line graph to see if they have been stretched or squeezed to create a particular impression. Avoid clutter that makes the data hard to see.

CHAPTER 10 EXERCISES

10.1 Lottery sales. States sell lots of lottery tickets. Table 10.2 shows where the money comes from. Make a bar graph that shows the distribution of lottery sales by type of game. Is it also proper to make a pie chart of these data?

10.2 Consistency? Table 10.2 shows how state lottery sales are divided among different types of games. What is the sum of the amounts spent on the five types of games? Why is this sum not exactly equal to the total given in the table?

Figure 10.9 Chart junk: this graph is so cluttered with unnecessary ink that it is hard to see the data.

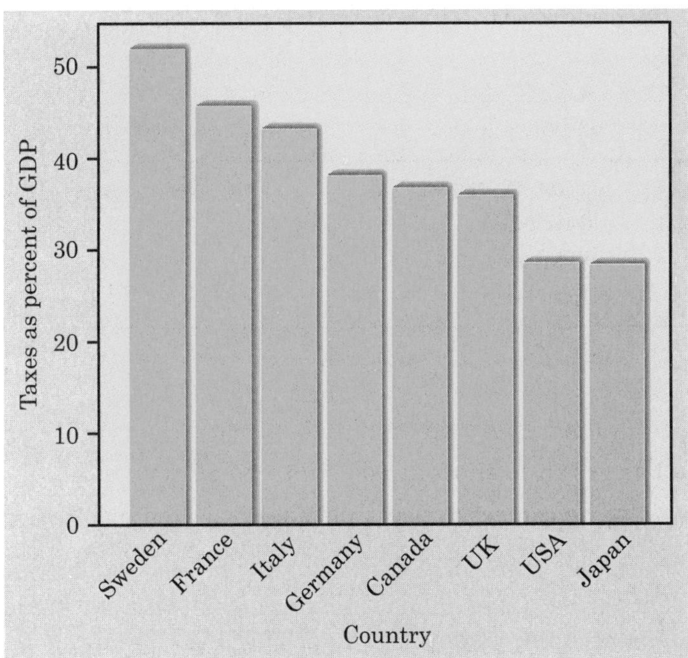

Figure 10.10 Taxes as a percent of gross domestic product in eight countries in 1996. Changing the order of the bars has improved the graph in Figure 10.5.

Table 10.2 State lottery sales by type of game, 1998

	Sales (millions of dollars)
Instant games	13,882
Three-digit games	5,643
Four-digit games	2,232
Lotto	9,854
Other games	3,978
Total	35,588

SOURCE: *Statistical Abstract of the United States, 1999.*

10.3 Marital status. In the *Statistical Abstract* we find these data on the marital status of adult American women as of 1998:

Marital status	Count (thousands)
Never married	21,043
Married	59,255
Widowed	11,027
Divorced	11,078

(a) How many women were not married in 1998?

(b) Make a bar graph to show the distribution of marital status.

(c) Would it also be correct to use a pie chart?

10.4 We pay high interest. Figure 10.11 shows a graph taken from an advertisement for an investment that promises to pay a higher interest rate than bank accounts and other competing investments. Is this graph a correct comparison of the four interest rates? Explain your answer.

10.5 We sell CDs. Who sold the most CDs on the Web just before listeners began to download their favorites rather than buy CDs? Figure 10.12 graphs

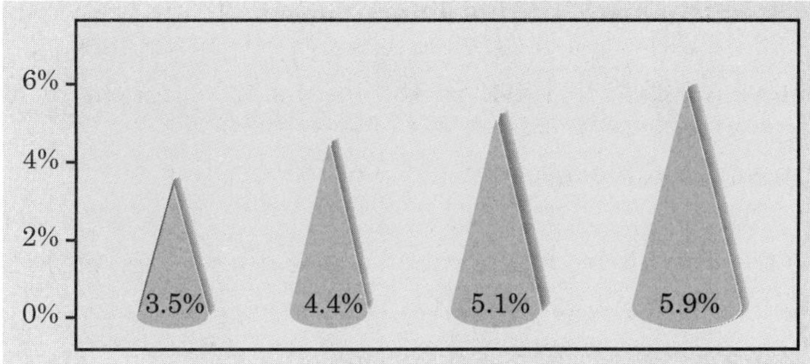

Figure 10.11 Comparing interest rates, for Exercise 10.4.

the shares of the market leaders in 1997 and says what percent of all online sales they had. Does the graph fairly represent the data? Explain your answer.

10.6 Murder weapons. The 1999 *Statistical Abstract of the United States* reports FBI data on murders for 1997. In that year, 53.3% of all murders were committed with handguns, 14.5% with other firearms, 13.0% with knives, 6.3% with a part of the body (usually the hands or feet), and 4.6% with blunt objects. Make a graph to display these data. Do you need an "other methods" category?

10.7 The cost of fresh oranges. Figure 10.13 is a line graph of the average cost of fresh oranges each month from January 1990 to January 2000. These data, from the Bureau of Labor Statistics' monthly survey of retail

Figure 10.12 Market share in percent for the three leading online sellers of compact discs in 1997, for Exercise 10.5. (Data from Jupiter Communications)

prices, are "index numbers" rather than prices in dollars and cents. That is, they give each month's price as a percent of the price in a base period, in this case the years 1982 to 1984. So 150 means "150% of the base price."
(a) The graph shows strong seasonal variation. How is this visible in the graph? Why would you expect the price of fresh oranges to show seasonal variation?

(b) What is the overall trend in orange prices during this period, after we take account of the seasonal variation?

10.8 College freshmen. A survey of college freshmen in 1998 asked what field they planned to study. The results: 10% arts and humanities, 17% business, 11% education, 16% engineering and science, 15% professional, and 8% social science. (From the 1999 *Statistical Abstract.*)
(a) What percent plan to study fields other than those listed?

(b) Make a graph that compares the percents of college freshmen planning to study various fields.

10.9 Exports. Figure 10.14 compares the value of exports (in billions of dollars) in 1997 from the world's leading exporters: Germany, Japan, and the United States.
(a) Explain why this is not an accurate graph.

(b) Make an accurate graph to compare the exports of these nations.

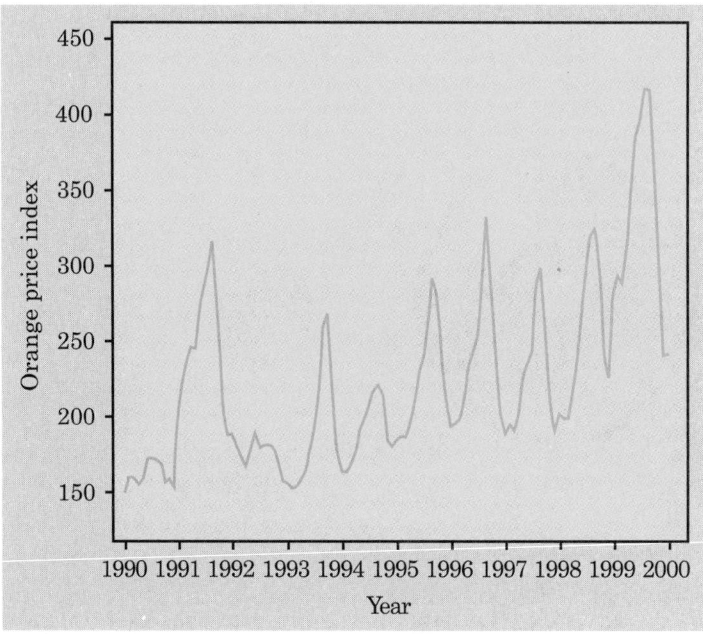

Figure 10.13 The price of fresh oranges, January 1990 to January 2000, for Exercise 10.7.

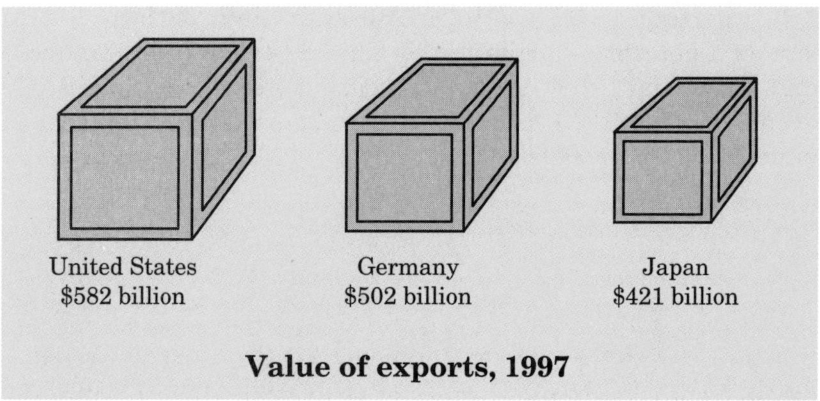

Figure 10.14 Value of goods exported by Germany, Japan, and the United States in 1997, for Exercise 10.9. (Data from the Organization for Economic Cooperation and Development.)

10.10 Civil disorders. The years around 1970 brought unrest to many U.S. cities. Here are government data on the number of civil disturbances in each 3-month period during the years 1968 to 1972.

Period	Count	Period	Count
1968, Jan.–Mar.	6	1970, July–Sept.	20
Apr.–June	46	Oct.–Dec.	6
July–Sept.	25	1971, Jan.–Mar.	12
Oct.–Dec.	3	Apr.–June	21
1969, Jan.–Mar.	5	July–Sept.	5
Apr.–June	27	Oct.–Dec.	1
July–Sept.	19	1972, Jan.–Mar.	3
Oct.–Dec.	6	Apr.–June	8
1970, Jan.–Mar.	26	July–Sept.	5
Apr.–June	24	Oct.–Dec.	5

(a) Make a line graph of these data.

(b) The data show both a longer-term trend and seasonal variation within years. Describe the nature of both patterns. Can you suggest an explanation for the seasonal variation in civil disorders?

10.11 Births to unwed mothers. Here are data on births to unwed mothers as a percent of births in the United States, from the *Statistical Abstract.* These data show a clear increasing trend over time. Make two line graphs of these data: one designed to show only a gradual increase over time, and a second designed to show a shockingly steep increase.

Year	1960	1965	1970	1975	1980	1985	1990	1995
Percent unwed mothers	5.3	7.7	10.7	14.2	18.4	22.0	28.0	32.2

10.12 A bad graph? Figure 10.15 shows a graph that appeared in the Lexington, Kentucky, *Herald-Leader* of October 5, 1975. Discuss the correctness of this graph.

10.13 Trends. Which of these series of data do you expect to show a clear trend? Will the trend be upward or downward? (All data are recorded annually.)

(a) The percent of students entering a university who bring a typewriter with them.

(b) The percent of students entering a university who bring a personal computer with them.

(c) The percent of adult women who do not work outside the home.

10.14 Seasonal variation. You examine the average temperature in Chicago each month for many years. Do you expect a line graph of the data

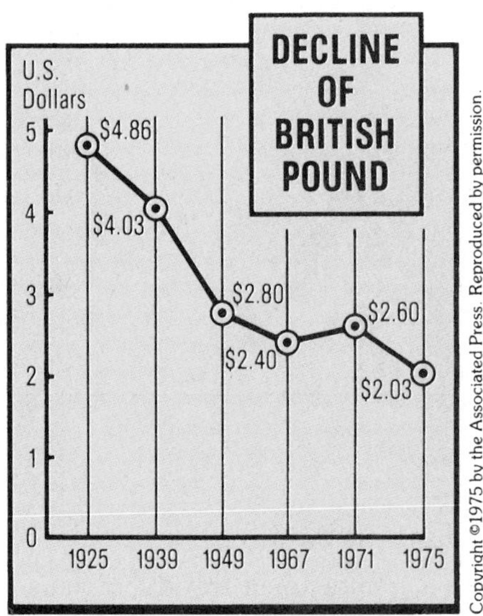

Figure 10.15 A newspaper's graph of the value of the British pound, for Exercise 10.12.

to show seasonal variation? Describe the kind of seasonal variation you expect to see.

10.15 Sales are up. The sales at your new gift shop in December are double the November value. Should you conclude that your shop is growing more popular and will soon make you rich? Explain your answer.

10.16 Counting employed people. A news article says, "More people were working in America in June than in any month since the end of 1990. If that's not what you thought you read in the papers last week, don't worry. It isn't. The report that employment plunged in June, with nonfarm payrolls declining by 117,000, helped to persuade the Federal Reserve to cut interest rates yet again. . . . In reality, however, there were 457,000 more people employed in June than in May." What explains the difference between the fact that employment went up by 457,000 and the official report that employment went down by 117,000?

10.17 The sunspot cycle. Some plots against time show **cycles** of up-and-down movements. Figure 10.16 is a line graph of the average number of sunspots on the sun's visible face for each month in the 20th century. What is the approximate length of the sunspot cycle? That is, how many

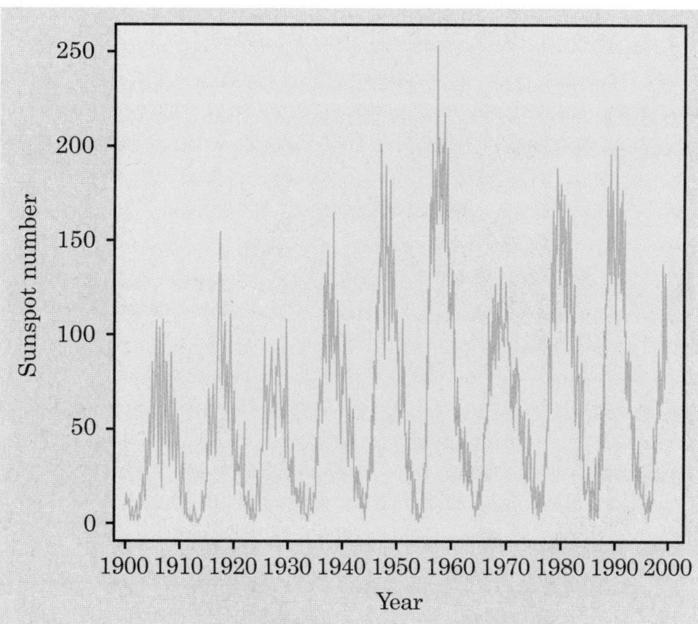

Figure 10.16 The sunspot cycle, for Exercise 10.17. This is a line graph of the monthly sunspot number for the years 1900 to 1999. (Data from the National Oceanic and Atmospheric Administration.)

years are there between the successive valleys in the graph? Is there any overall trend in the number of sunspots?

10.18 Trucks versus cars. Consumers are turning to trucks, SUVs, and minivans in place of passenger cars. Here are data on sales of new cars and trucks in the United States. (The definition of "truck" includes SUVs and minivans.) Plot two line graphs on the same axes to compare the change in car and truck sales over time. Describe the trend that you see.

Year	1981	1983	1985	1987	1989
Cars (1000s)	8536	9182	11,042	10,277	9772
Trucks (1000s)	2260	3129	4682	4912	4941

Year	1991	1993	1995	1997	1999
Cars (1000s)	8175	8518	8636	8273	8697
Trucks (1000s)	4365	5681	6481	7226	8717

10.19 Who sells cars? Figure 10.17 is a pie chart of the percent of passenger car sales in 1997 by various manufacturers. The artist has tried to make the graph attractive by using the wheel of a car for the "pie." Is the graph still a correct display of the data? Explain your answer.

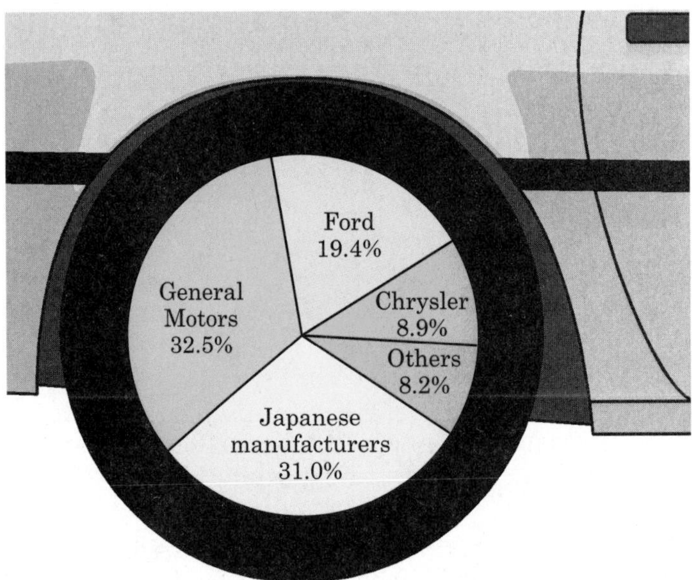

Figure 10.17 Passenger car sales by several manufacturers in 1997, for Exercise 10.19.

10.20 Who sells cars? Make a bar graph of the data in Exercise 10.19. What advantage does your new graph have over the pie chart in Figure 10.17?

10.21 The Border Patrol. Here are the numbers of deportable aliens caught by the U.S. Border Patrol for 1971 through 1997. Display these data in a graph. What are the most important facts that the data show?

Year	1971	1973	1975	1977	1979	1981	1983
Count (1000s)	420	656	767	1042	1076	976	1251

Year	1985	1987	1989	1991	1993	1995	1997
Count (1000s)	1349	1190	954	1198	1327	1395	1536

10.22 Sketch line graphs of a series of observations over time having each of these characteristics. Mark your time axis in years.

(a) A strong downward trend, but no seasonal variation.

(b) Seasonal variation each year, but no clear trend.

(c) A strong downward trend with yearly seasonal variation.

10.23 How much can a dollar buy? The buying power of a dollar changes over time. The Bureau of Labor Statistics measures the cost of a "market basket" of goods and services to compile its Consumer Price Index (CPI). If the CPI is 120, goods and services that cost \$100 in the base period now cost \$120. Here are the average values of the CPI for years between 1970 and 1998. The base period is the years 1982 to 1984.

(a) Make a graph that shows how the CPI has changed over time.

(b) What was the overall trend in prices during this period? Were there any years in which this trend was reversed?

(c) In which years were prices rising fastest? In what period were they rising slowest?

Year	CPI	Year	CPI	Year	CPI
1970	38.8	1980	82.4	1990	130.7
1972	41.8	1982	96.5	1992	140.3
1974	49.3	1984	103.9	1994	148.2
1976	56.9	1986	109.6	1996	156.9
1978	65.2	1988	118.3	1998	163.0

10.24 Bad habits. According to the National Household Survey on Drug Abuse, 20.5% of adolescents aged 12 to 17 years used alcohol in 1997, 9.4% used marijuana, 1.0% used cocaine, and 19.9% used cigarettes. Explain why it is *not* correct to display these data in a pie chart.

10.25 Accidental deaths. In 1997 there were 92,353 deaths from accidents in the United States. Among these were 42,340 deaths from motor vehicle accidents, 11,858 from falls, 10,163 from poisoning, 4051 from drowning, and 3601 from fires.

(a) Find the percent of accidental deaths from each of these causes, rounded to the nearest percent. What percent of accidental deaths were due to other causes?

(b) Make a well-labeled graph of the distribution of causes of accidental deaths.

10.26 Yields of money market funds. Many people invest in "money market funds." These are mutual funds that attempt to maintain a constant price of $1 per share while paying monthly interest. Table 10.3 gives the average annual interest rates (in percent) paid by all taxable money market funds since 1973, the first full year in which such funds were available.

(a) Make a line graph of the interest paid by money market funds for these years.

(b) Interest rates, like many economic variables, show **cycles,** clear but irregular up-and-down movements. In which years did the interest rate cycle reach temporary peaks?

Table 10.3 Average interest paid by money market funds, 1973–1996

Year	Rate	Year	Rate	Year	Rate	Year	Rate
1973	7.60	1979	10.92	1985	7.77	1991	5.70
1974	10.79	1980	12.88	1986	6.30	1992	3.31
1975	6.39	1981	17.16	1987	6.17	1993	2.62
1976	5.11	1982	12.55	1988	7.09	1994	3.65
1977	4.92	1983	8.69	1989	8.85	1995	5.37
1978	7.25	1984	10.21	1990	7.81	1996	4.80

SOURCE: Albert J. Fredman, "A closer look at money market funds," *American Association of Individual Investors Journal,* February 1997, pp. 22–27.

Table 10.4 Women's winning times (minutes) in the Boston Marathon

Year	Time	Year	Time	Year	Time
1972	190	1982	150	1992	144
1973	186	1983	143	1993	145
1974	167	1984	149	1994	142
1975	162	1985	154	1995	145
1976	167	1986	145	1996	147
1977	168	1987	146	1997	146
1978	165	1988	145	1998	143
1979	155	1989	144	1999	143
1980	154	1990	145	2000	146
1981	147	1991	144		

(c) A plot against time may show a consistent trend underneath cycles. When did interest rates reach their overall peak during these years? Has there been a general trend downward since that year?

10.27 The Boston Marathon. Women were allowed to enter the Boston Marathon in 1972. The times (in minutes, rounded to the nearest minute) for the winning woman from 1972 to 2000 appear in Table 10.4.
(a) Make a graph of the winning times.

(b) Give a brief description of the pattern of Boston Marathon winning times over these years. Have times stopped improving in recent years?

Chapter 11
Displaying Distributions with Graphs

Where does my school stand?

Tonya is a student at Grand Valley State University in her home state of Michigan. Grand Valley charges in-state students $4108 in tuition and fees. Tonya wonders how Grand Valley's tuition compares with the amount charged by other Michigan colleges and universities.

There are 81 colleges and universities in Michigan. Their tuition and fees for the 1999–2000 school year run from $1260 at Kalamazoo Valley Community College to $19,258 at Kalamazoo College. It isn't easy to compare Grand Valley State with a list of 80 other colleges. Let's make a graph. The histogram in Figure 11.1 shows that many colleges charge less than $2000. The distribution extends out to the right. At the upper extreme, two colleges charge between $18,000 and $20,000. Grand Valley State is one of 5 colleges in the $4000 to $6000 group.

Figure 11.2 is another, less familiar, graph of the same data. We have rounded the tuition charges to the nearest hundred dollars. Grand Valley State's $4108, for example, becomes just 41. Put the thousands digit or digits to the left of the line, and hang the hundreds one by one on these stems. Grand Valley State is the 1 on the 4 stem. This is a stemplot. The stemplot shows more detail than the histogram. Tonya can easily see, for example, that Grand Valley State's tuition and fees are 39th out of 81 colleges when we count up from the lowest charge.

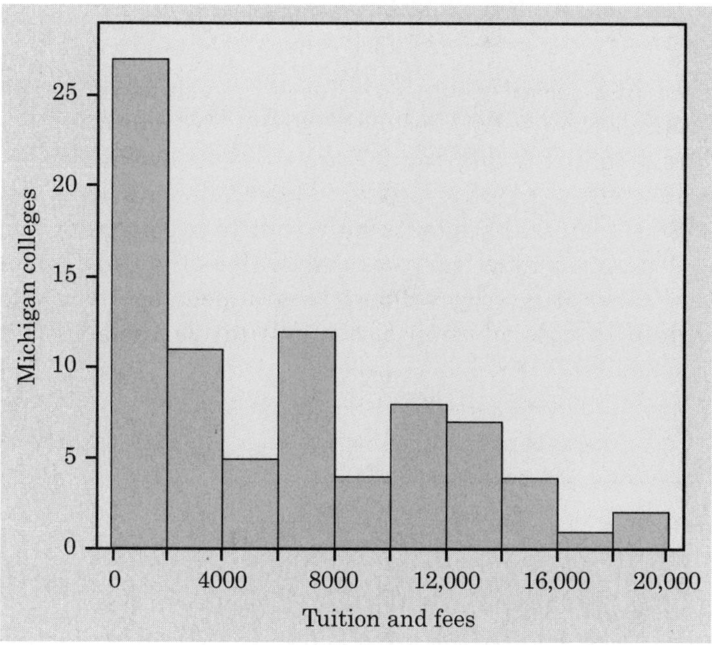

Figure 11.1 Histogram of the tuition and fees charged by 81 Michigan colleges and universities in the 1999–2000 academic year.

Stem	Leaves
1	355555556666667777788889999
2	8
3	155666899
4	0125
5	01
6	133466
7	014567
8	169
9	7
10	139
11	13678
12	39
13	24457
14	39
15	16
16	0
17	
18	2
19	3

Figure 11.2 Stemplot of the Michigan tuition and fee data.

Histograms

Categorical variables just record group membership, such as the marital status of a woman or the race of a college student. We can use a pie chart or bar graph to display the distribution of categorical variables because they have relatively few values. What about quantitative variables such as the SAT scores of students admitted to a college or the income of families? These variables take so many values that a graph of the distribution is clearer if nearby values are grouped together. The most common graph of the distribution of a quantitative variable is a *histogram.*

Example 1. How to make a histogram

Table 11.1 presents the percent of residents aged 65 years and over in each of the 50 states. To make a histogram of this distribution, proceed as follows.

Step 1: Divide the range of the data into classes of equal width. The data in Table 11.1 range from 5.5 to 18.3, so we choose as our classes

$$5.0 \leq \text{percent over } 65 < 6.0$$
$$6.0 \leq \text{percent over } 65 < 7.0$$
$$\vdots$$
$$18.0 \leq \text{percent over } 65 < 19.0$$

Be sure to specify the classes precisely so that each individual falls into exactly one class. A state with 5.9% of its residents aged 65 or older would fall into the first class, but 6.0% falls into the second.

Step 2: Count the number of individuals in each class. Here are the counts:

Class	Count	Class	Count	Class	Count
5.0 to 5.9	1	10.0 to 10.9	2	15.0 to 15.9	4
6.0 to 6.9	0	11.0 to 11.9	9	16.0 to 16.9	0
7.0 to 7.9	0	12.0 to 12.9	11	17.0 to 17.9	0
8.0 to 8.9	1	13.0 to 13.9	14	18.0 to 18.9	1
9.0 to 9.9	1	14.0 to 14.9	6		

Table 11.1 Percent of residents aged 65 and over in the states (1998)

State	Percent	State	Percent	State	Percent
Alabama	13.1	Louisiana	11.5	Ohio	13.4
Alaska	5.5	Maine	14.1	Oklahoma	13.4
Arizona	13.2	Maryland	11.5	Oregon	13.2
Arkansas	14.3	Massachusetts	14.0	Pennsylvania	15.9
California	11.1	Michigan	12.5	Rhode Island	15.6
Colorado	10.1	Minnesota	12.3	South Carolina	12.2
Connecticut	14.3	Mississippi	12.2	South Dakota	14.3
Delaware	13.0	Missouri	13.7	Tennessee	12.5
Florida	18.3	Montana	13.3	Texas	10.1
Georgia	9.9	Nebraska	13.8	Utah	8.8
Hawaii	13.3	Nevada	11.5	Vermont	12.3
Idaho	11.3	New Hampshire	12.0	Virginia	11.3
Illinois	12.4	New Jersey	13.6	Washington	11.5
Indiana	12.5	New Mexico	11.4	West Virginia	15.2
Iowa	15.1	New York	13.3	Wisconsin	13.2
Kansas	13.5	North Carolina	12.5	Wyoming	11.5
Kentucky	12.5	North Dakota	14.4		

Step 3: Draw the histogram. Mark on the horizontal axis the scale for the variable whose distribution you are displaying. That's "percent of residents aged 65 and over" in this example. The scale runs from 4 to 20 because that range spans the classes we chose. The vertical axis contains the scale of counts. Each bar represents a class. The base of the bar covers the class, and the bar height is the class count. There is no horizontal space between the bars unless a class is empty, so that its bar has height zero. Figure 11.3 is our histogram.

Just as with bar graphs, our eyes respond to the area of the bars in a histogram. Be sure that the classes for a histogram have equal widths. There is no one right choice of the classes. Too few classes will give a "skyscraper" histogram, with all values in a few classes with tall bars. Too many classes will produce a "pancake" graph, with most classes having one or no observations. Neither choice will give a good picture of the shape of the distribution. You must use your judgment in choosing classes to display the shape. Statistics software will choose the classes for you. The computer's choice is usually a good one, but you can change it if you want.

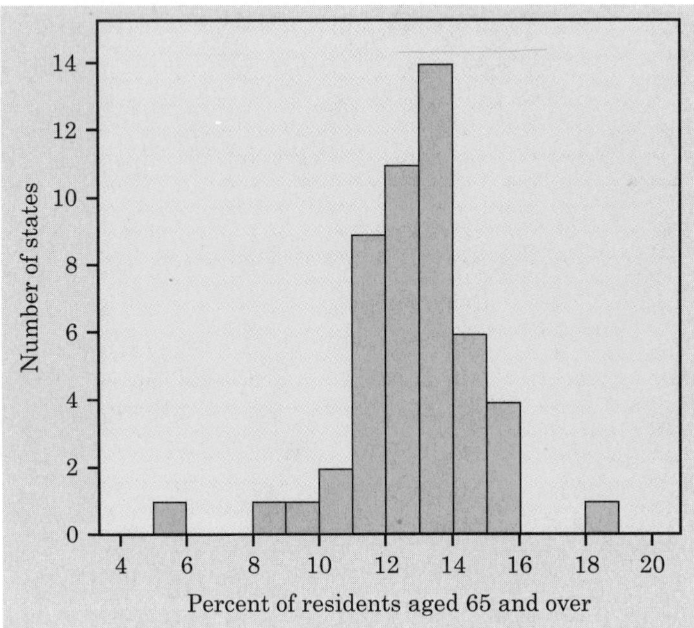

Figure 11.3 Histogram of the percent of residents aged 65 and older in the 50 states. Note the two outliers.

Interpreting histograms

Making a statistical graph is not an end in itself. The purpose of the graph is to help us understand the data. After you (or your computer) make a graph, always ask, "What do I see?" Here is a general strategy for looking at graphs:

Pattern and deviations

In any graph of data, look for an **overall pattern** and also for striking **deviations** from that pattern.

We have already applied this strategy to line graphs. Trend and seasonal variation are common overall patterns in a line graph. The upward spike in gasoline prices in Figure 10.7 (page 181) caused by the Iraqi invasion of Kuwait in 1990 is an example of a deviation from the general pattern. In the case of the histogram of Figure 11.3, it is easiest to begin with deviations from the overall pattern of the histogram. Two states stand out. You can find them in the table once the histogram has called attention to them. Florida has 18.3% of its residents over age 65, and Alaska has only 5.5%. These states are clear *outliers.*

Outliers

An **outlier** in any graph of data is an individual observation that falls outside the overall pattern of the graph.

Is Utah, with 8.8% of its population over 65, an outlier? Whether an observation is an outlier is to some extent a matter of judgment. Utah is the smallest of the main body of observations, not separated from the general pattern as Florida and Alaska are. I would not call it an outlier. Once you have spotted outliers, look for an explanation. Many outliers are due to mistakes, such as typing 4.0 as 40. Other outliers point to the special nature of some observations. Explaining outliers usually requires some background information. It is not surprising that Florida, with its many retired people, has many residents over 65 and that Alaska, the northern frontier, has few.

To see the *overall pattern* of a histogram, ignore any outliers. Here is a simple way to organize your thinking.

Overall pattern of a distribution

To describe the overall pattern of a distribution:

- Give the **center** and the **spread.**
- See if the distribution has a simple **shape** that you can describe in a few words.

We will learn how to describe center and spread numerically in Chapter 12. For now, we can describe the center of a distribution by its *midpoint,* the value with roughly half the observations taking smaller values and half taking larger values. We can describe the spread of a distribution by giving the *smallest and largest values,* ignoring any outliers.

Example 2. Describing distributions

Look again at the histogram in Figure 11.3. **Shape:** The distribution has a *single peak.* It is roughly *symmetric*—that is, the pattern is similar on both sides of the peak. **Center:** The midpoint of the distribution is close to the single peak, at about 13%. **Spread:** The spread is about 9% to 16% if we ignore the two outliers.

The distribution of tuition and fees at Michigan colleges, shown in Figure 11.1, has a quite different **shape.** There is a strong *peak* in the lowest cost class. Most colleges charge less than $8000, but there is a long right tail extending up to almost $20,000. We call a distribution with a long tail on one side *skewed.* The **center** is roughly $4500 (half the colleges charge less than this). The **spread** is large, from $1260 to more than $19,000. There are no outliers—the colleges with the highest tuition just continue the long right tail that is part of the overall pattern.

When you describe a distribution, concentrate on the main features. Look for major peaks, not for minor ups and downs in the bars of the histogram like those in Figure 11.1. Look for clear outliers, not just for the smallest and largest observations. Look for rough *symmetry* or clear *skewness.*

Symmetric and skewed distributions

A distribution is **symmetric** if the right and left sides of the histogram are approximately mirror images of each other.

A distribution is **skewed to the right** if the right side of the histogram (containing the half of the observations with larger values) extends much farther out than the left side. It is **skewed to the left** if the left side of the histogram extends much farther out than the right side.

In mathematics, symmetry means that the two sides of a figure like a histogram are exact mirror images of each other. Data are almost never exactly symmetric, so we are willing to call histograms like that in Figure 11.3 roughly symmetric as an overall description. The tuition distribution in Figure 11.1, on the other hand, is clearly skewed to the right. Here are more examples.

Example 3. Sampling again

The values a statistic takes in many random samples from the same population form a distribution with a regular pattern. The histogram in Figure 11.4 displays a distribution that we met in Chapter 3. Take a simple random sample of 1523 adults. Ask each whether they bought a lottery ticket in the last 12 months. The proportion who say "Yes" is the sample proportion $\hat{p}$. Do this 1000 times and collect the 1000 sample proportions $\hat{p}$ from the 1000 samples. Figure 11.4 shows the distribution of 1000 sample proportions when the truth about the population is that 60% have bought lottery tickets.

This distribution is quite symmetric about a single peak in the center. The center is at 0.60, reflecting the lack of bias of the statistic. The spread from the smallest to the largest of the 1000 values is from 0.556 to 0.643.

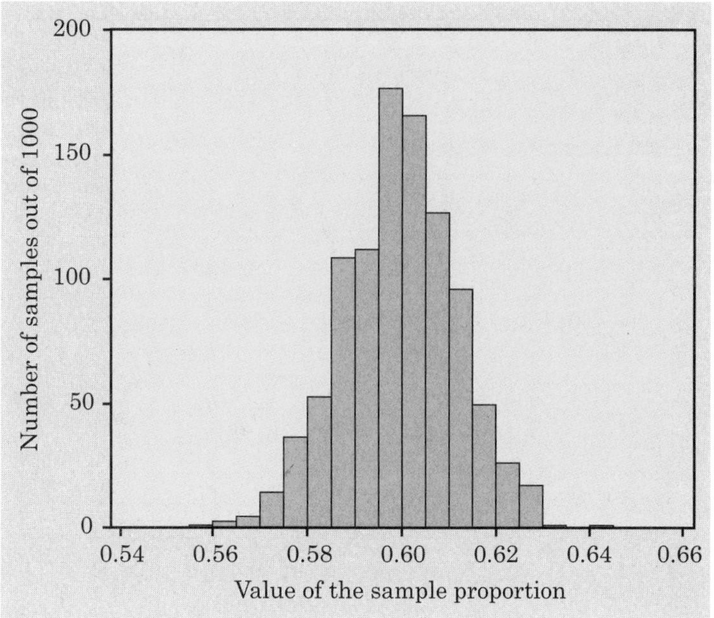

Figure 11.4 Histogram of the sample proportion $\hat{p}$ for 1000 simple random samples from the same population. This is a symmetric distribution.

Example 4. Shakespeare's words

Figure 11.5 shows the distribution of lengths of words used in Shakespeare's plays. This distribution has a single peak and is somewhat skewed to the right. There are many short words (3 and 4 letters) and few very long words (10, 11, or 12 letters), so that the right tail of the histogram extends out farther than the left tail. The center of the distribution is about 4. That is, about half of Shakespeare's words have 4 or fewer letters. The spread is from 1 letter to 12 letters.

Notice that the vertical scale in Figure 11.5 is not the *count* of words but the *percent* of all of Shakespeare's words that have each length. A histogram of percents rather than counts is convenient when the counts are very large or when we want to compare several distributions. Different kinds of writing have different distributions of word lengths, but all are right-skewed because short words are common and very long words are rare.

The overall shape of a distribution is important information about a variable. Some types of data regularly produce distributions that are symmetric or skewed. For example, the sizes of living things of the same species (like lengths of crickets) tend to be symmetric. Data on incomes (whether of individuals, companies, or nations) are usually strongly skewed to the right. There are many moderate incomes, some large incomes, and a few very large incomes. Do remember that many distributions have shapes that are neither symmetric nor skewed. Some data show other patterns. Scores on an exam, for

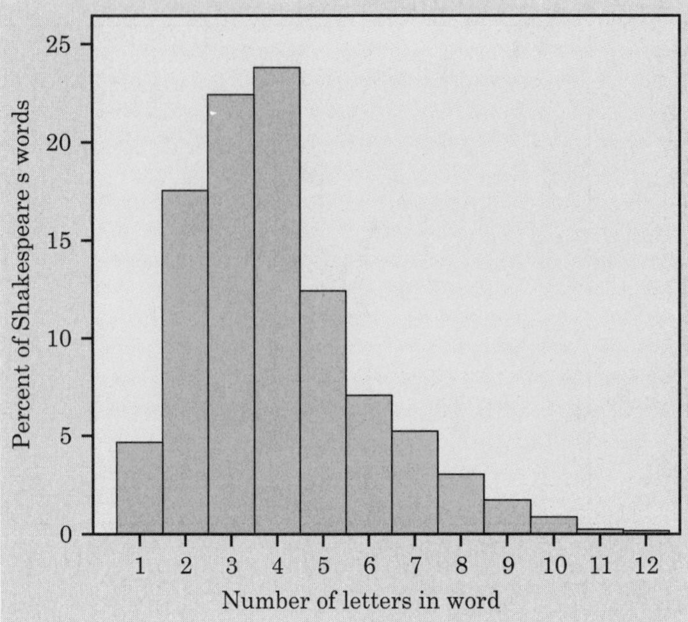

Figure 11.5 The distribution of word lengths used by Shakespeare in his plays. This distribution is skewed to the right.

example, may have a cluster near the top of the scale if many students did well. Or they may show two distinct peaks if a tough problem divided the class into those who did and didn't solve it. Use your eyes and describe what you see.

Stemplots

Histograms are not the only graphical display of distributions. For small data sets, a *stemplot* is quicker to make and presents more detailed information.

Stemplot

To make a **stemplot:**

1. Separate each observation into a **stem** consisting of all but the final (rightmost) digit and a **leaf,** the final digit. Stems may have as many digits as needed, but each leaf contains only a single digit.

2. Write the stems in a vertical column with the smallest at the top, and draw a vertical line at the right of this column.

3. Write each leaf in the row to the right of its stem, in increasing order out from the stem.

Example 5. Stemplot of the "65 and over" data

For the "65 and over" percents in Table 11.1, the whole-number part of the observation is the stem and the final digit (tenths) is the leaf. The Alabama entry, 13.1, has stem 13 and leaf 1. Stems can have as many digits as needed, but each leaf must consist of only a single digit. Figure 11.6 shows the steps in making a stemplot for the data in Table 11.1. First, draw the stems. Then go through the table adding a leaf for each observation. Finally, arrange the leaves in increasing order out from each stem.

A stemplot looks like a histogram turned on end. The stemplot in Figure 11.6 is just like the histogram in Figure 11.3 because the classes chosen for the histogram are the same as the stems in the stemplot. The stemplot in Figure 11.2, on the other hand, has almost twice as many classes as the histogram of the same data in Figure 11.1. We interpret stemplots like histograms, looking for the overall pattern and for any outliers.

You can choose the classes in a histogram. The classes (the stems) of a stemplot are given to you. You can get more flexibility by **rounding** the data so that the final digit after rounding is suitable as a leaf. Do this when the data have too many digits. For example, the tuition charges of Michigan colleges look like

$15,136 $1,940 $12,339 $6,960 . . .

A stemplot would have too many stems if we took all but the final digit as the stem and the final digit as the leaf. To make the stemplot in Figure 11.2, we round these data to the nearest hundred dollars:

151   19   123   70 . . .

These values appear in Figure 11.2 on the stems 15, 1, 12, and 7.

The chief advantage of a stemplot is that it displays the actual values of the observations. We can see from the stemplot in Figure 11.2, but not from the histogram in Figure 11.1, that Michigan's most expensive college charges $19,300 (rounded to the nearest hundred dollars). Stemplots are also faster to draw than histograms. A stemplot requires that we use the first digit or digits as stems. This amounts to an automatic choice of classes and can give a poor picture of the distribution. Stemplots do not work well with large data sets, because the stems then have too many leaves.

What the eye really sees

We make the bars in bar graphs and histograms equal in width because the eye responds to their area. That's roughly true. Careful study by statistician William Cleveland shows that our eyes "see" the size of a bar in proportion to the 0.7 power of its area. Suppose, for example, that one figure in a pictogram is both twice as high and twice as wide as another. The area of the bigger figure is 4 times that of the smaller. But we see the size of the bigger figure as 2.6 times the smaller, because 2.6 is 4 to the 0.7 power.

Stem	Stem	Leaves	Stem	Leaves
5	5	5	5	5
6	6		6	
7	7		7	
8	8	8	8	8
9	9	9	9	9
10	10	11	10	11
11	11	135554355	11	133455555
12	12	45553205253	12	02233455555
13	13	12035738634422	13	01222333445678
14	14	331043	14	013334
15	15	1962	15	1269
16	16		16	
17	17		17	
18	18	3	18	3
Step 1: Stems	Step 2: Add leaves		Step 3: Order leaves	

Figure 11.6 Making a stemplot of the data in Table 11.1. Whole percents form the stems and tenths of a percent form the leaves.

Statistcs in Summary The **distribution** of a variable tells us what values the variable takes and how often it takes each value. To display the distribution of a quantitative variable, use a **histogram** or a **stemplot.** We usually favor stemplots when we have a small number of observations and histograms for large amounts of data.

When you look at a graph, look for an **overall pattern** and for **deviations** from that pattern, such as **outliers.** We can describe the overall pattern of a histogram or stemplot by giving its **shape, center, and spread.** Some distributions have simple shapes such as **symmetric** or **skewed,** but others are too irregular to describe by a simple shape.

CHAPTER 11 EXERCISES

11.1 Lightning strikes. Figure 11.7 comes from a study of lightning storms in Colorado. It shows the distribution of the hour of the day during which the first lightning flash for that day occurred. Describe the shape, center, and spread of this distribution. Are there any outliers?

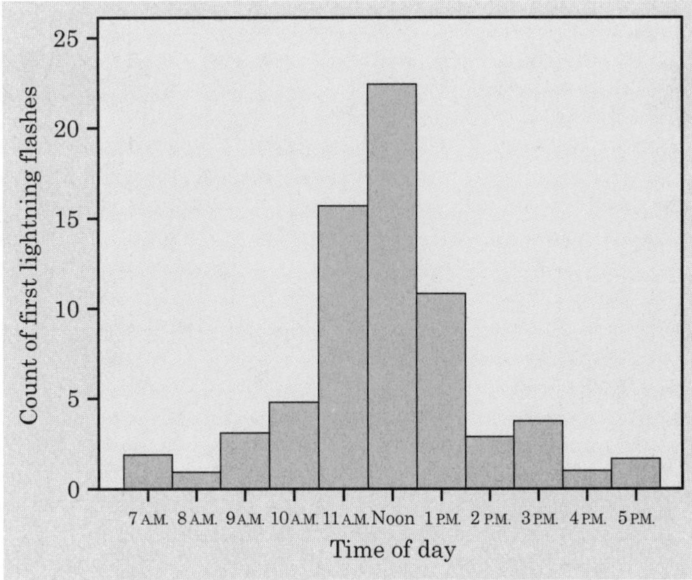

Figure 11.7 Histogram of the time of day at which the day's first lightning flash occurred (from a study in Colorado), for Exercise 11.1.

11.2 Where do the young live? Figure 11.8 is a stemplot of the percent of residents aged 25 to 34 in each of the 50 states. As in Figure 11.6 for older residents, the stems are whole percents and the leaves are tenths of a percent.

(a) The mountain states Montana and Wyoming have the smallest percent of young adults, perhaps because they lack job opportunities. What are the percents for these two states?

(b) Ignoring Montana and Wyoming, describe the shape, center, and spread of this distribution.

(c) Is the distribution for young adults more or less spread out than the distribution in Figure 11.6 for older adults?

11.3 Minority students in engineering. Figure 11.9 is a histogram of the number of minority students (black, Hispanic, Native American) who earned doctorate degrees in engineering from each of 115 universities in the

Stem	Leaves
10	9
11	0
12	1344677889
13	0012455566789999
14	11222344445789
15	24478999

Figure 11.8 Stemplot of the percent of each state's residents who are 25 to 34 years old, for Exercise 11.2.

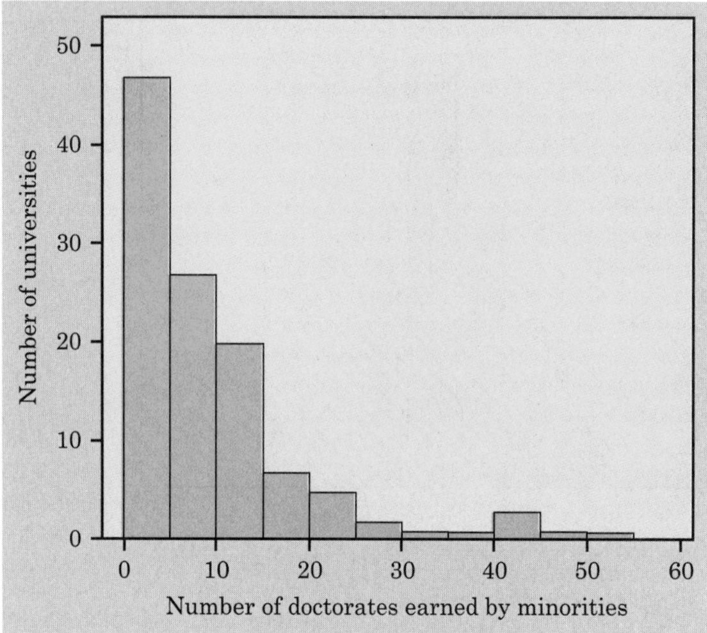

Figure 11.9 The distribution of number of engineering doctorates earned by minority students at 115 universities, 1992 to 1996, for Exercise 11.3.

years 1992 through 1996. Briefly describe the shape, center, and spread of this distribution.

11.4 Returns on common stocks. The total return on a stock is the change in its market price plus any dividend payments made. Total return is usually expressed as a percent of the beginning price. Figure 11.10 is a histogram of the distribution of total returns for all 1528 common stocks listed on the New York Stock Exchange in one year.

(a) Describe the overall shape of the distribution of total returns.

(b) What is the approximate center of this distribution? Approximately what were the smallest and largest total returns? (This describes the spread of the distribution.)

(c) A return less than zero means that owners of the stock lost money. About what percent of all stocks lost money?

11.5 Histogram or stemplot? Explain why we prefer a histogram to a stemplot for describing the returns on 1528 common stocks.

11.6 Automobile fuel economy. Government regulations require automakers to give the city and highway gas mileages for each model of car. Table 11.2 gives the highway mileages (miles per gallon) for 32 model year

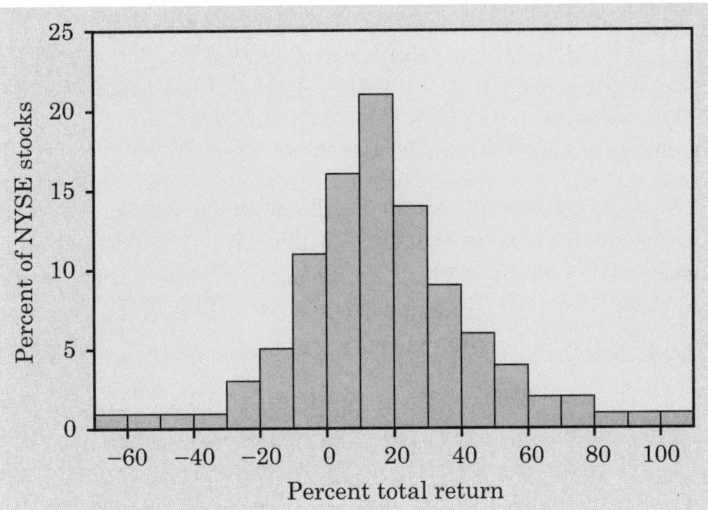

Figure 11.10 The distribution of total returns for all New York Stock Exchange common stocks in one year, for Exercise 11.4.

Table 11.2 Highway gas mileage for model year 2000 midsize cars

Model	MPG	Model	MPG
Acura 3.5RL	24	Lexus GS300	24
Audi A6 Quattro	24	Lexus LS400	25
BMW 740I Sport M	21	Lincoln-Mercury LS	25
Buick Regal	29	Lincoln-Mercury Sable	28
Cadillac Catera	24	Mazda 626	28
Cadillac Eldorado	28	Mercedes-Benz E320	30
Chevrolet Lumina	30	Mercedes-Benz E430	24
Chrysler Cirrus	28	Mitsubishi Diamante	25
Dodge Stratus	28	Mitsubishi Galant	28
Honda Accord	29	Nissan Maxima	28
Hyundai Sonata	28	Oldsmobile Intrigue	28
Infiniti I30	28	Saab 9-3	26
Infiniti Q45	23	Saturn LS	32
Jaguar Vanden Plas	24	Toyota Camry	30
Jaguar S/C	21	Volkswagen Passat	29
Jaguar X200	26	Volvo S70	27

2000 midsize passenger cars. Make a stemplot of the highway mileages of these cars. What can you say about the overall shape of the distribution? Where is the center (the value such that half the cars have better mileage and half have worse mileage)? Two of these cars are subject to the "gas guzzler tax" because of their low mileage. Which two?

11.7 The obesity epidemic. Medical authorities describe the spread of obesity in the United States as an epidemic. Table 11.3 gives the percent of adults who are obese in each of the 45 states that participated in a study of the problem. (NA marks the states for which data are not available.) Display the distribution in a graph and briefly describe its shape, center, and spread.

11.8 Yankee money. Table 11.4 gives the salaries of the players on the New York Yankees baseball team as of the opening day of the 1999 season. Make a histogram of these data. Is the distribution roughly symmetric, skewed to the right, or skewed to the left?

Table 11.3 *Percent of adult population who are obese (1998)*

State	Percent	State	Percent	State	Percent
Alabama	20.7	Louisiana	21.3	Ohio	19.5
Alaska	20.7	Maine	17.0	Oklahoma	18.7
Arizona	12.7	Maryland	19.8	Oregon	17.8
Arkansas	NA	Massachusetts	13.8	Pennsylvania	19.0
California	16.8	Michigan	20.7	Rhode Island	NA
Colorado	14.0	Minnesota	15.7	South Carolina	20.2
Connecticut	14.7	Mississippi	22.0	South Dakota	15.4
Delaware	16.6	Missouri	19.8	Tennessee	18.5
Florida	17.4	Montana	14.7	Texas	19.9
Georgia	18.7	Nebraska	17.5	Utah	15.3
Hawaii	15.3	Nevada	NA	Vermont	14.4
Idaho	16.0	New Hampshire	14.7	Virginia	18.2
Illinois	17.9	New Jersey	15.2	Washington	17.6
Indiana	19.5	New Mexico	14.7	West Virginia	22.9
Iowa	19.3	New York	15.9	Wisconsin	17.9
Kansas	NA	North Carolina	19.0	Wyoming	NA
Kentucky	19.9	North Dakota	18.7		

Table 11.4 Salaries of the New York Yankees, 1999

Player	Salary	Player	Salary
Bernie Williams	$9,857,143	Chad Curtis	$2,000,000
Roger Clemens	$8,250,000	Orlando Hernandez	$1,850,000
David Cone	$8,000,000	Jeff Nelson	$1,816,667
Paul O'Neill	$6,250,000	Luis Sojo	$800,000
Chuck Knoblauch	$6,000,000	Ramiro Mendoza	$375,000
Andy Pettitte	$5,950,000	Jason Grimsley	$370,000
Scott Brosius	$5,250,000	Jorge Posada	$350,000
Derek Jeter	$5,000,000	Dan Naulty	$300,000
Chili Davis	$4,333,333	Shane Spencer	$204,050
Tino Martinez	$4,300,000	Mike Figga	$203,650
Mariano Rivera	$4,250,000	Ricky Ledee	$202,850
Joe Girardi	$3,400,000	Clayton Bellinger	$200,000
Hideki Irabu	$3,125,000	Darrell Einertson	$200,000
Mike Stanton	$2,016,667	Mike Jerzembeck	$200,000

11.9 The statistics of writing style. Numerical data can distinguish different types of writing, and sometimes even individual authors. Here are data collected by students on the percentages of words of 1 to 15 letters used in articles in *Popular Science* magazine:

Length	1	2	3	4	5	6	7	8	9	10	11	12	13	14	15
Percent	3.6	14.8	18.7	16.0	12.5	8.2	8.1	5.9	4.4	3.6	2.1	0.9	0.6	0.4	0.2

(a) Make a histogram of this distribution. Describe its shape, center, and spread.

(b) How does the distribution of lengths of words used in *Popular Science* compare with the similar distribution in Figure 11.5 for Shakespeare's plays? Look in particular at short words (2, 3, and 4 letters) and very long words (more than 10 letters).

11.10 Skewed left. Sketch a histogram for a distribution that is skewed to the left. Suppose that you and your friends emptied your pockets of coins and recorded the year marked on each coin. The distribution of dates would be skewed to the left. Explain why.

11.11 What's my shape? Do you expect the distribution of the total player payroll for the 30 teams in major-league baseball to be roughly symmetric, clearly skewed to the right, or clearly skewed to the left? Why?

11.12 Hispanics in the eastern states. Here are the percents of the population that is of Hispanic origin in each state east of the Mississippi River in 1997:

Alabama	0.1	Maryland	3.5	Pennsylvania	2.5
Connecticut	7.9	Massachusetts	5.9	Rhode Island	6.2
Delaware	3.3	Michigan	2.6	South Carolina	1.2
Florida	14.4	Mississippi	0.8	Tennessee	1.1
Georgia	2.8	New Hampshire	1.4	Vermont	0.8
Illinois	9.9	New Jersey	11.9	Virginia	3.5
Indiana	2.3	New York	14.2	West Virginia	0.6
Kentucky	0.8	North Carolina	2.0	Wisconsin	2.5
Maine	0.7	Ohio	1.5		

Make a stemplot of these data. Describe the overall pattern of the distribution. Are there any outliers?

11.13 How many calories does a hot dog have? *Consumer Reports* magazine presented the following data on the number of calories in a hot dog for each of 17 brands of meat hot dogs:

173	191	182	190	172	147	146	139	175
136	179	153	107	195	135	140	138	

Make a stemplot of the distribution of calories in meat hot dogs and briefly describe the shape of the distribution. Most brands of meat hot dogs contain a mixture of beef and pork, with up to 15% poultry allowed by government regulations. The only brand with a different makeup was *Eat Slim Veal Hot Dogs*. Which point on your stemplot do you think represents this brand?

11.14 The changing age distribution of the United States. The distribution of the ages of a nation's population has a strong influence on economic and social conditions. Table 11.5 shows the age distribution of U.S. residents in 1950 and 2075, in millions of persons. The 1950 data come from that year's census. The 2075 data are projections made by the Census Bureau.

Table 11.5 Age distribution in the United States, 1950 and 2075 (in millions of persons)

Age group	1950	2075
Under 10 years	29.3	34.9
10 to 19 years	21.8	35.7
20 to 29 years	24.0	36.8
30 to 39 years	22.8	38.1
40 to 49 years	19.3	37.8
50 to 59 years	15.5	37.5
60 to 69 years	11.0	34.5
70 to 79 years	5.5	27.2
80 to 89 years	1.6	18.8
90 to 99 years	0.1	7.7
100 to 109 years	—	1.7
Total	151.1	310.6

(a) Because the total population in 2075 is much larger than the 1950 population, comparing percents in each age group is clearer than comparing counts. Make a table of the percent of the total population in each age group for both 1950 and 2075.

(b) Make a histogram of the 1950 age distribution (in percents). Then describe the main features of the distribution. In particular, look at the percent of children relative to the rest of the population.

(c) Make a histogram of the projected age distribution for the year 2075. Use the same scales as in (b) for easy comparison. What are the most important changes in the U.S. age distribution projected for the 125-year period between 1950 and 2075?

11.15 Babe Ruth's home runs. Here are the numbers of home runs that Babe Ruth hit in his 15 years with the New York Yankees, 1920 to 1934:

54 59 35 41 46 25 47 60 54 46 49 46 41 34 22

Make a stemplot of these data. Is the distribution roughly symmetric, clearly skewed, or neither? About how many home runs did Ruth hit in a typical year? Is his famous 60 home runs in 1927 an outlier?

11.16 Back-to-back stemplot. The current major-league single-season home run record is held by Mark McGwire of the St. Louis Cardinals. Here are McGwire's home run counts for 1987 to 1999:

49 32 33 39 22 42 9 9 39 52 58 70 65

A **back-to-back stemplot** helps us compare two distributions. Write the stems as usual, but with a vertical line both to their left and to their right. On the right, put leaves for Ruth (Exercise 11.15). On the left, put leaves for McGwire. Arrange the leaves on each stem in increasing order out from the stem. Now write a brief comparison of Ruth and McGwire as home run hitters. McGwire was injured in 1993 and there was a baseball strike in 1994. How do these events appear in the data?

11.17 When it rains, it pours. On July 6, 1994, 21.10 inches of rain fell on Americus, Georgia. That's the most rain ever recorded in Georgia for a 24-hour period. Table 11.6 gives the maximum precipitation ever recorded

Table 11.6 Record 24-hour precipitation amounts (inches) by state

State	Inches	State	Inches	State	Inches
Alabama	32.52	Louisiana	22.00	Ohio	10.75
Alaska	15.20	Maine	13.32	Oklahoma	15.68
Arizona	11.40	Maryland	14.75	Oregon	11.65
Arkansas	14.06	Massachusetts	18.15	Pennsylvania	34.50
California	26.12	Michigan	9.78	Rhode Island	12.13
Colorado	11.08	Minnesota	10.84	South Carolina	17.00
Connecticut	12.77	Mississippi	15.68	South Dakota	8.00
Delaware	8.50	Missouri	18.18	Tennessee	11.00
Florida	38.70	Montana	11.50	Texas	43.00
Georgia	21.10	Nebraska	13.15	Utah	6.00
Hawaii	38.00	Nevada	7.13	Vermont	8.77
Idaho	7.17	New Hampshire	10.38	Virginia	27.00
Illinois	16.91	New Jersey	14.81	Washington	14.26
Indiana	10.50	New Mexico	11.28	West Virginia	19.00
Iowa	16.70	New York	11.17	Wisconsin	11.72
Kansas	12.59	North Carolina	22.22	Wyoming	6.06
Kentucky	10.40	North Dakota	8.10		

in 24 hours (through 1998) at any weather station in each state. The record amount varies a great deal from state to state—hurricanes bring extreme rains on the Atlantic coast, and the mountain West is generally dry. Make a graph to display the distribution of records for the states. Mark where your state lies in this distribution. Briefly describe the distribution.

Chapter 12
Describing Distributions with Numbers

Does education pay?

People with more education earn more on the average than people with less education. How much more? A simple way to say is to compare *median* incomes, the amount such that half of a group of people earn more and half earn less. Here are the median incomes of adults with four different levels of education:

High school graduate only	Some college, no degree	Bachelor's degree	Advanced degree
\$16,297	\$18,988	\$32,581	\$47,000

These numbers come from 71,512 people interviewed by the Current Population Survey in March 1999. Every March, the CPS asks detailed questions about income. Boiling down 71,512 incomes into four numbers helps us get a quick picture of the relationship between education and income. The median college graduate, for example, earns about twice as much as the median high school graduate.

We know that incomes vary a lot. In fact, the highest income reported by the 31,621 people in the CPS sample who stopped with a high school diploma was $498,606. The medians compare the *centers* of the four income distributions. Can we also describe the *spreads* with just a few numbers? The largest and smallest incomes in a group of 31,621 people don't tell us much. Instead, let's give the range covered by the middle half of the incomes in each group. Here they are:

High school graduate only	Some college, no degree	Bachelor's degree	Advanced degree
$7,412 to $29,000	$7,803 to $33,150	$16,941 to $53,061	$28,075 to $75,162

The message of the 71,512 incomes is now clear: going to college but not getting a degree doesn't raise income much. People with bachelor's degrees, however, earn quite a bit more, and an advanced degree gives income another healthy boost. But there is a lot of variation among individuals—some very rich people never went to college. Finally, remember that this observational study says nothing about cause and effect. People who get advanced education are often smart, ambitious, and well-off to start with, so they might earn more even without their degrees.

In the summer of 1998, Mark McGwire and Sammy Sosa captured the public's imagination with their pursuit of baseball's single-season home run record. McGwire eventually set a new standard with 70 home runs. How does this accomplishment fit McGwire's career? Here are McGwire's home run counts for the years 1987 (his rookie year) to 1999:

1987	1988	1989	1990	1991	1992	1993	1994	1995	1996	1997	1998	1999
49	32	33	39	22	42	9	9	39	52	58	70	65

The stemplot in Figure 12.1 displays the data. The shape of the distribution is a bit irregular, but we might describe it as skewed to the right with a single peak and two low outliers. We can explain the outliers: McGwire was injured in 1993 and there was a players' strike in 1994.

0	99
1	
2	2
3	2399
4	29
5	28
6	5
7	0

Figure 12.1 Stemplot of the number of home runs hit by Mark McGwire in his first 13 seasons.

A graph and a few words give a good description of Mark McGwire's home run career. Words are less adequate to describe the incomes of 31,621 people with a high school education. We need *numbers* that summarize the center and spread of a distribution.

Median and quartiles

Our comparison of incomes for different levels of education introduced one simple and effective way to describe center and spread: give the *median* and the *quartiles*. The median is the midpoint, the value that separates the smaller half of the observations from the larger half. The first and third quartiles mark off the middle half of the observations. The quartiles get their name because with the median they divide the observations into quarters—one quarter lie below the first quartile, half the observations lie below the median, and three quarters lie below the third quartile. That's the idea. To actually get numbers, we need a rule that makes the idea exact.

Example 1. Finding the median

To find the median of Mark McGwire's 13 home run counts, first arrange them in order from smallest to largest:

9 9 22 32 33 39 **39** 42 49 52 58 65 70

The bold 39 is the center observation, with 6 observations to its left and 6 to its right. When the number of observations n is odd, there is always one observation in the center of the ordered list. This is the median, $M = 39$

We might compare McGwire's career with that of New York Yankee outfielder Roger Maris, the previous holder of the single-season record. Here are Maris's home run counts for his 10 years in the American League, arranged in order from smallest to largest:

8 13 14 16 **23** **26** 28 33 39 61

When n is even, there is no one middle observation. But there is a middle pair—the bold 23 and 26 have 4 observations on either side. We take the median to be halfway between this middle pair. So Maris's median is

$$M = \frac{23 + 26}{2} = \frac{49}{2} = 24.5$$

There is a fast way to locate the median in the ordered list: count up $(n + 1)/2$ places from the beginning of the list. Try it. For McGwire, $n = 13$

and $(n + 1)/2 = 7$, so the median is the 7th entry in the ordered list. For Maris, $n = 10$ and $(n + 1)/2 = 5.5$. This means "halfway between the 5th and 6th" entries, so M is the average of these two entries. This "$(n + 1)/2$ rule" is especially handy when you have many observations. The median of $n = 31{,}621$ incomes is the 15,811th in the ordered list. Be sure to note that $(n + 1)/2$ does *not* give the median M, just its position in the ordered list of observations.

The median M

The **median M** is the midpoint of a distribution, the number such that half the observations are smaller and the other half are larger. To find the median of a distribution:

1. Arrange all observations in order of size, from smallest to largest.

2. If the number of observations n is odd, the median M is the center observation in the ordered list. Find the location of the median by counting $(n + 1)/2$ observations up from the bottom of the list.

3. If the number of observations n is even, the median M is the average of the two center observations in the ordered list. The location of the median is again $(n + 1)/2$ from the bottom of the list.

The median income of people with a high school diploma in the Current Population Survey sample was $16,297. That's helpful but incomplete. Do most people with a high school education earn close to this amount, or are the incomes very spread out? The simplest useful description of a distribution consists of both a measure of *center* and a measure of *spread*. If we choose the median (the midpoint) to describe center, the quartiles are natural descriptions of spread. Again, the idea is clear: find the points one-quarter and three-quarters up the ordered list of observations. Again, we need a rule to make the idea precise. The rule for calculating the quartiles uses the rule for the median.

The quartiles Q_1 and Q_3

To calculate the **quartiles:**

1. Arrange the observations in increasing order and locate the median M in the ordered list of observations.

2. The **first quartile Q_1** is the median of the observations whose position in the ordered list is to the left of the location of the overall median.

3. The **third quartile Q_3** is the median of the observations whose position in the ordered list is to the right of the location of the overall median.

"Yup, Old Bob drowned due to being ignorant of statistics. He thought it was enough to know the average depth of the river."

Example 2. Finding the quartiles

Mark McGwire's home run counts (arranged in order) are

$$\begin{array}{ccccccccccccc} 9 & 9 & 22 & 32 & 33 & 39 & \mathbf{39} & 42 & 49 & 52 & 58 & 65 & 70 \\ & & \uparrow & & & & \uparrow & & & \uparrow & & & \\ & & Q_1 & & & & M & & & Q_3 & & & \end{array}$$

There is an odd number of observations, so the median is the one in the middle, the bold 39 in the list. To find the quartiles, ignore this central observation. The first quartile is the median of the 6 observations to the left of the bold 39 in the list. That's halfway between the 3rd and 4th, so

$$Q_1 = \frac{22 + 32}{2} = 27$$

The third quartile is the median of the 6 observations to the right of the bold 39. It is $Q_3 = 55$.

Roger Maris's 10 home run counts are

$$\begin{array}{cccccccccc} 8 & 13 & 14 & 16 & 23 & 26 & 28 & 33 & 39 & 61 \\ & & \uparrow & & & \uparrow & & \uparrow & & \\ & & Q_1 & & & M & & Q_3 & & \end{array}$$

The median lies halfway between the middle pair. There are 5 observations to the left of this location. The first quartile is the median of these 5 numbers. That is, $Q_1 = 14$. The third quartile is the median of the 5 observations to the right of the median's location, $Q_3 = 33$.

You can use the $(n + 1)/2$ rule to locate the quartiles when there are many observations. The median of our 31,621 incomes is the 15,811th in the list arranged in order from smallest to largest. So the first quartile is the median of the 15,810 incomes below this point in the list. Use the $(n + 1)/2$ rule with $n = 15{,}810$ to locate the quartile:

$$\frac{n+1}{2} = \frac{15{,}810 + 1}{2} = 7905.5$$

The average of the 7905th and 7906th incomes in the ordered list is $7412, as given in our opening example.

The five-number summary and boxplots

The smallest and largest observations tell us little about the distribution as a whole, but they give information about the tails of the distribution that is missing if we know only the median and the quartiles. To get a quick summary of both center and spread, combine all five numbers.

The five-number summary

The **five-number summary** of a distribution consists of the smallest observation, the first quartile, the median, the third quartile, and the largest observation, written in order from smallest to largest. In symbols, the five-number summary is

$$\text{Minimum} \quad Q_1 \quad M \quad Q_3 \quad \text{Maximum}$$

These five numbers offer a reasonably complete description of center and spread. The five-number summaries of home run counts are

9 27 39 55 70

for McGwire and

8 14 24.5 33 61

for Maris. The five-number summary of a distribution leads to a new graph, the *boxplot*. Figure 12.2 shows boxplots for the home run comparison.

Boxplot

A **boxplot** is a graph of the five-number summary.

- A central box spans the quartiles.
- A line in the box marks the median.
- Lines extend from the box out to the smallest and largest observations.

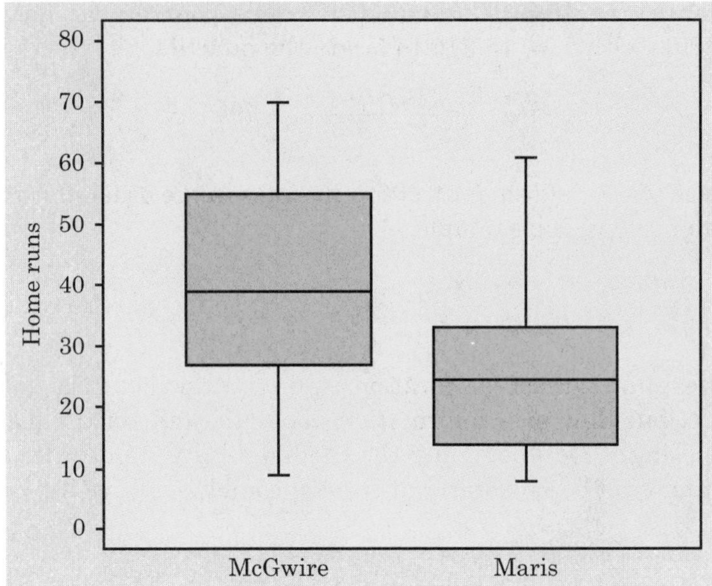

Figure 12.2 Boxplots comparing the yearly home run production of Mark McGwire and Roger Maris.

You can draw boxplots either horizontally or vertically. Be sure to include a numerical scale in the graph. When you look at a boxplot, first locate the median, which marks the center of the distribution. Then look at the spread. The quartiles show the spread of the middle half of the data, and the extremes (the smallest and largest observations) show the spread of the entire data set. We see from Figure 12.2 that McGwire's usual performance, as indicated by the median and the box that marks the middle half of the distribution, is better than that of Maris.

Because boxplots show less detail than histograms or stemplots, they are best used for side-by-side comparison of more than one distribution, as in Figure 12.2. For such small numbers of observations, a back-to-back stemplot is better yet (see Exercise 11.16, page 214). It would make clear, as the boxplot cannot, that Maris's record 61 home runs in 1961 is an outlier in his career. Let us look at an example where boxplots are more genuinely useful.

Example 3. Education and income

To see how income changes with level of education, we looked at the Current Population Survey annual sample of incomes. Figure 12.3 compares the income distributions for four levels of education, based on 71,512 observations. This figure is a variation on the boxplot idea. The largest income among tens of thousands of people will surely be very large. The highest single income for people who stopped after high school, for example, is $498,606. Only 5% of this group, however, have incomes above $54,481. Figure 12.3 uses the 5% and 95% points in the distribution in place of the single smallest and largest incomes. So the line above the box for the "High school only" group extends only to $54,481.

Figure 12.3 gives us a clear and simple visual comparison. We see how the median and middle half move up for people with bachelor's and advanced degrees. The income of the bottom 5% stays small because there are some people in each group with no income or even negative income, perhaps due to illness or disability. The 95% point, marking off the top 5% of incomes, shoots up among people with advanced degrees, who include doctors, lawyers, and holders of M.B.A. degrees.

Figure 12.3 also illustrates how boxplots often indicate the symmetry or skewness of a distribution. In a symmetric distribution, the first and third quartiles are equally distant from the median. In most distributions that are skewed to the right, on the other hand, the third quartile will be farther above the median than the first quartile is below it. The extremes behave the same way. Even with the top 5% not present, we can see the strong right-skewness of incomes among holders of advanced degrees.

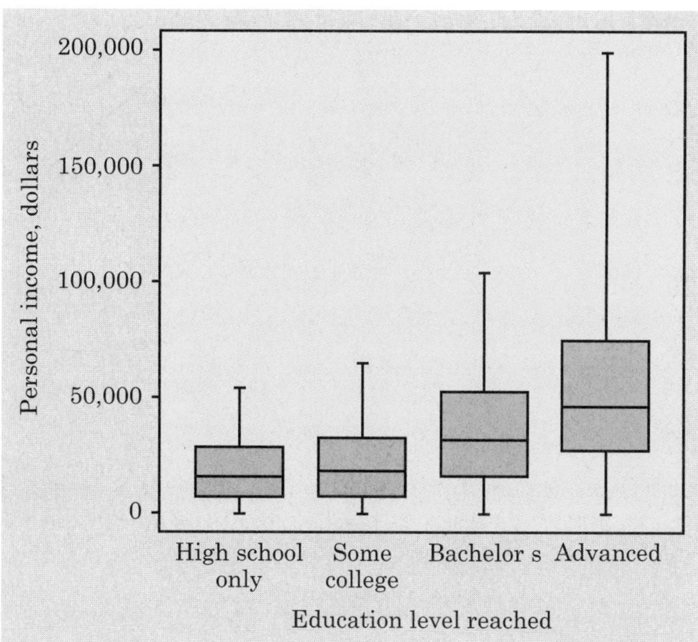

Figure 12.3 Boxplots comparing the distributions of income among adults with different levels of education. The ends of each plot are at the 5% and 95% points in the distribution.

Statistical Controversies

Income Inequality

During the prosperous 1980s and 1990s, the incomes of American households went up, but the gap between rich and poor grew. Figures 12.4 and 12.5 give two views of increasing inequality. Figure 12.4 is a line graph of household income, in dollars adjusted to have the same buying power every year. The lines show the median income and the 20th and 80th percentiles, which mark off the bottom fifth and the top fifth of households. The 80th percentile (up 41% between 1967 and 1998) is pulling away from the median and the 20th percentile, both up about 20%.

Figure 12.5 looks at the *share* of all income that goes to the top fifth and the bottom fifth. The bottom fifth's share has drifted down, to 3.6% of all income in 1998. The share of the top fifth grew to 49.2%. The share of the top 5% grew even faster, to more than 21% of the income of all households in the country. Income in-

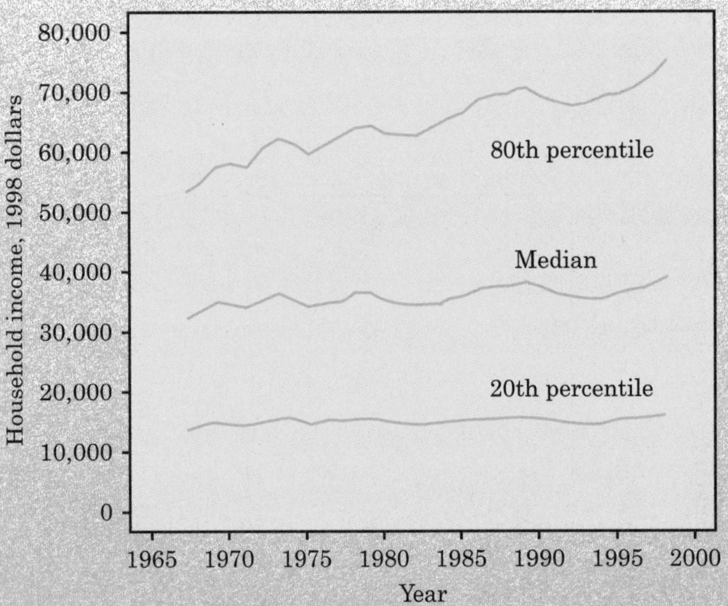

Figure 12.4 The change over time of three points in the distribution of incomes for American households. Eighty percent of households have incomes below the 80th percentile, half have incomes below the median, and 20% have incomes below the 20th percentile. In 1998, the 20th percentile was $16,116, the median household income was $38,885, and the 80th percentile was $75,000.

equality in the United States is greater than in other developed nations and has been increasing.

These are complicated issues, with much room for conflicting data and hidden agendas. The political left wants to reduce inequality, and the political right says the rich earn their high incomes. I want to point to just one important statistical twist. Figures 12.4 and 12.5 report "cross-section" data that give a snapshot of households in each year. "Longitudinal" data that follow households over time might paint a different picture. Consider a young married couple, Jamal and Tonya. They work part-time as students, then borrow to go to graduate school. They are down in the bottom fifth. When they get out of school, their income grows quickly. By age 40, they are happily in the top fifth. Many poor households are only temporarily poor.

Longitudinal studies are expensive because they must follow the same households for years. They are prone to bias because some households drop out over time. One study of income tax returns found that only 14% of the bottom fifth were still in the bottom fifth 10 years later. But really poor people don't file tax returns. Another study looked at children under 5 years old. Starting in both 1971 and 1981, it found that 60% of children who lived in households in the bottom fifth still lived in bottom-fifth households 10 years later. Many people do move from poor to rich as they grow older, but there are also many households that stay poor for years. Unfortunately, many children live in these households. Figure 12.6 shows one result of growing inequality: more children living in poverty.

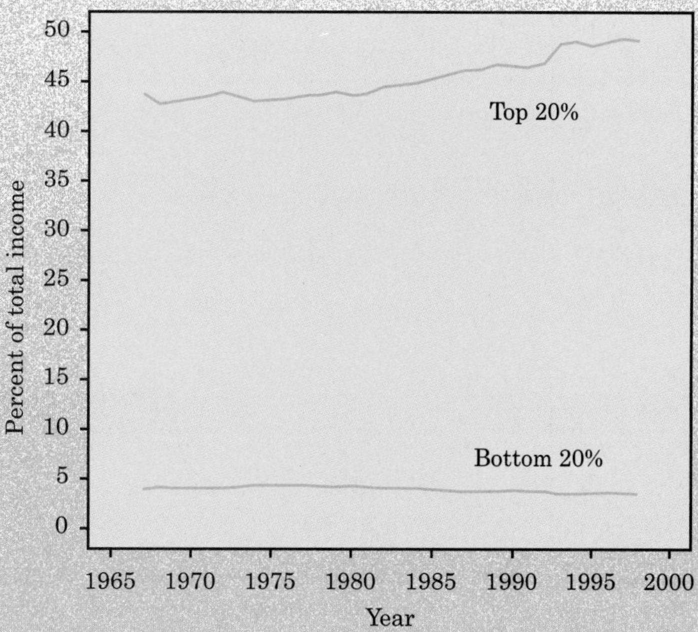

Figure 12.5 The change over time of the shares of total household income that go to the highest-income 20% and to the lowest-income 20% of households. In 1998, the top 20% of households received almost half of all income.

Jeff Greenberg/ The Image Works

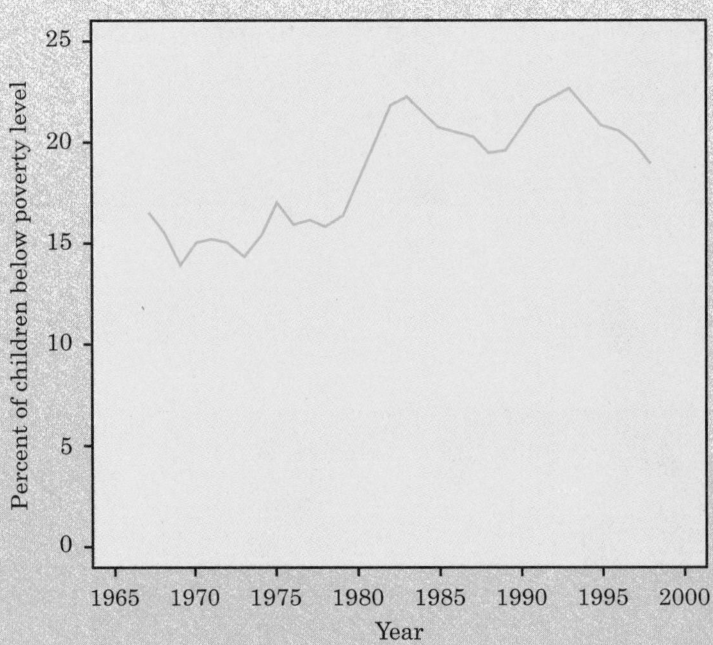

Figure 12.6 The change over time in the percent of children living in households with incomes below the poverty level set by the government. In 1998, 18.9% of American children lived in poverty.

Mean and standard deviation

The five-number summary is not the most common numerical description of a distribution. That distinction belongs to the combination of the *mean* to measure center and the *standard deviation* to measure spread. The mean is familiar—it is the ordinary average of the observations. The idea of the standard deviation is to give the average distance of observations from the mean. The "average distance" in the standard deviation is found in a rather obscure way. We will give the details, but you may want to just think of the standard deviation as "average distance from the mean" and leave the details to your calculator.

Mean and standard deviation

The **mean** $\bar{x}$ (pronounced "x-bar") of a set of observations is their average. To find the mean of n observations, add the values and divide by n:

$$\bar{x} = \frac{\text{sum of the observations}}{n}$$

The **standard deviation** s measures the average distance of the observations from their mean. It is calculated by finding an average of the squared distances and then taking the square root. To find the standard deviation of n observations:

1. Find the distance of each observation from the mean and square each of these distances.

2. Average the distances by dividing their sum by $n - 1$. This average squared distance is called the **variance.**

3. The standard deviation s is the square root of this average squared distance.

Example 4. Finding the mean and standard deviation

The numbers of home runs Mark McGwire hit in his first 13 major-league seasons are

49 32 33 39 22 42 9 9 39 52 58 70 65

To find the mean of these observations,

$$\bar{x} = \frac{\text{sum of observations}}{n}$$

$$= \frac{49 + 32 + \cdots + 65}{13}$$

$$= \frac{519}{13} = 39.92$$

Figure 12.7 displays the data as points above the number line, with their mean marked by an asterisk (*). The arrow shows one of the distances from the mean. The idea behind the standard deviation *s* is to average the 13 distances. To find the standard deviation by hand, you can use a table layout:

Observation	Distance from mean squared
49	$(49 - 39.92)^2 = (9.08)^2 = 82.45$
32	$(32 - 39.92)^2 = (-7.92)^2 = 62.73$
⋮	
65	$(65 - 39.92)^2 = (25.08)^2 = 629.01$
	sum = 4438.97

The average is

$$\frac{4438.97}{12} = 369.91$$

Notice that we "average" by dividing by *one less* than the number of observations. Finally, the standard deviation is the square root of this number:

$$s = \sqrt{369.91} = 19.23$$

In practice, you can key the data into your calculator and hit the mean key and the standard deviation key. Or you can enter the data into a spreadsheet or other software to find $\overline{x}$ and s. It is usual, for good but somewhat technical reasons, to average the squared distances by dividing their total by $n - 1$ rather than by n. Many calculators have two standard deviation buttons, giving you a choice between dividing by n and dividing by $n - 1$. Be sure to choose $n - 1$.

More important than the details of the calculation are the properties that show how the standard deviation measures spread.

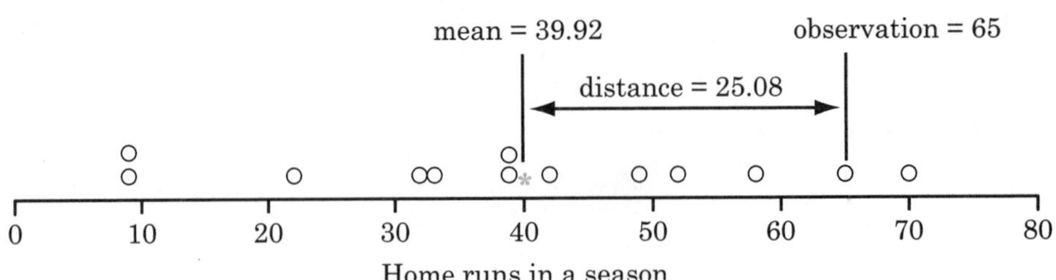

Figure 12.7 Mark McGwire's home run counts, with their mean (*) and the distance of one observation from the mean. Think of the standard deviation as an average of these distances.

Properties of the standard deviation s

- s measures spread about the mean $\bar{x}$. Use s to describe the spread of a distribution only when you use $\bar{x}$ to describe the center.
- $s = 0$ only when there is *no spread*. This happens only when all observations have the same value. So standard deviation zero means no spread at all. Otherwise $s > 0$. As the observations become more spread out about their mean, s gets larger.

Example 5. Investing 101

Enough examples about income. Here is an example about what to do with it once you've earned it. One of the first principles of investing is that taking more risk brings higher returns, at least on the average in the long run. Financial people measure risk by how unpredictable the return on an investment is. A bank account that is insured by the government and has a fixed rate of interest has no risk—its return is known exactly. Stock in a new company may soar one week and plunge the next. It has high risk because you can't predict what it will be worth when you want to sell.

Investors should think statistically. You can assess an investment by thinking about the distribution of (say) yearly returns. That means asking about both the center and spread of the pattern of returns. Only naive investors look for a high average return without asking about risk, that is, about how spread out or variable the returns are. Financial experts use the mean and standard deviation to describe returns on investments. The standard deviation was long considered too complicated to mention to the public, but now you will find standard deviations appearing regularly in mutual fund reports.

Here by way of illustration are the means and standard deviations of the yearly returns on three investments over the 50 years from 1950 to 1999:

Investment	Mean return	Standard deviation
Treasury bills	5.34%	2.96%
Treasury bonds	6.12%	10.73%
Common stocks	14.62%	16.32%

You can see that risk (variability) goes up as the mean return goes up, just as financial theory claims. Treasury bills and bonds are ways of loaning money to the U.S. government. Bills are paid back in one year, so their return changes from year to year depending on interest rates. Bonds are 30-year loans. They are riskier because the value of a bond you own will

Poor New York?

Is New York a rich state? New York's mean income per person ranks fourth among the states, right up there with its rich neighbors Connecticut and New Jersey, which rank first and second. But while Connecticut and New Jersey rank seventh and second in median household income, New York stands 29th, well below the national average. What's going on? Just another example of mean versus median. New York has many very highly paid people, who pull up its mean income per person. But it also has a higher proportion of poor households than do New Jersey and Connecticut, and this brings the median down. New York is not a rich state—it's a state with extremes of wealth and poverty.

Stem	Leaves
-2	6
-1	500
-0	98753
0	011456778
1	01224667999
2	01223344799
3	02222378
4	3
5	0

Figure 12.8 Stemplot of the yearly returns on common stocks for the 50 years 1950 to 1999. The returns are rounded to the nearest whole percent. The stems are 10s of percents and the leaves are single percents.

drop if interest rates go up. Stocks are yet riskier. They give higher returns (on the average in the long run), but at the cost of lots of sharp ups and downs on the way. As the stemplot in Figure 12.8 shows, stocks went up by as much as 50% and down by as much as 26% in one year during the 50 years covered by our data.

Choosing numerical descriptions

The five-number summary is easy to understand and is the best short description for most distributions. The mean and standard deviation are harder to understand but are more common. How can we decide which of these two descriptions of center and spread we should use? Let's start by comparing the mean and the median. "Midpoint" and "arithmetic average" are both reasonable ideas for describing the center of a set of data, but they are different ideas with different uses. The most important distinction is that the mean (the average) is strongly influenced by a few extreme observations and the median (the midpoint) is not.

Example 6. Mean versus median

Table 12.1 gives the approximate salaries (in millions of dollars) of the 14 members of the Los Angeles Lakers basketball team for the year 2000. You can calculate that the mean is $\overline{x}$ = \$4.1 million and that the median is M = \$2.6 million. No wonder professional basketball players have big houses.

Table 12.1 Year 2000 salaries for the Los Angeles Lakers

Player	Salary	Player	Salary
Shaquille O'Neal	$17.1 million	Ron Harper	$2.1 million
Kobe Bryant	$11.8 million	A. C. Green	$2.0 million
Robert Horry	$5.0 million	Devean George	$1.0 million
Glen Rice	$4.5 million	Brian Shaw	$1.0 million
Derek Fisher	$4.3 million	John Salley	$0.8 million
Rick Fox	$4.2 million	Tyronne Lue	$0.7 million
Travis Knight	$3.1 million	John Celestand	$0.3 million

Why is the mean so much higher than the median? Figure 12.9 is a stemplot of the salaries, with millions as stems. The distribution is skewed to the right and there are two high outliers. The very high salaries of Kobe Bryant and Shaquille O'Neal pull up the sum of the salaries and so pull up the mean. If we drop the outliers, the mean for the other 12 players is only $2.4 million. The median doesn't change nearly as much: it drops from $2.6 million to $2.05 million.

We can make the mean as large as we like by just increasing Shaquille O'Neal's salary. The mean will follow one outlier up and up. But to the median, Shaq's salary just counts as one observation at the upper end of the distribution. Moving it from $17.1 million to $171 million would not change the median at all.

The mean and median of a symmetric distribution are close to each other. In fact, $\bar{x}$ and M are exactly equal if the distribution is exactly symmetric. In skewed distributions, however, the mean runs away from the median toward the long tail. Many distributions of monetary values—incomes, house prices, wealth—are strongly skewed to the right. The mean may be much larger than the median. For example, we saw at the beginning of this chapter that the median income of people with advanced degrees in the Current Population Survey sample is $47,000. The distribution is skewed to the right, and the long right tail pulls the mean up to $65,220. Because monetary data often have a few extremely high observations, descriptions of these distributions usually employ the median.

You should think about more than symmetry versus skewness when choosing between the mean and the median. The distribution of selling

Figure 12.9 Stemplot of the salaries of Los Angeles Lakers players, from Table 12.1.

0	378
1	00
2	01
3	1
4	235
5	0
6	
7	
8	
9	
10	
11	8
12	
13	
14	
15	
16	
17	1

prices for homes in Middletown is no doubt skewed to the right—but if the Middletown city council wants to estimate the total market value of all houses in order to set tax rates, the mean and not the median helps them out. The total is just the number of houses times the mean but has no connection with the median.

The standard deviation is pulled up by outliers or the long tail of a skewed distribution even more strongly than the mean. The standard deviation of the Lakers' salaries is s = \$4.76 million for all 14 players and only s = \$1.72 million when the two outliers are removed. The quartiles are much less sensitive to a few extreme observations.There is another reason to avoid the standard deviation in describing skewed distributions. Because the two sides of a strongly skewed distribution have different spreads, no single number such as s describes the spread well. The five-number summary, with its two quartiles and two extremes, does a better job. In most situations, it is wise to use $\bar{x}$ and s only for distributions that are roughly symmetric.

Choosing a summary

The mean and standard deviation are strongly affected by outliers or by the long tail of a skewed distribution. The median and quartiles are less affected.

The five-number summary is usually better than the mean and standard deviation for describing a skewed distribution or a distribution with outliers. Use $\bar{x}$ and s only for reasonably symmetric distributions that are free of outliers.

Why do we bother with the standard deviation at all? One answer appears in the next chapter: the mean and standard deviation are the natural measures of center and spread for an important kind of symmetric distribution, called normal distributions.

Do remember that a graph gives the best overall picture of a distribution. Numerical measures of center and spread report specific facts about a distribution, but they do not describe its entire shape. Numerical summaries do not disclose the presence of multiple peaks or gaps, for example. *Always start with a graph of your data.*

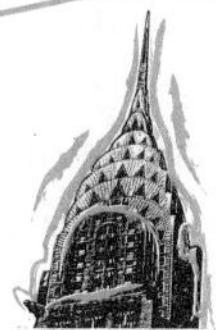

Exploring the Web

Many of our examples in this chapter have concerned income. At the Census Bureau Web site, www.census.gov, you can find the latest data on income and related issues. Look for *Money Income in the United States* (click on "Income" on the home page) and *Poverty in the United States* (click on "Poverty"). The full texts and the numerous tables in these annual publications are available online, along with many of the actual data, like the incomes and education of 71,512 people that lie behind Figure 12.3.

You can compare the behavior of the mean and median by using the "Mean and Median" applet at the *Statistics: Concepts and Controversies* Web site, www.whfreeman.com/scc. Click to enter data, then use the mouse to drag an outlier up and watch the mean chase after it.

Statistics in Summary To describe a set of data, always start with graphs. Then add well-chosen numbers that summarize specific aspects of the data. If we have data on a single quantitative variable, we start with a histogram or stemplot to display the distribution. Then we add numbers to describe the **center and spread** of the distribution.

There are two common descriptions of center and spread: **the five-number summary** and the **mean and standard deviation.** The five-number summary consists of the **median** to measure center and the two **quartiles** and the smallest and largest observations to describe spread. The median is the midpoint of the observations. The **mean** is the average of the observations. The **standard deviation** measures spread as a kind of average distance from the mean, so use it only with the mean.

The mean and standard deviation can be changed a lot by a few outliers. The mean and median agree for symmetric distributions, but the mean moves farther toward the long tail of a skewed distribution. In general, use the five-number summary to describe most distributions and the mean and standard deviation only for roughly symmetric distributions.

CHAPTER 12 EXERCISES

12.1 Median income. You read that the median income of U.S. households in 1998 was $38,885. Explain in plain language what "the median income" is.

12.2 What's the average? The Census Bureau publication *Money Income in the United States* gives several choices for "average income." In 1998, the median income of American households was $38,885. The mean household income was $51,855. The median income of families was $46,737, and the mean family income was $59,589. The Census Bureau says, "Households consist of all people who occupy a housing unit. The term 'family' refers to a group of two or more people related by birth, marriage, or adoption who reside together." Explain carefully why mean incomes are higher than median incomes and why family incomes are higher than household incomes.

12.3 Rich magazine readers. The business magazine *Forbes* reports (July 5, 1999) that the median household wealth of its readers is $956,000.
(a) Is the mean wealth of these households greater or less than $956,000? Why?

(b) The data were reported by a sample of *Forbes* readers contacted by telephone. I suspect that $956,000 is a bit higher than the true median wealth. Why?

12.4 College tuition. Figure 11.2 (page 197) is a stemplot of the tuition charged by 81 colleges in Michigan. The stems are thousands of dollars and the leaves are hundreds of dollars. For example, the highest tuition is $19,300 and appears as leaf 3 on stem 19.
(a) Find the five-number summary of Michigan college tuitions. You see that the stemplot already arranges the data in order.

(b) Would the mean tuition be clearly smaller than the median, about the same as the median, or clearly larger than the median? Why?

12.5 Where do the young live? Figure 11.8 (page 207) is a stemplot of the percent of residents aged 25 to 34 in each of the 50 states. The stems are whole percents and the leaves are tenths of a percent.
(a) The shape of the distribution suggests that the mean and median will not be far apart. Why?

(b) Find the mean and median of these data and verify that they are similar.

12.6 Gas mileage. Table 11.2 (page 209) gives the highway gas mileages for model year 2000 midsize cars.
(a) Make a stemplot of these data if you did not do so in Exercise 11.6.

(b) Find the five-number summary of gas mileages. Which cars are in the bottom quarter of gas mileages?

(c) The stemplot shows a fact about the overall shape of the distribution that the five-number summary cannot describe. What is it?

12.7 Yankee money. Table 11.4 (page 211) gives the salaries of the New York Yankees baseball team. What shape do you expect the distribution to have? Do you expect the mean salary to be close to the median, clearly higher, or clearly lower? Verify your choices by making a graph and calculating the mean and median.

12.8 The richest 1%. The distribution of individual incomes in the United States is strongly skewed to the right. In 1997, the mean and median incomes of the top 1% of Americans were $330,000 and $675,000. Which of these numbers is the mean and which is the median? Explain your reasoning.

12.9 How many calories does a hot dog have? *Consumer Reports* magazine presented the following data on the number of calories in a hot dog for each of 17 brands of meat hot dogs:

173	191	182	190	172	147	146	139	175
136	179	153	107	195	135	140	138	

Make a stemplot and find the five-number summary. The stemplot shows important facts about the distribution that the numerical summary does not tell us about. What are these facts?

12.10 Returns on common stocks. Example 5 informs us that financial theory uses the mean and standard deviation to describe the returns on investments. Figure 11.10 (page 209) is a histogram of the returns of all New York Stock Exchange common stocks in one year. Are the mean and standard deviation suitable as a brief description of this distribution? Why?

12.11 Minority students in engineering. Figure 11.9 (page 208) is a histogram of the number of minority students (black, Hispanic, Native American) who earned doctorate degrees in engineering from each of 115 universities in the years 1992 through 1996.

(a) What are the positions of each number in the five-number summary in a list of 115 observations arranged from smallest to largest?

(b) Even without the actual data, you can use your answer to (a) and the histogram to give the five-number summary approximately. Do this. About how many minority engineering Ph.D.'s must a university graduate to be in the top quarter?

12.12 The statistics of writing style. Here are data on the percents of words of 1 to 15 letters used in articles in *Popular Science* magazine. Exercise 11.9 asked you to make a histogram of these data.

Length	1	2	3	4	5	6	7	8	9	10	11	12	13	14	15
Percent	3.6	14.8	18.7	16.0	12.5	8.2	8.1	5.9	4.4	3.6	2.1	0.9	0.6	0.4	0.2

With a bit of thought, you can find the five-number summary of the distribution of word lengths from this table. Do it.

12.13 Immigrants in the eastern states. Here are the number of immigrants (in thousands) who settled in each state east of the Mississippi River in 1997:

Alabama	1.6	Connecticut	9.5	Delaware	1.4
Florida	82.3	Georgia	12.6	Illinois	38.1
Indiana	3.9	Kentucky	1.9	Maine	1.0
Maryland	19.0	Massachusetts	17.3	Michigan	14.7
Mississippi	1.1	New Hampshire	1.5	New Jersey	41.2
New York	123.7	North Carolina	5.9	Ohio	8.2
Pennsylvania	14.6	Rhode Island	2.5	South Carolina	2.4
Tennessee	4.4	Vermont	0.6	Virginia	19.3
West Virginia	0.6	Wisconsin	3.2		

Make a graph of the distribution. Describe its overall shape and any outliers. Then choose and calculate a suitable numerical summary.

12.14 Immigrants in the eastern states. New York State is a high outlier in the distribution of the previous exercise. Find the mean and the median for these data with and without New York. Which measure is more changed when we omit the outlier?

12.15 State SAT scores. Figure 12.10 is a histogram of the average scores on the mathematics part of the SAT exam for students in the 50 states and the District of Columbia. The distinctive overall shape of this distribution implies that a single measure of center such as the mean or the median is of little value in describing the distribution. Explain why this is true.

12.16 Highly paid athletes. A news article reports that of the 411 players on National Basketball Association rosters in February 1998, only 139 "made more than the league average salary" of \$2.36 million. Is \$2.36 million the mean or median salary for NBA players? How do you know?

12.17 Mean or median? Which measure of center, the mean or the median, should you use in each of the following situations? Why?

(a) Middletown is considering imposing an income tax on citizens. The city government wants to know the average income of citizens so that it can estimate the total tax base.

(b) In a study of the standard of living of typical families in Middletown, a sociologist estimates the average family income in that city.

12.18 Mean or median? You are planning a party and want to know how many cans of soda to buy. A genie offers to tell you either the mean

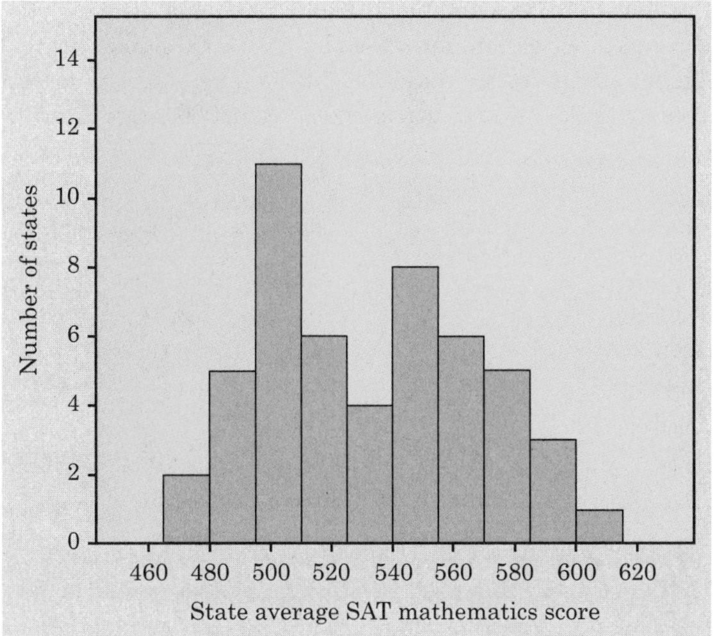

Figure 12.10 Histogram of the average score on the mathematics part of the SAT exam for students in the 50 states and the District of Columbia in 1998, for Exercise 12.15.

number of cans guests will drink or the median number of cans. Which measure of center should you ask for? Why? To make your answer concrete, suppose there will be 30 guests and the genie will tell you one of $\bar{x} = 5$ cans or $M = 3$ cans. How many cans should you have on hand?

12.19 State SAT scores. We want to compare the distributions of average SAT math and verbal scores for the states. We enter these data into a computer with the names SATM for math scores and SATV for verbal scores. Here is output from the statistical software package Minitab. (Other software produces similar output. Some software uses rules for finding the quartiles that differ slightly from ours. So software may not give exactly the answer you would get by hand.)

```
Variable     N          Mean        Median      StDev
SATV        51        532.22        525.00      33.86
SATM        51        533.47        528.00      35.13

Variable Minimum      Maximum         Q1          Q3
SATV       478.00      593.00      501.00      564.00
SATM       473.00      601.00      503.00      558.00
```

Use this output to make boxplots of SAT math and verbal scores for the states. Briefly compare the two distributions in words.

12.20 Do SUVs waste gas? Table 11.2 (page 209) gives the highway fuel consumption (in miles per gallon) for 32 model year 2000 midsize cars. You found the five-number summary for these data in Exercise 12.6. Here are the highway mileages for 26 four-wheel-drive model year 2000 sport utility vehicles:

Model	MPG	Model	MPG
BMW X5	17	Kia Sportage	22
Chevrolet Blazer	20	Land Rover	17
Chevrolet Tahoe	18	Lexus LX470	16
Dodge Durango	18	Lincoln Navigator	17
Ford Expedition	18	Mazda MPV	19
Ford Explorer	20	Mercedes-Benz ML320	20
Honda Passport	20	Mitsubishi Montero	20
Infiniti QX4	18	Nissan Pathfinder	19
Isuzu Amigo	19	Nissan Xterra	19
Isuzu Trooper	19	Subaru Forester	27
Jeep Cherokee	20	Suzuki Grand Vitara	20
Jeep Grand Cherokee	18	Toyota RAV4	26
Jeep Wrangler	19	Toyota 4Runner	21

(a) Give a graphical and numerical description of highway fuel consumption for SUVs. What are the main features of the distribution?

(b) Make boxplots to compare the highway fuel consumption of midsize cars and SUVs. What are the most important differences between the two distributions?

12.21 How many calories in a hot dog? Some people worry about how many calories they consume. *Consumer Reports* magazine, in a story on hot dogs, measured the calories in 20 brands of beef hot dogs, 17 brands of meat hot dogs, and 17 brands of poultry hot dogs. Here is computer output describing the beef hot dogs,

```
Mean = 156.8  Standard deviation = 22.64
Min = 111  Max = 190  N = 20
Median = 152.5  Quartiles = 140, 178.5
```

the meat hot dogs,

```
Mean = 158.7  Standard deviation = 25.24
Min = 107  Max = 195  N = 17
Median = 153  Quartiles = 139, 179
```

and the poultry hot dogs,

```
Mean = 122.5   Standard deviation = 25.48
Min = 87   Max = 170   N = 17
Median = 129   Quartiles = 102, 143
```

(Some software uses rules for finding the quartiles that differ slightly from ours. So software may not give exactly the answer you would get by hand.) Use this information to make boxplots of the calorie counts for the three types of hot dogs. Write a brief comparison of the distributions. Will eating poultry hot dogs usually lower your calorie consumption compared with eating beef or meat hot dogs?

12.22 Finding the standard deviation. The level of various substances in the blood influences our health. Here are measurements of the level of phosphate in the blood of a patient, in milligrams of phosphate per deciliter of blood, made on 6 consecutive visits to a clinic:

5.6 5.2 4.6 4.9 5.7 6.4

A graph of only 6 observations gives little information, so we proceed to compute the mean and standard deviation.

(a) Find the mean from its definition. That is, find the sum of the 6 observations and divide by 6.

(b) Find the standard deviation from its definition. That is, find the distance of each observation from the mean, square the distances, then calculate the standard deviation. Example 4 shows the method.

(c) Now enter the data into your calculator and use the mean and standard deviation buttons to obtain $\bar{x}$ and s. Do the results agree with your hand calculations?

12.23 What *s* measures. Use a calculator to find the mean and standard deviation of these two sets of numbers:

(a) 4 0 1 4 3 6

(b) 5 3 1 3 4 2

Make a back-to-back stemplot (see Exercise 11.16) to compare the two distributions. Which one is more spread out?

12.24 What *s* measures. Add 2 to each of the numbers in set (a) in the previous exercise. The data are now 6 2 3 6 5 8.

(a) Use a calculator to find the mean and standard deviation and compare your answers with those for data (a) in the previous exercise. How does adding 2 to each number change the mean? How does it change the standard deviation?

(b) Without doing the calculation, what would happen to $\bar{x}$ and s if we added 10 to each value in data (a) of the previous exercise? (This exercise demonstrates that the standard deviation measures only spread about the mean and ignores changes in where the data are centered.)

12.25 Cars and SUVs. Use the mean and standard deviation to compare the gas mileages of midsize cars (Table 11.2, page 209) and SUVs (Exercise 12.20). Do these numbers catch the main points of your more detailed comparison in Exercise 12.20?

12.26 A contest. This is a standard deviation contest. You must choose four numbers from the whole numbers 0 to 10, with repeats allowed.

(a) Choose four numbers that have the smallest possible standard deviation.

(b) Choose four numbers that have the largest possible standard deviation.

(c) Is more than one choice correct in either (a) or (b)? Explain.

12.27 $\bar{x}$ and s are not enough. The mean $\bar{x}$ and standard deviation s measure center and spread but are not a complete description of a distribution. Data sets with different shapes can have the same mean and standard deviation. To demonstrate this fact, use your calculator to find $\bar{x}$ and s for these two small data sets. Then make a stemplot of each and comment on the shape of each distribution.

Data A	9.14	8.14	8.74	8.77	9.26	8.10	6.13	3.10	9.13	7.26	4.74
Data B	6.58	5.76	7.71	8.84	8.47	7.04	5.25	5.56	7.91	6.89	12.50

12.28 Raising pay. A school system employs teachers at salaries between \$30,000 and \$60,000. The teachers' union and the school board are negotiating the form of next year's increase in the salary schedule. Suppose that every teacher is given a flat \$1000 raise.

(a) How much will the mean salary increase? The median salary?

(b) Will a flat \$1000 raise increase the spread as measured by the distance between the quartiles?

(c) Will a flat \$1000 raise increase the spread as measured by the standard deviation of the salaries? (See Exercise 12.24 if you need help.)

12.29 Raising pay. Suppose that the teachers in the previous exercise each receive a 5% raise. The amount of the raise will vary from \$1500 to \$3000, depending on present salary. Will a 5% across-the-board raise increase the spread of the distribution as measured by the distance between the quartiles? Do you think it will increase the standard deviation?

12.30 Making colleges look good. Colleges announce an "average" SAT score for their entering freshmen. Usually the college would like this "average" to be as high as possible. A *New York Times* article noted, "Private colleges that buy lots of top students with merit scholarships prefer the mean, while open-enrollment public institutions like medians." Use what you know about the behavior of means and medians to explain these preferences.

12.31 What graph to draw? We now understand three kinds of graphs to display distributions of quantitative variables: histograms, stemplots, and boxplots. Give an example (just words, no data) of a situation in which you would prefer each kind of graph.

Chapter 13
Normal Distributions

Technology strikes

Bar graphs and histograms are definitely old technology. Using bars to display data goes back to William Playfair (1759–1823), an English economist who was an early pioneer of data graphics. Histograms require that we choose classes, and their appearance can change with different choices. Surely modern software offers a better way to picture distributions?

Software can replace the separate bars of a histogram with a smooth curve that represents the overall shape of a distribution. Look at Figure 13.1. The data are the numbers of minority group members who earned doctorates in engineering from 115 universities between 1992 and 1996. We met these data in Chapter 11, and the histogram in Figure 13.1 repeats Figure 11.9. The curve is the new-technology replacement for the histogram. The software doesn't start from the histogram—it starts with the actual observations and cleverly draws a curve to describe their distribution.

In Figure 13.1, the software has caught the overall shape and shows the ripples in the long right tail more effectively than does the histogram. It struggles a bit with the peak. For example, it has extended the curve below zero in an attempt to smooth out the sharp peak. In Figure 13.2, we apply the same software to a larger set of data with a more regularly shaped distribution. These are the values of the sample proportion $\hat{p}$ for 1000 SRSs of size 1523 from a population in which the population proportion is $p = 0.6$. We also met these data in Chapter 11, and the histogram here repeats Figure 11.4. The software draws a curve that shows a distinctive symmetric, single-peaked, bell shape.

For the irregular distribution in Figure 13.1, we can't do better. In the case of the very symmetric sampling data in Figure 13.2, however, there is another way to get a smooth curve. It's a mathematical fact that this distribution can be described by a specific kind of smooth curve called a *normal curve.* Figure 13.3 shows the normal curve for these data. The curve looks a lot like the one in Figure 13.2, but a close look shows that it is smoother. The normal curve is much easier to work with and does not require clever software. We will see that normal curves have special properties that help us use them and think about them. Only some kinds of data fit normal curves, however, so keep the technology handy for those that don't.

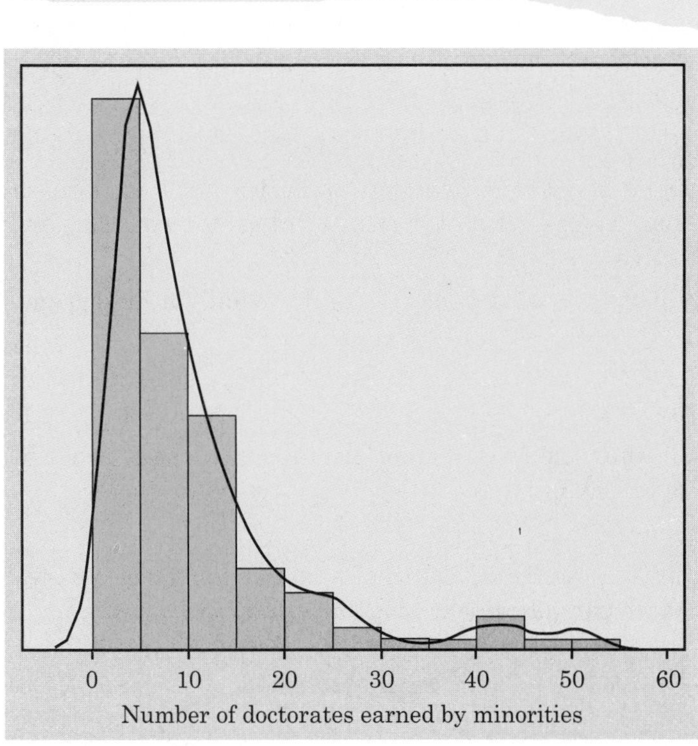

Figure 13.1 A histogram and a computer-drawn curve. Both picture the distribution of the number of doctorates in engineering earned by members of minority groups at 115 universities. This distribution is skewed to the right.

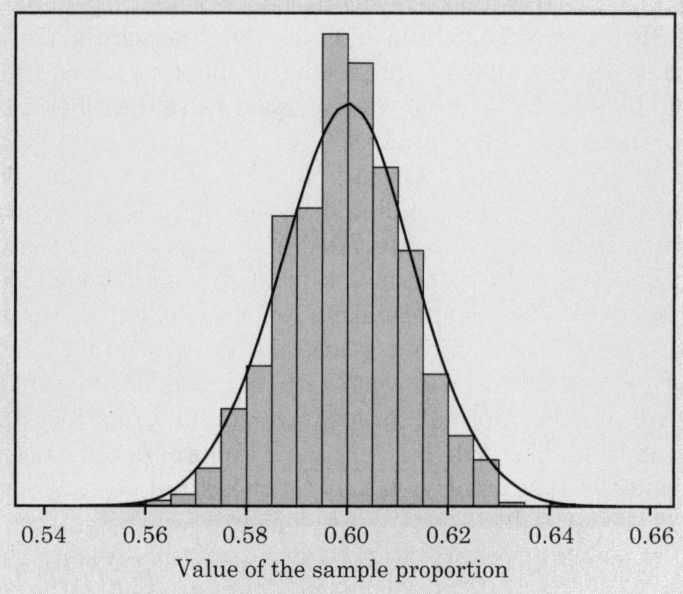

Figure 13.2 A histogram and a computer-drawn curve. Both picture the distribution of the sample proportion in 1000 simple random samples from the same population. This distribution is quite symmetric.

We now have a kit of graphical and numerical tools for describing distributions. What is more, we have a clear strategy for exploring data on a single quantitative variable:

1. Always plot your data: make a graph, usually a histogram or a stemplot.

2. Look for the overall pattern (shape, center, spread) and for striking deviations such as outliers.

3. Choose either the five-number summary or the mean and standard deviation to briefly describe center and spread in numbers.

Here is one more step to add to this strategy:

4. Sometimes the overall pattern of a large number of observations is so regular that we can describe it by a smooth curve.

Density curves

Figures 13.1 and 13.2 show curves used in place of histograms to picture the overall shape of a distribution of data. You can think of drawing a curve through the tops of the bars in a histogram, smoothing out the irregular ups and downs of the bars. There is one important distinction between histograms and these curves. Most histograms show the *counts* of observations in each class by the heights of their bars and therefore by the areas of the bars. We set up curves to show the *proportion* of observations in any region by areas under the curve. To do that, we choose the scale so that the total area under the curve is exactly 1. We then have a **density curve.**

Example 1. Using a density curve

Figure 13.4 copies Figure 13.3, showing the histogram and the normal density curve that describe this data set of 1000 sample proportions. What proportion of the observations are greater than 0.61? From the actual 1000 observations, we can count that exactly 195 are greater than 0.61. So the proportion is 195/1000 or 0.195. Because 0.61 is one of the break points between the classes in the histogram, the area of the shaded bars in Figure 13.4(a) makes up 0.195 of the total area of all the bars.

Now concentrate on the density curve drawn through the histogram. The total area under this curve is 1, and the shaded area in Figure 13.4(b) represents the proportion of observations that are greater than 0.61. This area is 0.208. You can see that the density curve is a quite good approximation—0.208 is quite close to 0.195.

The area under the density curve in Example 1 is not exactly equal to the true proportion because the curve is an idealized picture of the distribution. For example, the curve is exactly symmetric but the actual data

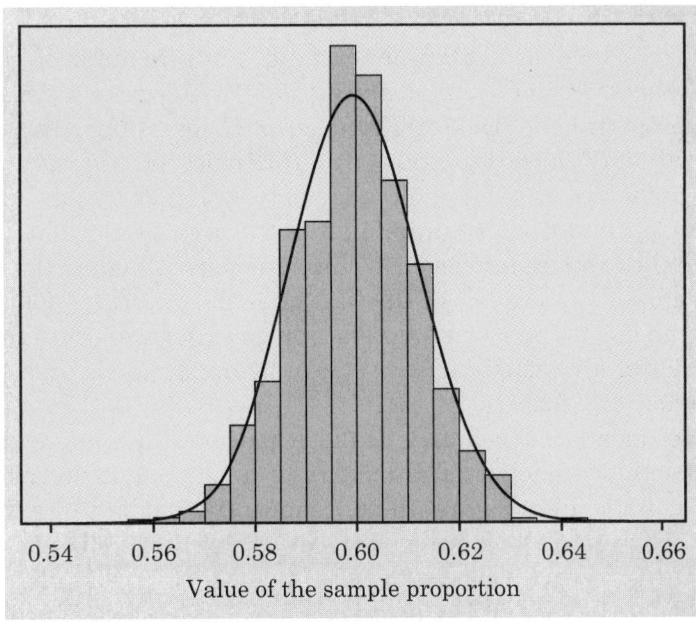

Figure 13.3 A perfectly symmetric normal curve used to describe the distribution of sample proportions.

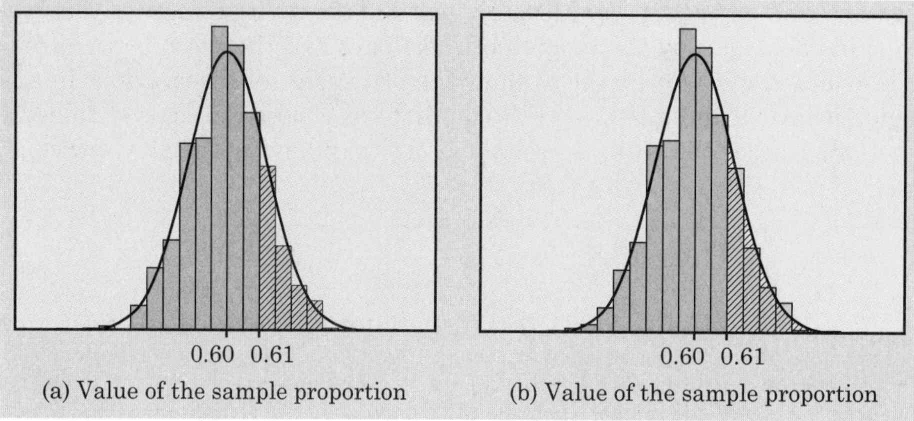

Figure 13.4 A normal density curve and a histogram. **(a)** The area of the shaded bars in the histogram represents observations greater than 0.61. These make up 195 of the 1000 observations. **(b)** The shaded area under the normal curve represents the proportion of observations greater than 0.61. This area is 0.208.

are only approximately symmetric. Because density curves are smoothed-out idealized pictures of the overall shapes of distributions, they are most useful for describing large numbers of observations.

The center and spread of a density curve

Density curves help us better understand our measures of center and spread. The median and quartiles are easy. Areas under a density curve represent proportions of the total number of observations. The median is the point with half the observations on either side. So *the median of a density curve is the equal-areas point,* the point with half the area under the curve to its left and the remaining half of the area to its right. The quartiles divide the area under the curve into quarters. One-fourth of the area under the curve is to the left of the first quartile, and three-fourths of the area is to the left of the third quartile. You can roughly locate the median and quartiles of any density curve by eye by dividing the area under the curve into four equal parts.

Because density curves are idealized patterns, a symmetric density curve is exactly symmetric. The median of a symmetric density curve is therefore at its center. Figure 13.5(a) shows the median of a symmetric curve. We can roughly locate the equal-areas point on a skewed curve like that in Figure 13.5(b) by eye.

What about the mean? The mean of a set of observations is their arithmetic average. If we think of the observations as weights stacked on a seesaw, the mean is the point at which the seesaw would balance. This fact is

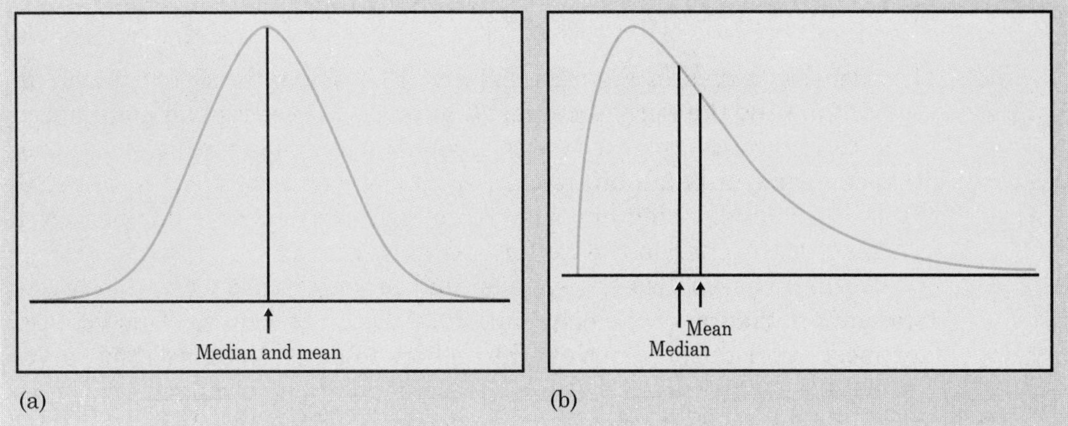

Figure 13.5 The median and mean for two density curves, a symmetric normal curve and a curve that is skewed to the right.

also true of density curves. *The mean is the point at which the curve would balance if made of solid material.* Figure 13.6 illustrates this fact about the mean. A symmetric curve balances at its center because the two sides are identical. *The mean and median of a symmetric density curve are equal,* as in Figure 13.5(a). We know that the mean of a skewed distribution is pulled toward the long tail. Figure 13.5(b) shows how the mean of a skewed density curve is pulled toward the long tail more than is the median.

Median and mean of a density curve

The **median** of a density curve is the equal-areas point, the point that divides the area under the curve in half.

The **mean** of a density curve is the balance point, at which the curve would balance if made of solid material.

The median and mean are the same for a symmetric density curve. They both lie at the center of the curve. The mean of a skewed curve is pulled away from the median in the direction of the long tail.

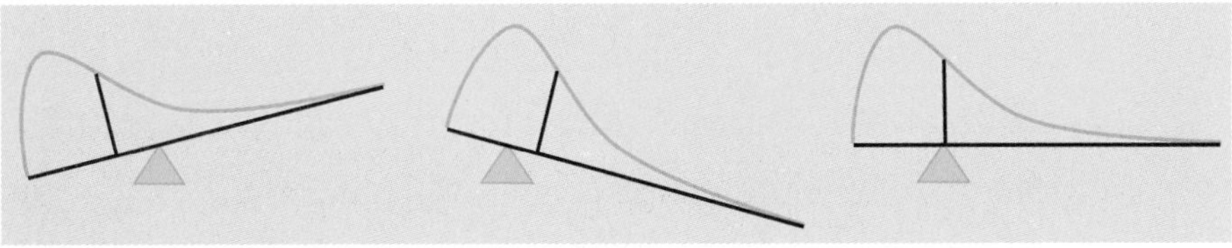

Figure 13.6 The mean of a density curve is the point at which it would balance.

Normal distributions

The density curves in Figures 13.3 and 13.4 belong to a particularly important family: the normal curves. Figure 13.7 presents two more normal density curves. Normal curves are symmetric, single-peaked, and bell-shaped. Their tails fall off quickly, so that we do not expect outliers. Because normal distributions are symmetric, the mean and median lie together at the peak in the center of the curve.

Normal curves also have the special property that we can locate the standard deviation of the distribution by eye on the curve. This isn't true for most other density curves. Here's how to do it. Imagine that you are skiing down a mountain that has the shape of a normal curve. At first, you descend at an ever-steeper angle as you go out from the peak:

Fortunately, before you find yourself going straight down, the slope begins to grow flatter rather than steeper as you go out and down:

The points at which this change of curvature takes place are located one standard deviation on either side of the mean. The standard deviations are marked on the two curves in Figure 13.7. You can feel the change as you run a pencil along a normal curve, and so find the standard deviation.

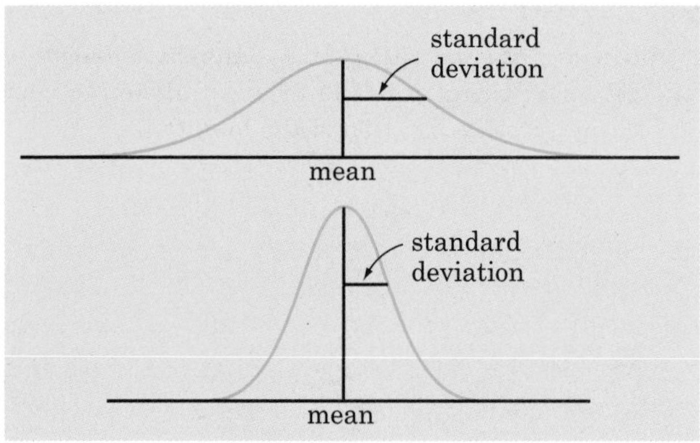

Figure 13.7 Two normal curves. The standard deviation fixes the spread of a normal curve.

Normal curves have the special property that giving the mean and the standard deviation completely specifies the curve. The mean fixes the center of the curve, and the standard deviation determines its shape. Changing the mean of a normal distribution does not change its shape, only its location on the axis. Changing the standard deviation does change the shape of a normal curve, as Figure 13.7 illustrates. The distribution with the smaller standard deviation is less spread out and more sharply peaked. Here is a summary of basic facts about normal curves:

Normal density curves

The **normal curves** are symmetric, bell-shaped curves that have these properties:

- A specific normal curve is completely described by giving its mean and its standard deviation.
- The mean determines the center of the distribution. It is located at the center of symmetry of the curve.
- The standard deviation determines the shape of the curve. It is the distance from the mean to the change-of-curvature points on either side.

Why are the normal distributions important in statistics? First, normal distributions are good descriptions for some distributions of *real data*. Normal curves were first applied to data by the great mathematician Carl Friedrich Gauss (1777–1855), who used them to describe the small errors made by astronomers and surveyors in repeated careful measurements of the same quantity. You will sometimes see normal distributions labeled "Gaussian" in honor of Gauss. For much of the 19th century normal curves were called "error curves" because they were first used to describe the distribution of measurement errors. As it became clear that the distributions of some biological and psychological variables were at least roughly normal, the "error curve" terminology was dropped. The curves were first called "normal" by Francis Galton in 1889. Galton, a cousin of Charles Darwin, pioneered statistical study of inheritance.

Normal curves also describe the distribution of *statistics such as sample proportions and sample means* when we take many samples from the same population. We used normal curves this way in Figures 13.3 and 13.4. The margins of error for the results of sample surveys are usually calculated from normal curves. However, even though many sets of data follow a normal distribution, many do not. Most income distributions, for example, are skewed to the right and so are not normal. Nonnormal data, like nonnormal people, not only are common but are sometimes more interesting than their normal counterparts.

The 68–95–99.7 rule

There are many normal curves, each described by its mean and standard deviation. All normal curves share many properties. In particular, the standard deviation is the natural unit of measurement for normal distributions. This fact is reflected in the following rule.

The 68–95–99.7 rule

In any normal distribution, approximately

- **68%** of the observations fall within one standard deviation of the mean.
- **95%** of the observations fall within two standard deviations of the mean.
- **99.7%** of the observations fall within three standard deviations of the mean.

Figure 13.8 illustrates the 68–95–99.7 rule. By remembering these three numbers, you can think about normal distributions without constantly making detailed calculations. Remember also, though, that no set of data is exactly described by a normal curve. The 68–95–99.7 rule will be only approximately true for SAT scores or the lengths of crickets.

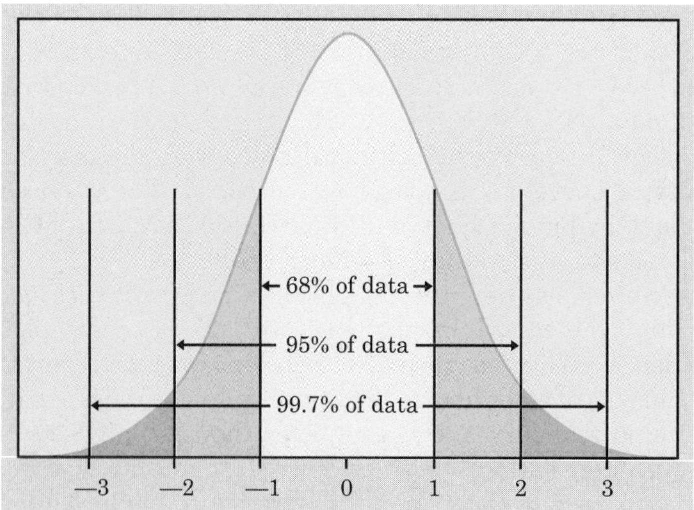

Figure 13.8 The 68–95–99.7 rule for normal distributions.

Example 2. Heights of young women

The distribution of heights of women aged 18 to 24 is approximately normal with mean 65 inches and standard deviation 2.5 inches. To use the 68–95–99.7 rule, always start by drawing a picture of the normal curve. Figure 13.9 shows what the rule says about women's heights.

Half of the observations in any normal distribution lie above the mean, so half of all young women are taller than 65 inches.

The central 68% of any normal distribution lies within one standard deviation of the mean. Half of this central 68%, or 34%, lie above the mean. So 34% of young women are between 65 inches and 67.5 inches tall. Adding the 50% who are shorter than 65 inches, we see that 84% of young women have heights less than 67.5 inches. That leaves 16% who are taller than 67.5 inches.

The central 95% of any normal distribution lies within two standard deviations of the mean. Two standard deviations is 5 inches here, so the middle 95% of young women's heights are between 60 inches (that's 65 − 5) and 70 inches (that's 65 + 5).

The other 5% of young women have heights outside the range from 60 to 70 inches. Because the normal distributions are symmetric, half of these women are on the short side. The shortest 2.5% of young women are less than 60 inches (5 feet) tall.

Almost all (99.7%) of the observations in any normal distribution lie within three standard deviations of the mean. Almost all young women are between 57.5 and 72.5 inches tall.

Standard scores

Jennie scored 600 on the verbal part of the SAT college entrance exam. How good a score is this? That depends on where a score of 600 lies in the distri-

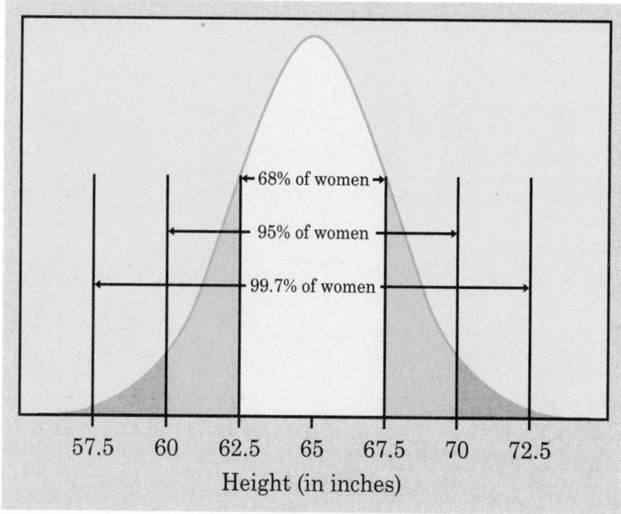

Figure 13.9 The 68–95–99.7 rule for heights of young women. This normal distribution has mean 65 inches and standard deviation 2.5 inches.

bution of all scores. The SAT exams are scaled so that scores should roughly follow the normal distribution with mean 500 and standard deviation 100. Jennie's 600 is one standard deviation above the mean. The 68–95–99.7 rule now tells us just where she stands (Figure 13.10). Half of all scores are below 500, and another 34% are between 500 and 600. So Jennie did better than 84% of the students who took the SAT. Her score report not only will say she scored 600 but will add that this is at the "84th percentile." That's statistics speak for "You did better than 84% of those who took the test."

Because the standard deviation is the natural unit of measurement for normal distributions, we restated Jennie's score of 600 as "one standard deviation above the mean." Observations expressed in standard deviations above or below the mean of a distribution are called *standard* scores.

Standard scores

The **standard score** for any observation is

$$\text{standard score} = \frac{\text{observation} - \text{mean}}{\text{standard deviation}}$$

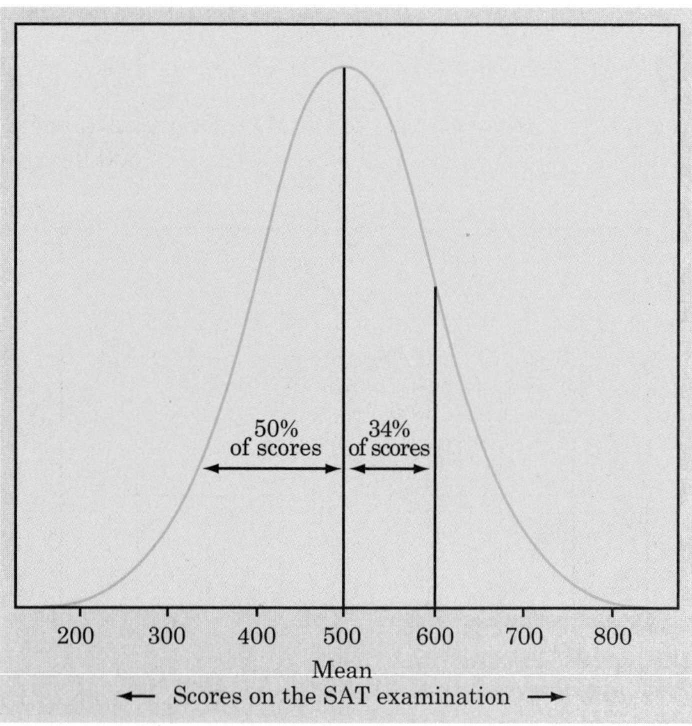

Figure 13.10 The 68–95–99.7 rule shows that 84% of any normal distribution lies to the left of the point one standard deviation above the mean. Here, this fact is applied to SAT scores.

A standard score of 1 says that the observation in question lies one standard deviation above the mean. An observation with standard score -2 is two standard deviations below the mean. Standard scores can be used to compare values in different distributions. Of course, you should not use standard scores unless you are willing to use the standard deviation to describe the spread of the distributions. That requires that the distributions be at least roughly symmetric.

Example 3. ACT versus SAT scores

Jennie scored 600 on the verbal part of the SAT. Her friend Gerald took the American College Testing (ACT) test and scored 21 on the verbal part. ACT scores are normally distributed with mean 18 and standard deviation 6. Assuming that both tests measure the same kind of ability, who has the higher score?

Jennie's standard score is

$$\frac{600-500}{100} = \frac{100}{100} = 1.0$$

Compare this with Gerald's standard score, which is

$$\frac{21-18}{6} = \frac{3}{6} = 0.5$$

Because Jennie's score is 1 standard deviation above the mean and Gerald's is only 0.5 standard deviation above the mean, Jennie's performance is better.

The bell curve?

Does the distribution of human intelligence follow the "bell curve" of a normal distribution? Scores on IQ tests do roughly follow a normal distribution. That is because a test score is calculated from a person's answers in a way that is designed to produce a normal distribution. To conclude that intelligence follows a bell curve, we must agree that the test scores directly measure intelligence. Many psychologists don't think there is one human characteristic that we can call "intelligence" and can measure by a single test score.

Percentiles of normal distributions*

For normal distributions, but not for other distributions, standard scores translate directly into *percentiles*.

The **cth percentile** of a distribution is a value such that *c* percent of the observations lie below it and the rest lie above.

The median of any distribution is the 50th percentile, and the quartiles are the 25th and 75th percentiles. In any normal distribution, the point one standard deviation above the mean (standard score 1) is the 84th percentile.

*This material is not needed to read the rest of the book.

Figure 13.10 shows why. Every standard score for a normal distribution translates into a specific percentile, which is the same no matter what the mean and standard deviation of the original normal distribution are. Table B at the end of this book gives the percentiles corresponding to various standard scores. This table enables us to do calculations in greater detail than the 68–95–99.7 rule.

Example 4. Percentiles for college entrance exams

Jennie's score of 600 on the SAT translates into a standard score of 1.0. We saw that the 68–95–99.7 rule says that this is the 84th percentile. Table B is a bit more precise: it says that standard score 1 is the 84.13 percentile of a normal distribution. Gerald's 21 on the ACT is a standard score of 0.5. Table B says that this is the 69.15 percentile. Gerald did well, but not as well as Jennie. The percentile is easier to understand than either the raw score or the standard score. That's why reports of exams such as the SAT usually give both the score and the percentile.

Example 5. Finding the observation that matches a percentile

How high must a student score on the SAT to fall in the top 20% of all scores? That requires a score at or above the 80th percentile. Look in the body of Table B for the percentiles closest to 80. You see that standard score 0.8 is the 78.81 percentile and standard score 0.9 is the 81.59 percentile. The percentile in the table closest to 80 is 78.81, so we conclude that a standard score of 0.8 is approximately the 80th percentile *of any normal distribution.*

To go from the standard score back to the scale of SAT scores, "undo" the standard score calculation as follows:

$$\begin{aligned} \text{observation} &= \text{mean} + \text{standard score} \times \text{standard deviation} \\ &= 500 + (0.8)(100) = 580 \end{aligned}$$

A score of 580 or higher will be in the top 20%. (More exactly, these scores are in the top 21.19% because 580 is exactly the 78.81 percentile, but we will be content to always use the closest entry in Table B.)

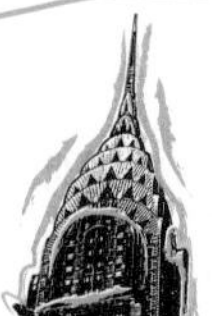

Exploring the Web

Have you ever seen a table of square roots? They were once common but have been replaced by the $\sqrt{x}$ button on a calculator. Tables of areas under a normal curve, like Table B at the back of this book, are still common but are also giving way to buttons and, better yet, to "applets" that let you find areas visually. Go to the *Statistics: Concepts and Controversies* Web site, www.whfreeman.com/scc, and look at the "Normal Curve" applet. You can even use this applet to do exercises that require normal curve calculations.

Statistics in Summary Stemplots, histograms, and boxplots all describe the distributions of quantitative variables. **Density curves** are another kind of graph that serves the same purpose. A density curve is a curve with area exactly 1 underneath it whose shape describes the overall pattern of a distribution. An area under the curve gives the proportion of the observations that fall in an interval of values. You can roughly locate the median (equal-areas point) and the mean (balance point) by eye on a density curve.

Normal curves are a special kind of density curve that describes the overall pattern of some sets of data. Normal curves are symmetric and bell-shaped. A specific normal curve is completely described by its mean and standard deviation. You can locate the mean (center point) and the standard deviation (distance from the mean to the change-of-curvature points) on a normal curve. All normal distributions obey the **68–95–99.7 rule. Standard scores** express observations in standard deviation units about the mean, which has standard score 0. A given standard score corresponds to the same **percentile** in any normal distribution. Table B gives percentiles for normal distributions.

CHAPTER 13 EXERCISES

13.1 Density curves.
(a) Sketch a density curve that is symmetric but has a shape different from that of the normal curves.

(b) Sketch a density curve that is strongly skewed to the left.

13.2 Mean and median. Figure 13.11 shows density curves of several shapes. Briefly describe the overall shape of each distribution. Several points are marked on each curve. The mean and the median are among these points. For each curve, which point is the median and which is the mean?

13.3 Random numbers. If you ask a computer to generate "random numbers" between 0 and 1, you will get observations from a *uniform distribution*. Figure 13.12 shows the density curve for a uniform distribution. This curve takes the constant value 1 between 0 and 1 and is zero outside that range. Use this density curve to answer these questions.
(a) Why is the total area under the curve equal to 1?

(b) The curve is symmetric. What is the value of the mean and median?

(c) What percent of the observations lie between 0 and 0.4?

IQ test scores. Figure 13.13 is a stemplot of the IQ test scores of 74 seventh-grade students. This distribution is very close to normal with mean 111 and standard deviation 11. It includes all the seventh graders in a rural Midwest school except for 4 low outliers who were dropped because they may have

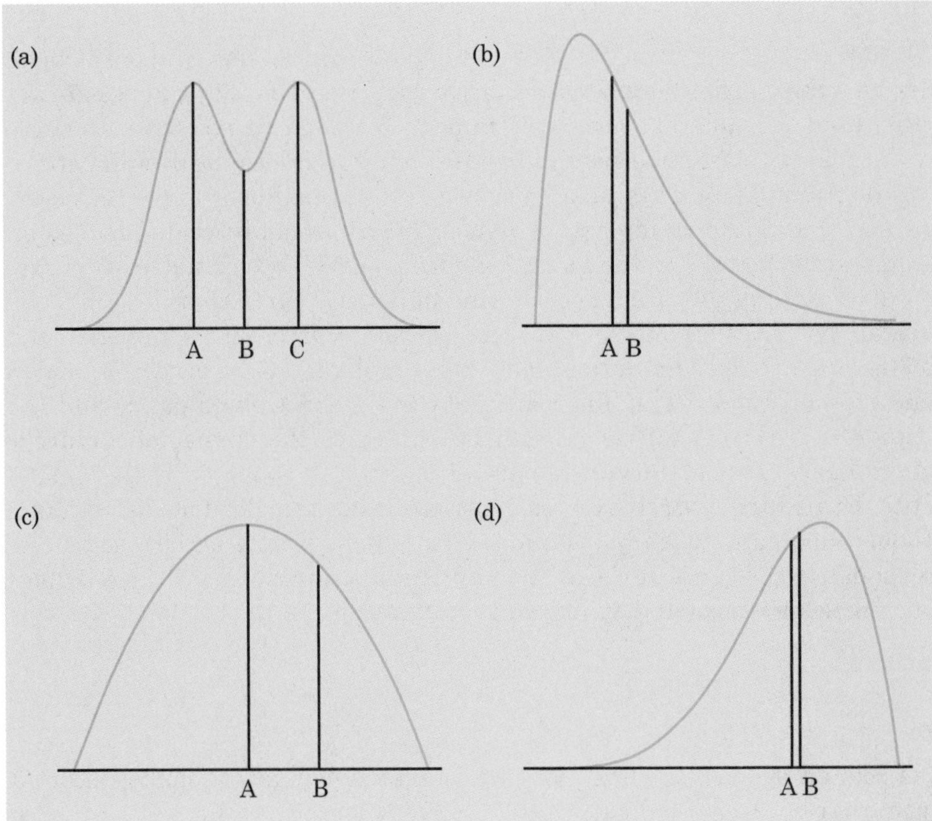

Figure 13.11 Four density curves of different shapes, for Exercise 13.2. In each case, the mean and the median are among the marked points.

been ill or otherwise not paying attention to the test. Take the normal distribution with mean 111 and standard deviation 11 as a description of the IQ test scores of all rural Midwest seventh-grade students. Use this distribution and the 68–95–99.7 rule to answer Exercises 13.4 to 13.6.

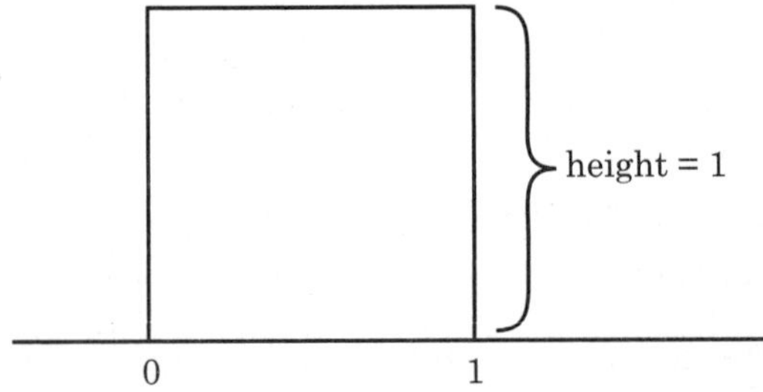

Figure 13.12 The density curve of a uniform distribution, for Exercise 13.3. Observations from this distribution are spread "at random" between 0 and 1.

8	69
9	0133
9	6778
10	0022333344
10	555666777789
10	0000111122223334444
11	55688999
11	003344
12	677888
12	02
13	6

Figure 13.13 Stemplot of the IQ scores of 74 seventh-grade students, for Exercises 13.4 to 13.6.

13.4 Between what values do the IQ scores of 95% of all rural Midwest seventh graders lie?

13.5 What percent of IQ scores for rural Midwest seventh graders are greater than 100?

13.6 What percent of all students have IQ scores 144 or higher? None of the 74 students in our sample school had scores this high. Are you surprised at this? Why?

13.7 **Length of pregnancies.** The length of human pregnancies from conception to birth varies according to a distribution that is approximately normal with mean 266 days and standard deviation 16 days. Use the 68–95–99.7 rule to answer the following questions.

(a) Between what values do the lengths of the middle 95% of all pregnancies fall?

(b) How short are the shortest 2.5% of all pregnancies?

13.8 **A normal curve.** What are the mean and standard deviation of the normal curve in Figure 13.14?

13.9 **Horse pregnancies.** Bigger animals tend to carry their young longer before birth. The length of horse pregnancies from conception to birth varies according to a roughly normal distribution with mean 336 days and standard deviation 3 days. Use the 68–95–99.7 rule to answer the following questions.

(a) Almost all (99.7%) of horse pregnancies fall in what range of lengths?

(b) What percent of horse pregnancies are longer than 339 days?

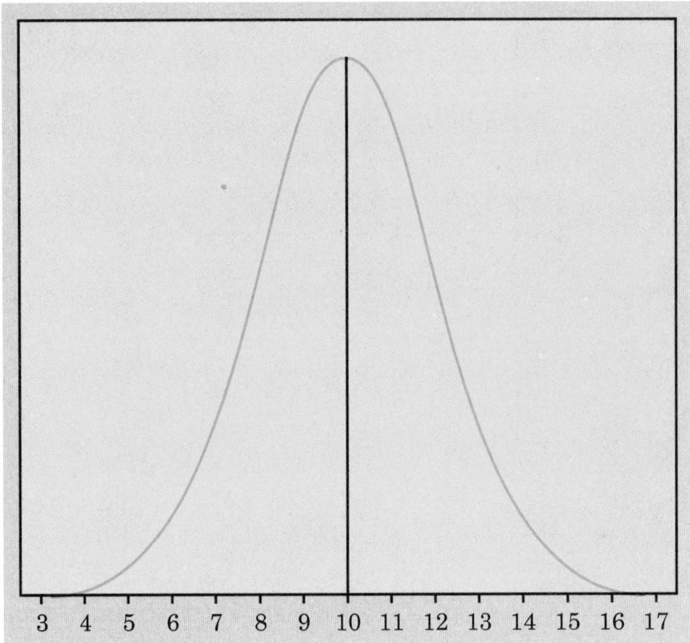

Figure 13.14 What are the mean and standard deviation of this normal density curve? For Exercise 13.8.

13.10 Three great hitters. Three landmarks of baseball achievement are Ty Cobb's batting average of .420 in 1911, Ted Williams's .406 in 1941, and George Brett's .390 in 1980. These batting averages cannot be compared directly because the distribution of major-league batting averages has changed over the years. The distributions are quite symmetric and (except for outliers such as Cobb, Williams, and Brett) reasonably normal. While the mean batting average has been held roughly constant by rule changes and the balance between hitting and pitching, the standard deviation has dropped over time. Here are the facts:

Decade	Mean	Std. dev.
1910s	.266	.0371
1940s	.267	.0326
1970s	.261	.0317

Compute the standard scores for the batting averages for Cobb, Williams, and Brett to compare how far each stood above his peers.

13.11 Comparing IQ scores. The Wechsler Adult Intelligence Scale (WAIS) is an "IQ test." Scores on the WAIS for the 20-to-34 age group are approximately normally distributed with mean 110 and standard deviation 25. Scores for the 60-to-64 age group are approximately normally distributed with mean 90 and standard deviation 25. Sarah, who is 30, scores 135 on the WAIS. Her mother, who is 60, also takes the test and scores 120.
(a) Express both scores as standard scores that show where each woman stands within her own age group.

(b) Who scored higher relative to her age group, Sarah or her mother? Who has the higher absolute level of the variable measured by the test?

13.12 Men's heights. The distribution of heights of young men is approximately normal with mean 70 inches and standard deviation 2.5 inches. Sketch a normal curve on which this mean and standard deviation are correctly located. (Hint: Draw the curve first, locate the points where the curvature changes, then mark the horizontal axis.)

13.13 More on men's heights. The distribution of heights of young men is approximately normal with mean 70 inches and standard deviation 2.5 inches. Use the 68–95–99.7 rule to answer the following questions.
(a) What percent of men are taller than 75 inches?

(b) Between what heights do the middle 95% of men fall?

(c) What percent of men are shorter than 67.5 inches?

13.14 Heights of men and women. The heights of young women are approximately normal with mean 65 inches and standard deviation 2.5 inches. The heights of men in the same age group have mean (and median) 70 inches. What percent of women are taller than a man of median height?

13.15 Heights of young adults. The mean height of men aged 18 to 24 is about 70 inches. Women that age have a mean height of about 65 inches. Do you think that the distribution of heights for all Americans aged 18 to 24 is approximately normal? Explain your answer.

13.16 Sampling. Suppose that the proportion of all adult Americans who are afraid to go out at night because of crime is $p = 0.4$. If we took many SRSs of size 1050, the sample proportion $\hat{p}$ would vary from sample to sample following a normal distribution with mean 0.4 and standard deviation 0.015. Use this fact and the 68–95–99.7 rule to answer these questions.
(a) In many samples, what percent of the values of $\hat{p}$ fall above 0.4? Above 0.43?

(b) In a large number of samples, what range contains the central 95% of values of $\hat{p}$?

13.17 Am I winning? Congressman Floyd commissions a sample survey of voters to learn what percent favor him in his race for reelection. To avoid spending too much, he samples only 400 voters. Suppose that in fact only 45% of the voters support Floyd. The percent favoring Floyd in a random sample of size 400 will vary from sample to sample according to a normal distribution with mean 45% and standard deviation 2.5%. What percent of all such samples will (wrongly) show that half or more of the voters favor Floyd?

The following optional exercises require use of Table B of normal distribution percentiles.

13.18 NCAA rules for athletes. The National Collegiate Athletic Association (NCAA) requires Division I athletes to score at least 820 on the combined mathematics and verbal parts of the SAT exam in order to compete in their first college year. (Higher scores are required for students with poor high school grades.) In 1999, the scores of the millions of students taking the SATs were approximately normal with mean 1017 and standard deviation 209. What percent of all students had scores less than 820?

13.19 More NCAA rules. The NCAA considers a student a "partial qualifier" eligible to practice and receive an athletic scholarship, but not to compete, if the combined SAT score is at least 720. Use the information in the previous exercise to find the percent of all SAT scores that are less than 720.

13.20 800 on the SAT. It is possible to score higher than 800 on either part of the SAT, but scores above 800 are reported as 800. (That is, a student can get a reported score of 800 without a perfect paper.) In 1999, the scores of men on the math part of the SAT followed a normal distribution with mean 531 and standard deviation 115. What percent of scores were above 800 (and so reported as 800)?

13.21 Women's SAT scores. The average performance of women on the SAT, especially the math part, is lower than that of men. The reasons for this gender gap are controversial. In 1999, women's scores on the math SAT followed a normal distribution with mean 495 and standard deviation 109. The mean for men was 531. What percent of women scored higher than the male mean?

13.22 Are we getting smarter? When the Stanford-Binet "IQ test" came into use in 1932, it was adjusted so that scores for each age group of children followed roughly the normal distribution with mean 100 and standard deviation 15. The test is readjusted from time to time to keep the mean at 100. If present-day American children took the 1932 Stanford-Binet test, their mean score would be about 120. The reasons for the increase in IQ over time are not known but probably include better childhood nutrition and more experience in taking tests.
(a) IQ scores above 130 are often called "very superior." What percent of children had very superior scores in 1932?

(b) If present-day children took the 1932 test, what percent would have very superior scores? (Assume that the standard deviation 15 does not change.)

13.23 Japanese IQ scores. The Wechsler Intelligence Scale for Children is used (in several languages) in the United States and Europe. Scores in each case are approximately normally distributed with mean 100 and standard deviation 15. When the test was standardized in Japan, the mean was 111. To what percentile of the American-European distribution of scores does the Japanese mean correspond?

13.24 The stock market. The annual rate of return on stock indexes (which combine many individual stocks) is very roughly normal. Since 1945, the Standard & Poor's 500 Index has had a mean yearly return of 12%, with a standard deviation of 16.5%. Take this normal distribution to be the distribution of yearly returns over a long period.

(a) In what range do the middle 95% of all yearly returns lie?

(b) The market is down for the year if the return on the index is less than zero. In what proportion of years is the market down?

(c) In what proportion of years does the index gain 25% or more?

13.25 Locating the quartiles. The quartiles of any distribution are the 25th and 75th percentiles. About how many standard deviations from the mean are the quartiles of any normal distribution?

13.26 Young women's heights. The heights of women aged 18 to 24 are approximately normal with mean 65 inches and standard deviation 2.5 inches. How tall are the tallest 10% of women? (Use the closest percent that appears in Table B.)

13.27 High IQ scores. Scores on the Wechsler Adult Intelligence Scale for the 20 to 34 age group are approximately normally distributed with mean 110 and standard deviation 25. How high must a person score to be in the top 25% of all scores?

Chapter 14
Describing Relationships: Scatterplots and Correlation

Ranking the states?

The news media have a weakness for lists. Best places to live, best colleges, healthiest foods, worst-dressed men . . . a list of best or worst is sure to find a place in the news. When the state-by-state SAT scores come out each year, it's therefore no surprise that we find news articles ranking the states from best (Iowa) to worst (South Carolina) according to the average SAT score achieved by their high school seniors.

The College Board, which sponsors the SAT exams, doesn't like this practice at all. "Comparing or ranking states on the basis of SAT scores alone is invalid and strongly discouraged by the College Board," says the heading on their table of state average SAT scores. To see why, let's look at the data.

Figure 14.1 shows the distribution of average scores on the mathematics part of the SAT exam for the 50 states and the District of Columbia. Iowa leads at 601, and South Carolina trails at 473 on the SAT scale of 400 to 800. The distribution has an unusual shape: it has two clear peaks. A double-peaked distribution is a clue that the data mix two distinct groups. Figure 14.2 unravels the mystery. This is a *scatterplot* that plots each state's average SAT score on the vertical axis against the percent of that state's high school graduates who take the SAT exam. The plot clearly shows two groups of states. In one group, no more than one-third of graduates take the exam and the average scores are high. In the other group of states, more than half take the SAT and the average scores are lower. The plot as a whole shows that average score goes down as more students take the exam.

There are in fact two common college entrance exams, the SAT and the ACT. Each state tends to favor one of these exams. In ACT states, only students applying to selective colleges take the SAT. They generally do well. No wonder that Iowa (home of the ACT), where only 5% of high school graduates take the SAT, does better than New York State, where 76% take the SAT. Iowa's higher score tells us nothing about the quality of education in Iowa relative to New York.

There are general lessons in this study of SAT scores. To understand one variable, we must often look at how it is related to other variables. And once again, to see what data say, start by making graphs.

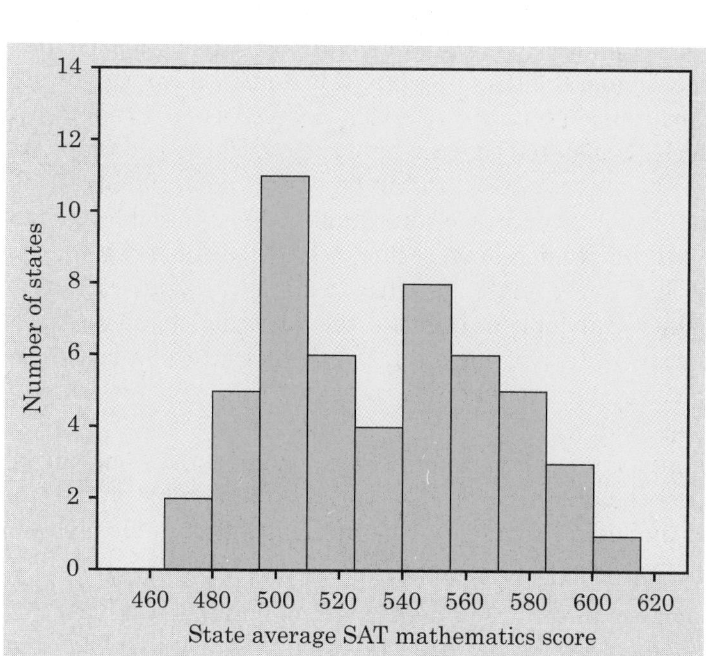

Figure 14.1 Histogram of the average scores of students in the 50 states and the District of Columbia on the mathematics part of the SAT examination.

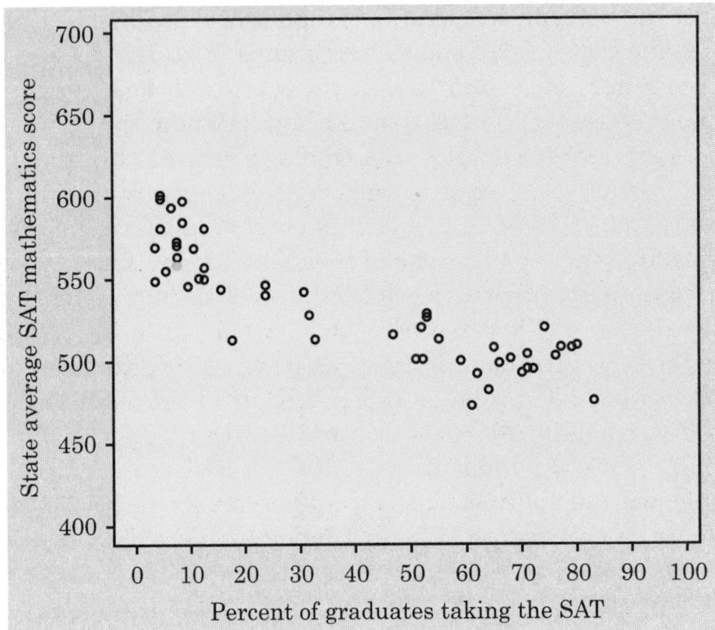

Figure 14.2 Scatterplot of average SAT mathematics score for each state against the percent of the state's high school graduates who take the SAT.

A medical study finds that short women are more likely to have heart attacks than women of average height, while tall women have the fewest heart attacks. An insurance group reports that heavier cars have fewer deaths per 10,000 vehicles registered than do lighter cars. These and many other statistical studies look at the relationship between two variables. To understand such a relationship, we must often examine other variables as well. To conclude that shorter women have higher risk of heart attacks, for example, the researchers had to eliminate the effect of other variables such as weight and exercise habits. Our topic in this and the following chapters is relationships between variables. One of our main themes is that the relationship between two variables can be strongly influenced by other variables that are lurking in the background.

Most statistical studies examine data on more than one variable. Fortunately, statistical analysis of several-variable data builds on the tools we used to examine individual variables. The principles that guide our work also remain the same:

- First plot the data, then add numerical summaries.
- Look for overall patterns and deviations from those patterns.
- When the overall pattern is quite regular, there is sometimes a way to describe it very briefly.

Scatterplots

The most common way to display the relation between two quantitative variables is a *scatterplot*. Figure 14.2 is a scatterplot that shows how the average SAT score in each state is related to the percent of high school graduates who take the SAT. We think that "percent taking" will help explain "average score." That is, "percent taking" is the *explanatory variable* and "average score" is the *response variable*. We want to see how average score changes when percent taking changes, so we put percent taking (the explanatory variable) on the horizontal axis. We can then see that as percent taking goes up, average score goes down. Each point on the plot represents one state. In Alabama, for example, 8% take the SAT, and the average SAT math score is 558. Find 8 on the x (horizontal) axis and 558 on the y (vertical) axis. Alabama appears as the orange point above 8 and to the right of 558.

Scatterplot

A **scatterplot** shows the relationship between two quantitative variables measured on the same individuals. The values of one variable appear on the horizontal axis, and the values of the other variable appear on the vertical axis. Each individual in the data appears as the point in the plot fixed by the values of both variables for that individual.

Always plot the explanatory variable, if there is one, on the horizontal axis (the x axis) of a scatterplot. As a reminder, we usually call the explanatory variable x and the response variable y. If there is no explanatory-response distinction, either variable can go on the horizontal axis.

Example 1. Health and wealth

Figure 14.3 is a scatterplot of data from the World Bank. The individuals are all the world's nations for which data are available. The explanatory variable is a measure of how rich a country is, the gross domestic product (GDP) per person. GDP is the total value of the goods and services produced in a country, converted into dollars. The response variable is life expectancy at birth.

We expect people in richer countries to live longer. The overall pattern of the scatterplot does show this, but the relationship has an interesting shape. Life expectancy rises very quickly as GDP increases, then levels off. People in very rich countries such as the United States live no longer than people in poorer but not extremely poor nations. Some of these countries, such as Costa Rica, even do better than the United States.

Three African nations are outliers. Their life expectancies are similar to those of their neighbors but their GDP is higher. Gabon produces oil, and Namibia and Botswana produce diamonds. It may be that income from mineral exports goes mainly to a few people and so pulls up GDP per person without much effect on either the income or the

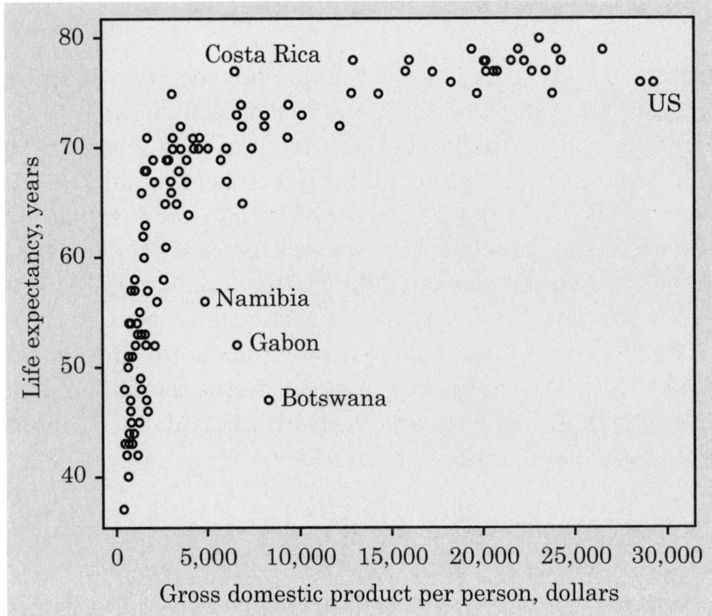

Figure 14.3 Scatterplot of the life expectancy of people in many nations against each nation's gross domestic product per person.

life expectancy of ordinary citizens. That is, GDP per person is a mean, and we know that mean income can be much higher than median income. In the case of Botswana, we may also have an incorrect value for GDP. The World Bank estimates $8310 per person, and this value is used in Figure 14.3, but the CIA estimates only $3600.

Interpreting scatterplots

To interpret a scatterplot, apply the usual strategies of data analysis.

Examining a scatterplot

In any graph of data, look for the **overall pattern** and for striking **deviations** from that pattern.

You can describe the overall pattern of a scatterplot by the **form, direction, and strength** of the relationship.

An important kind of deviation is an **outlier,** an individual value that falls outside the overall pattern of the relationship.

Both Figures 14.2 and 14.3 have a clear *direction:* average SAT scores go down as more students take the exam, and life expectancy generally goes up as GDP increases. We say that Figure 14.2 shows a *negative association* between the variables and that Figure 14.3 shows a *positive association.*

Positive association, negative association

Two variables are **positively associated** when above-average values of one tend to accompany above-average values of the other and below-average values also tend to occur together. The scatterplot slopes upward as we move from left to right.

Two variables are **negatively associated** when above-average values of one tend to accompany below-average values of the other, and vice versa. The scatterplot slopes downward from left to right.

Each of our scatterplots has a distinctive *form*. Figure 14.2 shows two *clusters* of states, and Figure 14.3 shows a *curved relationship*. The *strength* of a relationship in a scatterplot is determined by how closely the points follow a clear form. The relationships in Figures 14.2 and 14.3 are not strong. States with similar percents taking the SAT show quite a bit of scatter in their average scores, and nations with similar GDPs can have quite different life expectancies. Here is an example of a stronger relationship with a simple form.

Example 2. Classifying fossils

Archaeopteryx is an extinct beast having feathers like a bird but teeth and a long bony tail like a reptile. Only six fossil specimens are known. Because these fossils differ greatly in size, some scientists think they are different species rather than individuals from the same species. We will examine data on the lengths in centimeters of the femur (a leg bone) and the humerus (a bone in the upper arm) for the five fossils that preserve both bones. Here are the data:

Femur	38	56	59	64	74
Humerus	41	63	70	72	84

Because there is no explanatory/response distinction, we can put either measurement on the *x* axis of a scatterplot. The plot appears in Figure 14.4.

The plot shows a *strong positive straight-line association*. The straight-line form is important because it is common and simple. The association is strong because the points lie close to a line. It is positive because as the length of one bone increases, so does the length of the other bone. These data suggest that all five fossils belong to the same species and differ in size because some are younger than others. We expect that a different species would have a different relationship between the lengths of the two bones, so that it would appear as an outlier.

Correlation

A scatterplot displays the direction, form, and strength of the relationship between two variables. Straight-line relations are particularly important because a straight line is a simple pattern that is quite common. A straight-line

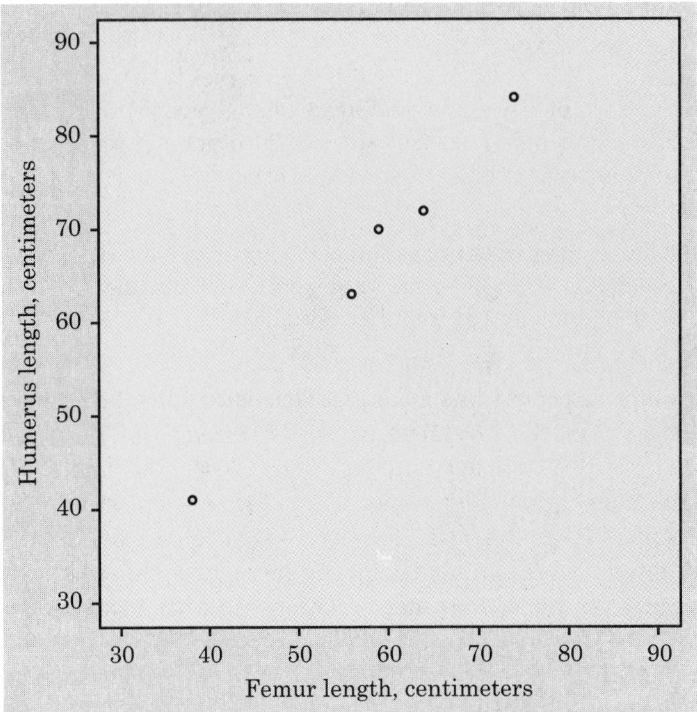

Figure 14.4 Scatterplot of the lengths of two bones in 5 fossil specimens of the extinct beast *Archaeopteryx*.

relation is strong if the points lie close to a straight line, and weak if they are widely scattered about a line. Our eyes are not good judges of how strong a relationship is. The two scatterplots in Figure 14.5 depict the same data, but the right-hand plot is drawn smaller in a large field. The right plot seems to show a stronger straight-line relationship. Our eyes can be fooled by changing the plotting scales or the amount of blank space around the cloud of points in a scatterplot. We need to follow our strategy for data analysis by using a numerical measure to supplement the graph. *Correlation* is the measure we use.

Correlation

The **correlation** describes the direction and strength of a straight-line relationship between two quantitative variables. Correlation is usually written as r.

Calculating a correlation takes a bit of work. You can usually think of r as the result of pushing a calculator button or giving a command in software and concentrate on understanding its properties and use. Knowing how we obtain r from data does help understand how correlation works, so here we go.

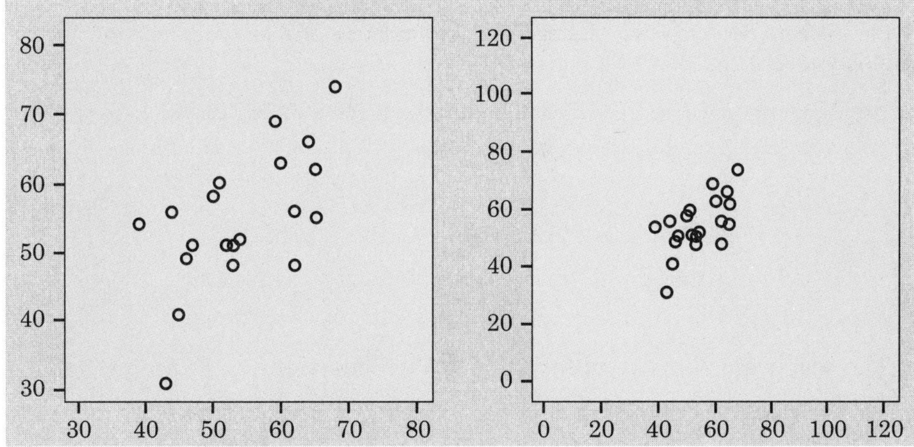

Figure 14.5 Two scatterplots of the same data. The right-hand plot suggests a stronger relationship between the variables because of the surrounding space.

Example 3. Calculating correlation

We have data on two variables, x and y, for n individuals. For the fossil data in Example 2, x is femur length, y is humerus length, and we have data for $n = 5$ fossils.

Step 1. Find the mean and standard deviation for both x and y. For the fossil data, a calculator tells us that

Femur:	$\overline{x} = 58.2\ \text{cm}$,	$s_x = 13.20\ \text{cm}$
Humerus:	$\overline{y} = 66.0\ \text{cm}$,	$s_y = 15.89\ \text{cm}$

We use s_x and s_y to remind ourselves that there are two standard deviations, one for the values of x and the other for the values of y.

Step 2. Using the means and standard deviations from Step 1, find the standard scores for each x-value and for each y-value:

Value of x	Standard score $(x - \overline{x})/s_x$	Value of y	Standard score $(y - \overline{y})/s_y$
38	(38 − 58.2)/13.20 = −1.530	41	(41 − 66.0)/15.89 = −1.573
56	(56 − 58.2)/13.20 = −0.167	63	(63 − 66.0)/15.89 = −0.189
59	(59 − 58.2)/13.20 = 0.061	70	(70 − 66.0)/15.89 = 0.252
64	(64 − 58.2)/13.20 = 0.439	72	(72 − 66.0)/15.89 = 0.378
74	(74 − 58.2)/13.20 = 1.197	84	(84 − 66.0)/15.89 = 1.133

Step 3. The correlation is the average of the products of these standard scores. As for the standard deviation, we "average" by dividing by $n - 1$, one fewer than the number of individuals:

$$r = \frac{1}{4}[(-1.530)(-1.573) + (-0.167)(-0.189) + (0.061)(0.252) + (0.439)(0.378) + (1.197)(1.133)]$$

$$= \frac{1}{4}(2.4067 + 0.0316 + 0.0154 + 0.1659 + 1.3562)$$

$$= \frac{3.9758}{4} = 0.994$$

The algebraic shorthand for the set of calculations in Example 3 is

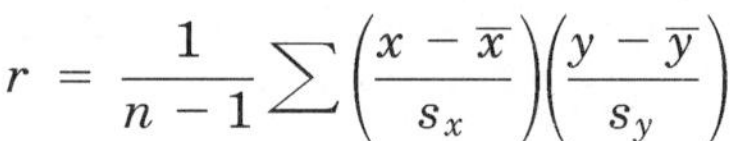

$$r = \frac{1}{n-1}\sum\left(\frac{x - \overline{x}}{s_x}\right)\left(\frac{y - \overline{y}}{s_y}\right)$$

The symbol Σ means "add them all up."

Understanding correlation

More important than calculating r (a task for a machine) is understanding how correlation measures association. Here are the facts:

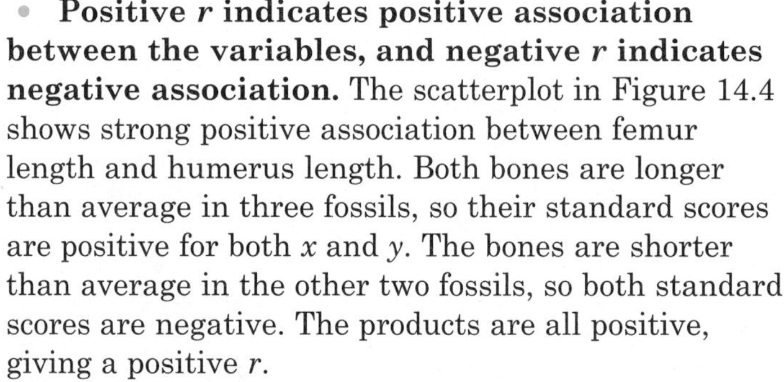

- **Positive r indicates positive association between the variables, and negative r indicates negative association.** The scatterplot in Figure 14.4 shows strong positive association between femur length and humerus length. Both bones are longer than average in three fossils, so their standard scores are positive for both x and y. The bones are shorter than average in the other two fossils, so both standard scores are negative. The products are all positive, giving a positive r.

- **The correlation r always falls between –1 and 1.** Values of r near 0 indicate a very weak straight-line relationship. The strength of the relationship increases as r moves away from 0 toward either –1 or 1. Values of r close to –1 or 1 indicate that the points lie close to a straight line. The extreme values $r = -1$ and $r = 1$ occur only when the points in a scatterplot lie exactly along a straight line.

  The result $r = 0.994$ in Example 3 reflects the strong positive straight-line pattern in Figure 14.4.

After you plot your data, think!

Abraham Wald (1902–1950), like many statisticians, worked on war problems during World War II. Wald invented some statistical methods that were military secrets until the war ended. Here is one of his simpler ideas. Asked where extra armor should be added to airplanes, Wald studied the location of enemy bullet holes in planes returning from combat. He plotted the locations on an outline of the plane. As data accumulated, most of the outline filled up. Put the armor in the few spots with no bullet holes, said Wald. That's where bullets hit the planes that didn't make it back.

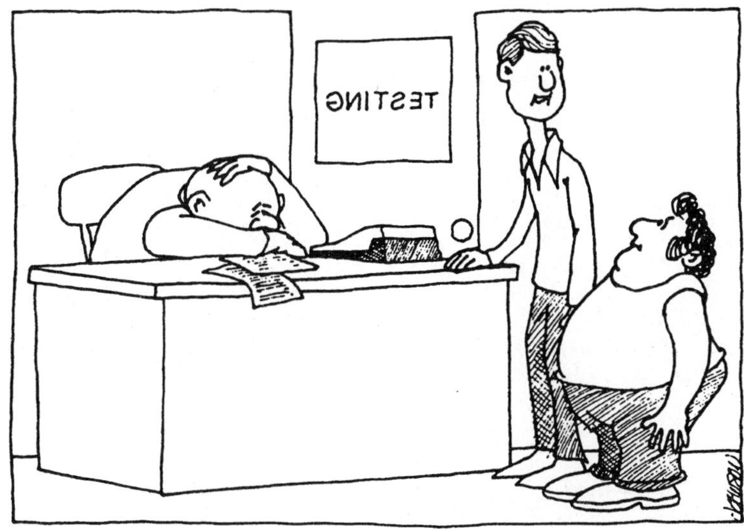

"He says we've ruined his positive correlation between height and weight."

The scatterplots in Figure 14.6 illustrate how r measures both the direction and strength of a straight-line relationship. Study them carefully. Note that the sign of r matches the direction of the slope in each plot, and that r approaches –1 or 1 as the pattern of the plot comes closer to a straight line.

- Because r uses the standard scores for the observations, **the correlation does not change when we change the units of measurement** of x, y, or both. Measuring length in inches rather than centimeters in Example 3 would not change the correlation $r = 0.994$.

  Our descriptive measures for one variable all share the same units as the original observations. If we measure length in centimeters, the median, quartiles, mean, and standard deviation are all in centimeters. The correlation between two variables, however, has no unit of measurement; it is just a number between –1 and 1.
- **Correlation ignores the distinction between explanatory and response variables**. If we reverse our choice of which variable to call x and which to call y, the correlation does not change.
- **Correlation measures the strength of only straight-line association between two variables.** Correlation does not describe curved relationships between variables, no matter how strong they are.
- Like the mean and standard deviation, **the correlation is strongly affected by a few outlying observations.** Use r with caution when

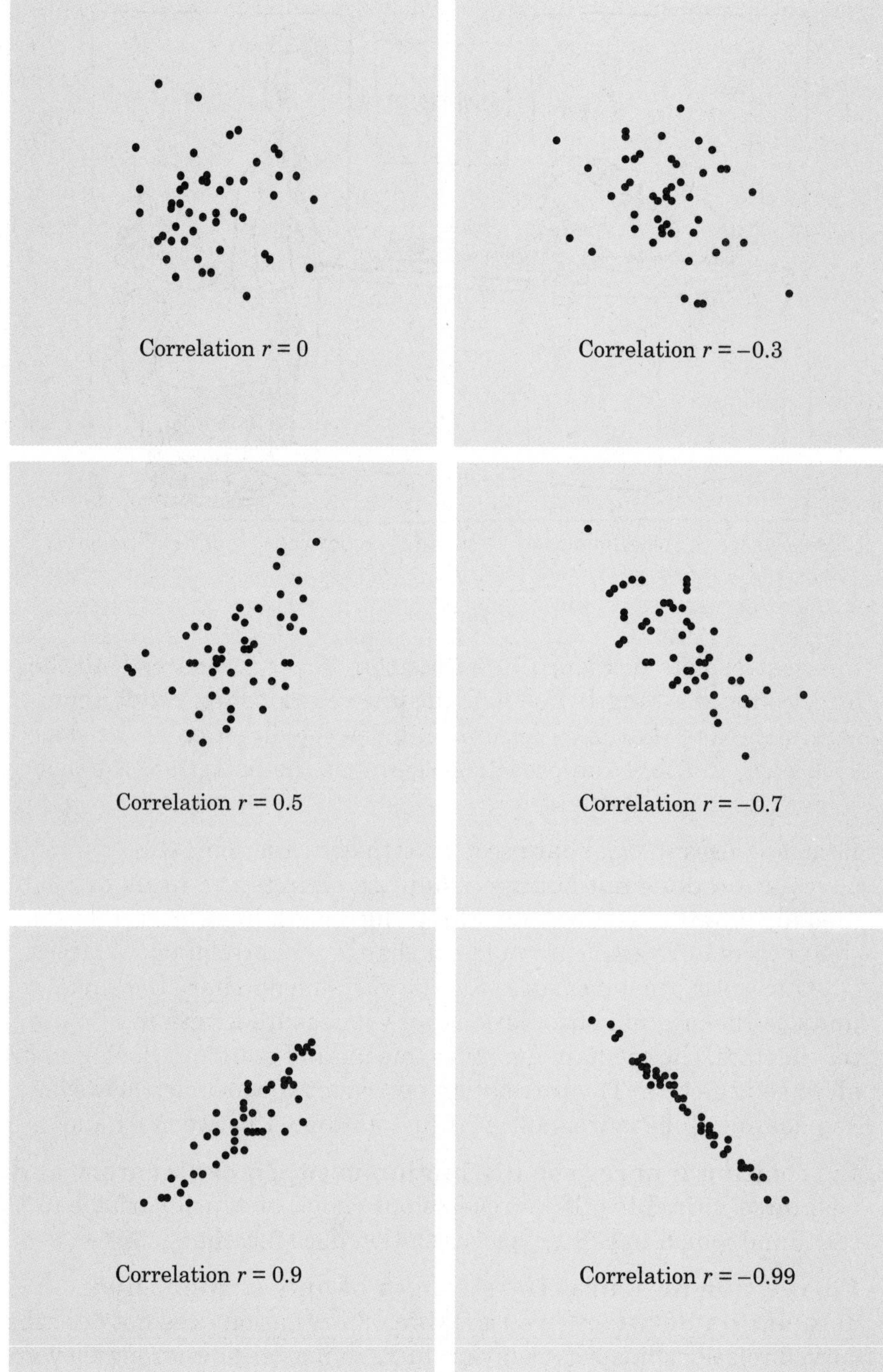

Figure 14.6 How correlation measures the strength of a straight-line relationship. Patterns closer to a straight line have correlations closer to 1 or −1.

outliers appear in the scatterplot. Look, for example, at Figure 14.7. We changed the femur length of the first fossil from 38 to 60 centimeters. Rather than falling in line with the other fossils, the first is now an outlier. The correlation drops from $r = 0.994$ for the original data to $r = 0.640$.

There are many kinds of relationships between variables, and many ways to measure them. Although correlation is very common, remember its limitations. Correlation makes sense only for quantitative variables—we can speak of the relationship between the sex of voters and the political party they prefer, but not of the correlation between these variables. Even for quantitative variables such as the length of bones, correlation measures only straight-line association.

Remember also that correlation is not a complete description of two-variable data, even when there is a straight-line relationship between the variables. You should give the means and standard deviations of both x and y along with the correlation. Because the formula for correlation uses the means and standard deviations, these measures are the proper choice to accompany a correlation.

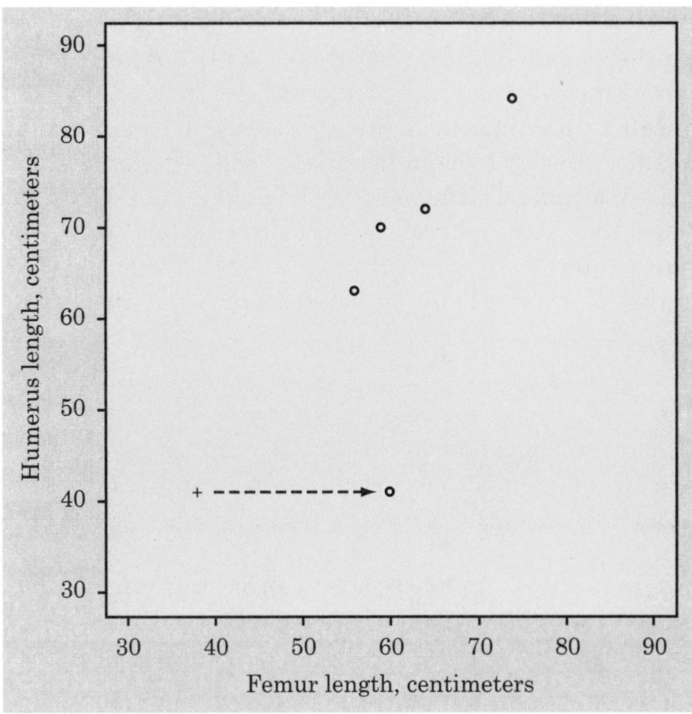

Figure 14.7 Moving one point reduces the correlation from $r = 0.994$ to $r = 0.640$.

Exploring the Web

The best way to grasp how the correlation reflects the pattern of the points on a scatterplot is to use an "applet" that allows you to plot and move data points and watch the correlation change. Go to the *Statistics: Concepts and Controversies* Web site, www.whfreeman.com/scc, and look at the "Correlation and Regression" applet. The correlation is constantly recalculated as you click to add points or use your mouse to move points.

Statistics in Summary Most statistical studies examine relationships between two or more variables. A **scatterplot** is a graph of the relationship between two quantitative variables. If you have an explanatory and a response variable, put the explanatory variable on the x (horizontal) axis of the scatterplot.

When you examine a scatterplot, look for the **direction, form, and strength** of the relationship and also for possible **outliers.** If there is a clear direction, is it positive (the scatterplot slopes upward from left to right) or negative (the plot slopes downward)? Is the form straight or curved? Are there clusters of observations? Is the relationship strong (a tight pattern in the plot) or weak (the points scatter widely)?

The **correlation r** measures the direction and strength of a straight-line relationship between two quantitative variables. Correlation is a number between −1 and 1. The sign of r shows whether the association is positive or negative. The value of r gets closer to −1 or 1 as the points cluster more tightly about a straight line. The extreme values −1 and 1 only occur when the scatterplot shows a perfectly straight line.

CHAPTER 14 EXERCISES

14.1 What number can I be?
(a) What are all the values that a correlation r can possibly take?

(b) What are all the values that a standard deviation s can possibly take?

14.2 Measuring crickets. For a biology project, you measure the length (centimeters) and weight (grams) of 12 crickets.
(a) Explain why you expect the correlation between length and weight to be positive.

(b) If you measured length in inches, how would the correlation change? (There are 2.54 centimeters in an inch.)

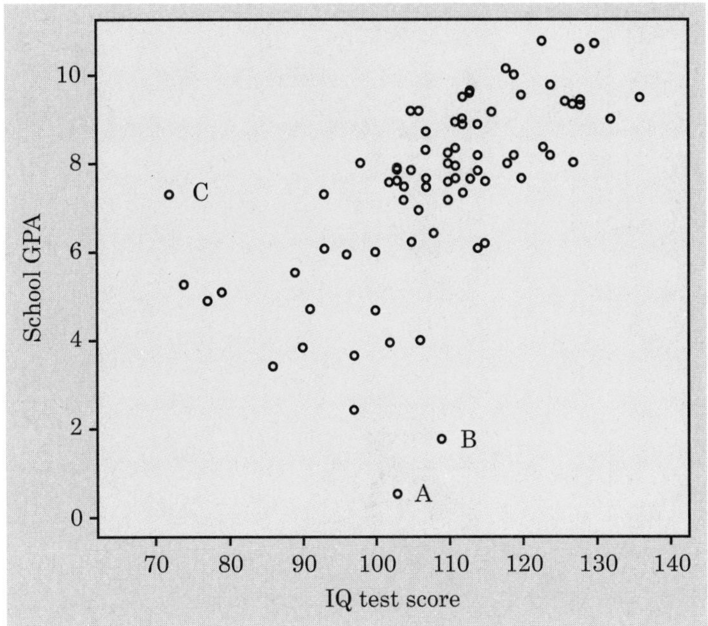

Figure 14.8 School grade point averages and IQ test scores for 78 seventh-grade students, for Exercise 14.3.

14.3 IQ and GPA. Figure 14.8 is a scatterplot of school grade point average versus IQ score for all 78 seventh-grade students in a rural Midwest school.
(a) Describe the overall pattern of the relationship in words. Points A, B, and C might be called outliers.

(b) About what are the IQ and GPA for student A?

(c) For each point A, B, and C, say how it is unusual (for example, "low GPA but a moderately high IQ score").

14.4 Calories and salt in hot dogs. Figure 14.9 shows the calories and sodium content in 17 brands of meat hot dogs. Describe the overall pattern of these data. In what way is the point marked A unusual?

14.5 IQ and GPA. Is the correlation r for the data in Figure 14.8 near –1, clearly negative but not near –1, near 0, clearly positive but not near 1, or near 1? Explain your answer.

14.6 Calories and salt in hot dogs. Is the correlation r for the data in Figure 14.9 near –1, clearly negative but not near –1, near 0, clearly positive but not near 1, or near 1? Explain your answer.

14.7 Comparing correlations. Which of Figures 14.8 and 14.9 has a correlation closer to 1? Explain your answer.

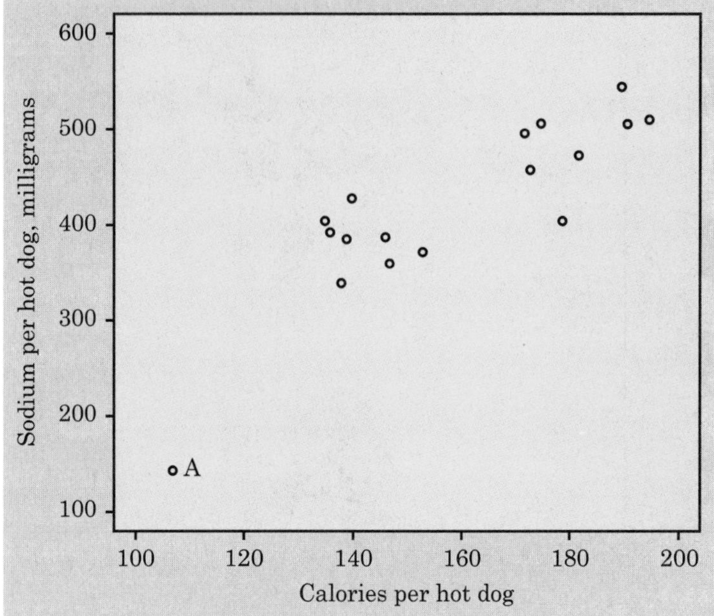

Figure 14.9 Calories and sodium content for 17 brands of meat hot dogs, for Exercise 14.4.

14.8 Outliers and correlation. Figure 14.8 contains outliers marked A, B, and C. The point marked A in Figure 14.9 is an outlier. Removing the outliers will *increase* the correlation r in one figure and *decrease* r in the other figure. What happens in each figure, and why?

14.9 The professor swims. Professor Moore swims 2000 yards regularly in a vain attempt to undo middle age. Here are his times (in minutes) and his pulse rate after swimming (in beats per minute) for 23 sessions in the pool:

Time	34.12	35.72	34.72	34.05	34.13	35.72	36.17	35.57
Pulse	152	124	140	152	146	128	136	144
Time	35.37	35.57	35.43	36.05	34.85	34.70	34.75	33.93
Pulse	148	144	136	124	148	144	140	156
Time	34.60	34.00	34.35	35.62	35.68	35.28	35.97	
Pulse	136	148	148	132	124	132	139	

(a) Make a scatterplot. (Which is the explanatory variable?)

(b) Is the association between these variables positive or negative? Explain why you expect the relationship to have this direction.

(c) Describe the form and strength of the relationship.

Table 14.1 Lean body mass and metabolic rate

Subject	Sex	Mass (kg)	Rate (cal)	Subject	Sex	Mass (kg)	Rate (cal)
1	M	62.0	1792	11	F	40.3	1189
2	M	62.9	1666	12	F	33.1	913
3	F	36.1	995	13	M	51.9	1460
4	F	54.6	1425	14	F	42.4	1124
5	F	48.5	1396	15	F	34.5	1052
6	F	42.0	1418	16	F	51.1	1347
7	M	47.4	1362	17	F	41.2	1204
8	F	50.6	1502	18	M	51.9	1867
9	F	42.0	1256	19	M	46.9	1439
10	M	48.7	1614				

14.10 Who burns more energy? Metabolic rate, the rate at which the body consumes energy, is important in studies of weight gain, dieting, and exercise. Table 14.1 gives data on the lean body mass and resting metabolic rate for 12 women and 7 men who are subjects in a study of dieting. Lean body mass, given in kilograms, is a person's weight leaving out all fat. Metabolic rate is measured in calories burned per 24 hours, the same calories used to describe the energy content of foods. The researchers believe that lean body mass is an important influence on metabolic rate.

(a) Make a scatterplot of the data for the female subjects. Which is the explanatory variable?

(b) Is the association between these variables positive or negative? What is the form of the relationship? How strong is the relationship?

(c) Now add the data for the male subjects to your graph, using a different color or a different plotting symbol. Does the pattern of relationship that you observed in (b) hold for men also? How do the male subjects as a group differ from the female subjects as a group?

14.11 Marriage. Suppose that women always married men 2 years older than themselves. Draw a scatterplot of the ages of 5 married couples, with the wife's age as the explanatory variable. What is the correlation r for your data? Why?

14.12 Stretching a scatterplot. Changing the units of measurement can greatly alter the appearance of a scatterplot. Return to the fossil data from Example 2:

Femur	38	56	59	64	74
Humerus	41	63	70	72	84

These measurements are in centimeters. Suppose a deranged scientist measured the femur in meters and the humerus in millimeters. The data would then be

Femur	0.38	0.56	0.59	0.64	0.74
Humerus	410	630	700	720	840

(a) Draw an x axis extending from 0 to 75 and a y axis extending from 0 to 850. Plot the original data on these axes. Then plot the new data on the same axes in a different color. The two plots look very different.

(b) Nonetheless, the correlation is exactly the same for the two sets of measurements. Why do you know that this is true without doing any calculations?

14.13 The professor swims. Exercise 14.9 gives data on the time to swim 2000 yards and the pulse rate after swimming for a middle-aged professor.
(a) Use a calculator to find the correlation r. Explain from looking at the scatterplot why this value of r is reasonable.

(b) Suppose that the times had been recorded in seconds. For example, the time 34.12 minutes would be 2047 seconds. How would the value of r change?

14.14 Who burns more energy? Table 14.1 gives data on the lean body mass and metabolic rate for 12 women and 7 men. You made a scatterplot of these data in Exercise 14.10.
(a) Do you think the correlation will be about the same for men and women or quite different for the two groups? Why?

(b) Calculate r for women alone and also for men alone. (Use your calculator.)

14.15 Strong association but no correlation. The gas mileage of an automobile first increases and then decreases as the speed increases. Suppose that this relationship is very regular, as shown by the following data on speed (miles per hour) and mileage (miles per gallon):

Speed	20	30	40	50	60
MPG	24	28	30	28	24

Make a scatterplot of mileage versus speed. Use a calculator to show that the correlation between speed and mileage is $r = 0$. Explain why the correlation is 0 even though there is a strong relationship between speed and mileage.

14.16 Body mass and metabolic rate. The body mass data in Table 14.1 are given in kilograms. There are 2.2 pounds in a kilogram. If we changed the data from kilograms to pounds, how would the mean body mass change? How would the correlation between body mass and metabolic rate change?

14.17 What are the units? Your data consist of observations on the age of several subjects (measured in years) and the reaction times of these subjects (measured in seconds). In what units are each of the following descriptive statistics measured?

(a) The mean age of the subjects.

(b) The standard deviation of the subjects' reaction times.

(c) The correlation between age and reaction time.

(d) The median age of the subjects.

14.18 Teaching and research. A college newspaper interviews a psychologist about student ratings of the teaching of faculty members. The psychologist says, "The evidence indicates that the correlation between the research productivity and teaching rating of faculty members is close to zero." The paper reports this as "Professor McDaniel said that good researchers tend to be poor teachers, and vice versa." Explain why the paper's report is wrong. Write a statement in plain language (don't use the word "correlation") to explain the psychologist's meaning.

14.19 Sloppy writing about correlation. Each of the following statements contains a blunder. Explain in each case what is wrong.

(a) "There is a high correlation between the gender of American workers and their income."

(b) "We found a high correlation ($r = 1.09$) between students' ratings of faculty teaching and ratings made by other faculty members."

(c) "The correlation between age and income was found to be $r = 0.53$ years."

14.20 Guess the correlation. Measurements in large samples show that the correlation

(a) between the heights of fathers and the heights of their adult sons is about ______________.

(b) between the heights of husbands and the heights of their wives is about ______________.

(c) between the heights of women at age 4 and their heights at age 18 is about ______________.

The answers (in scrambled order) are

$$r = 0.2 \quad r = 0.5 \quad r = 0.8$$

Match the answers to the statements and explain your choice.

14.21 Guess the correlation. For each of the following pairs of variables, would you expect a substantial negative correlation, a substantial positive correlation, or a small correlation?

(a) The age of secondhand cars and their prices.

(b) The weight of new cars and their gas mileages in miles per gallon.

(c) The heights and the weights of adult men.

(d) The heights and the IQ scores of adult men.

14.22 Investment diversification. A mutual fund company's newsletter says, "A well-diversified portfolio includes assets with low correlations." The newsletter includes a table of correlations between the returns on various classes of investments. For example, the correlation between municipal bonds and large-cap stocks is 0.50 and the correlation between municipal bonds and small-cap stocks is 0.21.

(a) Rachel invests heavily in municipal bonds. She wants to diversify by adding an investment whose returns do not closely follow the returns on her bonds. Should she choose large-cap stocks or small-cap stocks for this purpose? Explain your answer.

(b) If Rachel wants an investment that tends to increase when the return on her bonds drops, what kind of correlation should she look for?

Table 14.2 Hot dog and soda prices at major-league baseball stadiums

Team	Hot dog	Soda	Team	Hot dog	Soda	Team	Hot dog	Soda
Angels	2.50	1.75	Giants	2.75	2.17	Rangers	2.00	2.00
Astros	2.00	2.00	Indians	2.00	2.00	Red Sox	2.25	2.29
Braves	2.50	1.79	Marlins	2.25	1.80	Rockies	2.25	2.25
Brewers	2.00	2.00	Mets	2.50	2.50	Royals	1.75	1.99
Cardinals	3.50	2.00	Padres	1.75	2.25	Tigers	2.00	2.00
Dodgers	2.75	2.00	Phillies	2.75	2.20	Twins	2.50	2.22
Expos	1.75	2.00	Pirates	1.75	1.75	White Sox	2.00	2.00

14.23 Take me out to the ball game. What is the relationship between the price charged for a hot dog and the price charged for a 16-ounce soda in major-league baseball stadiums? Table 14.2 gives some data. Make a scatterplot appropriate for showing how soda price helps explain hot dog price. Describe the relationship that you see. Are there any outliers?

14.24 When it rains, it pours. Figure 14.10 plots the highest *yearly* precipitation ever recorded in each state against the highest *daily* precipitation ever recorded in that state. The points for Alaska (AK), Hawaii (HI), and Texas (TX) are marked on the scatterplot.

(a) About what are the highest daily and yearly precipitation values for Alaska?

(b) Alaska and Hawaii have very high yearly maximums relative to their daily maximums. Omit these two states as outliers. Describe the nature of the relationship for the other states. Would knowing a state's highest daily precipitation be a great help in predicting that state's highest yearly precipitation?

14.25 How many corn plants are too many? How much corn per acre should a farmer plant to obtain the highest yield? To find the best planting

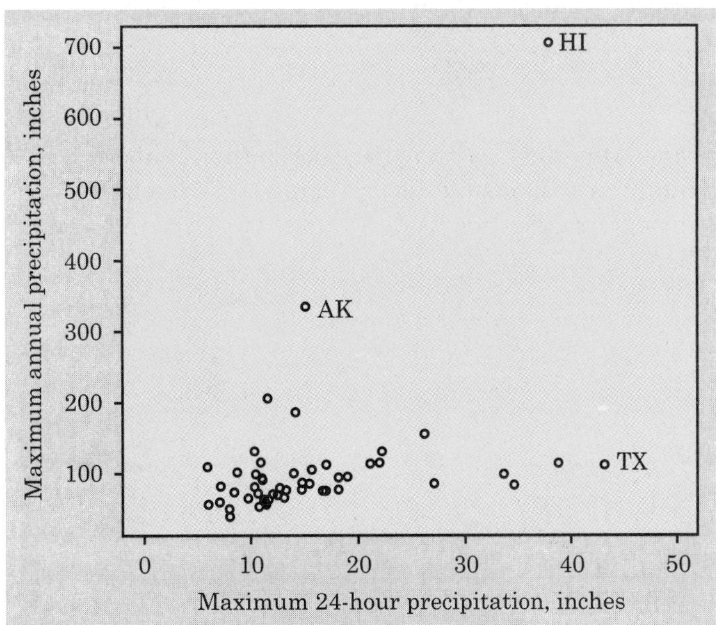

Figure 14.10 Record high yearly precipitation at any weather station in each state plotted against record high daily precipitation for the state, for Exercise 14.24.

rate, do an experiment: plant at different rates on several plots of ground and measure the harvest. Here are data from such an experiment:

Plants per acre	Yield (bushels per acre)			
12,000	150.1	113.0	118.4	142.6
16,000	166.9	120.7	135.2	149.8
20,000	165.3	130.1	139.6	149.9
24,000	134.7	138.4	156.1	
28,000	119.0	150.5		

(a) Is yield or planting rate the explanatory variable?

(b) Make a scatterplot of yield and planting rate.

(c) Describe the overall pattern of the relationship. Is it a straight line? Is there a positive or negative association, or neither? Explain why increasing the number of plants per acre of ground has the effect that your graph shows.

14.26 Why so small? Make a scatterplot of the following data:

x	1	2	3	4	10	10
y	1	3	3	5	1	11

Use your calculator to show that the correlation is about 0.5. What feature of the data is responsible for reducing the correlation to this value despite a strong straight-line association between x and y in most of the observations?

Chapter 15 Describing Relationships: Regression, Prediction, and Causation

Will stocks go up or down?

Predicting the future course of the stock market could make you rich. No wonder lots of people and lots of computers pore over market data looking for patterns.

Some popular patterns are a bit bizarre. The "Super Bowl Indicator" says that the football Super Bowl, played in January, predicts how stocks will behave each year. The current National Football League (NFL) was formed by merging the original NFL with the American Football League (AFL). The indicator claims that stocks go up in years when a team from the old NFL wins and down when an AFL team wins. The indicator was right in 28 of 33 years between the first Super Bowl in 1967 and 1999. Sounds impressive. But stocks went down only 6 times in those 33 years, so just predicting "up" every year would be right 27 times. Original NFL teams have won most Super Bowls for good reasons: there are 17 of them against only 11 old AFL teams, and the NFL was the established and stronger league. So "NFL wins" has been pretty much the same as "up every year." The great 1990s boom in stocks roared on unhindered by AFL wins in 1998 and 1999.

The Super Bowl predictor is simple-minded. There are statistical methods to predict one variable from others that go well beyond just counting ups and downs. *Regression*, which we will meet in this chapter, is the starting point for these methods. They go on quite a ways. One Web site sells stock price predictions based on "artificial neural networks, genetic algorithms, nearest-neighbor modeling, and other pattern classification techniques, combined with chaotic-, fractal-, and wavelet-based price-time series transforms." Sounds impressive, but does it work? Here's a hint: if these methods really predicted stock prices, their owners wouldn't be selling them to us—they would be quietly using them to get very rich.

If a pattern were found, says the accepted financial wisdom, so many people would try to take advantage of it that the pattern would quickly disappear. If, for example, stocks regularly went up between St. Patrick's Day (March 17) and April Fool's Day (April 1), people would buy stocks around March 17 and sell them around April 1. Result: buying would drive stock prices up at the beginning of the period and selling would push them down at the end, destroying the pattern. This logic is more convincing than claims to have discovered a way to beat the market. Prediction, as I will often remark in this chapter, is a subtle business.

Regression lines

If a scatterplot shows a straight-line relationship between two quantitative variables, we would like to summarize this overall pattern by drawing a line on the graph. A *regression line* summarizes the relationship between two variables, but only in a specific setting: one of the variables helps explain or predict the other. That is, regression describes a relationship between an explanatory variable and a response variable.

Regression line

A **regression line** is a straight line that describes how a response variable y changes as an explanatory variable x changes. We often use a regression line to predict the value of y for a given value of x.

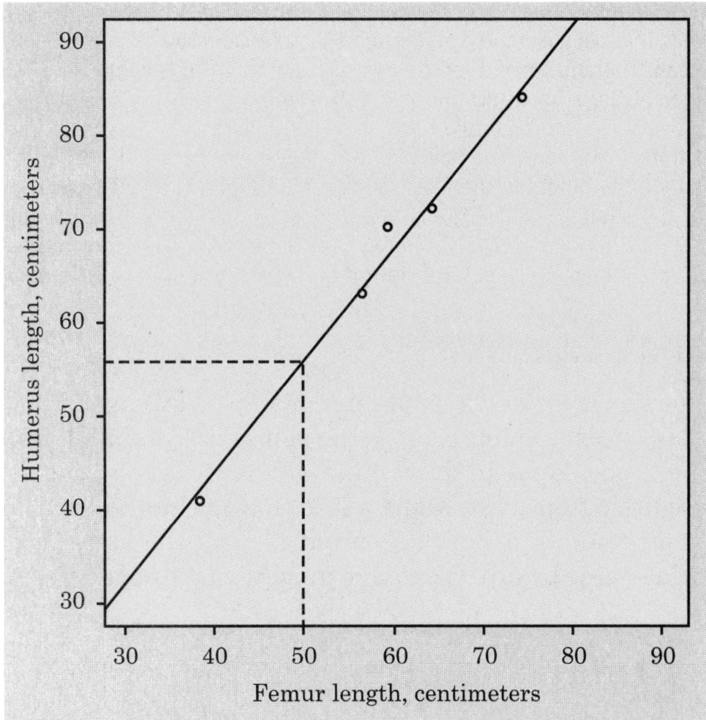

Figure 15.1 Using a straight-line pattern for prediction. The data are the lengths of two bones in 5 fossils of the extinct beast *Archaeopteryx*.

Example 1. Fossil bones

We saw that the lengths of two bones in fossils of the extinct beast *Archaeopteryx* closely follow a straight-line pattern. Figure 15.1 plots the lengths for the 5 available fossils. The line on the plot gives a quick summary of the overall pattern.

Another *Archaeopteryx* fossil is incomplete. Its femur is 50 centimeters long, but the humerus is missing. Can we predict how long the humerus is? The straight-line pattern connecting humerus length to femur length is so strong that we feel quite safe in using femur length to predict humerus length. Figure 15.1 shows how: starting at the femur length (50 cm), go up to the line, then over to the humerus length axis. We predict length about 56 cm. This is the length the humerus would have if this fossil's point lay exactly on the line. All the other points are close to the line, so we think the missing point would also be close to the line. That is, we think this prediction will be quite accurate.

Example 2. Presidential elections

Republican Ronald Reagan was elected president twice, in 1980 and in 1984. Figure 15.2 plots the percent who voted for Reagan's Democratic opponents: Jimmy Carter in

1980 and Walter Mondale in 1984. The plot shows a positive straight-line relationship. We expect this because some states tend to vote Democratic and others tend to vote Republican. There is one outlier: Georgia, President Carter's home state, voted 56% for the Democrat Carter in 1980 but only 40% Democratic in 1984.

We could use the regression line drawn in Figure 15.2 to predict a state's 1984 vote from its 1980 vote. The points in this figure are more widely scattered about the line than are the points in the fossil bone plot in Figure 15.1. The correlations, which measure the strength of the straight-line relationships, are $r = 0.994$ for Figure 15.1 and $r = 0.704$ for Figure 15.2. The scatter of the points makes it clear that predictions of voting will be generally less accurate than predictions of bone length.

Regression equations

When a plot shows a straight-line relationship as strong as that in Figure 15.1, it is easy to draw a line close to the points by eye. In Figure 15.2, however, different people might draw quite different lines by eye. Because we want to predict y from x, we want a line that is close to the points in the *vertical* (y) direction. It is hard to concentrate on just the vertical distances when drawing a line by eye. What is more, drawing by eye gives us a line on

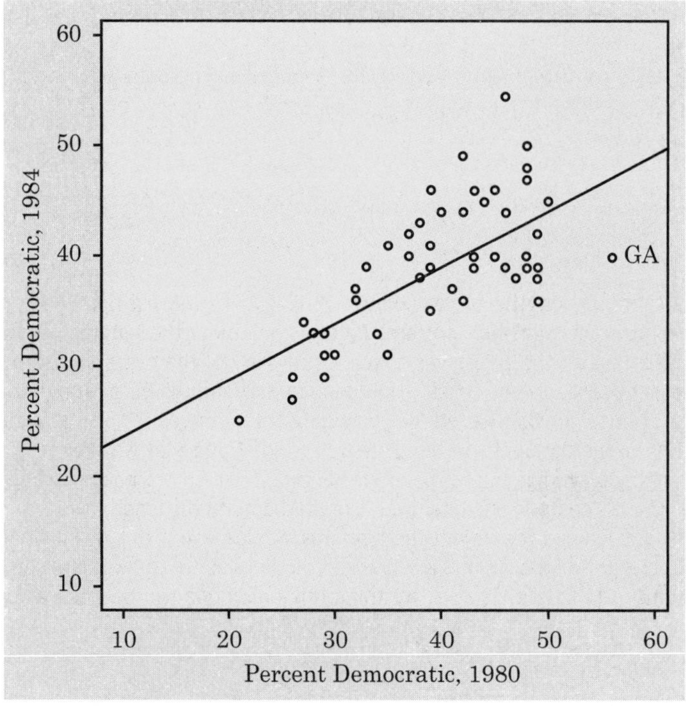

Figure 15.2 A weaker straight-line pattern. The data are the percent in each state who voted Democratic in the two Reagan presidential elections.

the graph, but not an equation for the line. We need a way to find from the data the equation of the line that comes closest to the points in the vertical direction. There are many ways to make the collection of vertical distances "as small as possible." The most common is the *least-squares* method.

Least-squares regression line

The **least-squares regression line** of y on x is the line that makes the sum of the squares of the vertical distances of the data points from the line as small as possible.

Figure 15.3 illustrates the least-squares idea. This figure magnifies the center part of Figure 15.1 to focus on 3 of the points. We see the vertical distances of these 3 points from the regression line. To find the least-squares line, look at these vertical distances (all 5 for the fossil data), square them, and move the line until the sum of the squares is the smallest it can be for

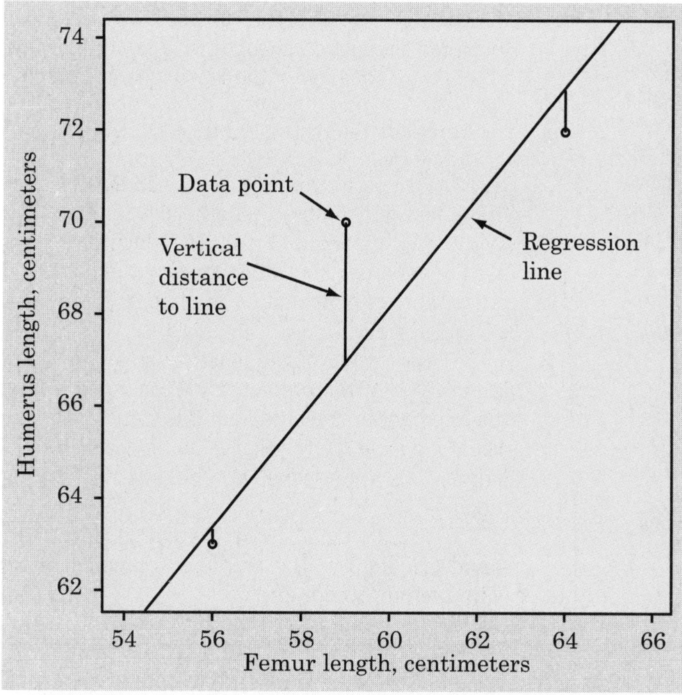

Figure 15.3 A regression line aims to predict y from x. So a good regression line makes the vertical distances from the data points to the line small.

Regression toward the mean

To "regress" means to go backward. Why are statistical methods for predicting a response from an explanatory variable called "regression"? Sir Francis Galton (1822–1911), who was the first to apply regression to biological and psychological data, looked at examples such as the heights of children versus the heights of their parents. He found that the taller-than-average parents tended to have children who were also taller than average, but not as tall as their parents. Galton called this fact "regression toward the mean" and the name came to be applied to the statistical method.

any line. The lines drawn on the scatterplots in Figures 15.1 and 15.2 are the least-squares regression lines. We won't give the formula for finding the least-squares line from data—that's a job for a calculator or computer. You should, however, be able to use the equation that the machine produces.

In writing the equation of a line, x stands as usual for the explanatory variable and y for the response variable. The equation of a line has the form

$$y = a + bx$$

The number b is the **slope** of the line, the amount by which y changes when x increases by one unit. The number a is the **intercept,** the value of y when $x = 0$. To use the equation for prediction, just substitute your x value into the equation and calculate the resulting y value.

Example 3. Using a regression equation

In Example 1, we used the "up and over" method in Figure 15.1 to predict the humerus length for a fossil whose femur length is 50 cm. The equation of the least-squares line is

$$\text{humerus length} = -3.66 + (1.197 \times \text{femur length})$$

The *slope* of this line is $b = 1.197$. This means that for these fossils, humerus length goes up by 1.197 cm when femur length goes up 1 cm. The slope of a regression line is usually important for understanding the data. The slope is the rate of change, the amount of change in the predicted y when x increases by 1.

The *intercept* of the least-squares line is $a = -3.66$. This is the value of the predicted y when $x = 0$. Although we need the intercept to draw the line, it is statistically meaningful only when x can actually take values close to zero. Here, femur length 0 is impossible, so the intercept has no statistical meaning.

To use the equation for *prediction,* substitute the value of x and calculate y. The predicted humerus length for a fossil with a femur 50 cm long is

$$\begin{aligned}\text{humerus length} &= -3.66 + (1.197)(50)\\ &= 56.2 \text{ cm}\end{aligned}$$

To *draw the line* on the scatterplot, predict y for two different values of x. This gives two points. Plot them and draw the line through them.

Understanding prediction

Computers make prediction easy and automatic, even from very large sets of data. Anything that can be done automatically is often done thoughtlessly. Regression software will happily fit a straight line to a curved relationship, for example. Also, the computer cannot decide which is the explanatory variable and which is the response variable. This is important, because the same data give two different lines depending on which is the explanatory variable.

In practice, we often use several explanatory variables to predict a response. A college might use SAT math and verbal scores and high school grades in English, math, and science (5 explanatory variables) to predict first-year college grades as part of its admissions process. Although the details are messy, all statistical methods of predicting a response share some basic properties of least-squares regression lines.

- **Prediction is based on fitting some "model" to a set of data.** In Figures 15.1 and 15.2, our model is a straight line which we draw through the points in a scatterplot. Other prediction methods use more elaborate models.

"How did I get into this business? Well, I couldn't understand regression and correlation in college, so I settled for this instead."

- **Prediction works best when the model fits the data closely.** Compare again Figure 15.1, where the data closely follow a line, with Figure 15.2, where they do not. Prediction is more trustworthy in Figure 15.1. It is not so easy to see patterns when there are many variables, but if the data do not have strong patterns, prediction may be very inaccurate.

- **Prediction outside the range of the available data is risky.** Suppose that you have data on a child's growth between 3 and 8 years of age. You find a strong straight-line relationship between age x and height y. If you fit a regression line to these data and use it to predict height at age 25 years, you will predict that the child will be 8 feet tall. Growth slows down and stops at maturity, so extending the straight line to adult ages is foolish. No one would make this mistake in predicting height. But almost all economic predictions try to tell us what will happen next quarter or next year. No wonder economic predictions are often wrong.

Example 4. Predicting the national surplus

The Congressional Budget Office is required to submit annual reports that predict the federal budget and its deficit or surplus for the next 5 years. These forecasts depend on future economic trends (unknown) and on what Congress will decide about taxes and spending (also unknown). Even the prediction of the state of the budget if current policies are not changed has been wildly inaccurate. The forecast made in 1996 for 1999, for example, missed by more than $300 billion. The 1997 forecast for the very next year was $102 billion off. As Senator Everett Dirksen once said, "A billion here and a billion there and pretty soon you are talking real money." By 1999, the Budget Office was predicting a surplus (ignoring Social Security) of $996 billion over the next 10 years. Politicians debated what to do with the money, but no one else believed the prediction.

Correlation and regression

Correlation measures the direction and strength of a straight-line relationship. Regression draws a line to describe the relationship. Correlation and regression are closely connected, even though regression requires choosing an explanatory variable and correlation does not.

Both correlation and regression are strongly affected by outliers. Be wary if your scatterplot shows strong outliers. Figure 15.4 plots the record high yearly precipitation in each state against that state's record high 24-hour precipitation. Hawaii is a high outlier, with a yearly record of 704.83 inches of rain recorded at Kukui in 1982. The correlation for all 50 states in Figure 15.4 is 0.408. If we leave out Hawaii, the correlation drops to r = 0.195 The solid line in the figure is the least-squares line for predicting the annual record from the 24-hour record. If we leave out Hawaii, the least-squares line drops down to the dotted line. This line is nearly flat—there is little relation between yearly and 24-hour record precipitation once we decide to ignore Hawaii.

The usefulness of the regression line for prediction depends on the strength of the association. That is, the usefulness of a regression line depends on the correlation between the variables. It turns out that the *square* of the correlation is the right measure.

R^2 in regression

The **square of the correlation, r^2**, is the fraction of the variation in the values of y that is explained by the least-squares regression of y on x.

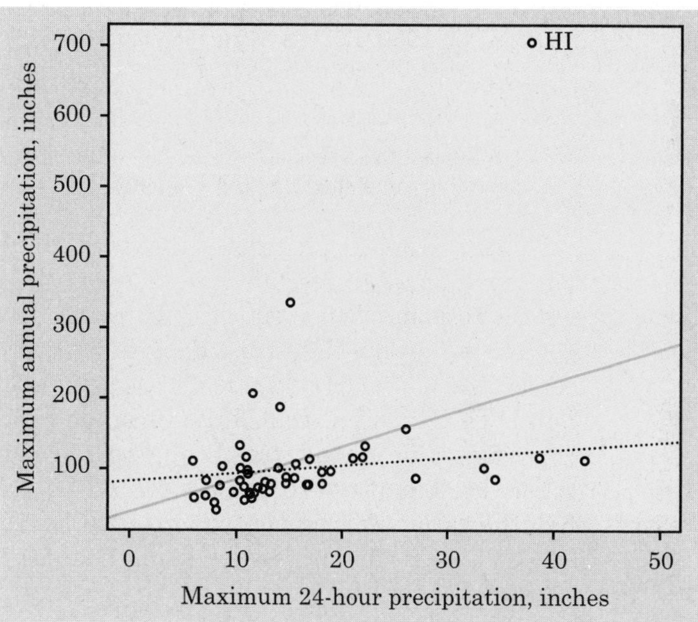

Figure 15.4 Least-squares regression lines are strongly influenced by outliers. The solid line is based on all 50 data points. The dotted line leaves out Hawaii.

The idea is that when there is a straight-line relationship, some of the variation in y is accounted for by the fact that as x changes it pulls y along with it.

Example 5. Using r^2

Look again at Figure 15.1. There is a lot of variation in the humerus lengths of these 5 fossils, from a low of 41 cm to a high of 84 cm. The scatterplot shows that we can explain almost all of this variation by looking at femur length and at the regression line. As femur length increases, it pulls humerus length up with it along the line. There is very little leftover variation in humerus length, which appears in the scatter of points about the line. Because $r = 0.994$ for these data, $r^2 = (0.994)^2 = 0.988$. So the variation "along the line" as femur length pulls humerus length with it accounts for 98.8% of all the variation in humerus length. The scatter of the points about the line accounts for only the remaining 1.2%. Little leftover scatter says that prediction will be accurate.

Contrast the voting data in Figure 15.2. There is still a straight-line relationship between the 1980 and 1984 Democratic votes, but there is also much more scatter of points about the regression line. Here, $r = 0.704$ and so $r^2 = 0.496$. Only about half the observed variation in the 1984 Democratic vote is explained by the straight-line pattern. You would still guess a higher 1984 Democratic vote for a state that was 45%

Did the vote counters cheat?

Republican Bruce Marks was ahead of Democrat William Stinson when the voting machines were tallied in their Pennsylvania election. But Stinson was ahead after absentee ballots were counted by the Democrats who controlled the election board. A court fight followed. The court called in a statistician, who used regression with data from past elections to predict the counts of absentee ballots from the voting-machine results. Marks's lead of 564 votes from the machines predicted that he would get 133 more absentee votes than Stinson. In fact, Stinson got 1025 more absentee votes than Marks. Did the vote counters cheat?

Democratic in 1980 than for a state that was only 30% Democratic in 1980. But lots of variation remains in the 1984 votes of states with the same 1980 vote. That is the other half of the total variation among the states in 1984. It is due to other factors, such as differences in the main issues in the two elections and the fact that President Reagan's two Democratic opponents came from different parts of the country.

In reporting a regression, it is usual to give r^2 as a measure of how successful the regression was in explaining the response. When you see a correlation, square it to get a better feel for the strength of the association. Perfect correlation ($r = -1$ or $r = 1$) means the points lie exactly on a line. Then $r^2 = 1$ and all of the variation in one variable is accounted for by the straight-line relationship with the other variable. If $r = -0.7$ or $r = 0.7$, $r^2 = 0.49$ and about half the variation is accounted for by the straight-line relationship. In the r^2 scale, correlation ± 0.7 is about halfway between 0 and ± 1.

The question of causation

There is a strong relationship between cigarette smoking and death rate from lung cancer. Does smoking cigarettes *cause* lung cancer? There is a strong association between the availability of handguns in a nation and that nation's homicide rate from guns. Does easy access to handguns *cause* more murders? It says right on the pack that cigarettes cause cancer. Whether more guns cause more murders is hotly debated. Why is the evidence for cigarettes and cancer better than the evidence for guns and homicide?

We already know three big facts about statistical evidence for cause and effect.

Statistics and causation

1. A strong relationship between two variables does not always mean that changes in one variable cause changes in the other.

2. The relationship between two variables is often influenced by other variables lurking in the background.

3. The best evidence for causation comes from randomized comparative experiments.

Example 6. Does television extend life?

Measure the number of television sets per person x and the life expectancy y for the world's nations. There is a high positive correlation: nations with many TV sets have higher life expectancies.

The basic meaning of causation is that by changing x we can bring about a change in y. Could we lengthen the lives of people in Botswana by shipping them TV sets? No. Rich nations have more TV sets than poor nations. Rich nations also have longer life expectancies because they offer better nutrition, clean water, and better health care. There is no cause-and-effect tie between TV sets and length of life.

Example 6 illustrates our first two big facts. Correlations such as this are sometimes called "nonsense correlations." The correlation is real. What is nonsense is the conclusion that changing one of the variables causes changes in the other. A lurking variable—such as national wealth in Example 6—that influences both x and y can create a high correlation even though there is no direct connection between x and y. We might call this *common response*: both the explanatory and the response variable are responding to some lurking variable.

"In a new attack on third-world poverty, aid organizations today began delivery of 100,000 television sets."

Example 7. Obesity in mothers and daughters

What causes obesity in children? Inheritance from parents, overeating, lack of physical activity, and too much television have all been named as explanatory variables.

The results of a study of Mexican-American girls aged 9 to 12 years are typical. Measure body mass index (BMI), a measure of weight relative to height, for both the girls and their mothers. People with high BMI are overweight or obese. Also measure hours of television, minutes of physical activity, and intake of several kinds of food. Result: the girls' BMIs were weakly correlated ($r = -0.18$) with physical activity and also with diet and television. The strongest correlation ($r = 0.506$) was between the BMI of daughters and the BMI of their mothers.

Body type is in part determined by heredity. Daughters inherit half their genes from their mothers. There is therefore a direct causal link between the BMI of mothers and daughters. Of course, the causal link is far from perfect. The mothers' BMIs explain only 25.6% (that's r^2 again) of the variation among the daughters' BMIs. Other factors, some measured in the study and others not measured, also influence BMI. *Even when direct causation is present, it is rarely a complete explanation of an association between two variables.*

Can we use r or r^2 from Example 7 to say how much inheritance contributes to the daughters' BMIs? No. Remember *confounding*. It may well be that mothers who are overweight also set an example of little exercise, poor eating habits, and lots of television. Their daughters pick up these habits to some extent, so the influence of heredity is mixed up with influences from the girls' environment. We can't say how much of the correlation between mother and daughter BMIs is due to inheritance.

Figure 15.5 shows in outline form how a variety of underlying links between variables can explain association. The dashed line represents an observed association between the variables x and y. Some associations are explained by a *direct cause-and-effect* link between the variables. The first diagram in Figure 15.5 shows "x causes y" by an arrow running from x to y. The second diagram illustrates *common response*. The observed association between the variables x and y is explained by a lurking variable z. Both x and y change in response to changes in z. This common response creates an association even though there may be no direct causal link between x and y. The third diagram in Figure 15.5 illustrates *confounding*. Both the explanatory variable x and the lurking variable z may influence the response variable y. Variables x and z are themselves associated, so we cannot distinguish the influence of x from the influence of z. We cannot say how strong the direct effect of x on y is. In fact, it can be hard to say if x influences y at all.

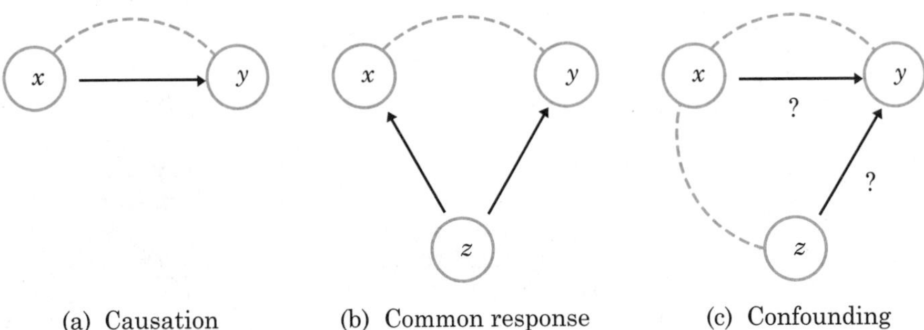

Figure 15.5 Some explanations for an observed association. A broken line shows an association. An arrow shows a cause-and-effect link. Variable x is explanatory, y is a response variable, and z is a lurking variable.

Statistical controversies

Gun Control and Crime

Do strict controls on guns, especially handguns, reduce crime? To many people, the answer must be "Yes." More than half of all murders in the United States are committed with handguns. The U.S. murder rate (per 100,000 population) is 1.7 times that of Canada, and the rate of murders with handguns is 15 times higher. Surely guns help bad things happen. Then John Lott, a University of Chicago economist, did an elaborate statistical study using data from all 3054 counties in the United States over the 18-year period from 1977 to 1994. Lott found that as states relaxed gun laws to allow adults to carry guns, the crime rate dropped. He argued that guns reduce crime by allowing citizens to defend themselves and by making criminals hesitate.

Lott used regression methods to determine the relationship between crime and many explanatory variables and to isolate the effect of permits to carry concealed guns after adjusting for other explanatory variables. The resulting debate, still going on, has been loud. People feel strongly about gun control. Most reacted to Lott's work based on whether or not they liked his conclusion. Gun supporters painted Lott as Moses revealing truth at last; opponents knew he must be both wrong and evil.

Is Lott right? I don't know. His work is more sophisticated than most older studies cited to support gun control. Yet large observational studies have many potential weaknesses, especially when they look for trends over time. Lots of things happen in 18 years, not all of which are in Lott's model—for example, police have become more aggressive in seizing illegal guns. Good data are hard to come by—for example, the number of people carrying guns legally is lower than the number of permits issued and is hard to estimate accurately. It would take very detailed study to reach an informed opinion on Lott's statistics.

The best reason to question Lott's findings combines awareness of the weaknesses of observational studies with the fact that statistical studies with stronger designs support reducing the presence of guns. Temporary bans on carrying guns in several cities in Colombia—highly publicized, and enforced by police checkpoints and searches—reduced murder rates compared with times when the bans were not in force. The Kansas City Gun Experiment compared two high-crime areas. In one, police seized guns by searches during traffic stops and after minor offenses. Gun crimes dropped by half in the treatment area and were unchanged in the control area. There seems good reason to think that reducing *illegal* carrying of guns reduces gun crime. Lott, of course, argues for *legal* carrying. Like many questions of causation, this one remains open.

Both common response and confounding involve the influence of a lurking variable or variables z on the response variable y. We won't labor the distinction between the two kinds of relationships. Just remember that "beware the lurking variable" is good advice in thinking about relationships between variables. Here is another example of common response, in a setting where we want to do prediction.

Example 8. SAT scores and college grades

High scores on the SAT examinations in high school certainly do not *cause* high grades in college. The moderate association (r^2 is about 27%) is no doubt explained by common response to such lurking variables as academic ability, study habits, and staying sober.

The ability of SAT scores to partly predict college performance doesn't depend on causation. We need only believe that the relationship between SAT scores and college grades that we see in past years will continue to hold for this year's high school graduates. Think once more of our fossils, where femur length predicts humerus length very well. The strong relationship is explained by common response to the overall age and size of the beasts whose fossils we now examine. *Prediction doesn't require causation.*

Discussion of these examples has brought to light two more big facts about causation:

More about statistics and causation

4. The observed relationship between two variables may be due to **direct causation, common response**, or **confounding**. Two or more of these factors may be present together.

5. An observed relationship can, however, be used for prediction without worrying about causation as long as the patterns found in past data continue to hold true.

Evidence for causation

Despite the difficulties, it is sometimes possible to build a strong case for causation in the absence of experiments. The evidence that smoking causes lung cancer is about as strong as nonexperimental evidence can be.

Doctors had long observed that most lung cancer patients were smokers. Observational studies comparing smokers and "similar" nonsmokers showed a strong association between smoking and death from lung cancer. Could the association be explained by lurking variables that the studies could not measure? Might there be, for example, a genetic factor that predisposes people both to nicotine addiction and to lung cancer? Smoking and lung cancer would then be positively associated even if smoking had no direct effect on the lungs. How were these objections overcome?

Let's answer this question in general terms: What are the criteria for establishing causation when we cannot do an experiment?

- **The association is strong.** The association between smoking and lung cancer is very strong.
- **The association is consistent.** Many studies of different kinds of people in many countries link smoking to lung cancer. That reduces the chance that a lurking variable specific to one group or one study explains the association.
- **Higher doses are associated with stronger responses.** People who smoke more cigarettes per day or who smoke over a longer period get lung cancer more often. People who stop smoking reduce their risk.
- **The alleged cause precedes the effect in time.** Lung cancer develops after years of smoking. The number of men dying of lung cancer rose as smoking became more common, with a lag of about 30 years. Lung cancer kills more men than any other form of cancer. Lung cancer was rare among women until women began to smoke. Lung cancer in women rose along with smoking, again with a lag of about 30 years, and has now passed breast cancer as the leading cause of cancer death among women.
- **The alleged cause is plausible.** Experiments with animals show that tars from cigarette smoke do cause cancer.

Medical authorities do not hesitate to say that smoking causes lung cancer. The U.S. Surgeon General has long stated that cigarette smoking is "the largest avoidable cause of death and disability in the United States." The evidence for causation is overwhelming—but it is not as strong as the evidence provided by well-designed experiments.

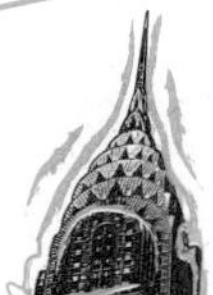

Exploring the Web

The best way to develop some feeling for how a regression line fits the points on a scatterplot is to use an applet that allows you to plot and move data points and watch the least-squares line move as the points change. Go to the *Statistics: Concepts and Controversies* Web site, www.whfreeman.com/scc, and look at the "Correlation and Regression" applet. Click in the "Show least-squares line" box to see the regression line.

Statistics in Summary

Regression is the name for statistical methods that fit some model to data in order to predict a response variable from one or more explanatory variables. The simplest kind of regression fits a straight line on a scatterplot for use in predicting y from x.

The most common way to fit a line is the **least-squares** method, which finds the line that makes the sum of the squared vertical distances of the data points from the line as small as possible.

Least-squares regression is closely related to correlation. In particular, the **squared correlation r^2** tells us what fraction of the variation in the responses is explained by the straight-line tie between y and x. It is generally true that the success of any statistical prediction depends on the presence of strong patterns in the data. Prediction outside the range of the data is risky because the pattern may be different there.

A strong relationship between two variables is not always evidence that changes in one variable **cause** changes in the other. Lurking variables can create relationships through **common response** or **confounding.** If we cannot do experiments, it is often difficult to get convincing evidence for causation.

CHAPTER 15 EXERCISES

15.1 Obesity in mothers and daughters. The study in Example 7 found that the correlation between the body mass index of young girls and their hours of physical activity in a day was $r = -0.18$. Why might we expect this correlation to be negative? What percent of the variation in BMI among the girls in the study can be explained by the straight-line relationship with hours of activity?

15.2 State SAT scores. Figure 14.2 (page 264) plots the average score of each state's high school seniors on the mathematics part of the SAT against the percent of seniors who take the exam. In addition to two clusters, the plot shows an overall rough straight-line pattern. The least-squares regression line for predicting SAT math score from percent taking is

$$\text{SAT score} = 574.6 - (1.102 \times \text{percent taking})$$

(a) What does the slope $b = -1.102$ tell us about the relationship between these variables?

(b) In New York State, 76% of high school seniors took the SAT. Predict their average score. (The actual average score in New York was 503.)

15.3 IQ and school GPA. Figure 14.8 (page 275) plots school grade point average (GPA) against IQ test score for 78 seventh-grade students. There is a rough straight-line pattern with quite a bit of scatter. The correlation between these variables is $r = 0.634$. What percent of the observed variation among the GPAs of these 78 students is explained by the straight-line relationship between GPA and IQ score? What percent of the variation is explained by differences in GPA among students with similar IQ scores?

15.4 State SAT scores. The correlation between the average SAT mathematics score in the states and the percent of high school seniors who take the SAT is $r = -0.879$.

(a) The correlation is negative. What does that tell us?

(b) How well does percent taking predict average score? (Use r^2 in your answer.)

15.5 IQ and school GPA. The least-squares line for predicting school GPA from IQ score, based on the 78 students plotted in Figure 14.8 (page 275) is

$$\text{GPA} = -3.56 + (0.101 \times \text{IQ})$$

Explain in words the meaning of the slope $b = 0.101$. Then predict the GPA of a student whose IQ score is 115.

15.6 The professor swims. Here are Professor Moore's times (in minutes) to swim 2000 yards and his pulse rate after swimming (in beats per minute) for 23 sessions in the pool:

Time	34.12	35.72	34.72	34.05	34.13	35.72	36.17	35.57
Pulse	152	124	140	152	146	128	136	144
Time	35.37	35.57	35.43	36.05	34.85	34.70	34.75	33.93
Pulse	148	144	136	124	148	144	140	156
Time	34.60	34.00	34.35	35.62	35.68	35.28	35.97	
Pulse	136	148	148	132	124	132	139	

You made a scatterplot of these data in Exercise 14.9 (page 276). The least-squares regression line is

$$\text{pulse rate} = 479.9 - (9.695 \times \text{time})$$

The next day's time is 34.30 minutes. Predict the professor's pulse rate. In fact, his pulse rate was 152. How accurate is your prediction?

15.7 Wine and heart disease. Drinking moderate amounts of wine may help prevent heart attacks. Let's look at data for entire nations. Table 15.1 gives data on yearly wine consumption (liters of alcohol from drinking wine, per person) and yearly deaths from heart disease (deaths per 100,000 people) in 19 developed countries.

(a) Make a scatterplot that shows how national wine consumption helps explain heart disease death rates.

Table 15.1 Wine consumption and heart disease

Country	Alcohol from wine*	Heart disease death rate†
Australia	2.5	211
Austria	3.9	167
Belgium/Lux.	2.9	131
Canada	2.4	191
Denmark	2.9	220
Finland	0.8	297
France	9.1	71
Iceland	0.8	211
Ireland	0.7	300
Italy	7.9	107
Netherlands	1.8	167
New Zealand	1.9	266
Norway	0.8	227
Spain	6.5	86
Sweden	1.6	207
Switzerland	5.8	115
United Kingdom	1.3	285
United States	1.2	199
West Germany	2.7	172

*Liters of alcohol from drinking wine, per person

†Deaths per 100,000 people

(b) Describe in words the direction, form, and strength of the relationship.

(c) The correlation for these variables is $r = -0.843$. Why does this value agree with your description in part (b)?

15.8 Beavers and beetles. Ecologists sometimes find rather strange relationships in our environment. One study seems to show that beavers benefit beetles. The researchers laid out 23 circular plots, each 4 meters in diameter, in an area where beavers were cutting down cottonwood trees.

In each plot, they counted the number of stumps from trees cut by beavers and the number of clusters of beetle larvae. Here are the data:

Stumps	2	2	1	3	3	4	3	1	2	5	1	3
Larvae clusters	10	30	12	24	36	40	43	11	27	56	18	40

Stumps	2	1	2	2	1	1	4	1	2	1	4
Larvae clusters	25	8	21	14	16	6	54	9	13	14	50

(a) Make a scatterplot that shows how the number of beaver-caused stumps influences the number of beetle larvae clusters. What does your plot show? (Ecologists think that the new sprouts from stumps are more tender than other cottonwood growth, so that beetles prefer them.)

(b) The least-squares regression line is

$$\text{larvae clusters} = -1.286 + (11.894 \times \text{stumps})$$

Draw this line on your plot. (To draw the line, use the equation to predict y for $x = 1$ and for $x = 5$. Plot the two (x, y) points and draw the line through them.)

(c) The correlation between these variables is $r = 0.916$. What percent of the observed variation in beetle larvae counts can be explained by straight-line dependence on stump counts?

(d) Based on your work in (a), (b), and (c), do you think that counting stumps offers a quick and reliable way to predict beetle larvae clusters?

15.9 Wine and heart disease. Table 15.1 gives data on wine consumption and heart disease death rates in 19 countries. A scatterplot (Exercise 15.7) shows a moderately strong relationship. The least-squares regression line for predicting heart disease death rate from wine consumption, calculated from the data in Table 15.1, is

$$y = 260.6 - 22.97x$$

Use this equation to predict the heart disease death rate in a country where adults average 1 liter of alcohol from wine each year and in a country that averages 8 liters per year. Use these two results to draw the least-squares line on your scatterplot.

15.10 Strong association but no correlation. Exercise 14.15 gives these data on the speed (miles per hour) and mileage (miles per gallon) of a car:

Speed	20	30	40	50	60
MPG	24	28	30	28	24

The least-squares line for predicting MPG from speed is

$$\text{MPG} = 26.8 + (0 \times \text{speed})$$

(a) Make a scatterplot of the data and draw this line on the plot.

(b) The correlation between MPG and speed is $r = 0$. What does this say about the usefulness of the regression line in predicting MPG?

15.11 Wine and heart disease. In Exercises 15.7 and 15.9, you examined data on wine consumption and heart disease deaths from Table 15.1. Suggest some differences among nations that may be confounded with wine-drinking habits. What is more, data about nations may tell us little about individual people. So these data alone are not evidence that you and I can lower our risk of heart disease by drinking more wine.

15.12 Correlation and regression. If the correlation between two variables x and y is $r = 0$, there is no straight-line relationship between the variables. It turns out that the correlation is 0 exactly when the slope of the least-squares regression line is 0. Explain why slope 0 means that there is no relationship between x and y. Start by drawing a line with slope 0 and explaining why in this situation x has no value for predicting y.

15.13 Acid rain. Researchers studying acid rain measured the acidity of precipitation in a Colorado wilderness area for 150 consecutive weeks. Acidity is measured by pH. Lower pH values show higher acidity. The acid rain researchers observed a straight-line pattern over time. They reported that the least-squares regression line

$$\text{pH} = 5.43 - (0.0053 \times \text{weeks})$$

fit the data well.

(a) Draw a graph of this line. Is the association positive or negative? Explain in plain language what this association means.

(b) According to the regression line, what was the pH at the beginning of the study (weeks = 1)? At the end (weeks = 150)?

(c) What is the slope of the regression line? Explain clearly what this slope says about the change in the pH of the precipitation in this wilderness area.

15.14 Review of straight lines. Fred keeps his savings in his mattress. He began with \$500 from his mother and adds \$100 each year. His total savings y after x years are given by the equation

$$y = 500 + 100x$$

(a) Draw a graph of this equation. (Choose two values of x, such as 0 and 10. Compute the corresponding values of y from the equation. Plot these two points on graph paper and draw the straight line joining them.)

(b) After 20 years, how much will Fred have in his mattress?

(c) If Fred had added \$200 instead of \$100 each year to his initial \$500, what is the equation that describes his savings after x years?

15.15 Review of straight lines. During the period after birth, a male white rat gains exactly 40 grams (g) per week. (This rat is unusually regular in his growth, but 40 g per week is a realistic rate.)
(a) If the rat weighed 100g at birth, give an equation for his weight after x weeks. What is the slope of this line?

(b) Draw a graph of this line between birth and 10 weeks of age.

(c) Would you be willing to use this line to predict the rat's weight at age 2 years? Do the prediction and think about the reasonableness of the result. (There are 454 grams in a pound. A large cat weighs about 10 pounds.)

15.16 More on correlation and regression. In Exercises 15.3 and 15.5, the correlation and the slope of the least-squares line for IQ and school GPA are both positive. In Exercises 15.7 and 15.9, both the correlation and the slope for wine consumption and heart disease deaths are negative. Is it possible for these two quantities to have opposite signs? Explain your answer.

15.17 Always plot your data! Table 15.2 presents four sets of data prepared by the statistician Frank Anscombe to illustrate the dangers of calculating without first plotting the data. *All four sets have the same correlation and the same least-squares regression line* to several decimal places. The regression equation is

$$y = 3 + 0.5x$$

(a) Make a scatterplot for each of the four data sets and draw the regression line on each of the plots. (To draw the regression line, substitute $x = 5$ and $x = 10$ into the equation. Find the predicted y for each x. Plot these two points and draw the line through them on all four plots.)

(b) In which of the four cases would you be willing to use the regression line to predict y given that $x = 10$? Explain your answer in each case.

15.18 Going to class helps. A study of class attendance and grades among first-year students at a state university showed that in general students who attended a higher percent of their classes earned higher grades. Class attendance explained 16% of the variation in grade index among the students. What is the numerical value of the correlation between percent of classes attended and grade index?

Table 15.2 Four data sets for exploring correlation and regression

Data Set A

x	10	8	13	9	11	14	6	4	12	7	5
y	8.04	6.95	7.58	8.81	8.33	9.96	7.24	4.26	10.84	4.82	5.68

Data Set B

x	10	8	13	9	11	14	6	4	12	7	5
y	9.14	8.14	8.74	8.77	9.26	8.10	6.13	3.10	9.13	7.26	4.74

Data Set C

x	10	8	13	9	11	14	6	4	12	7	5
y	7.46	6.77	12.74	7.11	7.81	8.84	6.08	5.39	8.15	6.42	5.73

Data Set D

x	8	8	8	8	8	8	8	8	8	8	19
y	6.58	5.76	7.71	8.84	8.47	7.04	5.25	5.56	7.91	6.89	12.50

SOURCE: Frank J. Anscombe, "Graphs in statistical analysis," *The American Statistician*, 27 (1973), pp. 17–21.

15.19 The declining farm population. The number of people living on American farms has declined steadily during this century. Here are data on the farm population (millions of persons) from 1935 to 1980:

Year	1935	1940	1945	1950	1955	1960	1965	1970	1975	1980
Population	32.1	30.5	24.4	23.0	19.1	15.6	12.4	9.7	8.9	7.2

(a) Make a scatterplot of these data. Draw by eye a regression line for predicting a year's farm population.

(b) Extend your line to predict the number of people living on farms in 1990. Is this result reasonable? Why?

15.20 Lots of wine. Exercise 15.9 gives us the least-squares line for predicting heart disease deaths per 100,000 people from liters of alcohol from wine, per person. The line is based on data from 19 rich countries. The equation is $y = 260.6 - 22.97x$. What is the predicted heart disease death rate for a country that drinks enough wine to supply 150 liters of alcohol per person? Explain why this result can't be true. Explain why using the regression line for this prediction is not intelligent.

15.21 Do firefighters make fires worse? Someone says, "There is a strong positive correlation between the number of firefighters at a fire and the amount of damage the fire does. So sending lots of firefighters just causes more damage." Explain why this reasoning is wrong.

15.22 How's your self-esteem? People who do well tend to feel good about themselves. Perhaps helping people feel good about themselves will help them do better in school and life. Raising self-esteem became for a time a goal in many schools. California even created a state commission to advance the cause. Can you think of explanations for the association between high self-esteem and good school performance other than "Self-esteem causes better work in school"?

15.23 Are big hospitals bad for you? A study shows that there is a positive correlation between the size of a hospital (measured by its number of beds x) and the median number of days y that patients remain in the hospital. Does this mean that you can shorten a hospital stay by choosing a small hospital? Why?

15.24 Health and wealth. An article entitled "The Health and Wealth of Nations" says:

> *The positive correlation between health and income per capita is one of the best-known relations in international development. This correlation is commonly thought to reflect a causal link running from income to health. . . . Recently, however, another intriguing possibility has emerged: that the health-income correlation is partly explained by a causal link running the other way—from health to income.*

Explain how higher income in a nation can cause better health. Then explain how better health can cause higher income. There is no simple way to determine the direction of the link.

15.25 Is math the key to success in college? Here is the opening of a newspaper account of a College Board study of 15,941 high school graduates:

> *Minority students who take high school algebra and geometry succeed in college at almost the same rate as whites, a new study says.*
>
> *The link between high school math and college graduation is "almost magical," says College Board President Donald Stewart, suggesting "math is the gatekeeper for success in college."*

> *"These findings," he says, "justify serious consideration of a national policy to ensure that all students take algebra and geometry."*

What lurking variables might explain the association between taking several math courses in high school and success in college? Explain why requiring algebra and geometry may have little effect on who succeeds in college.

15.26 Do artificial sweeteners cause weight gain? People who use artificial sweeteners in place of sugar tend to be heavier than people who use sugar. Does this mean that artificial sweeteners cause weight gain? Give a more plausible explanation for this association.

15.27 TV watching and school grades. Children who watch many hours of television get lower grades in school on the average than those who watch less TV. Suggest some lurking variables that may explain this relationship because they contribute to both heavy TV viewing and poor grades.

15.28 Correlation again. The correlation between IQ score and school GPA (Exercise 15.3) is $r = 0.634$. The correlation between wine consumption and heart disease deaths (Exercise 15.7) is $r = -0.843$. Which of these two correlations indicates a stronger straight-line relationship? Explain your answer.

15.29 Magic Mozart. In 1998, the Kalamazoo (Michigan) Symphony advertised a "Mozart for Minors" program with this statement: "Question: Which students scored 51 points higher in verbal skills and 39 points higher in math? Answer: Students who had experience in music." What do you think of the claim that "experience in music" causes higher test scores?

15.30 Calculating the least-squares line. Like to know the details when you study something? Here is the formula for the least-squares regression line for predicting y from x. Start with the means $\overline{x}$ and $\overline{y}$ and the standard deviations s_x and s_y of the two variables and the correlation r between them. The least-squares line has equation $y = a + bx$ with

$$b = r\frac{s_y}{s_x} \qquad \text{intercept:} \quad a = \overline{y} - b\overline{x}$$

Example 3 in Chapter 14 (page 269) gives the means, standard deviations, and correlation for the fossil bone length data. Use these values in the formulas just given to verify the equation of the least-squares line given on page 288:

$$\text{humerus length} = -3.66 + (1.197 \times \text{ femur length})$$

The remaining exercises require a two-variable statistics calculator or software that will calculate the least-squares regression line from data.

15.31 The professor swims. Return to the swimming data in Exercise 15.6.
(a) Verify the equation given for the least-squares line in that exercise.

(b) Suppose you were told only that the pulse rate was 152. You now want to predict swimming time. Find the equation of the least-squares regression line that is appropriate for this purpose. What is your prediction?

(c) The two lines in (a) and (b) are different. Explain clearly why there are two different regression lines.

15.32 Is wine good for your heart? Table 15.1 gives data on wine consumption and heart disease death rates in 19 countries. Verify the equation of the least-squares line given in Exercise 15.9.

15.33 Always plot your data! A skeptic might wonder if the four very different data sets in Table 15.2 really do have the same correlation and least-squares line. Verify that (to a close approximation) the least-squares line is $y = 3 + 0.5x$, as given in Exercise 15.17.

Chapter 16 The Consumer Price Index and Government Statistics

The government's most important statistic

Which of the many data series produced by government statistical offices in the United States is the most important? The unemployment rate, from the monthly Current Population Survey, is certainly a candidate. Rising unemployment can influence elections and change government policy. I suggest, however, another candidate: the monthly Consumer Price Index, or CPI.

The CPI measures changes in the prices of goods and services over time, so it measures the falling buying power of the dollar over time. If the same goods and services cost more, then the dollar is worth less. A dollar in 2000 buys less than a dollar bought in 1980, so it is a different dollar even though it looks the same. In fact, the CPI tells us that a dollar at the beginning of 2000 would buy only half as much as a dollar would buy in 1980. Someone who did not earn twice as many dollars in 2000 as in 1980 had lost buying power.

Why is the CPI so important? It does at times affect elections and government policies. It also has direct links to large parts of the economy shared by no other statistic. Nobody wants to lose buying power. So groups with enough clout tie their incomes directly to the CPI. Social Security payments rise automatically as the CPI increases, and so do the pensions of people retired from the military and the federal civil service. Over 2 million union members have contracts that tie their wages to the CPI. The Bureau of Labor Statistics says that the incomes of over 80 million people are directly affected by the CPI. When the CPI rises by 1%, government spending automatically goes up by $6 billion a year. Income tax brackets go up with the CPI. You can even buy U.S. savings bonds whose value is linked to the increase in the CPI.

The CPI is also important to anyone planning for the future. Saving for education or for retirement requires that we take account of the falling buying power of the dollar. We can use the CPI to compare 1980 dollars and 2000 dollars by turning them into "real dollars" that have the same buying power over time. We'll see how in this chapter.

We all notice the high salaries paid to professional athletes. In major-league baseball, for example, the mean salary rose from $143,756 in 1980 to $1,567,873 in 1999. That's a big jump. Not as big as it first appears, however. *A dollar in 1999 did not buy as much as a dollar in 1980, so 1980 salaries cannot be directly compared with 1999 salaries.* The hard fact that the dollar has steadily lost buying power over time means that we must make an adjustment whenever we compare dollar values from different years. The adjustment is easy. What is not easy is measuring the changing buying power of the dollar. The government's Consumer Price Index (CPI) is the tool we need.

"Now this here's a genuine 1980 dollar. They don't make 'em like that anymore."

Index numbers

The CPI is a new kind of numerical description, an *index number.* We can attach an index number to any quantitative variable that we measure repeatedly over time. The idea of the index number is to give a picture of changes in a variable much like that drawn by saying "The average cost of a day in the hospital rose 46% between 1990 and 1996." That is, an index number describes the percent change from a base period.

Index number

An **index number** measures the value of a variable relative to its value at a base period. To find the index number for any value of the variable:

$$\text{index number} = \frac{\text{value}}{\text{base value}} \times 100$$

Example 1. Calculating an index number

A gallon of unleaded regular gasoline cost $1.042 in January 1990 and $1.301 in January 2000. (These are national average prices collected by the Bureau of Labor Statistics.) The gasoline price index number for January 2000, with January 1990 as the base period, is

$$\begin{aligned}\text{index number} &= \frac{\text{value}}{\text{base value}} \times 100 \\ &= \frac{1.301}{1.042} \times 100 = 125\end{aligned}$$

The gasoline price index number for the base period, January 1990, is

$$\text{index number} = \frac{1.042}{1.042} \times 100 = 100$$

Knowing the base period is essential to making sense of an index number. Because the index number for the base period is always 100, it is usual to identify the base period as 1990 by writing "1990 = 100." In news reports concerning the Consumer Price Index, you will notice the mysterious equation "1982–84 = 100." That's shorthand for the fact that the years 1982 to 1984 are the base period for the CPI. An index number just gives the current value as a percent of the base value. Index number 125 means 125% of the base value, or a 25% increase from the base value. Index number 80 means that the current value is 80% of the base, a 20% decrease.

Fixed market basket price indexes

It may seem that index numbers are little more than a plot to disguise simple statements in complex language. Why say "The Consumer Price Index (1982–84 = 100) stood at 168.8 in January 2000" instead of "Consumer prices rose 68.8% between the 1982–84 average and January 2000"? In fact, the term "index number" usually means more than a measure of change relative to a base. It also tells us the kind of variable whose change we measure. That variable is a weighted average of several quantities, with fixed weights. Let's illustrate the idea by a simple price index.

Example 2. The Mountain Man Price Index

Bill Smith lives in a cabin in the mountains and strives for self-sufficiency. He buys only salt, kerosene, and the services of a professional welder. Here are Bill's purchases in 1990, the base period. His cost, in the last column, is the price per unit multiplied by the number of units he purchased.

Good or service	1990 quantity	1990 price	1990 cost
Salt	100 pounds	$0.50/pound	$50.00
Kerosene	50 gallons	1.00/gallon	50.00
Welding	10 hours	14.00/hour	140.00
		Total cost =	$240.00

The total cost of Bill's collection of goods and services in 1990 was $240. To find the "Mountain Man Price Index" for 2000, we use 2000 prices to calculate the 2000 cost of this *same* collection of goods and services. Here is the calculation:

Good or service	1990 quantity	2000 price	2000 cost
Salt	100 pounds	$0.80/pound	$80.00
Kerosene	50 gallons	1.00/gallon	50.00
Welding	10 hours	23.00/hour	230.00
		Total cost =	$360.00

The same goods and services that cost $240 in 1990 cost $360 in 2000. So the Mountain Man Price Index (1990 = 100) for 2000 is

$$\text{index number} = \frac{360}{240} \times 100 = 150$$

The point of Example 2 is that we follow the cost of the *same* collection of goods and services over time. It may be that Bill refused to hire the welder in 2000 because his hourly rate rose sharply. No matter—the index number uses the 1990 quantities, ignoring any changes in Bill's purchases between 1990 and 2000. We call the collection of goods and services whose total cost we follow a *market basket*. The index number is then a *fixed market basket price index*.

Fixed market basket price index

A **fixed market basket price index** is an index number for the total cost of a fixed collection of goods and services.

The basic idea of a fixed market basket price index is that the weight given to each component (salt, kerosene, welding) remains fixed over time. The CPI is in essence a fixed market basket price index, with several hundred items that represent all consumer purchases. Holding the market basket fixed allows a legitimate comparison of prices because we compare the prices of exactly the same items at each time. As we will see, it also poses severe problems for the CPI.

Using the CPI

For now, think of the CPI as an index number for the cost of everything that American consumers buy. That the CPI for January 2000 was 168.8 means that we must spend $168.80 in January 2000 to buy goods and services that cost $100 in the 1982 to 1984 base period. An index number for "the cost of everything" lets us compare dollar amounts from different years, by converting all the amounts into dollars of the same year. You will find tables in the *Statistical Abstract,* for example, with headings such as "Median Household Income, in Constant (2000) Dollars." That table has restated all incomes in dollars that will buy as much as the dollar would buy in 2000. Watch for the term *constant dollars* and for phrases like *real income.* They mean that all dollar amounts represent the same buying power even though they may describe different years.

Table 16.1 gives annual average CPIs from 1915 to 1999. Figure 16.1 is a line graph of the CPI values from the table. It shows that the 20th century was a time of inflation—prices rose throughout the century and increased rapidly after 1973. Faced with this depressing fact, it would be foolish to think about dollars without adjusting for their decline in buying power. Here is the recipe for converting dollars of one year into dollars of another year.

Adjusting for changes in buying power

To convert an amount in dollars at time A to the amount with the same buying power at time B:

$$\text{dollars at time B} = \text{dollars at time A} \times \frac{\text{CPI at time B}}{\text{CPI at time A}}$$

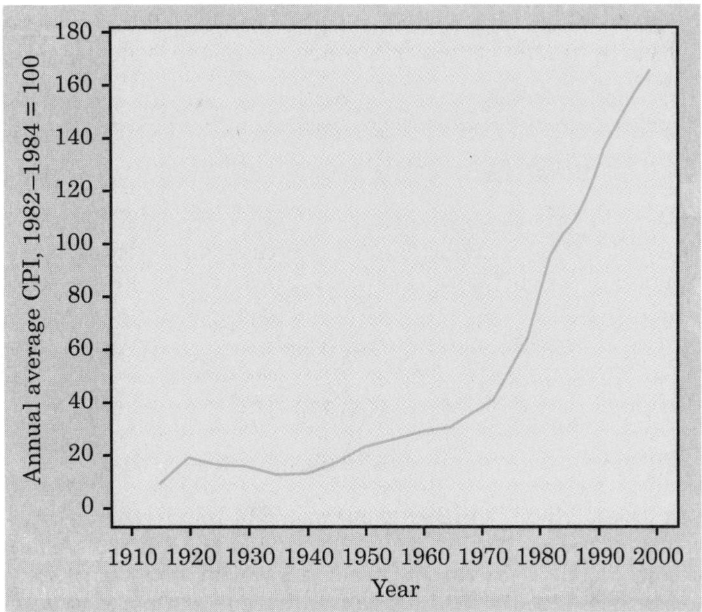

Figure 16.1 The Consumer Price Index (1982–84 = 100) from 1915 to 1999. Consumer prices in the United States rose sharply in the 20th century.

Table 16.1 Annual average Consumer Price Index, 1982–84 = 100

Year	CPI	Year	CPI	Year	CPI	Year	CPI
1915	10.1	1965	31.5	1982	96.5	1991	136.2
1920	20.0	1970	38.8	1983	99.6	1992	140.3
1925	17.5	1975	53.8	1984	103.9	1993	144.5
1930	16.7	1976	56.9	1985	107.6	1994	148.2
1935	13.7	1977	60.6	1986	109.6	1995	152.4
1940	14.0	1978	65.2	1987	113.6	1996	156.9
1945	18.0	1979	72.6	1988	118.3	1997	160.5
1950	24.1	1980	82.4	1989	124.0	1998	163.0
1955	26.8	1981	90.9	1990	130.7	1999	166.6
1960	29.6						

SOURCE: Bureau of Labor Statistics.

Notice that the CPI for the time you are *going to* appears on the top in the ratio of CPIs in this recipe. Here are some examples.

Example 3. Salaries of professional athletes

The mean salary of major-league baseball players rose from $143,756 in 1980 to $1,567,873 in 1999. How big was the increase in real terms? Let's convert the 1980 average into 1999 dollars. Table 16.1 gives the annual average CPIs that we need.

$$\text{1999 dollars} = \text{1980 dollars} \times \frac{\text{1999 CPI}}{\text{1980 CPI}}$$

$$= \$143{,}756 \times \frac{166.6}{82.4}$$

$$= \$290{,}652$$

That is, it took $290,652 in 1999 to buy what $143,756 would buy in 1980. We can now compare the 1980 mean salary of $290,652 *in 1999 dollars* with the actual 1999 mean salary, $1,567,873. Today's athletes earn much more than 1980 athletes even after adjusting for the fact that the dollar buys less now. (Of course, the mean salary is pulled up by the very high salaries paid to a few star players. The 1999 median salary was only $495,000.)

Example 4. Rising incomes?

For a more serious example, let's leave the pampered world of professional athletes and look at the incomes of ordinary people. The median annual income of all American households was $17,710 in 1980. By 1999, the median income had risen to $40,816. Dollar income more than doubled, but we know that much of that rise is an illusion because of the dollar's declining buying power. To compare these incomes, we must express them in dollars of the same year. Let's express the 1980 median household income in 1999 dollars:

$$\text{1999 dollars} = \$17{,}710 \times \frac{166.6}{82.4} = \$35{,}807$$

Real household incomes rose only from $35,807 to $40,816 in the 19 years between 1980 and 1999. That's a 14% increase.

The picture is different at the top. The 5% of households with the highest incomes earned $51,500 or more in 1980. In 1999 dollars, this is

$$\text{1999 dollars} = \$51{,}500 \times \frac{166.6}{82.4} = \$104{,}125$$

In fact, the top 5% of households earned $142,021 or more in 1999. That is, the real income of the highest earners increased by 36%.

Finally, let's look at production workers, the traditional "working men" (and women). Their average hourly earnings were $6.66 in 1980 and $13.28 in 1999. Restating the 1980 earnings in 1999 dollars,

$$1999 \text{ dollars} = \$6.66 \times \frac{166.6}{82.4} = \$13.47$$

The real earnings of production workers actually went down slightly between 1980 and 1999.

Example 4 illustrates how using the CPI to compare dollar amounts from different years brings out truths that are otherwise hidden. In this case, the truth is that the fruits of prosperity in the 1980s and 1990s went mostly to those at the top. Put another way, people with skills and education did much better than people like production workers who generally lack special skills and college educations. Economists suggest several reasons: the "new economy" that rewards knowledge, high immigration leading to competition for less-skilled jobs, more competition from abroad, and so on. Exactly why the rewards of skill and education have increased so much, and what we should do about the stagnant incomes of less-educated people, are controversial questions.

So you think that's inflation?

Americans were unhappy when oil price increases in 1973 set off a round of inflation that saw the CPI almost double in the following decade. That's nothing. In Argentina, prices rose 127% in a single month, July 1989. The Turkish lira went from 14 to the dollar in 1970 to 579,000 to the dollar in 2000. There were 65 German marks to a dollar in January 1920, and 4,200,000,000,000 marks to a dollar in November 1923. Now *that's* inflation.

Understanding the CPI

The idea of the CPI is that it is an index number for the cost of everything American consumers buy. That idea needs lots of fiddling to be practical. Much of the fiddling uses the results of large sample surveys.

Who is covered? The official name for the common version of the CPI (there are others, but we will ignore them) is the Consumer Price Index for All Urban Consumers. The CPI market basket represents the purchases of people living in urban areas. The official definition of "urban" is broad, so that about 80% of the U.S. population is covered. But if you live on a farm, the CPI doesn't apply to you.

How is the market basket chosen? Different households buy different things, so how can we get a single market basket? From a sample survey. The *Consumer Expenditure Survey* gathers detailed data on the spending of 29,000 households. The BLS breaks spending into categories such as "fresh fruits and vegetables," "new and used motor vehicles," and "hospital and related services." Then it chooses specific items such as "fresh oranges" to represent each group in the market basket. The items in the market basket get weights that represent their category's proportion of all spending. The weights, and even the specific market basket items, are updated regularly to keep up with changing buying habits. So the market basket isn't actually fixed.

How are the prices determined? From more sample surveys. The BLS must discover the price of "fresh oranges" every month. That price differs from city to city and from store to store in the same city. The *Point of Purchase Survey* of 16,800 households keeps the BLS up-to-date on where consumers shop for each category of goods and services (supermarkets, convenience stores, discount stores, and so on). Each month, the BLS records 80,000 prices in 85 cities at a sample of stores that represents actual buying habits.

Does the CPI measure changes in the cost of living? A fixed market basket price index measures the cost of *living the same* over time, as Example 2 illustrated. In fact, we don't keep buying the same market basket of goods and services over time. We switch from LP records to tapes and CDs and then to DVD-Audio disks. We don't buy new 1984 cars in 1995 or 2002. As prices change, we change what we buy—if beef becomes expensive, we buy less beef and more chicken or more tofu. A fixed market basket price index can't measure changes in the cost of living.

The BLS tries hard to keep its market basket up-to-date and to compensate for changes in quality. Every year, for example, the BLS must decide how much of the increase in new-car prices is paying for better quality. Only what's left counts as a genuine price increase in calculating the CPI. Between December 1967 and December 1994, actual car prices went up 313.4%, but the new-car price in the CPI went up only 172.1%. In 1995, adjustments for better quality reduced the overall rise in the prices of goods and services from 4.7% to only 2.2%. Prices of goods and services make up about 70% of the CPI. Most of the rest is the cost of shelter—renting an apartment or buying a house. House prices are another problem for the BLS. People buy houses partly to live in and partly because they think owning a house is a good investment. If we pay more for a house because we think it's a good investment, the full price should not go into the CPI.

By now it is clear that the CPI is *not* a fixed market basket price index, though that is the best way to start thinking about it. The BLS must constantly change the market basket as new products appear and our buying habits change. It must adjust the prices its sample surveys record to take account of better quality and the investment component of house prices. Yet the CPI still does not measure changes in our cost of living. It leaves out taxes, for example, which are certainly part of our cost of living.

Even if we agree that the CPI should only look at the goods and services we buy, it doesn't measure changes in our cost of living. In principle, a true "cost-of-living index" would measure the cost of the *same standard of living* over time. That's why we start with a fixed market basket price index, which also measures the cost of living the same over time but takes

Statistical Controversies

Does the CPI Overstate Inflation?

In 1995, Federal Reserve Chairman Alan Greenspan estimated that the CPI overstates inflation by somewhere between 0.5% and 1.5% per year. Mr. Greenspan was unhappy about this, because increases in the CPI automatically drive up federal spending. At the end of 1996, a group of outside experts appointed by the Senate Finance Committee estimated that the CPI had in the past overstated the rate of inflation by about 1.1% per year. The Bureau of Labor Statistics agreed that the CPI overstates inflation but thought that the experts' guess of 1.1% per year was too high.

The reasons the CPI shows the value of a dollar falling faster than is true are partly tied to the nature of the CPI and partly due to limits on how quickly the BLS can adjust the details of the enormous machine that lies behind the CPI. Think first about the details. The prices of new products, such as digital cameras and flat-screen computer monitors, often start high and drop rapidly. The CPI market basket changes too slowly to capture the early drop in price. Discount stores with lower prices also enter the CPI sample slowly. Although the BLS tries hard to adjust for better product quality, the outside experts thought these adjustments were often too little and too late. The BLS has made many improvements in these details. The improved CPI would have grown about 0.5% per year more slowly than the actual CPI between 1978 and 1998.

The wider issue is the nature of the CPI as essentially a fixed market basket index. Such an index has an upward bias because it can't track shifts from beef to tofu and back as consumers try to get the same quality of life from whatever products are cheaper this month. At the bottom of the outside experts' criticisms of the CPI was the fact that the CPI does not track the "cost of living." Their first recommendation was, "The BLS should establish a cost of living index as its objective in measuring consumer prices." The bureau said it agreed in principle but that neither it nor anyone else knows how to do this in practice. It also said, "Measurement of changes in 'quality of life' may require too many subjective judgments to furnish an acceptable basis for adjusting the CPI." Nonetheless, a new kind of index that in principle comes closer to measuring changes in the cost of living will appear beginning in 2002. Stay tuned for changes in the CPI—and for grumbling from groups such as Social Security recipients who would be hurt by less rapid increases.

the simple view that "the same" means buying exactly the same things. If we are just as satisfied after switching from beef to tofu to avoid paying more for beef, our standard of living hasn't changed and a cost-of-living index should ignore the higher price of beef. If we are willing to pay more for products that keep our environment clean, we are paying for a higher standard of living and the index should treat this just like an improvement in the quality of a new car. The BLS says that it would like the CPI to track changes in the cost of living, but that a true cost-of-living index isn't possible in the real world.

The place of government statistics

Modern nations run on statistics. Economic data in particular guide government policy and inform the decisions of private business and individuals. Price indexes and unemployment rates, along with many other less publicized series of data, are produced by government statistical offices.

Some countries have a single statistical office, such as Statistics Canada (www.statcan.ca). Others attach smaller offices to various branches of government. The United States is an extreme case: there are 72 federal statistical offices, with relatively weak coordination among them. The Census Bureau and the Bureau of Labor Statistics are the most important, but you may at times use the products of the Bureau of Economic Analysis, the National Center for Health Statistics, the Bureau of Justice Statistics, or others in the federal government's collection of statistical agencies.

A 1993 ranking of government statistical agencies by the heads of these agencies in several nations put Canada at the top, with the United States tied with Britain and Germany for sixth place. The top spots generally went to countries with a single, independent statistical office. In 1996, Britain combined its main statistical agencies to form a new Office for National Statistics (www.ons.gov.uk). American government statistics remains fragmented.

What do citizens need from their government statistical agencies? First of all, they need data that are *accurate, timely,* and *keep up with changes in society and the economy.* Producing accurate data quickly demands considerable resources. Think of the large-scale sample surveys that produce the unemployment rate and the CPI. The major U.S. statistical offices have a good reputation for accuracy and lead the world in getting data to the public quickly. Their record for keeping up with changes is less good. The struggle to adjust the CPI for changing buying habits and changing quality is one issue. Another is the failure of U.S. economic statistics to keep up with trends such as the shift from manufacturing to services as the center of economic activity. Business organizations have expressed strong dissatisfaction with the overall state of our economic data.

Much of the difficulty stems from lack of money. In the years after 1980, reducing federal spending was a political priority. Government statistical agencies lost staff and cut programs. Lower salaries made it hard to attract the best economists and statisticians to government. The level of government spending on data also depends on our view of what data the government should produce. In particular, should the government produce data that are used mainly by private business rather than by the government's own policy makers? Perhaps such data should be either compiled by private concerns or produced only for those who are willing to pay. This is a question of political philosophy rather than statistics, but it helps determine what level of government statistics we want to pay for.

"Yes sir, I know that we have to know where the economy is going. But do we have to publish the statistics so that everyone else does too?"

Freedom from political influence is as important to government statistics as accuracy and timeliness. When a statistical office is part of a government ministry, it can be influenced by the needs and desires of that ministry. The Census Bureau is in the Department of Commerce, which serves business interests. The BLS is in the Department of Labor. Thus business and labor each have "their own" statistical office. The professionals in the statistical offices successfully resist direct political interference—a poor unemployment report is never delayed until after an election, for example. But indirect influence is clearly present. The BLS must compete with other Department of Labor activities for its budget, for example. Political interference with statistical work seems to be increasing, as when Congress refused to allow the Census Bureau to use sample surveys to correct for undercounting in the 2000 census.

The 1996 reorganization of Britain's statistical offices was prompted in part by a widespread feeling that political influence was too strong. The details of how unemployment is measured in Britain were changed many times in the 1980s, for example, and almost all the changes had the effect of reducing the reported unemployment rate—just what the government wanted to see.

I favor a single "Statistics USA" office not attached to any other government ministry, as in Canada. Such unification might also help the money problem by eliminating duplication. It would at least allow a central decision about which programs deserve a larger share of limited resources. Unification is unlikely, but stronger coordination of the many federal statistical offices could achieve many of the same ends.

The question of social statistics

National economic statistics are well established with the government, the media, and the public. The government also produces many data on social issues such as education, health, housing, and crime. Social statistics are less complete than economic statistics. We have good data about how much money is spent on food but less information about how many people are poorly nourished. Social data are also less carefully produced than economic data. Economic statistics are generally based on larger samples, are compiled more often, and are published with a shorter time lag. The reason is clear: economic data are used by the government to guide economic policy month by month. Social data help us understand our society and address its problems but are not needed for short-term management.

There are other reasons the government is reluctant to produce social data. Many people don't want the government to ask about their sexual behavior or religion. Many feel that the government should avoid asking about our opinions—it's OK to ask "When did you last visit a doctor?" but not "How satisfied are you with the quality of your health care?" These hesitations reflect the American suspicion of government intrusion. Yet issues such as sexual behavior that contributes to the spread of AIDS and satisfaction with health care are important to citizens. Both facts and opinions on these issues can sway elections and influence policy. How can we get accurate information about social issues, collected consistently over time, and yet not entangle the government with sex, religion, and other touchy subjects?

The solution in the United States has been government funding of university sample surveys. After first deciding to undertake a sample survey asking people about their sexual behavior, in part to guide AIDS policy, the government backed away. Instead, it funded a much smaller survey of 3452 adults by the University of Chicago's National Opinion Research Center (NORC). NORC's General Social Survey (GSS), funded by the government's National Science Foundation, belongs with the Current Population Survey and the samples that undergird the CPI on any list of the most important sample surveys in the United States. The GSS includes both "fact" and "opinion" items. Respondents answer questions about their job security, their job satisfaction, their satisfaction with their city, their friends, and their families. They talk about race, religion, and sex. Many Americans would object if the government asked whether they had seen an X-rated movie in the past year, but they reply when the GSS asks this question.

This indirect system of government funding of a university-based sample survey fits the American feeling that the government itself should not be unduly invasive. It also insulates the survey from most political pressure. Alas, the government's budget-cutting extends to the GSS, which now describes itself as an "almost annual" survey because lack of funds has prevented taking samples in some years. The GSS is, I think, a bargain.

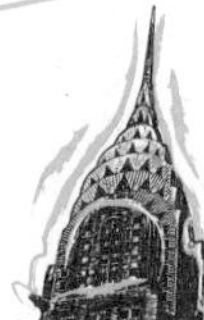

Exploring the Web

The Consumer Price Index has a home online at the Bureau of Labor Statistics Web site, stats.bls.gov/cpihome.htm. Look under "Frequently Asked Questions" for the bureau's explanation of the CPI and its uses. The most recent CPI appears in the latest news release posted at this site.

If you like data, you can go to the BLS data page, stats.bls.gov/datahome.htm, choose "Most Requested Series" and then "Average Price Data," and see for yourself how the prices of such things as white bread and gasoline have changed over time.

U.S. government statistical agencies are fragmented, but they have cooperated in providing a single FedStats Web site that offers access to all of them. To see the variety of U.S. government statistics, go to www.fedstats.gov and click on "Agencies." This is the mother lode of data about the United States.

Statistics in Summary

An **index number** describes the value of a variable relative to its value at some **base period**. A **fixed market basket price index** is an index number that describes the total cost of a collection of goods and services. Think of the government's **Consumer Price Index** as a fixed market basket price index for the collection of all the goods and services that consumers buy. Because the CPI shows how consumer prices change over time, we can use it to change a dollar amount at one time into the amount at another time that has the same buying power. This is needed to compare dollar values from different times in **real terms.**

The details of the CPI are complex. It uses data from several large sample surveys. It is not a true fixed market basket price index because of adjustments for changing buying habits, new products, and improved quality.

Government statistical offices produce data needed for government policy and decisions by businesses and individuals. The data should be accurate, timely, and free from political interference. Citizens therefore have a stake in the competence and independence of government statistical offices.

CHAPTER 16 EXERCISES

When you need the CPI for a year that does not appear in Table 16.1, use the table entry for the year that most closely follows the year you want.

16.1 The price of gasoline. The year-end average price of unleaded regular gasoline has fluctuated as follows:

1985	\$1.208 per gallon
1990	\$1.354 per gallon
1995	\$1.101 per gallon

Give the gasoline price index numbers (1990 = 100) for 1985, 1990, and 1995.

16.2 The cost of college. The part of the CPI that measures the cost of college tuition (1982–84 = 100) was 326.0 in January 2000. The overall CPI was 168.8 that month.

(a) Explain exactly what the index number 326.0 tells us about the rise in college tuition between the base period and the beginning of 2000.

(b) College tuition has risen much faster than consumer prices in general. How do you know this?

16.3 The price of gasoline. Use your results from Exercise 16.1 to answer these questions.

(a) By how many points did the gasoline price index number change between 1985 and 1995? What percent change was this?

(b) By how many points did the gasoline price index number change between 1990 and 1995? What percent change was this?

You see that the point change and the percent change in an index number are the same if we start in the base period, but not otherwise.

16.4 Toxic releases. The Environmental Protection Agency requires industry to report releases of any of a list of toxic chemicals. The total amounts released (in thousands of pounds) were 3,395,867 in 1988, 1,964,926 in 1995, and 1,941,870 in 1997. Give an index number for toxic chemical releases in each of these years, with 1988 as the base period. By what percent did releases increase or decrease between 1988 and 1997?

16.5 Los Angeles and New York. The Bureau of Labor Statistics publishes separate consumer price indexes for major metropolitan areas in addition to the national CPI. The CPI (1982–84 = 100) in January 2000 was 167.9 in Los Angeles and 179.2 in New York.

(a) These numbers tell us that prices rose faster in New York than in Los Angeles between the base period and the beginning of 2000. Explain how we know this.

(b) These numbers do *not* tell us that prices in January 2000 were higher in New York than in Los Angeles. Explain why.

16.6 The Food Faddist Price Index. A food faddist eats only steak, rice, and ice cream. In 1990, he bought:

Item	1990 quantity	1990 price
Steak	200 pounds	$5.45/pound
Rice	300 pounds	0.49/pound
Ice cream	50 gallons	5.08/gallon

After a visit from his mother, he adds oranges to his diet. Oranges cost $0.56/pound in 1990. Here are the food faddist's food purchases in 2000:

Item	2000 quantity	2000 price
Steak	175 pounds	$6.59/pound
Rice	325 pounds	0.53/pound
Ice cream	50 gallons	6.64/gallon
Oranges	100 pounds	0.61/pound

Find the fixed market basket Food Faddist Price Index (1990 = 100) for the year 2000.

16.7 The Guru Price Index. A guru purchases only olive oil, loincloths, and copies of the *Atharva Veda,* from which he selects mantras for his disciples. Here are the quantities and prices of his purchases in 1985 and 1995:

Item	1985 quantity	1985 price	1995 quantity	1995 price
Olive oil	20 pints	$2.50/pint	18 pints	$3.80/pint
Loincloth	2	$2.75 each	3	$2.80 each
Atharva Veda	1	$10.95	1	$12.95

From these data, find the fixed market basket Guru Price Index (1985 = 100) for 1995.

16.8 The curse of the Bambino. In 1920 the Boston Red Sox sold Babe Ruth to the New York Yankees for $125,000. Between 1920 and 2000, the Yankees won 26 World Series and the Red Sox won none. How much is $125,000 in 1999 dollars?

16.9 Dream on. When Julie started college in 1995, she set a goal of making $35,000 when she graduated. Julie graduated in 1999. What must Julie earn in 1999 in order to have the same buying power that $35,000 had in 1995?

16.10 Living too long? If both husband and wife are alive at age 65, in half the cases at least one will still be alive at age 93, 28 years later. This should frighten you. Myrna and Bill retired in 1970 with an income of $10,000 per year. They were quite comfortable—that was about the median family income in 1970. How much income did they need 28 years later, in 1998, to have the same buying power?

16.11 Microwaves on sale. The prices of new gadgets often start high and then fall rapidly. The first home microwave oven cost $1300 in 1955. You can now buy a better microwave oven for $100. Find the latest value of the CPI (it's on the BLS Web site) and use it to restate $100 in present-day dollars in 1955 dollars. Compare this with $1300 to see how much real microwave oven prices have come down.

16.12 Good golfers. In 1999, Tiger Woods won $6,620,970 on the Professional Golfers Association tour. The leading money winner in 1938 was Sam Snead, at $223,274. Tom Watson, the leader in 1980, won $1,041,002 that year. How do these amounts compare in real terms?

16.13 Joe DiMaggio. Yankee center fielder Joe DiMaggio was paid $32,000 in 1940 and $100,000 in 1950. Express his 1940 salary in 1950 dollars. By what percent did DiMaggio's real income change in the decade?

16.14 Calling London. A 10-minute telephone call to London via AT&T cost $12 in 1976 and $11 in 1999. Compare the real costs of these calls. By what percent did the real cost go down between 1976 and 1999?

16.15 Paying for Harvard. Harvard charged $5900 for tuition, room, and board in 1976. The 1999 charge was $32,164. Express Harvard's 1976 charges in 1999 dollars. Did the cost of going to Harvard go up faster or slower than consumer prices in general? How do you know?

16.16 The minimum wage. The federal government sets the minimum hourly wage that employers can pay a worker. Labor wants a high minimum wage, but many economists argue that too high a minimum makes employers reluctant to hire workers with low skills and so increases unemployment.

Here is information on changes in the federal minimum wage, in dollars per hour:

Year	1960	1965	1970	1975	1980	1985	1990	1995	1999
Minimum wage	$1.00	$1.25	$1.60	$2.10	$3.10	$3.35	$3.80	$4.25	$5.15

Use annual average CPIs from Table 16.1 to restate the minimum wage in constant 1960 dollars. Make two line graphs on the same axes, one showing the actual minimum wage during these years and the other showing the minimum wage in constant dollars. Explain carefully to someone who knows no statistics what your graphs show about the history of the minimum wage.

16.17 College tuition. Tuition for Indiana residents at Purdue University has increased as follows (use tuition at your own college if you have those data):

Year	1981	1983	1985	1987	1989
Tuition	$1158	$1432	$1629	$1816	$2032
Year	1991	1993	1995	1997	1999
Tuition	$2324	$2696	$3056	$3336	$3624

Use annual average CPIs from Table 16.1 to restate each year's tuition in constant 1981 dollars. Make two line graphs on the same axes, one showing actual dollar tuition for these years and the other showing constant dollar tuition. Then explain to someone who knows no statistics what your graphs show.

16.18 2000 versus 1980. The introduction to this chapter says that "a dollar at the beginning of 2000 would buy only half as much as a dollar would buy in 1980." The CPI for January 2000 was 168.8. Table 16.1 tells us that the average CPI in 1980 was 82.4. Explain carefully why these numbers justify the "half as much" claim.

16.19 Rising incomes? In Example 4, we see that the median real income (in 1999 dollars) of all households rose from $35,807 in 1980 to $40,816 in 1999. The real income that marks off the top 5% of households rose from $104,125 to $142,021 in the period. Verify the claim in Example 4 that median income rose 14% and that the real income of top earners rose 36%.

16.20 Cable TV. Suppose that cable television systems across the country add channels to their lineup and raise the monthly fee they charge subscribers. The part of the CPI that tracks cable TV prices might not go up at all, even though consumers must pay more. Explain why.

16.21 Item weights in the CPI. The cost of buying a house (with the investment component removed) makes up about 20% of the CPI. The cost of renting a place to live makes up about 6%. Where do the weights 20% and 6% come from? Why does the cost of buying get more weight in the index?

16.22 The CPI doesn't fit me. The CPI may not measure your personal experience with changing prices. Explain why the CPI will not fit each of these people:

(a) Marcia lives on a cattle ranch in Montana.

(b) Jim heats his home with a wood stove and does not have air-conditioning.

(c) Luis and Maria were in a serious auto accident and spent much of last year in a rehabilitation center.

16.23 Seasonal adjustment. Like many government data series, the CPI is published in both unadjusted and seasonally adjusted forms. The BLS says that it "strongly recommends using indexes unadjusted for seasonal variation (i.e., not seasonally adjusted indexes) for escalation." "Escalation" here means adjusting wage or other payments to keep up with changes in the CPI. Why is the unadjusted CPI preferred for this purpose?

16.24 More CPIs. In addition to the national CPI, the BLS publishes CPIs for 4 regions and for 26 local areas. Each regional and local CPI is based on just the part of the national sample of prices that takes place in the region or local area. The BLS says that the local CPIs should be used with caution because they are much more variable than national or regional CPIs. Why is this?

16.25 The poverty line. The federal government announces "poverty lines" each year for households of different sizes. Households with income below the announced levels are considered to be living in poverty. An economist looked at the poverty lines over time and said that they "show a pattern of getting higher in real terms as the real income of the general population rises." What does "getting higher in real terms" say about the official poverty level?

16.26 Real wages. In one of the many reports on stagnant incomes in the United States, we read this: "Practically every income group faced a decline in real wages during the 1980s. However, workers at the 33rd percentile experienced a 14 percent drop in the real wage, workers at the 66th percentile experienced only a 6 percent drop, and workers in the upper tail of the distribution experienced a 1 percent wage increase."

(a) What is meant by "the 33rd percentile of the income distribution"?

(b) What does "real wages" mean?

16.27 Saving money? One way to cut the cost of government statistics is to reduce the sizes of the samples. We might, for example, cut the Current Population Survey from 50,000 households to 20,000. Explain clearly, to someone who knows no statistics, why such cuts reduce the accuracy of the resulting data.

16.28 The General Social Survey. The General Social Survey places much emphasis on asking many of the same questions year after year. Why do you think it does this?

16.29 Measuring the effects of crime. We wish to include, as part of a set of social statistics, measures of the amount of crime and of the impact of crime on people's attitudes and activities. Suggest some possible measures in each of the following categories:
(a) Statistics to be compiled from official sources such as police records.

(b) Factual information to be collected using a sample survey of citizens.

(c) Information on opinions and attitudes to be collected using a sample survey.

16.30 Statistical agencies. Write a short description of the work of one of these government statistical agencies. You can find information by starting at the FedStats Web site (www.fedstats.gov) and going to "Agencies."
(a) Bureau of Economic Analysis (Department of Commerce).

(b) National Center for Education Statistics.

(c) National Center for Health Statistics.

Part II Review

Data analysis is the art of describing data using graphs and numerical summaries. The purpose of data analysis is to help us see and understand the most important features of a set of data. Chapter 10 commented on basic graphs, especially pie charts, bar graphs, and line graphs. Chapters 11 to 13 showed how data analysis works by presenting statistical ideas and tools for describing the distribution of one variable. Figure II.1 organizes the big ideas. We plot our data, then describe their center and spread using either the mean and standard deviation or the five-number summary. The last step, which makes sense only for some data, is to summarize the data in compact form by using a normal curve as a model for the overall pattern. The

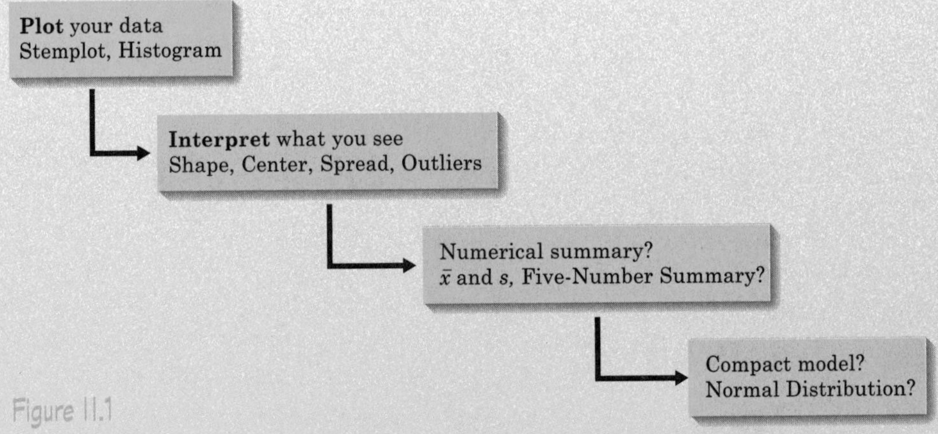

Figure II.1

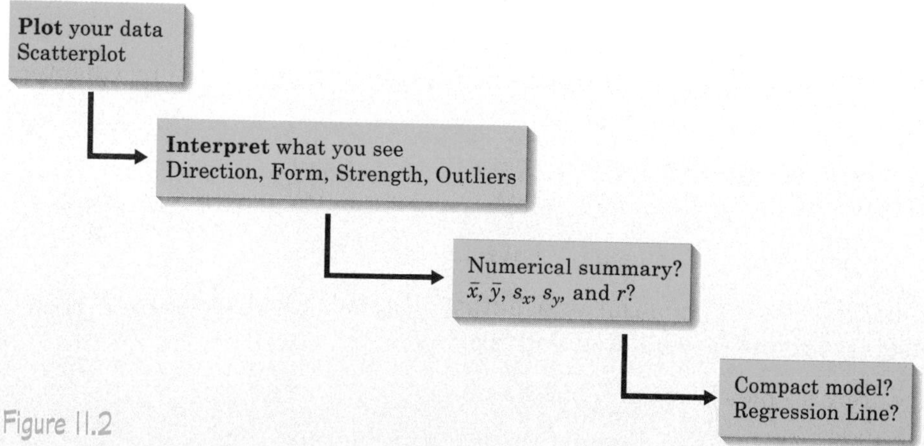

Figure II.2

question marks at the last two stages remind us that the usefulness of numerical summaries and normal distributions depends on what we find when we examine graphs of our data. No short summary does justice to irregular shapes or to data with several distinct clusters.

Chapters 14 and 15 applied the same ideas to relationships between two quantitative variables. Figure II.2 retraces the big ideas from Figure II.1, with details that fit the new setting. We always begin by making graphs of our data. In the case of a scatterplot, we have learned a numerical summary only for data that show a roughly straight-line pattern on the scatterplot. The summary is then the means and standard deviations of the two variables and their correlation. A regression line drawn on the plot gives us a compact model of the overall pattern that we can use for prediction. Once again there are question marks at the last two stages to remind us that correlation and regression only describe straight-line relationships.

Relationships often raise the question of causation. We know that evidence from randomized comparative experiments is the "gold standard" for deciding that one variable causes changes in another variable. Chapter 15 reminded us in more detail how strong associations can appear in data even when there is no direct causation. We must always think about the possible effects of variables lurking in the background. In Chapter 16 we met a new kind of description, index numbers, with the Consumer Price Index as the leading example. Chapter 16 also discussed government statistical offices, a quiet but important part of the statistical world.

PART II SUMMARY

Here are the most important skills you should have after reading Chapters 10 to 16.

A. DISPLAYING DISTRIBUTIONS

1. Recognize categorical and quantitative variables.
2. Recognize when a pie chart can and cannot be used.
3. Make a bar graph of the distribution of a categorical variable, or in general to compare related quantities.
4. Interpret pie charts and bar graphs.
5. Make a line graph of a quantitative variable over time.
6. Recognize patterns such as trends and seasonal variation in line graphs.
7. Be aware of graphical abuses, especially pictograms and distorted scales in line graphs.
8. Make a histogram of the distribution of a quantitative variable.
9. Make a stemplot of the distribution of a small set of observations. Round data as needed to make an effective stemplot.

B. DESCRIBING DISTRIBUTIONS (QUANTITATIVE VARIABLE)

1. Look for the overall pattern of a histogram or stemplot and for major deviations from the pattern.
2. Assess from a histogram or stemplot whether the shape of a distribution is roughly symmetric, distinctly skewed, or neither. Assess whether the distribution has one or more major peaks.
3. Describe the overall pattern by giving numerical measures of center and spread in addition to a verbal description of shape.
4. Decide which measures of center and spread are more appropriate: the mean and standard deviation (especially for symmetric distributions) or the five-number summary (especially for skewed distributions).
5. Recognize outliers and give plausible explanations for them.

C. NUMERICAL SUMMARIES OF DISTRIBUTIONS

1. Find the median M and the quartiles Q_1 and Q_3 for a set of observations.
2. Give the five-number summary and draw a boxplot; assess center, spread, symmetry, and skewness from a boxplot.
3. Find the mean $\bar{x}$ and (using a calculator) the standard deviation s for a small set of observations.

4. Understand that the median is less affected by extreme observations than the mean. Recognize that skewness in a distribution moves the mean away from the median toward the long tail.

5. Know the basic properties of the standard deviation: $s \geq 0$ always; $s = 0$ only when all observations are identical and increases as the spread increases; s has the same units as the original measurements; s is pulled strongly up by outliers or skewness.

D. NORMAL DISTRIBUTIONS

1. Interpret a density curve as a description of the distribution of a quantitative variable.

2. Recognize the shape of normal curves and estimate by eye both the mean and standard deviation from such a curve.

3. Use the 68–95–99.7 rule and symmetry to state what percent of the observations from a normal distribution fall between two points when both points lie at the mean or one, two, or three standard deviations on either side of the mean.

4. Find and interpret the standard score of an observation.

5. (Optional) Use Table B to find the percentile of a value from any normal distribution and the value that corresponds to a given percentile.

E. SCATTERPLOTS AND CORRELATION

1. Make a scatterplot to display the relationship between two quantitative variables measured on the same subjects. Place the explanatory variable (if any) on the horizontal scale of the plot.

2. Describe the form, direction, and strength of the overall pattern of a scatterplot. In particular, recognize positive or negative association and straight-line patterns. Recognize outliers in a scatterplot.

3. Judge whether it is appropriate to use correlation to describe the relationship between two quantitative variables. Use a calculator to find the correlation r.

4. Know the basic properties of correlation: r measures the strength and direction of only straight-line relationships; r is always a number between -1 and 1; $r = \pm 1$ only for perfect straight-line relations; r moves away from 0 toward ± 1 as the straight-line relation gets stronger.

F. REGRESSION LINES

1. Explain what the slope b and the intercept a mean in the equation $y = a + bx$ of a straight line.

2. Draw a graph of the straight line when you are given its equation.

3. Use a regression line, given on a graph or as an equation, to predict y for a given x. Recognize the danger of prediction outside the range of the available data.

4. Use r^2, the square of the correlation, to describe how much of the variation in one variable can be accounted for by a straight-line relationship with another variable.

G. STATISTICS AND CAUSATION

1. Give plausible explanations for an observed association between two variables: direct cause and effect, the influence of lurking variables, or both.

2. Assess the strength of statistical evidence for a claim of causation, especially when experiments are not possible.

H. THE CONSUMER PRICE INDEX (CPI) AND RELATED TOPICS

1. Calculate and interpret index numbers.

2. Calculate a fixed market basket price index for a small market basket.

3. Use the CPI to compare the buying power of dollar amounts from different years. Explain phrases such as "real income."

PART II REVIEW EXERCISES

Review exercises are short and straightforward exercises that help you solidify the basic ideas and skills in each part of this book.

II.1 Poverty in the states. Table II.1 gives the percents of people living below the poverty line in the 26 states east of the Mississippi River. Make a stemplot of these data. Is the distribution roughly symmetric, skewed to the right, or skewed to the left? Which states (if any) are outliers?

II.2 Quarterbacks. Table II.2 gives the total passing yards for National Football League starting quarterbacks during the 1999 season. (These are the starting quarterbacks at the end of the season. Some played fewer games than others.) Make a histogram of these data. Does the distribution have a clear shape: roughly symmetric, clearly skewed to the left, clearly skewed to the right, or none of these? Which quarterbacks (if any) are outliers?

II.3 Poverty in the states. Give the five-number summary for the data on poverty from Table II.1.

II.4 Quarterbacks. Give the five-number summary for the data on passing yards for NFL quarterbacks from Table II.2.

II.5 Poverty in the states. Find the mean percent of state residents living in poverty from the data in Table II.1. If we removed Mississippi from the data, would the mean increase or decrease? Why? Find the mean for the 25 remaining states to verify your answer.

Table II.1 Percent of state residents living in poverty, 1997

State	Percent	State	Percent	State	Percent
Alabama	15.7	Maryland	8.4	Pennsylvania	11.2
Connecticut	8.6	Massachusetts	12.2	Rhode Island	12.7
Delaware	9.6	Michigan	10.3	South Carolina	13.1
Florida	14.3	Mississippi	16.7	Tennessee	14.3
Georgia	14.5	New Hampshire	9.1	Vermont	9.3
Illinois	11.2	New Jersey	9.3	Virginia	12.7
Indiana	8.8	New York	16.5	West Virginia	16.4
Kentucky	15.9	North Carolina	11.4	Wisconsin	8.2
Maine	10.1	Ohio	11.0		

SOURCE: *Statistical Abstract of the United States.*

Table II.2 Passing yards for NFL quarterbacks in 1999

Quarterback	Yards	Quarterback	Yards	Quarterback	Yards
Troy Aikman	2964	Doug Flutie	3171	Ray Lucas	1678
Tony Banks	2136	Gus Frerotte	2117	Peyton Manning	4135
Steve Beuerlein	4436	Rich Gannon	3840	Dan Marino	2448
Jeff Blake	2670	Jeff Garcia	2544	Cade McNown	1465
Drew Bledsoe	3985	Jeff George	2816	Steve McNair	2179
Mark Brunell	3060	Elvis Grbac	3389	Doug Pederson	1276
Chris Chandler	2339	Brian Griese	3032	Jake Plummer	2111
Kerry Collins	2316	Jim Harbaugh	2761	Billy Joe Tolliver	1916
Tim Couch	2447	Brad Johnson	4005	Mike Tomczak	1625
Trent Dilfer	1619	Jon Kitna	3346	Kurt Warner	4353
Brett Favre	4091				

II.6 Big heads? The army reports that the distribution of head circumference among male soldiers is approximately normal with mean 22.8 inches and standard deviation 1.1 inches. Use the 68–95–99.7 rule to answer these questions.

(a) Between what values do the middle 95% of head circumferences fall?

(b) What percent of soldiers have head circumference greater than 23.9 inches?

II.7 SAT scores. The scale for SAT exam scores is set so that the distribution of scores is approximately normal with mean 500 and standard deviation 100. Answer these questions without using a table.

(a) What is the median SAT score?

(b) You run a tutoring service for students who score between 400 and 600 and hope to do better. What percent of SAT scores are between 400 and 600?

II.8 Explaining correlation. You have data on the current grade point average (GPA) of students in a basic statistics course and their scores on the first examination in the course. Say as specifically as you can what the correlation r between GPA and exam score measures.

II.9 Data on mice. For a biology project, you measure the tail length (centimeters) and weight (grams) of 12 mice of the same variety. What units of measurement do each of the following have?

(a) The mean length of the tails.

(b) The first quartile of the tail lengths.

(c) The standard deviation of the tail lengths.

(d) The correlation between tail length and weight.

II.10 More data on mice. For a biology project, you measure the tail length (centimeters) and weight (grams) of 12 mice of the same variety.

(a) Explain why you expect the correlation between tail length and weight to be positive.

(b) The mean tail length turns out to be 9.8 centimeters. What is the mean length in inches? (There are 2.54 centimeters in an inch.)

(c) The correlation between tail length and weight turns out to be $r = 0.6$. If you measured length in inches instead of centimeters, what would be the new value of r?

Figure II.3 plots the average brain weight in grams versus average body weight in kilograms for many species of mammals. There are many small mammals whose points at the lower left overlap. Exercises II.11 to II.16 are based on this scatterplot.

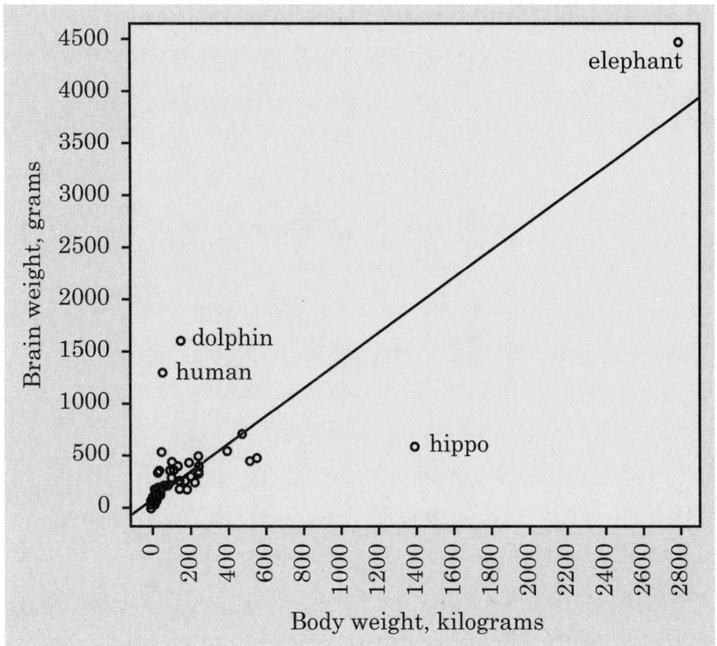

Figure II.3 Scatterplot of the average brain weight (grams) against the average body weight (kilograms) for 96 species of mammals, for Exercises II.11 to II.16.

II.11 Dolphins and hippos. The points for the dolphin and hippopotamus are labeled in Figure II.3. Read from the graph the approximate body weight and brain weight for these two species.

II.12 Dolphins and hippos. One reaction to this scatterplot is "Dolphins are smart, hippos are dumb." What feature of the plot lies behind this reaction?

II.13 Outliers. The African elephant is much larger than any other mammal in the data set but lies roughly in the overall straight-line pattern. Dolphins, humans, and hippos lie outside the overall pattern. The correlation between body weight and brain weight for the entire data set is $r = 0.86$.

(a) If we removed the elephant, would this correlation increase or decrease or not change much? Explain your answer.

(b) If we removed dolphins, hippos, and humans, would this correlation increase or decrease or not change much? Explain your answer.

II.14 Brain and body. The correlation between body weight and brain weight is $r = 0.86$. How well does body weight explain brain weight for mammals? Give a number to answer this question, and briefly explain what the number tells us.

II.15 Prediction. The line on the scatterplot in Figure II.3 is the least-squares regression line for predicting brain weight from body weight. Suppose that a new mammal species is discovered hidden in the rain forest with body weight 600 kilograms. Predict the brain weight for this species.

II.16 Slope. The line on the scatterplot in Figure II.3 is the least-squares regression line for predicting brain weight from body weight. The slope of this line is one of the numbers below. Which number is the slope? Why?

(a) $b = 0.5$.

(b) $b = 1.3$.

(c) $b = 3.2$.

From Rex Boggs in Australia comes an unusual data set: before showering in the morning, he weighed the bar of soap in his shower stall. The weight goes down as the soap is used. The data appear in Table II.3 (weights in grams). Notice that Mr. Boggs forgot to weigh the soap on some days. Exercises II.17 to II.19 are based on the soap data set.

II.17 Scatterplot. Plot the weight of the bar of soap against day. Is the overall pattern roughly straight-line? Based on your scatterplot, is the correlation between day and weight close to 1, positive but not close to 1, close to 0, negative but not close to –1, or close to –1? Explain your answer.

II.18 Regression. The equation for the least-squares regression line for the data in Table II.3 is

$$\text{weight} = 133.2 - 6.31 \times \text{day}$$

(a) Explain carefully what the slope $b = -6.31$ tells us about how fast the soap lost weight.

Table II.3 Weight (grams) of a bar of soap used to shower

Day	Weight	Day	Weight	Day	Weight
1	124	8	84	16	27
2	121	9	78	18	16
5	103	10	71	19	12
6	96	12	58	20	8
7	90	13	50	21	6

SOURCE: Rex Boggs.

(b) Mr. Boggs did not measure the weight of the soap on day 4. Use the regression equation to predict that weight.

(c) Draw the regression line on your scatterplot from the previous exercise.

II.19 Prediction? Use the regression equation in the previous exercise to predict the weight of the soap after 30 days. Why is it clear that your answer makes no sense? What's wrong with using the regression line to predict weight after 30 days?

II.20 Keeping up with the Joneses. The Jones family had a household income of $30,000 in 1980, when the CPI (1982–84 = 100) was 82.4. The CPI at the beginning of 2000 was 168.8. How much must the Joneses earn in 2000 to have the same buying power they had in 1980?

II.21 Affording a Mercedes. A Mercedes-Benz 190 cost $24,000 in 1981, when the CPI (1982–84 = 100) was 90.9. The average CPI for 1999 was 166.6. How many 1999 dollars must you earn to have the same buying power as $24,000 had in 1981?

II.22 Steinway. A Steinway concert grand piano cost $13,500 in 1976. A similar Steinway cost $79,900 in 1999. Has the cost of the piano gone up or down in real terms? Give a calculation to justify your answer.

II.23 The price of gold. Some people recommend that investors buy gold "to protect against inflation." Here are the prices of an ounce of gold at the end of the year for the years between 1983 and 1999. Make a graph that shows how the price of gold changed in real terms over this period. Would an investment in gold have protected against inflation by holding its value in real terms?

Year	1983	1985	1987	1989	1991	1993	1995	1997	1999
Gold price	$385	$329	$486	$403	$354	$391	$392	$368	$295

II.24 Never on Sunday? The Canadian Province of Ontario carries out statistical studies of the working of Canada's national health care system in the province. The bar graphs in Figure II.4 come from a study of admissions and discharges from community hospitals in Ontario. They show the number of heart attack patients admitted and discharged on each day of the week during a 2-year period.

(a) Explain why you expect the number of patients admitted with heart attacks to be roughly the same for all days of the week. Do the data show that this is true?

(b) Describe how the distribution of the day on which patients are discharged from the hospital differs from that of the day on which they are admitted. What do you think explains the difference?

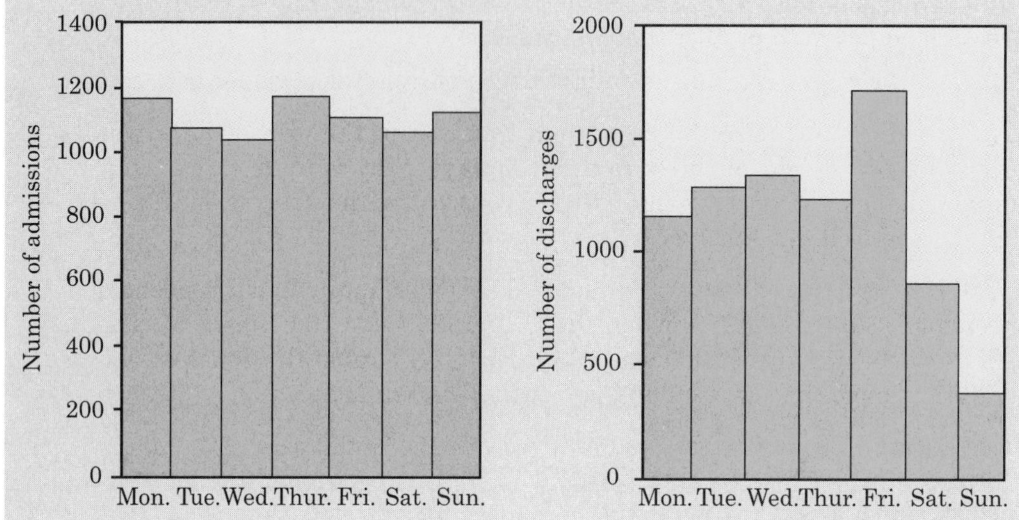

Figure II.4 Bar graphs of the number of heart attack victims admitted and discharged from hospitals in Ontario, Canada, on each day of the week, for Exercise II.24.

II.25 Drive time. Professor Moore, who lives a few miles outside a college town, records the time he takes to drive to the college each morning. Here are the times (in minutes) for 42 consecutive weekdays, with the dates in order along the rows:

8.25	7.83	8.30	8.42	8.50	8.67	8.17	9.00	9.00	8.17	7.92
9.00	8.50	9.00	7.75	7.92	8.00	8.08	8.42	8.75	8.08	9.75
8.33	7.83	7.92	8.58	7.83	8.42	7.75	7.42	6.75	7.42	8.50
8.67	10.17	8.75	8.58	8.67	9.17	9.08	8.83	8.67		

(a) Make a histogram of these drive times. Is the distribution roughly symmetric, clearly skewed, or neither? Are there any clear outliers?

(b) Make a line graph of the drive times. (Label the horizontal axis in days, 1 to 42.) The plot shows no clear trend, but it does show one unusually low drive time and two unusually high drive times. Circle these observations on your plot.

II.26 Drive time outliers. In the previous exercise, there are three outliers in Professor Moore's drive times to work. All three can be explained. The low time is the day after Thanksgiving (no traffic on campus). The two high times reflect delays due to an accident and icy roads. Remove these three observations. To summarize normal drive times, use a calculator to find the mean $\bar{x}$ and standard deviation s of the remaining 39 times.

II.27 House prices. The median selling price of existing single-family homes was $133,400 at the beginning of the year 2000. Would the mean selling price be higher, about the same, or lower? Why?

II.28 The 1996 election. Bill Clinton was reelected president in 1996 with 49.2% of the popular vote. His Republican opponent, Bob Dole, received 40.7% of the vote, with minor candidates taking the remaining votes. Table II.4 gives the percent of the popular vote won by President Clinton in each state. Describe these data with a graph, a numerical summary, and a brief verbal description.

Table II.4 Percent of votes for President Clinton, 1996

State	Percent	State	Percent	State	Percent
Alabama	43.2	Louisiana	52.9	Ohio	47.4
Alaska	33.3	Maine	51.6	Oklahoma	40.5
Arizona	46.5	Maryland	54.2	Oregon	47.2
Arkansas	53.7	Massachusetts	61.5	Pennsylvania	49.2
California	51.1	Michigan	51.7	Rhode Island	59.7
Colorado	44.4	Minnesota	51.1	South Carolina	44.0
Connecticut	52.8	Mississippi	44.1	South Dakota	43.0
Delaware	51.8	Missouri	47.5	Tennessee	48.0
Florida	48.0	Montana	41.2	Texas	43.8
Georgia	45.8	Nebraska	35.0	Utah	33.3
Hawaii	56.9	Nevada	43.9	Vermont	53.4
Idaho	33.6	New Hampshire	49.3	Virginia	45.2
Illinois	54.3	New Jersey	53.7	Washington	49.8
Indiana	41.6	New Mexico	49.2	West Virginia	51.5
Iowa	50.3	New York	59.4	Wisconsin	48.8
Kansas	36.1	North Carolina	44.0	Wyoming	36.8
Kentucky	45.8	North Dakota	40.1		

SOURCE: Federal Election Commission.

II.29 Statistics for investing. Joe's retirement plan invests in stocks through an "index fund" that follows the behavior of the the stock market as a whole, as measured by the Standard & Poor's 500 Index. Joe wants to buy a mutual fund that does not track the index closely. He reads that monthly returns from Fidelity Technology Fund have correlation $r = 0.77$ with the S&P 500 Index and that Fidelity Real Estate Fund has correlation $r = 0.37$ with the index.
(a) Which of these funds has the closer relationship to returns from the stock market as a whole? How do you know?
(b) Does the information given tell Joe anything about which fund has had higher returns?

PART II PROJECTS

Projects are longer exercises that require gathering information or producing data and emphasize writing a short essay to describe your work. Many are suitable for teams of students.

Project 1. Statistical graphics in the press. Graphs good and bad fill the news media. Some publications, such as *USA Today*, make particularly heavy use of graphs to present data. Collect several graphs (at least five) from newspapers and magazines (not from advertisements). Include some graphs that, in your opinion, represent good style and some that represent poor style or are misleading. Use your collection as examples in a brief essay about the clarity, accuracy, and attractiveness of graphs in the press.

Project 2. Roll your own regression. Choose two quantitative variables that you think have a roughly straight-line relationship. Gather data on these variables and do a statistical analysis: make a scatterplot, find the correlation, find the regression line (use a statistical calculator or software), and draw the line on your plot. Then write a report on your work. Some examples of suitable pairs of variables are:
(a) The height and arm span of a group of people.

(b) The height and walking stride length of a group of people.

(c) The price per ounce and bottle size in ounces for several brands of shampoo and several bottle sizes for each brand.

Project 3. High school dropouts. Write a factual report on high school dropouts in the United States. Some examples of questions you might address are: What states have the highest percentages of adults who did not finish high school? How do the earnings and employment rates of dropouts compare with those of other adults? Is the percentage who fail to finish high school higher among blacks and Hispanics than among whites?

The *Statistical Abstract* will supply you with data. Look in the index under "education" for an entry on "high school dropouts." You may want to look at other parts of the *Statistical Abstract* for data on earnings and other variables broken down by education.

Project 4. Association is not causation. Write a snappy, attention-getting article on the theme that "association is not causation." Use pointed but not-too-serious examples like those in Example 6 (page 293) and Exercise 15.21 (page 305) of Chapter 15, or this one: there is an association between long hair and being female, but cutting a woman's hair will not turn her into a man. Be clear, but don't be technical. Imagine that you are writing for high school students.

Project 5. Military spending. Here are data on U.S. spending for national defense for fiscal years between 1940 and 1999 from the *Statistical Abstract*. You may want to look in the latest volume for data from the most recent year. The units are millions of dollars (this is serious money).

Year	1940	1945	1950	1955	1960	1965	1970
Military spending	1,660	82,965	13,724	42,729	48,130	50,620	81,692

Year	1975	1980	1985	1990	1995	1999
Military spending	86,509	133,995	252,748	299,331	272,066	276,730

Write an essay that describes the changes in military spending in real terms during this period from just before World War II until a decade after the end of the cold war. Do the necessary calculations and write a brief description that ties military spending to the major military events of this period: World War II (1941–1945), the Korean War (1950–1953), the Vietnam War (roughly 1964–1975), and the end of the cold war after the fall of the Berlin Wall in 1989.

Project 6. Your pulse rate. What is your "resting pulse rate"? Of course, even if you measure your pulse rate while resting, it may vary from day to day and with the time of day. Measure your resting pulse rate at least six times each day (spaced through the day) for at least four days. Write a discussion that includes a description of how you made your measurements and an analysis of your data. Based on the data, what would you say when someone asks your resting pulse rate? (If several students do this prcject, you can then discuss variation in pulse rate among a group of individuals as well.)

Project 7. The dates of coins. Coins are stamped with the year in which they were minted. Collect data from at least 50 coins of each denomination: pennies, nickels, dimes, and quarters. Write a description of the distribution of dates on coins now in circulation, including graphs and numerical descriptions. Are there differences among the denominations? Did you find any outliers?

Part III

"What kind of childish nonsense are you working on now?"

Chance

"If chance will have me king, why, chance will crown me." So said Macbeth in Shakespeare's great play. Chance does indeed play with us all, and we can do little to understand or manage it. Sometimes, however, chance is tamed. A roll of dice, a simple random sample, even the inheritance of eye color or blood type, represent chance tied down so that we can understand and manage it. Unlike Macbeth's life or ours, we can roll the dice again. And again, and again. The outcomes are governed by chance, but in many repetitions a pattern emerges. Chance is no longer mysterious, because we can describe its pattern.

We humans use mathematics to describe regular patterns, whether the circles and triangles of geometry or the movements of the planets. We use mathematics to understand the regular patterns of chance behavior when chance is tamed in a setting where we can repeat the same chance phenomena again and again. The mathematics of chance is called probability. Probability is the topic of this part of the book, though we will go light on the math in favor of experimenting and thinking.

Chapter 17
Thinking about Chance

What's the chance?

On January 28, 1986, the space shuttle *Challenger* exploded soon after takeoff. A presidential commission investigated. What did those involved estimate the chance of such a failure to be? Some working engineers said roughly 1 in 100. Management said roughly 1 in 100,000. On hearing the latter estimate, physicist Richard Feynman, a member of the commission, asked, "You mean to tell me that if you launched a rocket every day for 300 years, you'd only expect one failure?" Feynman was good at mental arithmetic: 300 years is 109,500 days if we ignore leap years.

Feynman did two important things in this little interchange. Management was clearly just guessing. (OK, they thought they were giving their "informed judgment" of risk, but it comes to the same thing.) That is, they were using the language of chance to express their personal opinion or judgment. Feynman turned this vague personal opinion into the much more concrete image of trying the same thing many times: If we launched a very large number of shuttles, how often would one fail? That should sound familiar, because it's the way we think about sampling: "If we choose many samples from the same population, the truth about the population will be within the margin of error 95% of the time."

Feynman's second tactic was to make the idea of trying something 100,000 times more understandable by bringing it into the real world—do it every day for 300 years. That's still not easy to grasp. Our minds don't deal well with really large numbers or with really small probabilities, like the 1-in-80-million chance of winning the lottery or the 1-in-7-million chance of dying in an airplane crash on your next flight.

Chance is a slippery subject. We will follow Feynman in starting with "what would happen if we did this many times" before we try to think about using the language of chance to express personal opinions. We will also start with examples like the 1-in-2 chance of a head in tossing a coin before we try to think about the lottery.

The idea of probability

Even the rules of football agree that tossing a coin avoids favoritism. Favoritism in choosing subjects for a sample survey or allotting patients to treatment and placebo groups in a medical experiment is as undesirable as it is in awarding first possession of the ball in football. That's why statisticians recommend random samples and randomized experiments, which are fancy versions of tossing a coin. A big fact emerges when we watch coin tosses or the results of random samples closely: **Chance behavior is unpredictable in the short run but has a regular and predictable pattern in the long run.**

Toss a coin, or choose a simple random sample. The result can't be predicted in advance, because the result will vary when you toss the coin or choose the sample repeatedly. But there is still a regular pattern in the results, a pattern that emerges clearly only after many repetitions. This remarkable fact is the basis for the idea of probability.

Example 1. Coin tossing

When you toss a coin, there are only two possible outcomes, heads or tails. Figure 17.1 shows the results of tossing a coin 1000 times. For each number of tosses from 1 to 1000, I have plotted the proportion of those tosses that gave a head. The first toss was a head, so the proportion of heads starts at 1. The second toss was a tail, reducing the proportion of heads to 0.5 after two tosses. The next three tosses gave a tail followed by two heads, so the proportion of heads after five tosses is 3/5, or 0.6.

The proportion of tosses that produce heads is quite variable at first, but it settles down as we make more and more tosses. Eventually this proportion gets close to 0.5 and stays there. We say that 0.5 is the *probability* of a head. The probability 0.5 appears as a horizontal line on the graph.

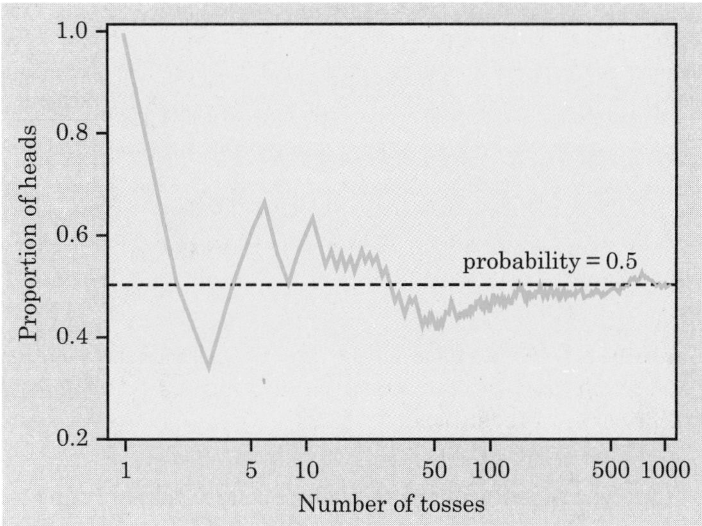

Figure 17.1 Toss a coin many times. The proportion of heads changes as we make more tosses but eventually gets very close to 0.5. This is what we mean when we say, "The probability of a head is one-half."

"Random" in statistics is not a synonym for "haphazard" but a description of a kind of order that emerges only in the long run. We encounter the unpredictable side of randomness in our everyday experience, but we rarely see enough repetitions of the same random phenomenon to observe the long-term regularity that probability describes. You can see that regularity emerging in Figure 17.1. In the very long run, the proportion of tosses that give a head is 0.5. This is the intuitive idea of probability. Probability 0.5 means "occurs half the time in a very large number of trials."

We might suspect that a coin has probability 0.5 of coming up heads just because the coin has two sides. But babies must have one of the two sexes, and the probabilities aren't equal—the probability of a boy is about 0.51, not 0.50. The idea of probability is empirical. That is, it is based on data rather than theorizing. Probability describes what happens in very many trials, and we must actually observe many coin tosses or many babies to pin down a probability. In the case of tossing a coin, some diligent people have in fact made thousands of tosses.

Example 2. Some coin tossers

The French naturalist Count Buffon (1707–1788) tossed a coin 4040 times. Result: 2048 heads, or proportion 2048/4040 = 0.5069 for heads.

Around 1900, the English statistician Karl Pearson heroically tossed a coin 24,000 times. Result: 12,012 heads, a proportion of 0.5005.

While imprisoned by the Germans during World War II, the South African mathematician John Kerrich tossed a coin 10,000 times. Result: 5067 heads, a proportion of 0.5067.

Randomness and probability

We call a phenomenon **random** if individual outcomes are uncertain but there is nonetheless a regular distribution of outcomes in a large number of repetitions.

The **probability** of any outcome of a random phenomenon is a number between 0 and 1 that describes the proportion of times the outcome would occur in a very long series of repetitions.

An outcome with probability 0 never occurs. An outcome with probability 1 happens on every repetition. An outcome with probability 1/2 happens half the time in a very long series of trials. Of course, we can never observe a probability exactly. We could always continue tossing the coin, for example. Mathematical probability is an idealization based on imagining what would happen in an indefinitely long series of trials.

We aren't thinking deeply here. That some things are random is simply an observed fact about the world. Probability just gives us a language to describe the long-term regularity of random behavior. The outcome of a coin toss, the time between emissions of particles by a radioactive source, and the sexes of the next litter of lab rats are all random. So is the outcome of a random sample or a randomized experiment. The behavior of large groups of individuals is often as random as the behavior of many coin tosses or many random samples. Life insurance, for example, is based on the fact that deaths occur at random among many individuals.

Example 3. The probability of dying

We can't predict whether a particular person will die in the next year. But if we observe millions of people, deaths are random. The National Center for Health Statistics says that the proportion of men aged 20 to 24 years who die in any one year is 0.0015. This is the *probability* that a young man will die next year. For women that age, the probability of death is about 0.0005.

If an insurance company sells many policies to people aged 20 to 24, it knows that it will have to pay off next year on about 0.15% of the policies sold to men and on about 0.05% of the policies sold to women. It will charge more to insure a man because the probability of having to pay is higher.

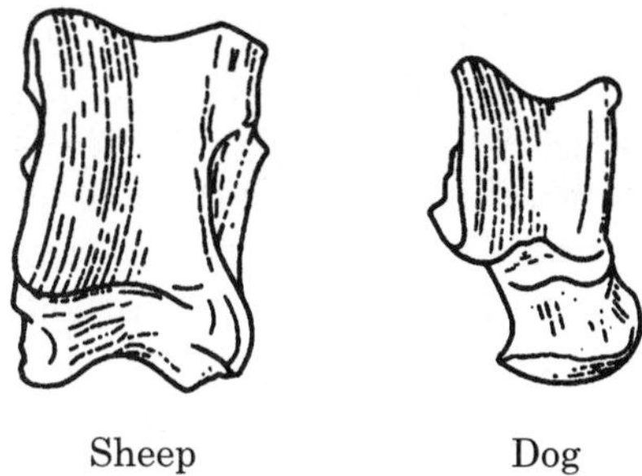

Figure 17.2 Animal heel bones (*astragali*), actual size. (From F. N. David, *Games, Gods, and Gambling,* Charles Griffin & Company, 1962. Reproduced by permission of the publishers.)

The ancient history of chance

Randomness is most easily noticed in many repetitions of games of chance—rolling dice, dealing shuffled cards, spinning a roulette wheel. Chance devices similar to these have been used from remote antiquity to discover the will of the gods. The most common method of randomization in ancient times was "rolling the bones," tossing several *astragali.* The astragalus (Figure 17.2) is a solid, quite regular bone from the heel of animals that, when thrown, will come to rest on any of four sides. (The other two sides are rounded.) Cubical dice, made of pottery or bone, came later, but even dice existed before 2000 B.C. Gambling on the throw of astragali or dice is, compared with divination, almost a modern development. There is no clear record of this vice before about 300 B.C. Gambling reached flood tide in Roman times, then temporarily receded (along with divination) in the face of Christian displeasure.

Chance devices such as astragali have been used from the beginning of recorded history. Yet none of the great mathematicians of antiquity studied the regular pattern of many throws of bones or dice. Perhaps this is because astragali and most ancient dice were so irregular that each had a different pattern of

Does God play dice?

Few things in the world are truly random in the sense that no amount of information will allow us to predict the outcome. We could in principle apply the laws of physics to a tossed coin, for example, and calculate whether it will land heads or tails. But randomness does rule events inside individual atoms. Albert Einstein didn't like this feature of the new quantum theory. "God does not play dice with the universe," said the great scientist. Eighty years later, it appears that Einstein was wrong.

outcomes. Or perhaps the reasons lie deeper, in the classical reluctance to engage in systematic experimentation.

Professional gamblers, who are not as inhibited as philosophers and mathematicians, did notice the regular pattern of outcomes of dice or cards and tried to adjust their bets to the odds of success. "How should I bet?" is the question that launched mathematical probability. The systematic study of randomness began (I oversimplify, but not too much) when 17th-century French gamblers asked French mathematicians for help in figuring out the "fair value" of bets on games of chance. *Probability theory,* the mathematical study of randomness, originated with Pierre de Fermat and Blaise Pascal in the 17th century and was well developed by the time statisticians took it over in the 20th century.

Myths about chance behavior

The idea of probability seems straightforward. It answers the question "What would happen if we did this many times?" In fact, both the behavior of random phenomena and the idea of probability are a bit subtle. We meet chance behavior constantly, and psychologists tell us that we deal with it poorly.

The myth of short-run regularity. The idea of probability is that randomness is regular *in the long run.* Unfortunately, our intuition about randomness tries to tell us that random phenomena should also be regular in the short run. When they aren't, we look for some explanation other than chance variation.

Example 4. What looks random?

Toss a coin six times and record heads (H) or tails (T) on each toss. Which of these outcomes is more probable?

HTHTTH TTTHHH

Almost everyone says that HTHTTH is more probable, because TTTHHH does not "look random." In fact, both are equally probable. That heads and tails are equally probable says only that about half of a very long sequence of tosses will be heads. It doesn't say that heads and tails must come close to alternating in the short run. The coin has no memory. It doesn't know what past outcomes were, and it can't try to create a balanced sequence.

The outcome TTTHHH in tossing six coins looks unusual because of the runs of 3 straight heads and 3 straight tails. Runs seem "not random" to our intuition but are quite common. Here's an example more striking than tossing coins.

Example 5. The hot hand in basketball

Belief that runs must result from something other than "just chance" influences behavior. If a basketball player makes several consecutive shots, both the fans and his teammates believe that he has a "hot hand" and is more likely to make the next shot. This is wrong. Careful study has shown that runs of baskets made or missed are no more frequent in basketball than would be expected if each shot is independent of the player's previous shots. Players perform consistently, not in streaks. If a player makes half her shots in the long run, her hits and misses behave just like tosses of a coin—and that means that runs of hits and misses are more common than our intuition expects.

The myth of the surprise meeting. Julie is spending the summer in London. One day, on the third floor of the Victoria and Albert Museum, she runs into Jim, a casual friend from college. "How unusual! Maybe we were fated to meet."

Well, maybe not. It is certainly unlikely that Julie would run into *this particular* acquaintance that day, but it is not at all unlikely that Julie will meet *some* acquaintance during her summer in London. After all, a typical adult has about 1500 casual acquaintances. When something unusual happens, we look back and say, "Wasn't that unlikely?" We would have said the same if any of 1500 other unlikely things had happened. Here's an example where we can actually calculate the probabilities.

Example 6. Winning the lottery twice

In 1986, Evelyn Marie Adams won the New Jersey state lottery for the second time, adding $1.5 million to her previous $3.9 million jackpot. The *New York Times* (February 14, 1986) claimed that the odds of one person winning the big prize twice were about 1 in 17 trillion. Nonsense, said two statistics professors in a letter that appeared in the *Times* two weeks later. The chance that Evelyn Marie Adams would win twice in her lifetime is indeed tiny, but it is almost certain that *someone* among the millions of regular lottery players in the United States would win two jackpot prizes. The statisticians estimated even odds of another double winner within seven years. Sure enough, Robert Humphries won his second Pennsylvania lottery jackpot ($6.8 million total) in May 1988.

Unusual events—especially distressing events—bring out the human desire to pinpoint a reason, a *cause*. Here's a sequel to our earlier discussion of causation: sometimes it's just the play of chance.

The probability of rain is . . .

You work all week. Then it rains on the weekend. Can there really be a statistical truth behind our perception that the weather is against us? At least on the east coast of the United States, the answer is "Yes." Going back to 1946, it seems that Sundays receive 22% more precipitation than Mondays. The likely explanation is that the pollution from all those workday cars and trucks forms the seeds for raindrops—with just enough delay to cause rain on the weekend.

Example 7. Cancer clusters

In 1984, residents of a neighborhood in Randolph, Massachusetts, counted 67 cancer cases in their 250 residences. This cluster of cancer cases seemed unusual, and the residents expressed concern that runoff from a nearby chemical plant was contaminating their water supply and causing cancer.

In 1979, two of the eight town wells serving Woburn, Massachusetts, were found to be contaminated with organic chemicals. Alarmed citizens began counting cancer cases. Between 1964 and 1983, 20 cases of childhood leukemia were reported in Woburn. This is an unusual number of cases of this rather rare disease. The residents believed that the well water had caused the leukemia and proceeded to sue two companies held responsible for the contamination.

Cancer is a common disease, accounting for more than 23% of all deaths in the United States. That cancer cases sometimes occur in clusters in the same neighborhood is not surprising; there are bound to be clusters *somewhere* simply by chance. But when a cancer cluster occurs in *our* neighborhood, we tend to suspect the worst and look for someone to blame. State authorities get several thousand calls a year from people worried about "too much cancer" in their area. But, as the National Cancer Institute says, "The majority of cancer clusters are simply the result of chance."

Both of the Massachusetts cancer clusters were investigated by statisticians from the Harvard School of Public Health. The investigators tried to obtain complete data on everyone who had lived in the neighborhoods in the periods in question and to estimate their exposure to the suspect drinking water. They also tried to obtain data on other factors that might explain cancer, such as smoking and occupational exposure to toxic substances. The verdict: chance is the likely explanation of the Randolph cluster, but there is evidence of an association between drinking water from the two Woburn wells and developing childhood leukemia.

The myth of the law of averages. Once at a convention in Las Vegas I roamed the gambling floors, watching money disappear into the drop boxes under the tables. You can see some interesting human behavior in a casino. When the shooter in the dice game craps rolls several winners in a row, some gamblers think she has a "hot hand" and bet that she will keep on winning. Others say that "the law of averages" means that she must now lose so that wins and losses will balance out. Believers in the law of averages think that if you toss a coin six times and get TTTTTT, the next toss must be more likely to give a head. It's true that in the long run heads must appear half the time. What is myth is that future outcomes must make up for an imbalance like six straight tails.

Coins and dice have no memories. A coin doesn't know that the first six outcomes were tails, and it can't try to get a head on the next toss to even things out. Of course, things do even out *in the long run*. After 10,000 tosses, the results of the first six tosses don't matter. They are overwhelmed by the results of the next 9994 tosses, not compensated for.

"So the law of averages doesn't guarantee me a girl after seven straight boys, but can't I at least get a group discount on the delivery fee?"

Example 8. We want a boy

Belief in this phony "law of averages" can lead to consequences close to disastrous. A few years ago, "Dear Abby" published in her advice column a letter from a distraught mother of eight girls. It seems that she and her husband had planned to limit their family to four children. When all four were girls, they tried again—and again, and again. After seven straight girls, even her doctor had assured her that "the law of averages was in our favor 100 to 1." Unfortunately for this couple, having children is like tossing coins. Eight girls in a row is highly unlikely, but once seven girls have been born, it is not at all unlikely that the next child will be a girl—and it was.

Personal probabilities

Joe sits staring into his beer as his favorite baseball team, the Chicago Cubs, lose another game. The Cubbies have some good young players, so let's ask Joe, "What's the chance that the Cubs will go to the World Series next year?" Joe brightens up. "Oh, about 10%," he says.

Does Joe assign probability 0.10 to the Cubs' appearing in the World Series? The outcome of next year's pennant race is certainly unpredictable, but we can't reasonably ask what would happen in many repetitions. Next year's baseball season will happen only once and will differ from all other seasons in players, weather, and many other ways. The answer to our question seems clear: if probability measures "what would happen if we did this many times," Joe's 0.10 is not a probability. Probability is based on data about many repetitions of the same random phenomenon. Joe is giving us something else, his personal judgment.

Yet we often use "probability" in a way that includes personal judgments of how likely it is that some event will happen. We make decisions based on these judgments—we take the bus downtown because we think the probability of finding a parking spot is low. More serious decisions also take "how likely" judgments into account. A company deciding whether to build a new plant must judge how likely it is that there will be high demand for its products three years from now when the plant is ready. Many companies express "How likely is it?" judgments as numbers—probabilities—and use these numbers in their calculations. High demand in three years, like the Cubs' winning next year's pennant, is a one-time event that doesn't fit the "do it many times" way of thinking. What is more, several company officers may give several different probabilities, reflecting differences in their individual judgment. We need another kind of probability, *personal probability.*

Personal probability

A **personal probability** of an outcome is a number between 0 and 1 that expresses an individual's judgment of how likely the outcome is.

Personal probabilities have the great advantage that they aren't limited to repeatable settings. They are useful because we base decisions on them: "I think the probability that the Vikings will win the Super Bowl is 0.75, so I'm going to bet on the game." Just remember that personal probabilities are different in kind from probabilities as "proportions in many repetitions." Because they express individual opinion, they can't be said to be right or wrong.

This is true even in a "many repetitions" setting. If Craig has a gut feeling that the probability of a head on the next toss of this coin is 0.7, that's what Craig thinks and that's all there is to it. Tossing the coin many times may show that the proportion of heads is very close to 0.5, but that's another matter. **There is no reason why a person's degree of confidence in the outcome of one try must agree with the results of many tries.** I stress this because it is common to say that "personal probability" and "what happens in many trials" are somehow two interpretations of the same idea. In fact, they are quite different ideas.

Why do we even use the word "probability" for personal opinions? There are two good reasons. First, we usually do base our personal opinions on data from many trials when we have such data. Data from Buffon, Pearson, and Kerrich (Example 2) and perhaps from our own experience convince us that coins come up heads very close to half the time in many tosses. When we say that a coin has probability 1/2 of coming up heads *on this toss,* we are applying to a single toss a measure of the chance of a head based on what would happen in a long series of tosses. Second, personal probability and probability as long-term proportion both obey the same mathematical rules. To start, both kinds of probabilities are numbers between 0 and 1. This isn't as important to us as it is to mathematicians, but we will look at some of the rules of probability in the next chapter. These rules apply to both kinds of probability.

What are the odds?

Gamblers often express chance in terms of *odds* rather than probability. Odds of A to B against an outcome means that the probability of that outcome is $B/(A + B)$. So "odds of 5 to 1" is another way of saying "probability 1/6." A probability is always between 0 and 1, but odds range from 0 to infinity. Although odds are mainly used in gambling, they give us a way to make very small probabilities clearer. "Odds of 999 to 1" may be easier to understand than "probability 0.001."

Probability and risk

Once we understand that "personal judgment of how likely" and "what happens in many repetitions" are different ideas, we have a good start toward understanding why the public and the experts disagree so strongly about what is risky and what isn't. The experts use probabilities from data to describe the risk of an unpleasant event. Individuals and society, however, seem to ignore data. We worry about some risks that almost never occur while ignoring others that are much more probable.

Example 9. Asbestos in the schools

High exposures to asbestos are dangerous. Low exposures, such as that experienced by teachers and students in schools where asbestos is present in the insulation around pipes, are not very risky. The probability that a teacher who works for 30 years in a school with typical asbestos levels will get cancer from the asbestos is around 15/1,000,000. The risk of dying in a car accident during a lifetime of driving is about 15,000/1,000,000. That is, driving regularly is 1000 times more risky than teaching in a school where asbestos is present.

Risk does not stop us from driving. Yet the much smaller risk from asbestos launched massive cleanup campaigns and a federal requirement that every school inspect for asbestos and make the findings public.

Why do we take asbestos so much more seriously than driving? Why do we worry about very unlikely threats such as tornados and terrorists more than we worry about heart attacks?

- We feel safer when a risk seems under our control than when we cannot control it. We are in control (or so we imagine) when we are driving, but we can't control the risk from asbestos or tornados or terrorists.
- It is hard to comprehend very small probabilities. Probabilities of 15 per million and 15,000 per million are both so small that our intuition

cannot distinguish between them. Psychologists have shown that we generally overestimate very small risks and underestimate higher risks. Perhaps this is part of the general weakness of our intuition about how probability operates.

- The probabilities for risks like asbestos in the schools are not as certain as probabilities for tossing coins. They must be estimated by experts from complicated statistical studies. Perhaps it is safest to suspect that the experts may have underestimated the level of risk.

Our reactions to risk depend on more than probability, even if our personal probabilities are higher than the experts' data-based probabilities. We are influenced by our psychological makeup and by social standards. As one writer noted, "Few of us would leave a baby sleeping alone in a house while we drove off on a 10-minute errand, even though car-crash risks are much greater than home risks."

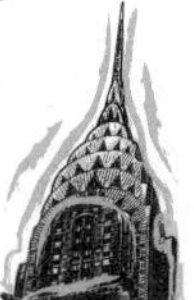

Exploring the Web

One of the best ways to grasp the idea of probability is to watch the proportion of trials on which an outcome occurs gradually settle down at the outcome's probability. Computer simulations can show this. Go to the *Statistics: Concepts and Controversies* Web site, www.whfreeman.com/scc, and look at the "Probability" applet.

Cancer clusters cause so much public concern that the National Cancer Institute has devoted an entire page to them: cancernet.nci.nih.gov/clinpdq/risk/Cancer_Clusters.html.

Statistics in Summary Some things in the world, both natural and of human design, are **random.** That is, their outcomes have a clear pattern in very many repetitions even though the outcome of any one trial is unpredictable. **Probability** describes the long-term regularity of random phenomena. The probability of an outcome is the proportion of very many repetitions on which that outcome occurs. A probability is a number between 0 (never occurs) and 1 (always occurs). I emphasize this kind of probability because it is based on data.

Probabilities describe only what happens in the long run. Short runs of random phenomena like tossing coins or shooting a basketball often don't look random to us because they do not show the regularity that in fact only emerges in very many repetitions.

Personal probabilities express an individual's personal judgment of how likely outcomes are. Personal probabilities are also numbers between

0 and 1. Different people can have different personal probabilities, and a personal probability need not agree with a proportion based on data about similar cases.

CHAPTER 17 EXERCISES

17.1 Pennies spinning. Hold a penny upright on its edge under your forefinger on a hard surface, then snap it with your other forefinger so that it spins for some time before falling. Based on 50 spins, estimate the probability of heads.

17.2 Pennies falling over. You may feel that it is obvious that the probability of a head in tossing a coin is about 1/2 because the coin has two faces. Such opinions are not always correct. The previous exercise asked you to spin a penny rather than toss it—that changes the probability of a head. Now try another variation. Stand a penny on edge on a hard, flat surface. Pound the surface with your hand so that the penny falls over. What is the probability that it falls with heads upward? Make at least 50 trials to estimate the probability of a head.

17.3 Random digits. The table of random digits (Table A) was produced by a random mechanism that gives each digit probability 0.1 of being a 0. What proportion of the first 200 digits in the table are 0s? This proportion is an estimate, based on 200 repetitions, of the true probability, which in this case is known to be 0.1.

17.4 How many tosses to get a head? When we toss a penny, experience shows that the probability (long-term proportion) of a head is close to 1/2. Suppose now that we toss the penny repeatedly until we get a head. What is the probability that the first head comes up in an odd number of tosses (1, 3, 5, and so on)? To find out, repeat this experiment 50 times, and keep a record of the number of tosses needed to get a head on each of your 50 trials.
(a) From your experiment, estimate the probability of a head on the first toss. What value should we expect this probability to have?

(b) Use your results to estimate the probability that the first head appears on an odd-numbered toss.

17.5 Tossing a thumbtack. Toss a thumbtack on a hard surface 100 times. How many times did it land with the point up? What is the approximate probability of landing point up?

17.6 Three of a kind. You read in a book on poker that the probability of being dealt three of a kind in a five-card poker hand is 1/50. Explain in simple language what this means.

17.7 From words to probabilities. Probability is a measure of how likely an event is to occur. Match one of the probabilities that follow with each statement of likelihood given. (The probability is usually a more exact measure of likelihood than is the verbal statement.)

0 0.01 0.3 0.6 0.99 1

(a) This event is impossible. It can never occur.

(b) This event is certain. It will occur on every trial.

(c) This event is very unlikely, but it will occur once in a while in a long sequence of trials.

(d) This event will occur more often than not.

17.8 Winning a baseball game. Over the period from 1969 to 1989 the champions of baseball's two major leagues won 63% of their home games during the regular season. At the end of the season, the two league champions meet in the baseball World Series. Would you use the study results to assign probability 0.63 to the event that the home team wins a World Series game? Explain your answer.

17.9 Will you have an accident? The probability that a randomly chosen driver will be involved in an accident in the next year is about 0.2. This is based on the proportion of millions of drivers who have accidents. "Accident" includes things like crumpling a fender in your own driveway, not just highway accidents.

(a) What do you think is your own probability of being in an accident in the next year? This is a personal probability.

(b) Give some reasons why your personal probability might be a more accurate prediction of your "true chance" of having an accident than the probability for a random driver.

(c) Almost everyone says their personal probability is lower than the random driver probability. Why do you think this is true?

17.10 Marital status. The probability that a randomly chosen woman 40 years old is divorced is about 0.16. This probability is a long-run proportion based on all the millions of women aged 40. Let's suppose that the proportion stays at 0.16 for the next 20 years. Bridget is now 20 years old and is not married.

(a) Bridget thinks her own chances of being divorced at age 40 are about 5%. Explain why this is a personal probability.

(b) Give some good reasons why Bridget's personal probability might differ from the proportion of all women aged 40 who are divorced.

(c) You are a government official charged with looking into the impact of the Social Security system on middle-aged divorced women. You care only about the probability 0.16, not about anyone's personal probability. Why?

17.11 Personal probability versus data. Give an example in which you would rely on a probability found as a long-term proportion from data on many trials. Give an example in which you would rely on your own personal probability.

17.12 Personal probability? When there are few data, we often fall back on personal probability. There had been just 24 space shuttle launches, all successful, before the *Challenger* disaster. The shuttle program management thought the chances of such a failure were only 1 in 100,000. Give some reasons why such an estimate is likely to be too optimistic.

17.13 Personal random numbers? Ask several of your friends (at least 10 people) to choose a four-digit number "at random." How many of the numbers chosen start with 1 or 2? How many start with 8 or 9? (There is strong evidence that people in general tend to choose numbers starting with low digits.)

17.14 Playing "pick four." The "pick four" games in many state lotteries announce a four-digit winning number each day. The winning number is essentially a four-digit group from a table of random digits. You win if your choice matches the winning digits, in exact order. The winnings are divided among all players who matched the winning digits. That suggests a way to get an edge.

(a) The winning number might be, for example, either 2873 or 9999. Explain why these two outcomes have exactly the same probability. (It is 1 in 10,000.)

(b) If you asked many people which outcome is more likely to be the randomly chosen winning number, most would favor one of them. Use the information in this chapter to say which one and to explain why. If you choose a number that people think is unlikely, you have the same chance to win, but you will win a larger amount because few other people will choose your number.

17.15 Surprising? You are getting to know your new roommate, assigned to you by the college. In the course of a long conversation, you find that both of you have sisters named Deborah. Should you be surprised? Explain your answer.

17.16 Shaq's free throws. The basketball player Shaquille O'Neal makes about half of his free throws over an entire season. In today's game, Shaq makes his first three free throws. The TV commentator says, "Shaq's technique really looks good today." Explain why the claim that Shaq has improved his shooting technique is not justified.

17.17 In the long run. Probability works not by compensating for imbalances but by overwhelming them. Suppose that the first six tosses of a coin give six tails and that tosses after that are exactly half heads and half tails. (Exact balance is unlikely, but it illustrates how the first six outcomes are swamped by later outcomes.) What is the proportion of heads after the first six tosses? What is the proportion of heads after 100 tosses if the last 94 produce 47 heads? What is the proportion of heads after 1000 tosses if the last 994 produce half heads? What is the proportion of heads after 10,000 tosses if the last 9994 produce half heads?

17.18 The "law of averages." The baseball player Tony Gwynn gets a hit about 35% of the time over an entire season. After he has failed to hit safely in six straight at-bats, the TV commentator says, "Tony is due for a hit by the law of averages." Is that right? Why?

17.19 Cold weather coming. A meteorologist, predicting a colder-than-normal winter, said, "First, in looking at the past few winters, there has been a lack of really cold weather. Even though we are not supposed to use the law of averages, we are due." Do you think that "due by the law of averages" makes sense in talking about the weather?

17.20 An unenlightened gambler.
(a) A gambler knows that red and black are equally likely to occur on each spin of a roulette wheel. He observes five consecutive reds occur and bets heavily on black at the next spin. Asked why, he explains that black is "due by the law of averages." Explain to the gambler what is wrong with this reasoning.

(b) After hearing you explain why red and black are still equally likely after five reds on the roulette wheel, the gambler moves to a poker game. He is dealt five straight red cards. He remembers what you said and assumes that the next card dealt in the same hand is equally likely to be red or black. Is the gambler right or wrong, and why?

17.21 Reacting to risks. The probability of dying if you play high school football is about 10 per million each year you play. The risk of getting cancer from asbestos if you attend a school in which asbestos is present for 10 years is about 5 per million. If we ban asbestos from schools, should we also ban high school football? Briefly explain your position.

17.22 Reacting to risks. National newspapers such as *USA Today* and the *New York Times* carry many more stories about deaths from airplane crashes than about deaths from automobile crashes. Auto accidents kill about 40,000 people in the United States each year. Crashes of all scheduled air carriers, including commuter carriers, have killed between 44 and 394 people per year in recent years.
(a) Why do the news media give more attention to airplane crashes?

(b) How does news coverage help explain why many people consider flying more dangerous than driving?

17.23 What probability doesn't say. The probability of a head in tossing a coin is 1/2. This means that as we make more tosses, the *proportion* of heads will eventually get close to 0.5. It does not mean that the *count* of heads will get close to 1/2 the number of tosses. To see why, imagine that the proportion of heads is 0.51 in 100 tosses, 1000 tosses, 10,000 tosses, and 100,000 tosses of a coin. How many heads came up in each set of tosses? How close is the number of heads to half the number of tosses?

Chapter 18

Probability Models

Heads, tails, or confusion

Martha: OK, Tom, here's a new Sacagawea dollar coin. What do you think is the probability of a head if I toss it?

Tom: Oh, I'd say 60%.

Martha: And what's your probability of a tail?

Tom: Make that 50%.

Martha: 60% chance of a head and 50% chance of a tail add up to 110% for a head or a tail. That doesn't make sense.

Tom: Well, those are my personal probabilities, so they can be anything I want them to be.

Martha: No they can't. They have to make sense when you look at them both together.

Martha says that personal probabilities must obey common-sense rules if they are to make sense. Probabilities that "make sense" may not describe what will happen if we toss the coin many times. We know that Tom's personal probabilities may not agree with how often Sacagawea really comes up in many tosses. But just as we check to see if data are consistent with each other, we insist that probabilities be consistent with each other. The probability of heads and the probability of tails for the same coin can't add to more than 1. In fact, they must add to exactly 1 unless we allow some other outcome, such as the coin landing on edge and staying upright. That's because all possible outcomes together must have probability exactly 1. And that's one of the "rules of probability" we will explore in this chapter.

Probability models

Choose a woman aged 25 to 29 years old at random and record her marital status. "At random" means that we give every such woman the same chance to be the one we choose. That is, we choose a random sample of size 1. The probability of any marital status is just the proportion of all women aged 25 to 29 who have that status—if we chose many women, this is the proportion we would get. Here is the set of probabilities:

Marital status	Never married	Married	Widowed	Divorced
Probability	0.386	0.555	0.004	0.055

This table gives a *probability model* for drawing a young woman at random and finding out her marital status. It tells us what are the possible outcomes (there are only four) and it assigns probabilities to these outcomes. The probabilities here are the proportions of all women who are in each marital class. That makes it clear that the probability that a woman is not married is just the sum of the probabilities of the three classes of unmarried women:

$$\begin{aligned} P(\text{not married}) &= P(\text{never married}) + P(\text{widowed}) + P(\text{divorced}) \\ &= 0.386 + 0.004 + 0.055 = 0.445 \end{aligned}$$

As a shorthand, we often write P(not married) for "the probability that the woman we choose is not married." You see that our model does more than assign a probability to each individual outcome—we can find the probability of any collection of outcomes by adding up individual outcome probabilities.

Probability model

A **probability model** for a random phenomenon describes all the possible outcomes and says how to assign probabilities to any collection of outcomes. We sometimes call a collection of outcomes an **event.**

Probability rules

Because the probabilities in this example are just the proportions of all women who have each marital status, they follow rules that say how proportions behave. Here are some basic rules that any probability model must obey:

A. **Any probability is a number between 0 and 1.** Any proportion is a number between 0 and 1, so any probability is also a number between 0 and 1. An event with probability 0 never occurs, and an event with probability

1 occurs on every trial. An event with probability 0.5 occurs in half the trials in the long run.

B. **All possible outcomes together must have probability 1.** Because some outcome must occur on every trial, the sum of the probabilities for all possible outcomes must be exactly 1.

C. **The probability that an event does not occur is 1 minus the probability that the event does occur.** If an event occurs in (say) 70% of all trials, it fails to occur in the other 30%. The probability that an event occurs and the probability that it does not occur always add to 100%, or 1.

D. **If two events have no outcomes in common, the probability that one or the other occurs is the sum of their individual probabilities.** If one event occurs in 40% of all trials, a different event occurs in 25% of all trials, and the two can never occur together, then one or the other occurs on 65% of all trials because 40% + 25% = 65%.

Politically correct

In 1950, the Russian mathematician B. V. Gnedenko (1912–1995) wrote a text on *The Theory of Probability* that was popular around the world. The introduction contains a mystifying paragraph that begins, "We note that the entire development of probability theory shows evidence of how its concepts and ideas were crystallized in a severe struggle between materialistic and idealistic conceptions." It turns out that "materialistic" is jargon for "Marxist-Leninist." It was good for the health of Russian scientists in the Stalin era to add such statements to their books.

Example 1. Marital status of young women

Look again at the probabilities for the marital status of young women. Each of the four probabilities is a number between 0 and 1. Their sum is

$$0.386 + 0.555 + 0.004 + 0.055 = 1$$

This assignment of probabilities satisfies Rules A and B. **Any assignment of probabilities to all individual outcomes that satisfies Rules A and B is legitimate.** That is, it makes sense as a set of probabilities. Rules C and D are then automatically true. Here is an example of the use of Rule C.

The probability that the woman we draw is not married is, by Rule C,

$$\begin{aligned} P(\text{not married}) &= 1 - P(\text{married}) \\ &= 1 - 0.555 = 0.445 \end{aligned}$$

That is, if 55.5% are married, then the remaining 44.5% are not married. Rule D says that you can also find the probability that a woman is not married by adding the probabilities of the three distinct ways of being not married, as we did earlier. This gives the same result.

Example 2. Rolling two dice

Rolling two dice is a common way to lose money in casinos. There are 36 possible outcomes when we roll two dice and record the up faces in order (first die, second die). Figure 18.1 displays these outcomes. What probabilities should we assign?

Casino dice are carefully made. Their spots are not hollowed out, which would give the faces different weights, but are filled with white plastic of the same density as the red plastic of the body. For casino dice it is reasonable to assign the same probability to each of the 36 outcomes in Figure 18.1. Because these 36 probabilities must have sum 1 (Rule B), each outcome must have probability 1/36.

We are interested in the sum of the spots on the up faces of the dice. What is the probability that this sum is 5? The event "roll a 5" contains four outcomes, and its probability is the sum of the probabilities of these outcomes:

$$\begin{aligned} P(\text{roll a 5}) &= P(\text{⚀ ⚃}) + P(\text{⚁ ⚂}) + P(\text{⚂ ⚁}) + P(\text{⚃ ⚀}) \\ &= \frac{1}{36} + \frac{1}{36} + \frac{1}{36} + \frac{1}{36} \\ &= \frac{4}{36} = 0.111 \end{aligned}$$

The rules tell us only what probability models *make sense.* They don't tell us whether the probabilities are *correct,* that is, whether they describe what actually happens in the long run. The probabilities in Example 2 are correct for casino dice. Inexpensive dice with hollowed-out spots are not balanced, and this probability model does not describe their behavior.

What about personal probabilities? As Tom said, "Well, those are my personal probabilities, so they can be anything I want them to be." We can't say that personal probabilities that don't obey Rules A and B are wrong, but we can say that they are **incoherent.** That is, they don't go together in a way that makes sense. So we usually insist that personal probabilities for all the outcomes of a random phenomenon obey Rules A and B. That is, the same rules govern both kinds of probability.

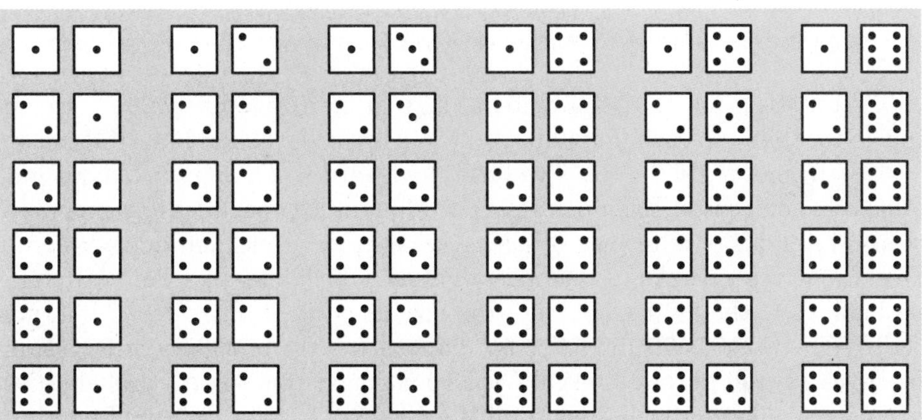

Figure 18.1 The 36 possible outcomes from rolling two dice.

Probability models for sampling

Choosing a random sample from a population and calculating a statistic such as the sample proportion is certainly a random phenomenon. The *distribution* of the statistic tells us what values it can take and how often it takes those values. That sounds a lot like a probability model.

Example 3. A sampling distribution

Take a simple random sample (SRS) of 1523 adults. Ask each whether they bought a lottery ticket in the last 12 months. The proportion who say "Yes,"

$$\hat{p} = \frac{\text{number who say "Yes"}}{1523}$$

is the sample proportion $\hat{p}$. Do this 1000 times and collect the 1000 sample proportions $\hat{p}$ from the 1000 samples. The histogram in Figure 18.2 shows the distribution of 1000 sample proportions when the truth about the population is that 60% have bought lottery tickets. The results of random sampling are of course random: we can't predict the outcome of one sample, but the figure shows that the outcomes of many samples have a regular pattern.

We have seen Figure 18.2 before, in Chapter 13. In fact, we saw the histogram part of this figure earlier, in Chapters 3 and 11. The repetition reminds us that the regular pattern of repeated random samples is one of the big ideas of statistics. The normal curve in the figure is a good approximation to the histogram. The histogram is the result of these particular 1000 SRSs. Think of the normal curve as the idealized pattern we would get if we kept on taking SRSs from this population forever. That's exactly the idea of probability—the pattern we would see in the very long run. *The normal curve assigns probabilities to the outcomes of random sampling.*

This normal curve has mean 0.6 and standard deviation about 0.0125. The "95" part of the 68–95–99.7 rule says that 95% of all samples will give a $\hat{p}$ falling within 2 standard deviations of the mean. That's within 0.025 of 0.6, or between 0.575 and 0.625. We now have more concise language for this fact: the *probability* is 0.95 that between 57.5% and 62.5% of the people in a sample will say "Yes." The word "probability" says we are talking about what would happen in the long run, in very many samples.

A statistic from a large sample has a great many possible values. Assigning a probability to each individual outcome worked well for 4 marital classes or 36 outcomes of rolling two dice but is awkward when there are thousands of possible outcomes. Example 3 uses a different approach: assign probabilities to intervals of outcomes by using areas under a normal density curve. Density curves have area 1 underneath them, which lines up nicely with total probability 1. The total area under the normal curve in Figure 18.2 is 1, and the area between 0.575 and 0.625 is 0.95, which is the probability that a sample gives a result in that interval. When a normal curve assigns probabilities, you can calculate probabilities from the 68–95–99.7 rule or from Table B of percentiles of normal distributions. These probabilities satisfy Rules A to D.

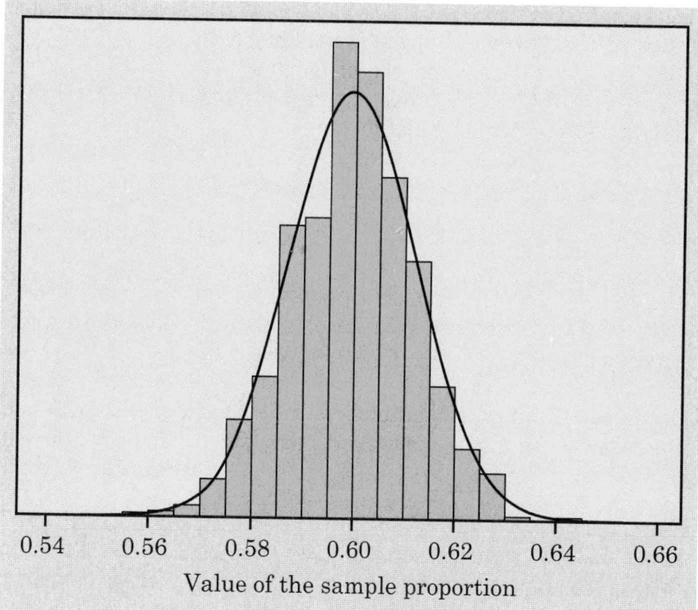

Figure 18.2 The sampling distribution of a sample proportion $\hat{p}$ from SRSs of size 1523 drawn from a population in which 60% of the members would give positive answers. The histogram shows the distribution from 1000 samples. The normal curve is the ideal pattern that describes the results of a very large number of samples.

Sampling distribution

The **sampling distribution** of a statistic tells us what values the statistic takes in repeated samples from the same population and how often it takes those values.

We think of a sampling distribution as assigning probabilities to the values the statistic can take. Because there are usually many possible values, sampling distributions are often described by a density curve such as a normal curve.

Example 4. Do you approve of gambling?

An opinion poll asks an SRS of 501 teens, "Generally speaking, do you approve or disapprove of legal gambling or betting?" Suppose that in fact exactly 50% of all teens would say "Yes" if asked. (This is close to what polls show to be true.) The poll's statisticians tell us that the sample proportion who say "Yes" will vary in repeated samples according to a normal distribution with mean 0.5 and standard deviation about 0.022. This is the *sampling distribution* of the sample proportion $\hat{p}$.

The 68–95–99.7 rule says that the probability is 0.16 that the poll gets a sample in which fewer than 47.8% say "Yes." Figure 18.3 shows how to get this result from the normal curve of the sampling distribution.

Example 5. Using normal percentiles*

What is the probability that the opinion poll in Example 4 will get a sample in which 52% or more say "Yes"? Because 0.52 is not 1, 2, or 3 standard deviations away from the mean, we can't use the 68–95–99.7 rule. We will use Table B of percentiles of normal distributions.

To use Table B, first turn the outcome $\hat{p} = 0.52$ into a standard score by subtracting the mean of the distribution and dividing by its standard deviation:

$$\frac{0.52 - 0.5}{0.022} = 0.9$$

Now look in Table B. A standard score of 0.9 is the 81.59 percentile of a normal distribution. This means that the probability is 0.8159 that the poll gets a smaller result. By Rule C (or just the fact that the total area under the curve is 1), this leaves probability 0.1841 for outcomes 52% or more "Yes." Figure 18.4 shows the probabilities as areas under the normal curve.

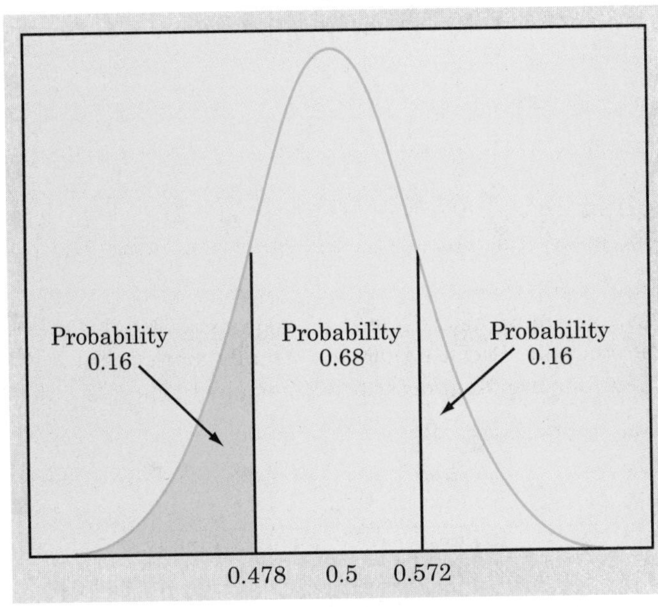

Figure 18.3 The normal sampling distribution for Example 4. Because 0.478 is one standard deviation below the mean, the area under the curve to the left of 0.478 is 0.16.

*Example 5 is optional.

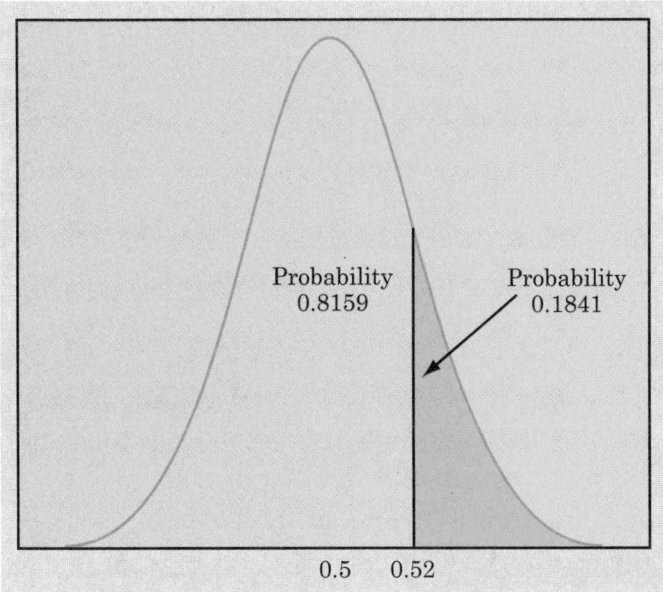

Figure 18.4 The normal sampling distribution for Example 5. The outcome 0.52 has standard score 0.9, so Table B tells us that the area under the curve to the left of 0.52 is 0.8159.

Statistics in Summary A **probability model** describes a random phenomenon by telling what outcomes are possible and how to assign probabilities to them. There are two simple ways to give a probability model. The first assigns a probability to each individual outcome. These probabilities must be numbers between 0 and 1 (Rule A) and they must add to exactly 1 (Rule B). To find the probability of any **event**, add the probabilities of the outcomes that make up the event.

The second kind of probability model assigns probabilities as areas under a **density curve**, such as a normal curve. The total probability is 1 because the total area under the curve is 1. This kind of probability model is often used to describe the **sampling distribution** of a statistic. This is the pattern of values of the statistic in many samples from the same population.

All legitimate assignments of probability, whether data-based or personal, obey the same **probability rules.** So the mathematics of probability is always the same.

CHAPTER 18 EXERCISES

18.1 Moving up. A sociologist studying social mobility in Denmark finds that the probability that the son of a lower-class father remains in the lower class is 0.46. What is the probability that the son moves to one of the higher classes?

18.2 Causes of death. Government data assign a single cause for each death that occurs in the United States. The data show that the probability is 0.45 that a randomly chosen death was due to cardiovascular (mainly heart) disease, and 0.23 that it was due to cancer. What is the probability that a death was due either to cardiovascular disease or to cancer? What is the probability that the death was due to some other cause?

18.3 Land in Canada. Choose an acre of land in Canada at random. The probability is 0.35 that it is forest and 0.03 that it is pasture.
(a) What is the probability that the acre chosen is not forested?

(b) What is the probability that it is either forest or pasture?

(c) What is the probability that a randomly chosen acre in Canada is something other than forest or pasture?

18.4 Do husbands do their share? An opinion poll interviewed a random sample of 1025 women. The married women in the sample were asked whether their husbands did their fair share of household chores. Here are the results:

Outcome	Probability
Does more than his fair share	0.12
Does his fair share	0.61
Does less than his fair share	?

These proportions are probabilities for the random phenomenon of choosing a married woman at random and asking her opinion.
(a) What must be the probability that the woman chosen says that her husband does less than his fair share? Why?

(b) The event "I think my husband does at least his fair share" contains the first two outcomes. What is its probability?

18.5 Rolling a die. Figure 18.5 displays several assignments of probabilities to the six faces of a die. We can learn which assignment is actually *correct* for a particular die only by rolling the die many times. However, some of the assignments are not *legitimate* assignments of probability. That is, they do not obey the rules. Which are legitimate and which are not? In the case of the illegitimate models, explain what is wrong.

Outcome	Probability: Model 1	Model 2	Model 3	Model 4
⚀	1/7	1/3	1/3	1
⚁	1/7	1/6	1/6	1
⚂	1/7	1/6	1/6	2
⚃	1/7	0	1/6	1
⚄	1/7	1/6	1/6	1
⚅	1/7	1/6	1/6	2

Figure 18.5 Four probability models for rolling a die, for Exercise 18.5.

18.6 High school academic rank. Select a first-year college student at random and ask what his or her academic rank was in high school. Here are the probabilities, based on proportions from a large sample survey of first-year students:

Rank	Top 20%	Second 20%	Third 20%	Fourth 20%	Lowest 20%
Probability	0.41	0.23	0.29	0.06	0.01

(a) What is the sum of these probabilities? Why do you expect the sum to have this value?

(b) What is the probability that a randomly chosen first-year college student was not in the top 20% of his or her high school class?

(c) What is the probability that a first-year student was in the top 40% in high school?

18.7 Tetrahedral dice. Psychologists sometimes use tetrahedral dice to study our intuition about chance behavior. A tetrahedron (Figure 18.6) is a pyramid with 4 faces, each a triangle with all sides equal in length. Label the 4 faces of a tetrahedral die with 1, 2, 3, and 4 spots. Give a probability model for rolling such a die and recording the number of spots on the down face. Explain why you think your model is at least close to correct.

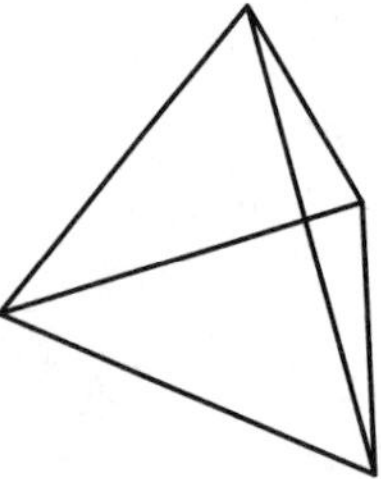

Figure 18.6 A tetrahedron. Exercises 18.7 and 18.9 concern dice with this shape.

18.8 Birth order. A couple plan to have three children. There are 8 possible arrangements of girls and boys. For example, GGB means the first two children are girls and the third child is a boy. All 8 arrangements are (approximately) equally likely.

(a) Write down all 8 arrangements of the sexes of three children. What is the probability of any one of these arrangements?

(b) What is the probability that the couple's children are 2 girls and 1 boy?

18.9 More tetrahedral dice. Tetrahedral dice are described in Exercise 18.7. Give a probability model for rolling two such dice. That is, write down all possible outcomes and give a probability to each. (Example 2 and Figure 18.1 may help you.) What is the probability that the sum of the down faces is 5?

18.10 Roulette. A roulette wheel has 38 slots, numbered 0, 00, and 1 to 36. The slots 0 and 00 are colored green, 18 of the others are red, and 18 are black. The dealer spins the wheel and at the same time rolls a small ball along the wheel in the opposite direction. The wheel is carefully balanced so that the ball is equally likely to land in any slot when the wheel slows. Gamblers can bet on various combinations of numbers and colors.

(a) What is the probability of any one of the 38 possible outcomes? Explain your answer.

(b) If you bet on "red," you win if the ball lands in a red slot. What is the probability of winning?

(c) The slot numbers are laid out on a board on which gamblers place their bets. One column of numbers on the board contains all multiples of 3, that is, 3, 6, 9, . . . , 36. You place a "column bet" that wins if any of these numbers comes up. What is your probability of winning?

18.11 Colors of M&M's. If you draw an M&M candy at random from a bag of the candies, the candy you draw will have one of six colors. The probability of drawing each color depends on the proportion of each color among all candies made.

(a) Here are the probabilities of each color for a randomly chosen plain M&M:

Color	Brown	Red	Yellow	Green	Orange	Blue
Probability	0.3	0.2	0.2	0.1	0.1	?

What must be the probability of drawing a blue candy?

(b) The probabilities for peanut M&M's are a bit different. Here they are:

Color	Brown	Red	Yellow	Green	Orange	Blue
Probability	0.2	0.1	0.2	0.1	0.1	?

What is the probability that a peanut M&M chosen at random is blue?

(c) What is the probability that a plain M&M is any of red, yellow, or orange? What is the probability that a peanut M&M has one of these colors?

18.12 Legitimate probabilities? In each of the following situations, state whether or not the given assignment of probabilities to individual outcomes is legitimate, that is, satisfies the rules of probability. If not, give specific reasons for your answer.

(a) When a coin is spun, $P(\text{H}) = 0.55$ and $P(\text{T}) = 0.45$.

(b) When two coins are tossed, $P(\text{HH}) = 0.4$, $P(\text{HT}) = 0.4$, and $P(\text{TH}) = 0.4$ and $P(\text{TT}) = 0.4$.

(c) Plain M&M's have not always had the mixture of colors given in Exercise 18.11. In the past there were no red candies and no blue candies. Tan had probability 0.10 and the other four colors had the same probabilities that are given in Exercise 18.11.

18.13 Polling women. Suppose that 47% of all adult women think they do not get enough time for themselves. An opinion poll interviews 1025 randomly chosen women and records the sample proportion who don't feel they get enough time for themselves. This statistic will vary from sample to sample if the poll is repeated. The sampling distribution is approximately normal with mean 0.47 and standard deviation about 0.016. Sketch this normal curve and use it to answer the following questions.

(a) The truth about the population is 0.47. In what range will the middle 95% of all sample results fall?

(b) What is the probability that the poll gets a sample in which fewer than 45.4% say they do not get enough time for themselves?

18.14 What will we die of? Suppose that 70% of all adults think that deaths from guns will increase in the future. An opinion poll plans an SRS of 1009 adults to ask about gun violence. The proportion of the sample who think gun deaths will increase will vary if we take many samples from this same population. The sampling distribution of the sample proportion is approximately normal with mean 0.70 and standard deviation about 0.014. Sketch this normal curve and use it to answer these questions:

(a) What is the probability that the poll gets a sample in which more than 72.8% of the people think gun deaths will increase?

(b) What is the probability of getting a sample that misses the truth (70%) by 2.8% or more?

18.15 Polling women (optional). In the setting of Exercise 18.13, what is the probability of getting a sample in which more than 51% of the women think they do not get enough time for themselves? (Use Table B.)

18.16 What will we die of (optional)? In the setting of Exercise 18.14, what is the probability of getting a sample in which fewer than 66.5% think that gun deaths will increase in the future? (Use Table B.)

18.17 Do you jog? An opinion poll asks an SRS of 1500 adults, "Do you happen to jog?" Suppose (as is approximately correct) that the population proportion who jog is $p = 0.15$ In a large number of samples, the proportion $\hat{p}$ who answer "Yes" will be approximately normally distributed with mean 0.15 and standard deviation 0.009. Sketch this normal curve and use it to answer these questions:

(a) What percent of many samples will have a sample proportion who jog that is 0.15 or less? Explain clearly why this percent is the probability that $\hat{p}$ is 0.15 or less.

(b) What is the probability that $\hat{p}$ will take a value between 0.141 and 0.159? (Use the 68–95–99.7 rule.)

(c) Now use Rule C for probability: What is the probability that $\hat{p}$ does not lie between 0.141 and 0.159?

18.18 Applying to college. You ask an SRS of 1500 college students whether they applied for admission to any other college. Suppose that in fact 35% of all college students applied to colleges besides the one they are attending. (That's close to the truth.) The sampling distribution of the proportion $\hat{p}$ of your sample who say "Yes" is approximately normal with mean 0.35 and standard deviation 0.01. Sketch this normal curve and use it to answer these questions:

(a) Explain in simple language what the sampling distribution tells us about the results of our sample.

(b) What percent of many samples would have a $\hat{p}$ larger than 0.37? (Use the 68–95–99.7 rule.) Explain in simple language why this percent is the probability of an outcome larger than 0.37.

(c) What is the probability that your sample will have a $\hat{p}$ less than 0.33?

(d) Use Rule D: What is the probability that your sample result will be either less than 0.33 or greater than 0.35?

18.19 Generating a sampling distribution. Let us illustrate the idea of a sampling distribution in the case of a very small sample from a very small population. The population is the scores of 10 students on an exam:

Student	0	1	2	3	4	5	6	7	8	9
Score	82	62	80	58	72	73	65	66	74	62

The parameter of interest is the mean score in this population. The sample is an SRS of size $n = 4$ drawn from the population. Because the students are labeled 0 to 9, a single random digit from Table A chooses one student for the sample.

(a) Find the mean of the 10 scores in the population. This is the population mean.

(b) Use Table A to draw an SRS of size 4 from this population. Write the four scores in your sample and calculate the mean $\bar{x}$ of the sample scores. This statistic is an estimate of the population mean.

(c) Repeat this process 10 times using different parts of Table A. Make a histogram of the 10 values of $\bar{x}$. You are constructing the sampling distribution of $\bar{x}$. Is the center of your histogram close to the population mean you found in (a)?

Chapter 19
Simulation

Getting through toll booths faster

We all know and hate the toll booth lines at bridges, tunnels, and toll roads. Statistics can't reduce the number of vehicles trying to get through, but it can help them get through faster.

Suppose that we must design the toll plaza for a new tunnel. Traffic in both directions will pay at the same location. Drivers may use cash, credit cards, or an electronic pay-while-moving system. How many toll booths do we need? Should each booth accept all types of payment, or should they specialize? What will happen in tourist season, when fewer drivers use the local electronic payment system and more pay cash? When traffic at a booth doubles, waiting times often quadruple or worse—will backups at peak times block the main road?

Settings like this are too complicated to just "think about." They aren't described by any set of mathematical equations that we can try to solve. What we do is roll out our random number generators and try to imitate the behavior of drivers. The formal term for this imitation is *simulation.* Cars and trucks arrive at random times, but we can see the probability distributions of times and vehicle types from past data. Drivers choose different payment systems, described by more probability distributions (different for local cars, tourist cars, and trucks). Different drivers are more or less aggressive in diving for the shortest line—another probability distribution. The time it takes to serve a driver at a booth is also random. The fellow with a $50 bill who wants a receipt takes longer.

We can now watch the toll booths run on the computer screen. The waiting time and lines of vehicles grow and shrink as the random flow of traffic arrives. Are lines often too long at rush hour in tourist season? What happens if we add a booth? What happens if we make that booth "cash only"? What happens if we make the approach longer so drivers have more time to see the shortest line? What happens if, alas, the amount of traffic doubles? Nothing good will happen, but perhaps we can prevent overflow from the cash lanes blocking access to the electronic lanes as well.

Simulation of complex interconnected systems like our toll plaza is a standard tool. The details are awful, but the ideas are pretty simple. The ideas behind simulation are our topic in this chapter.

Where do probabilities come from?

The probabilities of heads and tails in tossing a coin are very close to 1/2. In principle, these probabilities come from *data* on many coin tosses. Joe's personal probabilities for the winner of next year's Super Bowl come from Joe's own *individual judgment*. What about the probability that we get a run of three straight heads somewhere in 10 tosses of a coin? We can find this probability by *calculation from a model* that describes tossing coins. That is, once we have used data to give a probability model for the random fall of coins, we don't have to go back to the beginning every time we want the probability of a new event.

The big advantage of probability models is that they allow us to calculate the probabilities of complicated events starting from an assignment of probabilities to simple events like "heads on one toss." This is true whether the model reflects probabilities from data or personal probabilities. Unfortunately, the math needed to do probability calculations is often tough. Technology rides to the rescue: once we have a probability model, we can use a computer to *simulate* many repetitions. This is easier than math and much faster than actually running many repetitions in the real world. You might compare finding probabilities by simulation to practicing flying in a computer-controlled flight simulator. Both kinds of simulation are in wide use. Both have similar drawbacks: they are only as good as the model you start with. Flight simulators use a software "model" of how an airplane reacts. Simulations of probabilities use a probability model. We set the model in motion by using our old friends the random digits from Table A.

Simulation

Using random digits from a table or from computer software to imitate chance behavior is called **simulation.**

We look at simulation partly because it is how engineers designing a toll plaza really do find probabilities and partly because simulation forces us to think clearly about probability models. We'll do the hard part—setting up the model—and leave the easy part—telling a computer to do 10,000 repetitions—to those who really need the right probability at the end.

"We've done a computer simulation of your projected performance in five years. You're fired."

Simulation basics

Simulation is an effective tool for finding probabilities of complex events once we have a trustworthy probability model. We can use random digits to simulate many repetitions quickly. The proportion of repetitions on which an event occurs will eventually be close to its probability, so simulation can give good estimates of probabilities. The art of simulation is best learned from a series of examples.

Example 1. Doing a simulation

Toss a coin 10 times. What is the probability of a run of at least 3 consecutive heads or 3 consecutive tails?

Step 1. Give a probability model. Our model for coin tossing has two parts:

- Each toss has probabilities 0.5 for a head and 0.5 for a tail.
- Tosses are *independent* of each other. That is, knowing the outcome of one toss does not change the probabilities for the outcomes of any other toss.

Step 2. Assign digits to represent outcomes. Digits in Table A of random digits will stand for the outcomes, in a way that matches the probabilities from Step 1. We know that each digit in Table A has probability 0.1 of being any one of 0, 1, 2, 3, 4, 5, 6, 7, 8, or 9, and that successive digits in the table are independent. Here is one assignment of digits for coin tossing:

- One digit simulates one toss of the coin.
- Odd digits represent heads; even digits represent tails.

This works because the 5 odd digits give probability 5/10 to heads. Successive digits in the table simulate independent tosses.

Step 3. Simulate many repetitions. Ten digits simulate 10 tosses, so looking at 10 consecutive digits in Table A simulates one repetition. Read many groups of 10 digits from the table to simulate many repetitions. Be sure to keep track of whether or not the event we want (a run of 3 heads or 3 tails) occurs on each repetition.

Here are the first three repetitions, starting at line 101 in Table A. I have underlined all runs of 3 or more heads or tails.

Really random digits

For purists, the RAND Corporation long ago published a book titled *One Million Random Digits*. The book lists 1,000,000 digits that were produced by a very elaborate physical randomization and really are random. An employee of RAND once told me that this is not the most boring book that RAND has ever published. . .

	Repetition 1	Repetition 2
Digits	1 9 2 2 3 9 5 0 3 4	0 5 7 5 6 2 8 7 1 3
Heads/tails	H H T T <u>H H H</u> T H T	T <u>H H H</u> <u>T T T</u> <u>H H H</u>
Run of 3?	YES	YES

	Repetition 3
Digits	9 6 4 0 9 1 2 5 3 1
Heads/tails	H <u>T T T</u> H H T <u>H H H</u>
Run of 3?	YES

Continuing in Table A, I did 25 repetitions; 23 of them did have a run of 3 or more heads or tails. So we estimate the probability of a run by the proportion

$$\text{estimated probability} = \frac{23}{25} = 0.92$$

Of course, 25 repetitions are not enough to be confident that my estimate is accurate. Now that we understand how to do the simulation, we can tell a computer to do many thousands of repetitions. A long simulation (or hard mathematics) finds that the true probability is about 0.826. Most people think runs are somewhat unlikely, so even my short simulation corrects our intuition by showing that runs of 3 occur most of the time in 10 tosses.

Once you have gained some experience in simulation, setting up the probability model (Step 1) is usually the hardest part of the process. Although coin tossing may not fascinate you, the model in Example 1 is typical of many probability problems because it consists of *independent trials* (the tosses) all having the *same possible outcomes* with the *same probabilities.* Shooting 10 free throws and observing the sexes of 10 children have similar models and are simulated in much the same way. The new part of the model is independence, which simplifies our work because it allows us to simulate each of the 10 tosses in exactly the same way.

Independence

Two random phenomena are **independent** if knowing the outcome of one does not change the probabilities for outcomes of the other.

Independence, like all aspects of probability, can be verified only by observing many repetitions. It is plausible that repeated tosses of a coin are independent (the coin has no memory), and observation shows that they are. It seems less plausible that successive shots by a basketball player are independent, but observation shows that they are at least very close to independent.

Step 2 (assigning digits) rests on the properties of the random digit table. Here are some examples of this step.

Example 2. Assigning digits for simulation

(a) Choose a person at random from a group of which 70% are employed. One digit simulates one person:

$$0, 1, 2, 3, 4, 5, 6 = \text{employed}$$
$$7, 8, 9 = \text{not employed}$$

(b) Choose one person at random from a group of which 73% are employed. Now *two* digits simulate one person:

$$00, 01, 02, \ldots, 72 = \text{employed}$$
$$73, 74, 75, \ldots, 99 = \text{not employed}$$

"I've had it! Simulated wood, simulated leather, simulated coffee, and now simulated probabilities!"

We assigned 73 of the 100 two-digit pairs to "employed" to get probability 0.73. Representing "employed" by 01, 02, . . . , 73 would also be correct.

(c) Choose one person at random from a group of which 50% are employed, 20% are unemployed, and 30% are not in the labor force. There are now three possible outcomes, but the principle is the same. One digit simulates one person:

$$0, 1, 2, 3, 4 = \text{employed}$$
$$5, 6 = \text{unemployed}$$
$$7, 8, 9 = \text{not in the labor force}$$

Was he good or was he lucky?

When a baseball player hits .300, everyone applauds. A .300 hitter gets a hit in 30% of times at bat. Could a .300 year just be luck? Typical major leaguers bat about 500 times a season and hit about .260. A hitter's successive tries seem to be independent. From this model, we can calculate or simulate the probability of hitting .300. It is about 0.025. Out of 100 run-of-the-mill major league hitters, two or three each year will bat .300 because they were lucky.

More elaborate simulations

The building and simulation of random models is a powerful tool of contemporary science, yet a tool that can be understood without advanced mathematics. What is more, doing a few simulations will increase your understanding of probability more than many pages of my prose. Having in mind these two goals of understanding simulation for itself and understanding simulation to understand probability, let us look at two more elaborate examples. The first still has independent trials, but there is no longer a fixed number of trials as there was when we tossed a coin 10 times.

Example 3. We want a girl

A couple plan to have children until they have a girl or until they have three children, whichever comes first. What is the probability that they will have a girl among their children?

Step 1. The probability model is like that for coin tossing:

- Each child has probability 0.49 of being a girl and 0.51 of being a boy. (Yes, more boys than girls are born. Boys have higher infant mortality, so the sexes even out by the time it matters.)
- The sexes of successive children are independent.

Step 2. Assigning digits is also easy. Two digits simulate the sex of one child. We assign 49 of the 100 pairs to "girl" and the remaining 51 to "boy":

$$00, 01, 02, \ldots, 48 = \text{girl}$$
$$49, 50, 51, \ldots, 99 = \text{boy}$$

Step 3. To simulate one repetition of this childbearing strategy, read pairs of digits from Table A until the couple have either a girl or three children. The number of pairs needed to simulate one repetition depends on how quickly the couple get a girl. Here are 10 repetitions, simulated using line 130 of Table A. To interpret the pairs of digits, I have written G for girl and B for boy under them, have added space to separate repetitions, and under each repetition have written "+" if a girl was born and "–" if not.

6905	16	48	17	8717	40	9517	845340	648987	20
B G	G	G	G	B G	G	B G	B B G	B B B	G
+	+	+	+	+	+	+	+	–	+

In these 10 repetitions, a girl was born 9 times. Our estimate of the probability that this strategy will produce a girl is therefore

$$\text{estimated probability} = \frac{9}{10} = 0.9$$

Some mathematics shows that if our probability model is correct, the true probability of having a girl is 0.867. Our simulated answer came quite close. Unless the couple are unlucky, they will succeed in having a girl.

Our final example has stages that are *not independent*. That is, the probabilities at one stage depend on the outcome of the preceding stage.

Example 4. A kidney transplant

Morris's kidneys have failed and he is awaiting a kidney transplant. His doctor gives him this information for patients in his condition: 90% survive the transplant, and 10% die. The transplant succeeds in 60% of those who survive, and the other 40% must return to kidney dialysis. The proportions who survive five years are 70% for those with a new kidney and 50% for those who return to dialysis. Morris wants to know the probability that he will survive for five years.

Step 1. The **tree diagram** in Figure 19.1 organizes this information to give a probability model in graphical form. The tree shows the three stages and the possible outcomes and probabilities at each stage. Each path through the tree leads to either survival for five years or to death in less than five years. To simulate Morris's fate, we must simulate each of the three stages. The probabilities at Stage 3 depend on the outcome of Stage 2.

Step 2. Here is our assignment of digits to outcomes:

Stage 1:

$$0 = \text{die}$$
$$1, 2, 3, 4, 5, 6, 7, 8, 9 = \text{survive}$$

Stage 2:

$$0, 1, 2, 3, 4, 5 = \text{transplant succeeds}$$
$$6, 7, 8, 9 = \text{return to dialysis}$$

Stage 3 with kidney:

$$0, 1, 2, 3, 4, 5, 6 = \text{survive five years}$$
$$7, 8, 9 = \text{fail to survive}$$

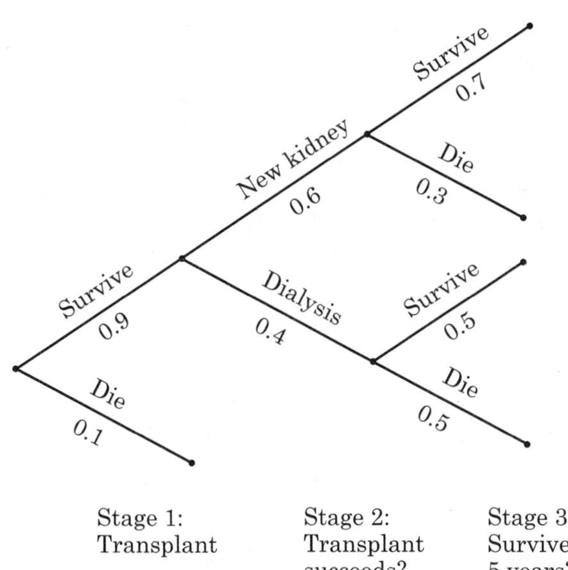

Figure 19.1 A tree diagram for the probability model of Example 4. Each branch point starts a new stage, with outcomes and their probabilities written on the branches. One repetition of the model follows the tree to one of its end points.

Stage 3 with dialysis:

0, 1, 2, 3, 4 = survive five years

5, 6, 7, 8, 9 = fail to survive

The assignment of digits at Stage 3 depends on the outcome of Stage 2. That's lack of independence.

Step 3. Here are simulations of several repetitions, each arranged vertically. I used random digits from line 110 of Table A.

	Repetition 1	Repetition 2	Repetition 3	Repetition 4
Stage 1	3 survive	4 survive	8 survive	9 survive
Stage 2	8 dialysis	8 dialysis	7 dialysis	1 kidney
Stage 3	4 survive	4 survive	8 die	8 die

Morris survives five years in 2 of our 4 repetitions. Now that we understand how to arrange the simulation, we should turn it over to a computer to do many repetitions. From a long simulation or from mathematics, we find that Morris has probability 0.558 of living for five years.

Exploring the Web

The Web abounds in applets that simulate various random phenomena. One amusing probability problem is named *Buffon's needle*. Draw lines 1 inch apart on a piece of paper, then drop a 1-inch-long needle on the paper. What is the probability that the needle crosses a line? You can find both a solution by mathematics and a simulation at George Reese's site, www.mste.uiuc.edu/reese/buffon/buffon.html. Personal sites sometimes vanish; a search on "Buffon's needle" will turn up alternative sites. The probability turns out to be $2/\pi$, where any circle's circumference is π times its radius. So the simulation is also a way to calculate π, one of the most famous numbers in mathematics.

Statistics in Summary We can use random digits to **simulate** random outcomes if we know the probabilities of the outcomes. Use the fact that each random digit has probability 0.1 of taking any one of the 10 possible digits and that all digits in the random number table are **independent** of each other. To simulate more complicated random phenomena, string together simulations of each stage. A common situation is several independent trials with the same possible outcomes and probabilities on each trial. Think of several tosses of a coin or several rolls of a die. Other simulations may require varying numbers of trials or different probabilities at each stage or may have stages that are not independent so that the probabilities at some stage depend on the outcome of earlier stages. The key to successful simulation is thinking carefully about the probability model.

CHAPTER 19 EXERCISES

19.1 Which party does it better? An opinion poll selects adult Americans at random and asks them, "Which political party, Democratic or Republican, do you think is better able to manage the economy?" Explain carefully how you would assign digits from Table A to simulate the response of one person in each of the following situations.
(a) Of all adult Americans, 50% would choose the Democrats and 50% the Republicans.

(b) Of all adult Americans, 60% would choose the Democrats and 40% the Republicans.

(c) Of all adult Americans, 40% would choose the Democrats, 40% would choose the Republicans, and 20% are undecided.

(d) Of all adult Americans, 53% would choose the Democrats and 47% the Republicans.

19.2 A small opinion poll. Suppose that 80% of a university's students favor abolishing evening exams. You ask 10 students chosen at random. What is the probability that all 10 favor abolishing evening exams?
(a) Give a probability model for asking 10 students independently of each other.

(b) Assign digits to represent the answers "Yes" and "No."

(c) Simulate 25 repetitions, starting at line 129 of Table A. What is your estimate of the probability?

19.3 Basic simulation. Use Table A to simulate the responses of 10 independently chosen adults in each of the four situations of Exercise 19.1.
(a) For situation (a), use line 110.

(b) For situation (b), use line 111.

(c) For situation (c), use line 112.

(d) For situation (d), use line 113.

19.4 Simulating an opinion poll. A recent opinion poll showed that about 70% of married women agree that their husbands do at least their fair share of household chores. Suppose that this is exactly true. Choosing a married woman at random then has probability 0.7 of getting one who agrees that her husband does his share. If we interview women separately, we can assume that their responses are independent. We want to know the probability that a simple random sample of 100 women will contain at least 80 who say that their husbands do their share. Explain carefully how to do this simulation and simulate *one* repetition of the poll using line 112 of Table A. How many of the 100 women said their husbands do their share? Explain how you would estimate the probability by simulating many repetitions.

19.5 Course grades. Choose a student at random from all who took beginning statistics at Upper Wabash Tech in recent years. The probabilities for the student's grade are

Grade	A	B	C	D or F
Probability	0.2	0.3	0.3	?

(a) What must be the probability of getting a D or an F?

(b) To simulate the grades of randomly chosen students, how would you assign digits to represent the four possible outcomes listed?

19.6 Class rank. Choose a college student at random and ask his or her class rank in high school. Probabilities for the outcomes are

Class rank	Top 10%	Top quarter but not top 10%	Top half but not top quarter	Bottom half
Probability	0.3	0.3	0.3	?

(a) What must be the probability that a randomly chosen student was in the bottom half of his or her high school class?

(b) To simulate the class standing of randomly chosen students, how would you assign digits to represent the four possible outcomes listed?

19.7 More on course grades. In Exercise 19.5 you explained how to simulate the grade of a randomly chosen student in a statistics course. Five students on the same floor of a dormitory are taking this course. They don't study together, so their grades are independent. Use simulation to estimate the probability that all five get a C or better in the course. (Simulate 20 repetitions.)

19.8 More on class rank. In Exercise 19.6 you explained how to simulate the high school class rank of a randomly chosen college student. The

Random Foundation decides to offer 8 randomly chosen students full college scholarships. What is the probability that no more than 3 of the 8 students chosen are in the bottom half of their high school class? Simulate 10 repetitions of the foundation's choices to estimate this probability.

19.9 Shaq's free throws. The basketball player Shaquille O'Neal makes about half of his free throws over an entire season. Take his probability of a hit to be 0.5 on each shot. Simulate 25 repetitions of his performance in a game in which he shoots 12 free throws, using line 122 of Table A. The probability model is exactly that for tossing a coin 12 times.

(a) Estimate the probability that Shaq makes at least 8 of 12 free throws.

(b) Examine the sequence of hits and misses in your 25 repetitions. How long was the longest run of shots made? Of shots missed?

19.10 Tonya's free throws. Tonya makes 70% of her free throws in a long season. In a tournament game she shoots 5 free throws late in the game and misses 3 of them. The fans think she was nervous, but the misses may simply be chance. Let's shed some light by estimating a probability.

(a) Describe how to simulate a single shot if the probability of making each shot is 0.7. Then describe how to simulate 5 independent shots.

(b) Simulate 50 repetitions of the 5 shots and record the number missed on each repetition. Use Table A, starting at line 125. What is the approximate probability that Tonya will miss 3 or more of the 5 shots?

19.11 Repeating an exam. Elaine is enrolled in a self-paced course that allows three attempts to pass an examination on the material. She does not study and has probability 2/10 of passing on any one attempt by luck. What is Elaine's probability of passing in three attempts? (Assume the attempts are independent because she takes a different examination on each attempt.)

(a) Explain how you would use random digits to simulate one attempt at the exam.

(b) Elaine will stop taking the exam as soon as she passes. (This is much like Example 3.) Simulate 50 repetitions, starting at line 120 of Table A. What is your estimate of Elaine's probability of passing the course?

(c) Do you think the assumption that Elaine's probability of passing the exam is the same on each trial is realistic? Why?

19.12 A better model for repeating an exam. A more realistic probability model for Elaine's attempts to pass an exam in the previous exercise is as follows. On the first try she has probability 0.2 of passing. If she fails on the first try, her probability on the second try increases to 0.3 because she learned something from her first attempt. If she fails on two attempts, the probability of passing on a third attempt is 0.4. She will stop as soon as she passes. The course rules force her to stop after three attempts in any case.

(a) Make a tree diagram of Elaine's progress. Notice that she has different probabilities of passing on each successive try.

(b) Explain how to simulate one repetition of Elaine's tries at the exam.

(c) Simulate 50 repetitions and estimate the probability that Elaine eventually passes the exam. Use Table A, starting at line 130.

19.13 Gambling in ancient Rome. Tossing four astragali was the most popular game of chance in Roman times. Many throws of a present-day sheep's astragalus show that the approximate probability distribution for the four sides of the bone that can land uppermost are

Outcome	Probability
Narrow flat side of bone	1/10
Broad concave side of bone	4/10
Broad convex side of bone	4/10
Narrow hollow side of bone	1/10

The best throw of four astragali was the "Venus," when all the four uppermost sides were different.

(a) Explain how to simulate the throw of a single astragalus. Then explain how to simulate throwing four astragali independently of each other.

(b) Simulate 25 throws of four astragali. Estimate the probability of throwing a "Venus." Be sure to say what part of Table A you used.

19.14 The Asian stochastic beetle. We can use simulation to examine the fate of populations of living creatures. Consider the Asian stochastic beetle. Females of this insect have the following pattern of reproduction:

- 20% of females die without female offspring, 30% have 1 female offspring, and 50% have 2 female offspring.
- Different females reproduce independently.

What will happen to the population of Asian stochastic beetles: will they increase rapidly, barely hold their own, or die out? It's enough to look at the female beetles, as long as there are some males around.

(a) Assign digits to simulate the offspring of one female beetle.

(b) Make a tree diagram for the female descendants of one beetle through three generations. The second generation, for example, can have 0, 1, or 2 females. If it has 0, we stop. Otherwise, we simulate the offspring of each second-generation female. What are the possible numbers of beetles after three generations?

(c) Use line 105 of Table A to simulate the offspring of 5 beetles to the third generation. How many descendants does each have after three generations? Does it appear that the beetle population will grow?

19.15 Two warning systems. An airliner has two independent automatic systems that sound a warning if there is terrain ahead (that means the airplane is about to fly into a mountain). Neither system is perfect. System A signals in time with probability 0.9. System B does so with probability 0.8. The pilots are alerted if either system works.
(a) Explain how to simulate the response of System A to terrain.

(b) Explain how to simulate the response of System B.

(c) Both systems are in operation simultaneously. Draw a tree diagram with System A as the first stage and System B as the second stage. Simulate 100 trials of the reaction to terrain ahead. Estimate the probability that a warning will sound. The probability for the combined system is higher than the probability for either A or B alone.

19.16 Playing craps. The game of craps is played with two dice. The player rolls both dice and wins immediately if the outcome (the sum of the faces) is 7 or 11. If the outcome is 2, 3, or 12, the player loses immediately. If he rolls any other outcome, he continues to throw the dice until he either wins by repeating the first outcome or loses by rolling a 7.
(a) Explain how to simulate the roll of a single fair die. (Hint: Just use digits 1 to 6 and ignore the others.) Then explain how to simulate a roll of two fair dice.

(b) Draw a tree diagram for one play of craps. In principle, a player could continue forever, but stop your diagram after four rolls of the dice. Use Table A, beginning at line 114, to simulate plays and estimate the probability that the player wins.

19.17 The airport van. Your company operates a van service from the airport to downtown hotels. Each van carries 7 passengers. Many passengers who reserve seats don't show up—in fact, the probability is 0.25 that a randomly chosen passenger will fail to appear. Passengers are independent. If you allow 9 reservations for each van, what is the probability that more than 7 passengers will appear? Do a simulation to estimate this probability.

19.18 A multiple-choice exam. Matt has lots of experience taking multiple-choice exams without doing much studying. He is about to take a quiz that has 10 multiple-choice questions, each with four possible answers. Here

is Matt's personal probability model. He thinks that in 60% of questions he can eliminate one answer as obviously wrong; then he guesses from the remaining three. He then has probability 1/3 of guessing the right answer. For the other 40% of questions, he must guess from all four answers, with probability 1/4 of guessing correctly.
(a) Make a tree diagram for the outcome of a single question. Explain how to simulate Matt's success or failure on one question.

(b) Questions are independent. To simulate the quiz, just simulate 10 questions. Matt needs to get at least 5 questions right to pass the quiz. You could find his probability of passing by simulating many tries at the quiz, but I ask you to simulate just one try. Did Matt pass this quiz?

19.19 More on the airport van. Let's continue the simulation of Exercise 19.17. You have a backup van, but it serves several stations. The probability that it is available to go to the airport at any one time is 0.6. You want to know the probability that some passengers with reservations will be left stranded because the first van is full and the backup van is not available. Draw a tree diagram with the first van (full or not) as the first stage and the backup (available or not) as the second stage. In Exercise 19-17 you simulated a number of repetitions of the first stage. Add simulations of the second stage whenever the first van is full. What is your estimate of the probability of stranded passengers?

19.20 The birthday problem. A famous example in probability theory shows that the probability that at least two people in a room have the same birthday is already greater than 1/2 when 23 people are in the room. The probability model is

- The birth date of a randomly chosen person is equally likely to be any of the 365 dates of the year.
- The birth dates of different people in the room are independent.

To simulate birthdays, let each three-digit group in Table A stand for one person's birth date. That is, 001 is January 1 and 365 is December 31. Ignore leap years and skip groups that don't label birth dates. Use line 139 of Table A to simulate birthdays of randomly chosen people until you hit the same date a second time. How many people did you look at to find two with the same birthday?

With a computer, you could easily repeat this simulation many times. You could find the probability that at least 2 out of 23 people have the same birthday, or you could find the expected number of people you must question to find two with the same birthday. These problems are a bit tricky to do by math, so they show the power of simulation.

19.21 The multiplication rule. Here is another basic rule of probability: *If several events are independent, the probability that all of the events happen is the product of their individual probabilities.* We know, for example, that a child has probability 0.49 of being a girl and probability 0.51 of being a boy, and that the sexes of successive children are independent. So the probability that a couple's two children are two girls is (0.49)(0.49) = 0.2401. You can use this multiplication rule to calculate the probability that we simulated in Example 3.

(a) Write down all 8 possible arrangements of the sexes of three children, for example, BBB and BBG. Use the multiplication rule to find the probability of each outcome. Check your work by verifying that your 8 probabilities add to 1.

(b) The couple in Example 3 plan to stop when they have a girl or to stop at 3 children even if all are boys. Use your work from (a) to find the probability that they get a girl.

Chapter 20

The House Edge: Expected Values

The house edge, and the real house edge

If you gamble, you care about how often you will win. The probability of winning tells you what proportion of a large number of bets will be winners. You care even more about *how much* you will win, because winning a lot is better than winning a little. If you play a game with a 50% chance of winning \$1, in the long run you will win half the time and average 50 cents winnings per bet. A 10% chance of winning \$100 does better—now you win only one time in 10 in the long run, but you average \$10 winnings per bet. Those "average winnings per bet" numbers are *expected values*. Like probabilities, expected values only tell us what will happen on the average in very many bets.

It's no secret that the expected values of games of chance are set so that the house wins on the average in the long run. In roulette, for example, the expected value of a \$1 bet is \$0.947. The house keeps the other 5.3 cents per dollar bet. Because the house plays hundreds of thousands of times each month, it will keep almost exactly 5.3 cents for every dollar bet.

What *is* a secret, at least to naive gamblers, is that in the real world a casino does much better than expected values suggest. In fact, casinos keep a bit over 20% of the money gamblers spend on roulette chips. That's because players who win keep on playing. Think of a player who gets back exactly 95% of each dollar bet. After one bet, he has 95 cents; after two bets he has 95% of that, or 90.25 cents; after three bets he has 95% of that, or 85.7 cents. The longer he keeps recycling his original dollar, the more of it the casino keeps. Real gamblers don't get a fixed percent back on each bet, but even the luckiest will lose his stake if he plays long enough. The casino keeps 5.3 cents of every dollar bet but 20 cents of every dollar that walks in the door.

That brings us to baccarat, the preferred game of the idle rich. Baccarat is a card game that requires no decisions and no skill, but it looks elegant. Some players bet $100,000 per hand. There are only a few high-stakes players, and they don't play many hands by the standards of more plebeian games like roulette. So "the long run" that the casino profits from is not as long in baccarat. Runs of luck average out over a month or so in roulette but not in baccarat. The quarterly profit statements of some casino companies have suffered from bad luck at the baccarat tables.

Expected values

Gambling on chance outcomes goes back to ancient times. Both public and private lotteries were common in the early years of the United States. After disappearing for a century or so, government-run gambling reappeared in 1964, when New Hampshire caused a furor by introducing a lottery to raise public revenue without raising taxes. The furor subsided quickly as larger states adopted the idea. Thirty-seven states and all Canadian provinces now sponsor lotteries. State lotteries made gambling acceptable as entertainment. Some form of legal gambling is allowed in 48 of the 50 states. Over half of all adult Americans have gambled legally. They spend more betting than on spectator sports, video games, theme parks, and movie tickets combined. If you are going to bet, you should understand what makes a

bet good or bad. As our introductory case study says, we care about how much we win as well as about our probability of winning.

Example 1. The Tri-State Daily Numbers

Here is a simple lottery wager, the "Straight" from the *Pick 3* game of the Tri-State Daily Numbers offered by New Hampshire, Maine, and Vermont. You pay $1 and choose a three-digit number. The state chooses a three-digit winning number at random and pays you $500 if your number is chosen. Because there are 1000 three-digit numbers, you have probability 1/1000 of winning. Here is the probability model for your winnings:

Outcome	$0	$500
Probability	0.999	0.001

What are your average winnings? The ordinary average of the two possible outcomes $0 and $500 is $250, but that makes no sense as the average winnings because $500 is much less likely than $0. In the long run you win $500 once in every 1000 bets and $0 on the remaining 999 of 1000 bets. (Of course, if you buy exactly 1000 *Pick 3* tickets, there is no guarantee that you will win exactly once. Probabilities are only *long-run* proportions.) Your long-run average winnings from a ticket are

$$\$500\,\frac{1}{1000} + \$0\,\frac{999}{1000} = \$0.50$$

or 50 cents. You see that in the long run the state pays out half the money bet and keeps the other half.

Here is a general definition of the kind of "average outcome" we used to evaluate the bets in Example 1.

Expected value

The **expected value** of a random phenomenon that has numerical outcomes is found by multiplying each outcome by its probability and then summing over all possible outcomes.

In symbols, if the possible outcomes are $a_1, a_2, \ldots, a_k$ and their probabilities are $p_1, p_2, \ldots, p_k$, the expected value is

$$\text{expected value} = a_1p_1 + a_2p_2 + \cdots + a_kp_k$$

An expected value is an average of the possible outcomes, but it is not an ordinary average in which all outcomes get the same weight. Instead, each outcome is weighted by its probability, so that outcomes that occur more often get higher weights.

Example 2. The Tri-State Daily Numbers, continued

The Straight wager in Example 1 pays off if you match the three-digit winning number exactly. You can choose instead to make a $1 StraightBox wager. You again choose a three-digit number, but you now have two ways to win. You win $292 if you exactly match the winning number, and you win $42 if your number has the same digits as the winning number, in any order. For example, if your number is 123, you win $292 if the winning number is 123 and $42 if the winning number is any of 132, 213, 231, 312, and 321. In the long run, you win $292 once every 1000 bets and $42 five times for every 1000 bets.

The probability model for the amount you win is

Outcome	$0	$42	$292
Probability	0.994	0.005	0.001

The expected value is

$$\text{expected value} = (\$0)(0.994) + (\$42)(0.005) + (\$292)(0.001) = \$0.502$$

We see that the StraightBox is a slightly better bet than the Straight bet, because the state pays out slightly more than half the money bet.

The Tri-State Daily Numbers is unusual among state lottery games in that it pays a fixed amount for each type of bet. Most states pay off on the "pari-mutuel" system. New Jersey's *Pick-3* game is typical: the state pools the money bet and pays out half of it, equally divided among the winning tickets. You still have probability 1/1000 of winning a Straight bet, but the amount your number 123 wins depends both on how much was bet on *Pick-3* that day and on how many other players chose the number 123. Without fixed amounts, we can't find the expected value of today's bet on 123, but one thing is constant: the state keeps half the money bet.

The idea of expected value as an average applies to random outcomes other than games of chance. It is used, for example, to describe the uncertain return from buying stocks or building a new factory. Here is a different example.

Example 3. How many vehicles per household?

What is the average number of motor vehicles in American households? The Census Bureau tells us that the distribution of vehicles per household (as of 1997) is as follows:

Number of vehicles	0	1	2	3	4	5
Proportion of households	0.04	0.25	0.45	0.18	0.06	0.02

This is a probability model for choosing a household at random and counting its vehicles. (I ignored the very few households with more than 5 vehicles.) The expected value for this model is the average number of vehicles per household. This average is

$$\text{expected value} = (0)(0.04) + (1)(0.25) + (2)(0.45) + (3)(0.18) + (4)(0.06) + (5)(0.02)$$
$$= 2.03$$

The law of large numbers

The definition of expected value says that it is an average of the possible outcomes, but an average in which outcomes with higher probability count more. I argued that the expected value is also the average outcome in another sense—it represents the long-run average we will actually see if we repeat a bet many times or choose many households at random. This is more than intuition. Mathematicians can prove, starting from just the basic rules of probability, that the expected value calculated from a probability model really is the "long-run average." This famous fact is called the *law of large numbers.*

Rigging the lottery

We have all seen the televised live lottery drawings, in which numbered balls bubble about and are randomly popped out by air pressure. How might we rig such a drawing? In 1980, when only three balls popped out, the Pennsylvania lottery was in fact rigged by the smiling host and several stagehands. They injected paint into all balls bearing 8 of the 10 digits. This weighed them down and guaranteed that all 3 balls for the winning number would have the remaining 2 digits. The perps then bet on all combinations of these digits. When 6-6-6 popped out, they won $1.2 million. Yes, they were caught.

The law of large numbers

If a random phenomenon with numerical outcomes is repeated many times independently, the mean of the actually observed outcomes approaches the expected value.

The law of large numbers is closely related to the idea of probability. In many independent repetitions, the proportion of each possible outcome will be close to its probability, and the average outcome obtained will be close to the expected value. These facts express the long-run regularity of chance events. They are the true version of the "law of averages."

The law of large numbers explains why gambling, which is a recreation or an addiction for individuals, is a business for a casino. The "house" in a gambling operation is not gambling at all. The average winnings of a large number of customers will be quite close to the expected value. The house has calculated the expected value ahead of time and knows what its take will be in the long run. There is no need to load the dice or stack the cards to guarantee a profit. Casinos concentrate on inexpensive entertainment and cheap bus trips to keep the customers flowing in. If enough bets are placed, the law of large numbers guarantees the house a profit. Life insurance companies operate much like casinos—they bet that the people who buy insurance will not die. Some do die, of course, but the insurance company knows the probabilities and relies on the law of large numbers to predict the average amount it will have to pay out. Then the company sets its premiums high enough to guarantee a profit.

High-tech gambling

There are more than 450,000 slot machines in the United States. Once upon a time, you put in a coin and pulled the lever to spin three wheels, each with 20 symbols. No longer. Now the machines are video games with flashy graphics and outcomes produced by random number generators. Machines can accept many coins at once, can pay off on a bewildering variety of outcomes, and can be networked to allow common jackpots. Gamblers still search for systems, but in the long run the random number generator guarantees the house its 5% profit.

Thinking about expected values

As with probability, it is worth exploring a few fine points about expected values and the law of large numbers.

How large is a large number? The law of large numbers says that the actual average outcome of many trials gets closer to the expected value as more trials are made. It doesn't say how many trials are needed to guarantee an average outcome close to the expected value. That depends on the *variability* of the random outcomes.

The more variable the outcomes, the more trials are needed to ensure that the mean outcome is close to the expected value. Games of chance must be quite variable if they are to hold the interest of gamblers. Even a long evening in a casino has an unpredictable outcome. Gambles with extremely variable outcomes, like state lottos with their very large but very improbable jackpots, require impossibly large numbers of trials to ensure that the average outcome is close to the expected value. (The state doesn't rely on the law of large numbers—lotto payoffs, unlike casino games, use the pari-mutuel system.)

Though most forms of gambling are less variable than lotto, the practical answer to the applicability of the law of large numbers is usually that the house plays often enough to rely on it, but you don't. Much of the psychological allure of gambling is its unpredictability for the player. The business of gambling rests on the fact that the result is not unpredictable for the house.

Is there a winning system? Serious gamblers often follow a system of betting in which the amount bet on each play depends on the outcome of previous plays. You might, for example, double your bet on each spin of the roulette wheel until you win—or, of course, until your fortune is exhausted. Such a system tries to take advantage of the fact that you have a memory even though the roulette wheel does not. Can you beat the odds with a system? No. Mathematicians have established a stronger version of the law of large numbers that says that if you do not have an infinite fortune to gamble with, your average winnings (the expected value) remains the same as long as successive trials of the game (such as spins of the roulette wheel) are independent. Sorry.

Finding expected values by simulation

How can we calculate expected values in practice? You know the mathematical recipe, but that requires that you start with the probability of each outcome. Expected values too difficult to compute in this way can be found by simulation. The procedure is as before: give a probability model, use random digits to imitate it, and simulate many repetitions. By the law of large numbers, the average outcome of these repetitions will be close to the expected value.

Example 4. We want a girl, again

A couple plan to have children until they have a girl or until they have three children, whichever comes first. We simulated 10 repetitions of this scheme in Example 3 of Chapter 19 (page 382). There, we estimated the probability that they will have a girl among their children. Now we ask a different question: how many children, on the average, will couples who follow this plan have? That is, we want the expected number of children.

The simulation is exactly as before. The probability model says that the sexes of successive children are independent and that each child has probability 0.49 of being a girl. Here are our earlier simulation results—but rather than noting whether the couple did have a girl, we now record the number of children they have. Recall that a pair of digits simulates one child, with 00 to 48 (probability 0.49) standing for a girl.

6905	16	48	17	8717	40	9517	845340	648987	20
B G	G	G	G	B G	G	B G	B B G	B B B	G
2	1	1	1	2	1	2	3	3	1

The mean number of children in these 10 repetitions is

$$\bar{x} = \frac{2+1+1+1+2+1+2+3+3+1}{10} = \frac{17}{10} = 1.7$$

We estimate that if many couples follow this plan, they will average 1.7 children each. This simulation is too short to be trustworthy. Math or a long simulation shows that the actual expected value is 1.77 children.

Statistical Controversies

The State of Legalized Gambling

Most voters think some forms of gambling should be legal, and the majority has its way: lotteries and casinos are common both in the United States and in other nations. The arguments in favor of allowing gambling are straightforward. Many people find betting entertaining and are willing to lose a bit of money in exchange for some excitement. Gambling doesn't harm other people, at least directly. A democracy should allow entertainments that a majority supports and that don't do harm. State lotteries raise money for good causes such as education and are a kind of voluntary tax that no one is forced to pay.

Opponents of gambling nevertheless have good arguments that may yet lead to more regulation. Some people find betting addictive. A study by the National Opinion Research Center estimated that pathological gamblers account for 15% of gambling revenue, and that each such person costs the rest of us $12,000 over his lifetime for social and police work. Gambling does ruin some lives, and it does indirectly harm others.

State-run lotteries involve governments in trying to persuade their citizens to gamble. In the early days of the New York lottery, I recall billboards that said, "Support education—play the lottery." That didn't work, and the ads quickly changed to "Get rich—play the lottery." Lotteries typically pay out only about half the money bet, so they are a lousy way to get rich even when compared with the slots at the local casino. Professional gamblers and statisticians avoid them, not wanting to waste money on so bad a bargain. Poor people spend a larger proportion of their income on lotteries than do the rich and are the main players of daily numbers games. The lottery may be a voluntary tax, but it hits the poor hardest, and states spend hundreds of millions on advertising to persuade the poor to lose yet more money. Some modest suggestions: states should cut out the advertising and pay out more of what is bet.

States license casinos because they pay taxes and attract tourists—and of course because many citizens want them. In fact, most casinos outside Las Vegas draw gamblers mainly from nearby areas. Crime is higher in counties with casinos—but lots of lurking variables may explain this association. Pathological gamblers do have high rates of arrest, but again the causal link is not clear.

The debate continues. Meanwhile, technology in the form of Internet gambling is bypassing governments and creating a new gambling economy that makes many of the old arguments outdated.

"I think the lottery is a great idea. If they raised taxes instead, we'd have to pay them."

Exploring the Web

The debate over legalized gambling continues. For the case against, visit the National Coalition Against Legalized Gambling, www.ncalg.org. For the defense by the casino industry, visit the American Gaming Association, www.americangaming.org. State lotteries make their case via the National Association of State and Provincial Lotteries, www.naspl.org. The National Indian Gaming Association, www.indiangaming.org, is more assertive; click on "Myths," for example. The report of a commission established by Congress to study the impact of gambling is at www.ngisc.gov. You'll find lots of facts and figures at all of these sites.

Statistics in Summary Probability tells us how often (in the very long run) a random phenomenon takes each of its possible outcomes. When the outcomes are numbers, as in games of chance, we are also interested in the long-run average outcome. The **law of large numbers** says that the mean outcome in many repetitions eventually gets close to the **expected value.** The expected value is found as an average of all the possible outcomes, each weighted by its probability. If you don't know the outcome probabilities, you can estimate the expected value (along with the probabilities) by simulation.

CHAPTER 20 EXERCISES

20.1 The numbers racket. *Pick 3* lotteries (Example 1) copy the numbers racket, an illegal gambling operation common in the poorer areas of large cities. One version works as follows. You choose any one of the 1000 three-digit numbers 000 to 999 and pay your local numbers runner \$1 to enter your bet. Each day, one three-digit number is chosen at random and pays off \$600. What is the expected value of a bet on the numbers? Is the numbers racket more or less favorable to gamblers than the *Pick 3* game in Example 1?

20.2 Pick 4. The Tri-State Daily Numbers *Pick 4* is much like the *Pick 3* game of Example 1. You pay \$1 and pick a four-digit number. The state chooses a four-digit number at random and pays you \$5000 if your number is chosen. What are the expected winnings from a \$1 *Pick 4* wager?

20.3 Red or black in roulette. An American roulette wheel has 38 slots, of which 18 are black, 18 are red, and 2 are green. When the wheel is spun, the ball is equally likely to come to rest in any of the slots. A bet of $1 on red will win $2 if the ball lands in a red slot. (When gamblers bet on red or black, the two green slots belong to the house.) Give a probability model for the winnings of a $1 bet on red and find the expected value of this bet.

20.4 More Pick 4. Just as with *Pick 3* (Example 2), you can make more elaborate bets in *Pick 4*. In the $1 StraightBox bet, if you choose 1234 you win $2604 if the randomly chosen winning number is 1234 and you win $104 if the winning number has the digits 1, 2, 3, and 4 in any other order. What is the expected amount you win?

20.5 More roulette. Gamblers bet on roulette by placing chips on a table that lays out the numbers and colors of the 38 slots in the roulette wheel. The red and black slots are arranged on the table in three columns of 12 slots each. A $1 column bet wins $3 if the ball lands in one of the 12 slots in that column. What is the expected amount such a bet wins? Is a column bet more or less favorable to a gambler than a bet on red or black (Exercise 20.3)?

20.6 Making decisions. The psychologist Amos Tversky did many studies of our perception of chance behavior. In its obituary of Tversky, the *New York Times* cited the following example.

(a) Tversky asked subjects to choose between two public health programs that affect 600 people. One has probability 1/2 of saving all 600 and probability 1/2 that all 600 will die. The other is guaranteed to save exactly 400 of the 600 people. Find the expected number of people saved by the first program.

(b) Tversky then offered a different choice. One program has probability 1/2 of saving all 600 and probability 1/2 of losing all 600, while the other will definitely lose exactly 200 lives. What is the difference between this choice and that in (a)?

(c) Given option (a), most subjects choose the second program. Given option (b), most subjects choose the first program. Do the subjects appear to use expected values in their choice? Why do you think the choices differ in the two cases?

20.7 Making decisions. A six-sided die has four green and two red faces and is balanced so that each face is equally likely to come up. You must choose one of the following three sequences of colors:

RGRRR

RGRRRG

GRRRRR

Now start rolling the die. You will win \$25 if the first rolls give the sequence you chose.

(a) Which sequence has the highest probability? Why? (You can see which is most probable without actually finding the probabilities.) Because the \$25 payoff is fixed, the most probable sequence has the highest expected value.

(b) In a psychological experiment, 63% of 260 students who had not studied probability chose the second sequence. Based on the discussion of "myths about chance behavior" in Chapter 17, explain why most students did not choose the sequence with the best chance of winning.

20.8 Estimating sales. Gain Communications sells aircraft communications units. Next year's sales depend on market conditions that cannot be predicted exactly. Gain follows the modern practice of using probability estimates of sales. The sales manager estimates next year's sales as follows:

Units sold	1000	3000	5000	10,000
Probability	0.1	0.3	0.4	0.2

These are personal probabilities that express the informed opinion of the sales manager. What is the expected value of next year's sales?

20.9 Keno. Keno is a popular game in casinos. Balls numbered 1 to 80 are tumbled in a machine as the bets are placed, then 20 of the balls are chosen at random. Players select numbers by marking a card. Here are two of the simpler Keno bets. Give the expected winnings for each.

(a) A \$1 bet on "Mark 1 number" pays \$3 if the single number you mark is one of the 20 chosen; otherwise you lose your dollar.

(b) A \$1 bet on "Mark 2 numbers" pays \$12 if both your numbers are among the 20 chosen. The probability of this is about 0.06. Is Mark 2 a more or a less favorable bet than Mark 1?

20.10 Rolling two dice. Example 2 of Chapter 18 (page 364) gives a probability model for rolling two casino dice and recording the number of spots on the two up faces. That example also shows how to find the probability that the total number of spots showing is 5. Follow that method to give a probability model for the total number of spots. The possible outcomes are 2, 3, 4, . . . , 12. Then use the probabilities to find the expected value of the total.

20.11 The Asian stochastic beetle. We met this insect in Exercise 19.14 (page 388). Females have this probability model for their number of female offspring:

Offspring	0	1	2
Probability	0.2	0.3	0.5

(a) What is the expected number of female offspring?

(b) Use the law of large numbers to explain why the population should grow if the expected number of female offspring is greater than 1 and die out if this expected value is less than 1.

20.12 An expected rip-off? A "psychic" runs the following ad in a magazine:

> *Expecting a baby? Renowned psychic will tell you the sex of the unborn child from any photograph of the mother. Cost $10. Moneyback guarantee.*

This may be a profitable con game. Suppose that the psychic simply replies "boy" to all inquiries. In the worst case, everyone who has a girl will ask for her money back. Find the expected value of the psychic's profit by filling in the table below.

Sex of child	Probability	The psychic's profit
Boy	0.51	
Girl	0.49	

20.13 The Asian stochastic beetle again. In Exercise 20.11 you found the expected number of female offspring of the Asian stochastic beetle. Simulate the offspring of 100 beetles and find the mean number of offspring for these 100 beetles. Compare this mean with the expected value from Exercise 20.11. (The law of large numbers says that the mean will be very close to the expected value if we simulate enough beetles.)

20.14 Life insurance. You might sell insurance to a 21-year-old friend. The probability that a man aged 21 will die in the next year is about 0.0015. You decide to charge $200 for a policy that will pay $100,000 if your friend dies.

(a) What is your expected profit on this policy?

(b) Although you expect to make a good profit, you would be foolish to sell your friend the policy. Why?

(c) A life insurance company that sells thousands of policies, on the other hand, would do very well selling policies on exactly these same terms. Explain why.

20.15 Household size. The Census Bureau gives this distribution for the number of people in American households:

Household size	1	2	3	4	5	6	7
Proportion	0.26	0.32	0.17	0.15	0.07	0.02	0.01

This is also the probability distribution for the size of a randomly chosen household. The expected value of this distribution is the average number of people in a household. What is this expected value?

20.16 Course grades. The distribution of grades in a large statistics course is as follows:

Grade	A	B	C	D	F
Probability	0.2	0.3	0.3	0.1	0.1

To calculate student grade point averages, grades are expressed in a numerical scale with $A = 4$, $B = 3$ and so on down to $F = 0$.

(a) Find the expected value. This is the average grade in this course.

(b) Explain how to simulate choosing students at random and recording their grades. Simulate 50 students and find the mean of their 50 grades. Compare this estimate of the expected value with the exact expected value from (a). (The law of large numbers says that the estimate will be very accurate if we simulate a very large number of students.)

20.17 We really want a girl. Example 4 estimates the expected number of children a couple will have if they keep going until they get a girl or until they have three children. Suppose that they set no limit on the number of children but just keep going until they get a girl. Their expected number of children must now be higher than in Example 4.

How would you simulate such a couple's children? Simulate 25 repetitions. What is your estimate of the expected number of children?

20.18 Play this game, please. OK, friends, I've got a little deal for you. We have a fair coin (heads and tails each have probability 1/2). Toss it twice. If two heads come up, you win right there. If you get any result other than two heads, I'll give you another chance: toss the coin twice more, and if you get two heads, you win. (Of course, if you fail to get two heads on the second try, I win.) Pay me a dollar to play. If you win, I'll give you your dollar back plus another dollar.

(a) Make a tree diagram for this game. Use the diagram to explain how to simulate one play of this game.

(b) Your dollar bet can win one of two amounts, 0 if I win and $2 if you win. Simulate 50 plays, using Table A, starting at line 125. Use your simulation to estimate the expected value of the game.

20.19 A multiple-choice exam. Charlene takes a quiz with 10 multiple-choice questions, each with five answer choices. If she just guesses independently at each question, she has probability 0.2 of guessing right on each. Use simulation to estimate Charlene's expected number of correct answers. (Simulate 20 repetitions.)

20.20 Repeating an exam. Exercise 19.12 (page 387) gives a model for up to three attempts at an exam in a self-paced course. In that exercise, you simulated 50 repetitions to estimate Elaine's probability of passing the exam. Use those simulations (or do 50 new repetitions) to estimate the expected number of tries Elaine will make.

20.21 A common expected value. Here is a common setting that we simulated in Chapter 19: there are a fixed number of independent trials with the same two outcomes and the same probabilities on each trial. Tossing a coin, shooting basketball free throws, and observing the sex of newborn babies are all examples of this setting. Call the outcomes "hit" and "miss." We can see what the expected number of hits should be. If Shaquille O'Neal shoots 12 free throws and has probability 1/2 of hitting each one, the expected number of hits is 1/2 of 12, or 6. By the same reasoning, if we have n trials with probability p of a hit on each trial, the expected number of hits is np. This fact can be proved mathematically. Can we verify it by simulation?

Simulate 10 tosses of a fair coin 100 times. (To do this quickly, use the first 10 digits and the last 10 digits in each of the 50 rows of Table A, with odd digits meaning a head and even digits a tail.) What is the expected number of heads by the np formula? What is the mean number of heads in your 100 repetitions?

20.22 Your state's lottery. Most states have a lotto game that offers large prizes for choosing (say) 6 out of 51 numbers. If your state has a lotto game, find out what percent of the money bet is returned to the bettors in the form of prizes. What percent of the money bet is used by the state to pay lottery expenses? What percent is net revenue to the state? For what purposes does the state use lottery revenue?

Part III Review

Some phenomena are random. Although their individual outcomes are unpredictable, there is a regular pattern in the long run. Gambling devices and taking an SRS are examples of random phenomena. Probability and expected value give us a language to describe randomness. Random phenomena are not haphazard any more than random sampling is haphazard. Randomness is instead a kind of order, a long-run regularity as opposed to either chaos or a determinism that fixes events in advance. Chapter 17 discusses randomness, Chapter 18 presents facts about probability, and Chapter 20 discusses expected values.

When randomness is present, probability answers the question "How often in the long run?" and expected value answers the question "How much on the average in the long run?" The answers are tied together by the definition of expected value in terms of probabilities. Some "probability models" assign probabilities to outcomes. Any such model must obey the rules of probability. Another kind of probability model uses a density curve such as a normal curve to assign probabilities as areas under the curve. Personal probabilities express an individual's judgment of how likely some event is. Personal probabilities must follow the rules of probability if they are to be consistent with each other.

To calculate the probability of a complicated event without using complicated math, we use random digits to simulate many repetitions. You can also find expected values by simulation. Chapter 19 shows how to do simulations. First give a probability model for the outcomes, then assign random digits to imitate the assignment of probabilities. The table of random digits now imitates repetitions. Keep track of the proportion of repetitions on which an event occurs to estimate its probability. Keep track of the mean outcome to estimate an expected value.

PART III SUMMARY

Here are the most important skills you should have after reading Chapters 17 to 20.

A. RANDOMNESS AND PROBABILITY

1. Recognize that some phenomena are random. Probability describes the long-run regularity of random phenomena.

2. Understand the idea of the probability of an event as the proportion of times the event occurs in very many repetitions of a random phenomenon. Use the idea of probability as long-run proportion to think about probability.

3. Recognize that short runs of random phenomena do not display the regularity described by probability. Accept that randomness is unpredictable in the short run, and avoid seeking causal explanations for random occurrences.

B. PROBABILITY MODELS

1. Use basic probability facts to detect illegitimate assignments of probability: Any probability must be a number between 0 and 1, and the total probability assigned to all possible outcomes must be 1.

2. Use basic probability facts to find the probabilities of events that are formed from other events: The probability that an event does not occur is 1 minus its probability. If two events cannot occur at the same time, the probability that one or the other occurs is the sum of their individual probabilities.

3. When probabilities are assigned to individual outcomes, find the probability of an event by adding the probabilities of the outcomes that make it up.

4. When probabilities are assigned by a normal curve, find the probability of an event by finding an area under the curve.

C. EXPECTED VALUE

1. Understand the idea of expected value as the average of numerical outcomes in very many repetitions of a random phenomenon.

2. Find the expected value from a probability model that lists all outcomes and their probabilities (when the outcomes are numerical).

D. SIMULATION

1. Specify simple probability models that assign probabilities to each of several stages and take the stages to be independent of each other.

2. Assign random digits to simulate such models.

3. Estimate either a probability or an expected value by repeating a simulation many times.

PART III REVIEW EXERCISES

Review exercises are short and straightforward exercises that help you solidify the basic ideas and skills in each part of this book.

III.1 What's the probability? Open your local telephone directory to any page in the residential listing. Look at the last four digits of each telephone number, the digits that specify an individual number within an exchange given by the first three digits. Note the first of these four digits in each of the first 100 telephone numbers on the page.
(a) How many of the digits were 1, 2, or 3? What is the approximate probability that the first of the four "individual digits" in a telephone number is 1, 2, or 3?

(b) If all 10 possible digits had the same probability, what would be the probability of getting a 1, 2, or 3? Based on your work in (a), do you think the first of the four "individual digits" in telephone numbers is equally likely to be any of the 10 possible digits?

III.2 Course grades. Choose a student at random from all who took Math 101 in recent years. The probabilities for the student's grade are

Grade	A	B	C	D	F
Probability	0.2	0.3	0.3	0.1	?

(a) What must be the probability of getting an F?

(b) To simulate the grades of randomly chosen students, how would you assign digits to represent the four possible outcomes listed?

III.3 Blood types. Choose a person at random and record his or her blood type. Here are the probabilities for each blood type:

Blood Type	Type O	Type A	Type B	Type AB
Probability	0.4	0.3	0.2	?

(a) What must be the probability that a randomly chosen person has Type AB blood?

(b) To simulate the blood types of randomly chosen people, how would you assign digits to represent the four types?

III.4 Course grades. If you choose 5 students at random from all those who have taken the course described in Exercise III.2, what is the probability that all the students chosen got a C or better? Simulate 10 repetitions of this random choosing and use your results to estimate the probability. (Your estimate from only 10 repetitions isn't reliable, but if you can do 10, you could do 10,000.)

III.5 Blood types. People with Type B blood can receive blood donations from other people with either Type B or Type O blood. Tyra has Type B blood. What is the probability that 2 or more of Tyra's 6 close friends can donate blood to her? Using your work in Exercise III.3, simulate 10 repetitions and estimate this probability. (Your estimate from just 10 repetitions isn't reliable, but you have shown in principle how to find the probability.)

III.6 Course grades. Choose a student at random from the course of Exercise III.2 and observe what grade that student earns ($A = 4, B = 3, C = 2, D = 1, F = 0$).

(a) What is the expected grade of a randomly chosen student?

(b) The expected grade is not one of the 5 grades possible for one student. Explain why your result nevertheless makes sense as an expected value.

III.7 Dice. What is the expected number of spots observed in rolling a carefully balanced die once?

III.8 Profit from a risky investment. Rotter Partners is planning a major investment. The amount of profit X is uncertain, but a probabilistic estimate gives the following distribution (in millions of dollars):

Profit	1	1.5	2	4	10
Probability	0.1	0.2	0.4	0.2	0.1

What is the expected value of the profit?

III.9 Poker. Deal a five-card poker hand from a shuffled deck. The probabilities of several types of hand are approximately as follows:

Hand	Worthless	One pair	Two pairs	Better hands
Probability	0.50	0.42	0.05	?

(a) What must be the probability of getting a hand better than two pairs?

(b) What is the expected number of hands a player is dealt before the first hand better than one pair appears? Explain how you would use simulation to answer this question, then simulate just 2 repetitions.

III.10 How much education? The *Statistical Abstract* gives this distribution of education for a randomly chosen American over 25 years old:

Education	Less than high school	High school graduate	College, no bachelor's	Bachelor's degree	Advanced degree
Probability	0.17	0.34	0.25	0.16	0.08

(a) How do you know that this is a legitimate probability model?

(b) What is the probability that a randomly chosen person over age 25 has at least a high school education?

(c) What is the probability that a randomly chosen person over age 25 has at least a bachelor's degree?

III.11 Language study. Choose a student in grades 9 to 12 at random and ask if he or she is studying a language other than English. Here is the distribution of results:

Language	Spanish	French	German	All others	None
Probability	0.26	0.09	0.03	0.03	0.59

(a) Explain why this is a legitimate probability model.

(b) What is the probability that a randomly chosen student is studying a language other than English?

(c) What is the probability that a randomly chosen student is studying French, German, or Spanish?

III.12 Choosing at random. Abby, Deborah, Mei-Ling, Sam, and Roberto work in a firm's public relations office. Their employer must choose two of them to attend a conference in Paris. To avoid unfairness, the choice will be made by drawing two names from a hat. (This is an SRS of size 2.)

(a) Write down all possible choices of two of the five names. These are the possible outcomes.

(b) The random drawing makes all outcomes equally likely. What is the probability of each outcome?

(c) What is the probability that Mei-Ling is chosen?

(d) What is the probability that neither of the two men (Sam and Roberto) is chosen?

III.13 Satisfaction with colleges. Ask a randomly chosen adult, "Are the colleges in your state doing an excellent, good, fair, or poor job, or don't you know enough to say?" Here is the distribution of opinions:

Opinion	Excellent	Good	Fair	Poor	Don't know
Probability	0.15	0.42	0.13	0.03	?

(a) What is the probability that a randomly chosen adult doesn't know enough to say?

(b) What is the probability that a randomly chosen adult thinks that colleges are doing a good or excellent job?

III.14 An IQ test. The Wechsler Adult Intelligence Scale (WAIS) is a common "IQ test" for adults. The distribution of WAIS scores for persons over 16 years of age is approximately normal with mean 100 and standard deviation 15. Use the 68–95–99.7 rule to answer these questions.
(a) What is the probability that a randomly chosen individual has a WAIS score of 115 or higher?

(b) In what range do the scores of the middle 95% of the adult population lie?

III.15 We like opinion polls. Are Americans interested in opinion polls about the major issues of the day? Suppose that 40% of all adults are very interested in such polls. (According to sample surveys that ask this question, 40% is about right.) A polling firm chooses an SRS of 1015 people. If they do this many times, the percent of the sample who say they are very interested will vary from sample to sample following a normal distribution with mean 40% and standard deviation 1.5%. Use the 68–95–99.7 rule to answer these questions.
(a) What is the probability that one such sample gives a result within ±1.5% of the truth about the population?

(b) What is the probability that one such sample gives a result within ±3% of the truth about the population?

III.16 An IQ test (optional). Use the information in Exercise III.14 and Table B to find the probability that a randomly chosen person has a WAIS score 112 or higher.

III.17 We like opinion polls (optional). Use the information in Exercise III.15 and Table B to find the probability that one sample misses the truth about the population by 4% or more. (This is the probability that the sample result is either less than 36% or greater than 44%.)

III.18 An IQ test (optional). How high must a person score on the WAIS test to be in the top 10% of all scores? Use the information in Exercise III.14 and Table B to answer this question.

III.19 Models legitimate and not. A bridge deck contains 52 cards, four of each of the 13 face values ace, king, queen, jack, ten, nine, . . . , two. You deal a single card from such a deck and record the face value of the card dealt. Give an assignment of probabilities to these outcomes that should be correct if the deck is thoroughly shuffled. Give a second assignment of probabilities that is legitimate (that is, obeys the rules of probability) but differs from your first choice. Then give a third assignment of probabilities that is *not* legitimate, and explain what is wrong with this choice.

III.20 Mendel's peas. Gregor Mendel used garden peas in some of the experiments that revealed that inheritance operates randomly. The seed color of Mendel's peas can be either green or yellow. Suppose we produce seeds by "crossing" two plants, both of which carry the G (green) and Y (yellow) genes. Each parent has probability 1/2 of passing each of its genes to a seed, independently of the other parent. A seed will be yellow unless both parents contribute the G gene. Seeds that get two G genes are green.

What is the probability that a seed from this cross will be green? Set up a simulation to answer this question, and estimate the probability from 25 repetitions.

III.21 Predicting the winner. There are 11 teams in the Big Ten athletic conference. Here's one set of personal probabilities for next year's basketball champion: Michigan State has probability 0.3 of winning. Iowa, Minnesota, Northwestern, and Penn State have no chance. That leaves 6 teams. Michigan, Ohio State, and Purdue all have the same probability of winning. Illinois, Indiana, and Wisconsin also have the same probability, but that probability is one-half that of the first 3. What probability does each of the 11 teams have?

III.22 Selling cars. Bill sells new cars for a living. On a weekday afternoon, he will deal with 1 customer with probability 0.2, 2 customers with probability 0.4, and 3 customers with probability 0.4. Each customer has probability 0.2 of buying a car. Customers buy independently of each other.

Describe how you would simulate the number of cars Bill sells in an afternoon. You must first simulate the number of customers, then simulate the buying decisions of 1, 2, or 3 customers. Simulate one afternoon to demonstrate your procedure.

PART III PROJECTS

Projects are longer exercises that require gathering information or producing data and emphasize writing a short essay to describe your work. Many are suitable for teams of students.

Project 1. A bit of history. On page 350, I said, "The systematic study of randomness began when 17th-century French gamblers asked French mathematicians for help in figuring out the 'fair value' of bets on games of chance." Pierre de Fermat and Blaise Pascal were two of the mathematicians who responded. Both are interesting characters. Choose one of these men. Write a brief essay giving his dates, some anecdotes you find noteworthy from his life, and at least one example of a probability problem he studied. (A Web search on the name will produce abundant information. Remember to use your own words in writing your essay.)

Project 2. Reacting to risks. On page 356, I quoted a writer as saying, "Few of us would leave a baby sleeping alone in a house while we drove off on a 10-minute errand, even though car-crash risks are much greater than home risks." Take it as a fact that the probability that the baby will be injured in the car is very much higher than the probability of any harm occurring at home in the same time period. Would you leave the baby alone? Explain your reasons in a short essay. If you would not leave the baby alone, be sure to explain why you choose to ignore the probabilities.

Project 3. First digits. Here is a remarkable fact: the first digits of the numbers in long tables are usually *not* equally likely to have any of the 10 possible values 0, 1, 2, 3, 4, 5, 6, 7, 8, and 9. The digit 1 tends to occur with probability roughly 0.3, the digit 2 with probability about 0.17, and so on. You can find more information about this fact, called "Benford's law," on the Web or in two articles by Theodore P. Hill, "The difficulty of faking data," *Chance*, 12 (1999), No. 3, pp. 27–31, and "The first digit phenomenon," *American Scientist*, 86 (1998), pp. 358–363. You don't have to read these articles for this project.

Locate at least two long tables whose entries could plausibly begin with any digit. You may choose data tables, such as populations of many cities or the number of shares traded on the New York Stock Exchange on many days, or mathematical tables such as logarithms or square roots. I hope it's clear that you can't use the table of random digits. Let's require that your examples each contain at least 300 numbers. Tally the first digits of all entries in each table. Report the distributions (in percents) and compare them with each other, with Benford's law, and with the equally likely distribution.

Project 4. Personal probability. Personal probabilities are personal, so we expect them to vary from person to person. Choose an event that most students at your school should have an opinion about, such as rain next Friday or a victory in your team's next game. Ask many students (at least 50) to tell you what probability they would assign to rain or a victory. Then analyze the data with a graph and numbers—shape, center, spread, and all that. What do your data show about personal probabilities for this future event?

Project 5. Making decisions. Exercise 20.6 (page 401) reported the results of a study by the psychologist Amos Tversky on the effect of wording on people's decisions about chance outcomes. His subjects were college students. Repeat Tversky's study at your school. Prepare two typed cards. One says:

> *You are responsible for treating 600 people who have been exposed to a fatal virus. Treatment A has probability 1/2 of saving all 600 and probability 1/2 that all 600 will die. Treatment B is guaranteed to save exactly 400 of the 600 people. Which treatment will you give?*

The second card says:

> *You are responsible for treating 600 people who have been exposed to a fatal virus. Treatment A has probability 1/2 of saving all 600 and probability 1/2 that all 600 will die. Treatment B will definitely lose exactly 200 of the lives. Which treatment will you give?*

Show each card to at least 25 people (25 different people for each, chosen as randomly as you can conveniently manage, and chosen from people who have not studied probability). Record the choices. Tversky claims that people shown the first card tend to choose B, while those shown the second card tend to choose A. Do your results agree with this claim? Write a brief summary of your findings: Do people use expected values in their decisions? Does the frame in which a decision is presented (the wording, for example) influence choices? (You can find lots of background on similar quirks of the mind in Thomas Gilovich, *How We Know What Isn't So: The Fallibility of Reason in Everyday Life,* Free Press, 1991.)

DILBERT reprinted by permission of United Features Syndicate.

Inference

To *infer* is to draw a conclusion from evidence. Statistical *inference* draws a conclusion about a population from evidence provided by a sample. Drawing conclusions in mathematics is a matter of starting from a hypothesis and using logical argument to prove without doubt that the conclusion follows. Statistics isn't like that. Statistical conclusions are uncertain, because the sample isn't the entire population. So statistical inference has to state conclusions and also say how uncertain they are. We use the language of probability to express uncertainty.

Because inference must both give conclusions and say how uncertain they are, it is the most technical part of statistics. Texts and courses intended to *train people to do* statistics spend most of their time on inference. My aim in this book is to *help you understand* statistics, which takes less technique but often more thought. We will look only at a few basic techniques of inference. The techniques are simple, but the ideas are subtle, so prepare to think. To start, think about what you already know and don't be too impressed by elaborate statistical techniques: even the fanciest inference cannot remedy fundamental flaws such as voluntary response samples or uncontrolled experiments.

Chapter 21
What Is a Confidence Interval?

Don't get mad

Know someone who is prone to anger? Nature has a way to calm such people: they get heart disease more often. Several observational studies have discovered a link between anger and heart disease. The best study looked at 12,986 people, both black and white, chosen at random from four communities. When first examined, all subjects were between the ages of 45 and 64 and were free of heart disease. Let's focus on the 8474 people in this sample who had normal blood pressure.

A short psychological test, the Spielberger Trait Anger Scale, measured how easily each person became angry. There were 633 people in the high range of the anger scale, 4731 in the middle, and 3110 in the low range. Now follow these people forward in time for almost six years and compare the rate of heart disease in the high and low groups. There are some lurking variables: people in the high-anger group are somewhat more likely to be male, to have less than a high school education, and to be smokers and drinkers. After adjusting for these differences, high-anger people were 2.2 times as likely to get heart disease and 2.7 times as likely to have an acute heart attack as low-anger people.

That makes anger sound serious. But only 53 people in the low group and 27 in the high group got heart disease during the period of the study. We know that the numbers 2.2 and 2.7 won't be exactly right for the population of all people aged 45 to 64 years with normal blood pressure. News reports of the study cited those numbers, but the full report in the medical journal *Circulation* gave confidence intervals. With 95%

confidence, high-anger people are between 1.36 and 3.55 times as likely to get heart disease as low-anger people. They are between 1.48 and 4.90 times as likely to have an acute heart attack. The intervals remind us that any statement we make about the population is uncertain because we have data only from a sample. For the sample, we can say "exactly 2.2 times as likely." For the whole population, the sample data allow us to say only "between 1.36 and 3.55 times as likely," and only with 95% confidence. To go behind the news to the real thing, in medicine and in other areas, we must speak the language of confidence intervals.

Estimating

Statistical inference draws conclusions about a population on the basis of data about a sample. One kind of conclusion answers questions like "What percent of employed women have a college degree?" or "What is the mean survival time for patients with this type of cancer?" These questions ask about a number (a percent, a mean) that describes a population. Numbers that describe a population are **parameters**. To estimate a population parameter, choose a sample from the population and use a **statistic**, a number calculated from the sample, as your estimate. Here's an example.

Example 1. Risky behavior in the age of AIDS

How common is behavior that puts people at risk of AIDS? The National AIDS Behavioral Surveys interviewed a random sample of 2673 adult heterosexuals. Of these, 170 said they had more than one sexual partner in the past year. That's 6.36% of the sample. This result may be biased by reluctance to tell the truth about sexual behavior (see Exercise 21.10). For now, assume that the people in the sample told the truth. Based on these data, what can we say about the percent of all adult heterosexuals who had multiple partners?

Our population is adult heterosexuals. The parameter is the proportion who have had more than one sexual partner in the past year. Call this unknown parameter p, for "proportion." The statistic that estimates the parameter p is the **sample proportion**

$$\hat{p} = \frac{\text{count in the sample}}{\text{size of the sample}} = \frac{170}{2673} = 0.0636$$

A basic move in statistical inference is to use a sample statistic to estimate a population parameter. Once we have the sample in hand, we estimate that the proportion of all adult heterosexuals with multiple partners is "about

6.36%" because the proportion in the sample was exactly 6.36%. We can only estimate that the truth about the population is "about" 6.36% because we know that the sample result is unlikely to be exactly the same as the true population proportion. A confidence interval makes that "about" precise.

95% confidence interval

A **95% confidence interval** is an interval calculated from sample data that is guaranteed to capture the true population parameter in 95% of all samples.

We will first march straight through to the interval for a population proportion, then reflect on what we have done and generalize a bit.

Estimating with confidence

We want to estimate the proportion p of the individuals in a population who have some characteristic—they are employed, or they approve the president's performance, for example. Let's call the characteristic we are looking for a "success." We use the proportion $\hat{p}$ of successes in a simple random sample (SRS) to estimate the proportion p of successes in the population. How good is the statistic $\hat{p}$ as an estimate of the parameter p? To find out, we ask, "What would happen if we took many samples?" Well, we know that $\hat{p}$ would vary from sample to sample. We also know that this sampling variability isn't haphazard. It has a clear pattern in the long run, a pattern that is pretty well described by a normal curve. Here are the facts.

Sampling distribution of a sample proportion

The **sampling distribution** of a statistic is the distribution of values taken by the statistic in all possible samples of the same size from the same population.

Take an SRS of size n from a large population that contains proportion p of successes. Let $\hat{p}$ be the **sample proportion** of successes,

$$\hat{p} = \frac{\text{count of successes in the sample}}{n}$$

Then, if the sample is large enough:

- The sampling distribution of $\hat{p}$ is **approximately normal**.
- The **mean** of the sampling distribution is p.
- The **standard deviation** of the sampling distribution is

$$\sqrt{\frac{p(1-p)}{n}}$$

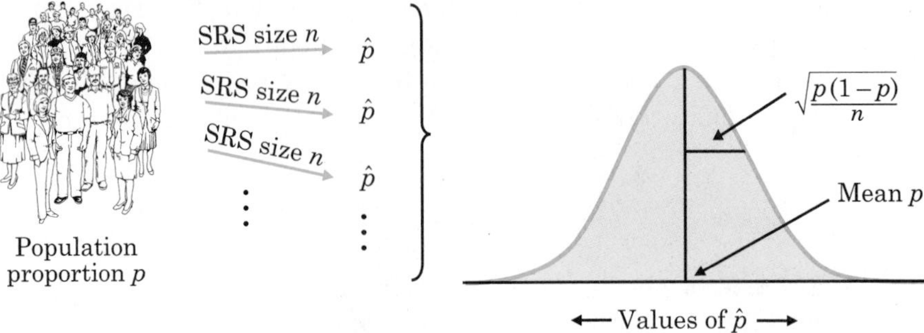

Figure 21.1 Repeat many times the process of selecting an SRS of size n from a population of which the proportion p are successes.The values of the sample proportion of successes $\hat{p}$ have this normal sampling distribution.

These facts can be proved by mathematics, so they are a solid starting point. Figure 21.1 summarizes them in a form that also reminds us that a sampling distribution describes the results of lots of samples from the same population.

Example 2. Risky behavior

Suppose, for example, that the truth is that 6% of adult heterosexuals have multiple partners. Then in the setting of Example 1, $p = 0.06$ The National AIDS Behavioral Surveys sample of size $n = 2673$ would, if repeated many times, produce sample proportions $\hat{p}$ that closely follow the normal distribution with

$$\text{mean} = p = 0.06$$

and

$$\begin{aligned} \text{standard deviation} &= \sqrt{\frac{p(1-p)}{n}} \\ &= \sqrt{\frac{(0.06)(0.94)}{2673}} \\ &= \sqrt{0.0000211} = 0.0046 \end{aligned}$$

The center of this normal distribution is at the truth about the population. That's the absence of bias in random sampling once again. The standard deviation is small because the sample is quite large. So almost all samples will produce a statistic $\hat{p}$ that is close to the true p. In fact, the 95 part of the 68–95–99.7 rule says that 95% of all sample outcomes will fall between

$$\text{mean} - 2 \text{ standard deviations} = 0.06 - 0.0092 = 0.0508$$

and

$$\text{mean} + 2 \text{ standard deviations} = 0.06 + 0.0092 = 0.0692$$

Figure 21.2 displays these facts.

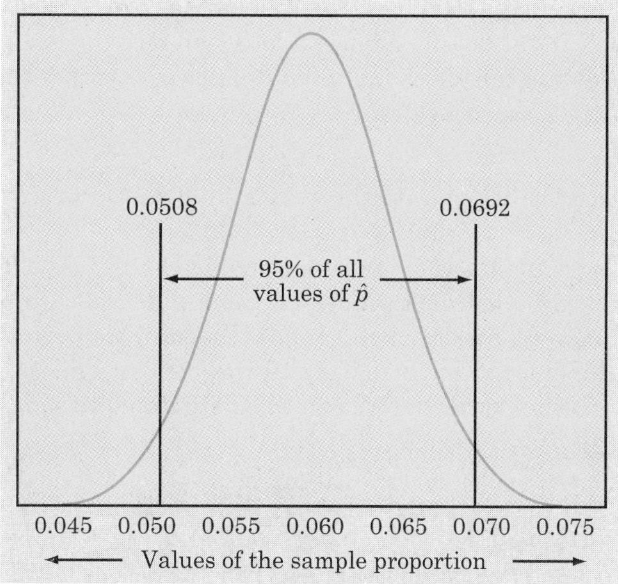

Figure 21.2 Repeat many times the process of selecting an SRS of size 2673 from a population of which the proportion $p = 0.06$ are successes. The middle 95% of the values of the sample proportion $\hat{p}$ will lie between 0.0508 and 0.0692.

So far, we have just put numbers on what we already knew: we can trust the results of large random samples because almost all such samples give results that are close to the truth about the population. The numbers say that in 95% of all samples of size 2673, the statistic $\hat{p}$ and the parameter p are within 0.0092 of each other. We can put this another way: 95% of all samples give an outcome $\hat{p}$ such that the population truth p is captured by the interval from $\hat{p} - 0.0092$ to $\hat{p} + 0.0092$.

The 0.0092 came from substituting $p = 0.06$ into the formula for the standard deviation of $\hat{p}$. For any value of p, the general fact is:

> When the population proportion has the value p, 95% of all samples catch p in the interval extending 2 standard deviations on either side of $\hat{p}$.

That's the interval

$$\hat{p} \pm 2\sqrt{\frac{p(1-p)}{n}}$$

Is this the 95% confidence interval we want? Not quite. The interval can't be found just from the data because the standard deviation involves the population proportion p, and in practice we don't know p. In Example 2, we applied the formula for $p = 0.06$, but this may not be the true p.

What to do? Well, the standard deviation of the statistic $\hat{p}$ does depend on the parameter p, but it doesn't change a lot when p changes. Go back to Example 2 and redo the calculation for other values of p. Here's the result:

Value of p	0.04	0.05	0.06	0.07	0.08
Standard deviation	0.0038	0.0042	0.0046	0.0049	0.0052

We see that if we guess a value of p reasonably close to the true value, the standard deviation found from the guessed value will be about right. We know that when we take a large random sample, the statistic $\hat{p}$ is almost always close to the parameter p. So we will use $\hat{p}$ as the guessed value of the unknown p. Now we have an interval that we can calculate from the sample data.

95% confidence interval for a proportion

Choose an SRS of size n from a large population that contains an unknown proportion p of successes. Call the proportion of successes in this sample $\hat{p}$. An approximate 95% confidence interval for the parameter p is

$$\hat{p} \pm 2\sqrt{\frac{\hat{p}(1-\hat{p})}{n}}$$

Example 3. A confidence interval for risky behavior

The National AIDS Behavioral Surveys random sample of 2673 adult heterosexuals found that 170 had multiple partners in the past year, a sample proportion $\hat{p} = 0.0636$. The 95% confidence interval for the proportion of all adult heterosexuals with multiple partners is

$$\begin{aligned}\hat{p} \pm 2\sqrt{\frac{\hat{p}(1-\hat{p})}{n}} &= 0.0636 \pm 2\sqrt{\frac{(0.0636)(0.9364)}{2673}} \\ &= 0.0636 \pm (2)(0.0047) \\ &= 0.0636 \pm 0.0094 \\ &= 0.0542 \text{ to } 0.0730\end{aligned}$$

Interpret this result as follows: we got this interval by using a recipe that catches the true unknown population proportion in 95% of all samples. The shorthand is: we are **95% confident** that the true proportion of heterosexuals with multiple partners lies between 5.42% and 7.30%.

Understanding confidence intervals

Our 95% confidence interval for a population proportion has the familiar form

estimate ± margin of error

We know that news reports of sample surveys, for example, usually give the estimate and the margin of error separately: "A new Gallup poll shows that 66% of women favor new laws restricting guns. The margin of error is plus or minus four percentage points." We also know that news reports usually leave out the level of confidence.

Not all confidence intervals have this form. The confidence intervals for the added risk of a heart attack in angry people that open this chapter have a different form, for example. Here's a complete description of a confidence interval.

Confidence interval

A **level *C* confidence interval** for a parameter has two parts:

- An **interval** calculated from the data.
- A **confidence level** *C*, which gives the probability that the interval will capture the true parameter value in repeated samples.

There are many recipes for statistical confidence intervals for use in many situations. Be sure you understand how to interpret a confidence interval. The interpretation is the same for any recipe, and you can't use a calculator or a computer to do the interpretation for you.

Confidence intervals use the central idea of probability: ask what would happen if we repeated the sampling many times. The 95% in a 95% confidence interval is a probability, the probability that the method produces an interval that does capture the true parameter.

"It was a numbers explosion."

Example 4. How confidence intervals behave

The National AIDS Behavioral Surveys sample of 2673 heterosexuals found 170 with multiple partners, so the sample proportion was

$$\hat{p} = \frac{170}{2673} = 0.0636$$

and the 95% confidence interval was

$$\hat{p} \pm 2\sqrt{\frac{\hat{p}(1-\hat{p})}{n}} = 0.0636 \pm 0.0094$$

Who is a smoker?

When estimating a proportion p, be sure you know what counts as a "success." The news says that 20% of adolescents smoke. Shocking. It turns out that this is the percent who smoked at least once in the past month. If we say that a smoker is someone who smoked in at least 20 of the past 30 days and smoked at least half a pack on those days, fewer than 4% of adolescents qualify.

Draw a second sample from the same population. It finds 148 of its 2673 respondents who have multiple partners. For this sample,

$$\hat{p} = \frac{148}{2673} = 0.0554$$

$$\hat{p} \pm 2\sqrt{\frac{\hat{p}(1-\hat{p})}{n}} = 0.0554 \pm 0.0088$$

Draw another sample. Now the count is 152 and the sample proportion and confidence interval are

$$\hat{p} = \frac{152}{2673} = 0.0569$$

$$\hat{p} \pm 2\sqrt{\frac{\hat{p}(1-\hat{p})}{n}} = 0.0569 \pm 0.0090$$

Keep sampling. Each sample yields a new estimate $\hat{p}$ and a new confidence interval. *If we sample forever, 95% of these intervals capture the true parameter.* This is true no matter what the true value is. Figure 21.3 summarizes the behavior of the confidence interval in graphical form.

On the ground that two pictures are better than one, Figure 21.4 gives a different view of how confidence intervals behave. Example 4 and Figure 21.3 remind us that repeated samples give different results and that we are guaranteed only that 95% of the samples give a correct result. Figure 21.4 goes behind the scenes. The vertical line is the true value of the population proportion p. The normal curve at the top of the figure is the sampling distribution of the sample statistic $\hat{p}$, which is centered at the true p. We are behind the scenes because in real-world statistics we don't know p.

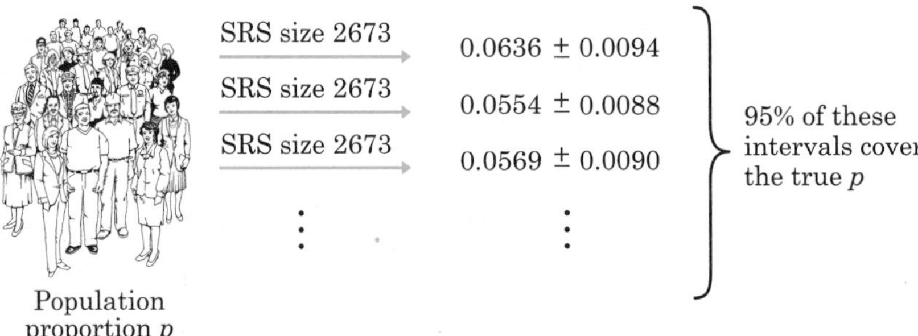

Figure 21.3 Repeated samples from the same population give different 95% confidence intervals, but 95% of these intervals capture the true population proportion p.

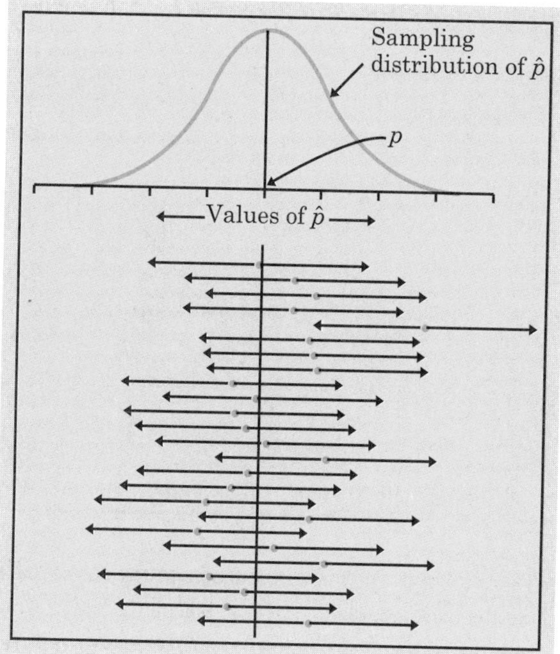

Figure 21.4 Twenty-five samples from the same population give these 95% confidence intervals. In the long run, 95% of all such intervals cover the true population proportion, marked by the vertical line.

The 95% confidence intervals from 25 SRSs appear below, one after the other. The central dots are the values of $\hat{p}$, the centers of the intervals. The arrows on either side span the confidence interval. In the long run, 95% of the intervals will cover the true p and 5% will miss. Of the 25 intervals in Figure 21.4, 24 hit and 1 misses. (Remember that probability describes only what happens in the long run—we don't expect exactly 95% of 25 intervals to capture the true parameter.)

Don't forget that our interval is only *approximately* a 95% confidence interval. It isn't exact for two reasons. The sampling distribution of the sample proportion $\hat{p}$ isn't exactly normal. And we don't get the standard deviation of $\hat{p}$ exactly right because we used $\hat{p}$ in place of the unknown p. Both of these difficulties go away as the sample size n gets larger. So our recipe is good only for large samples. What is more, the recipe assumes that the population is really big—at least 10 times the size of the sample. Professional statisticians use more elaborate methods that take the size of the population into account and work even for small samples. But our method works well enough for many practical uses. More important, it shows how we get a confidence interval from the sampling distribution of a statistic. That's the reasoning behind any confidence interval.

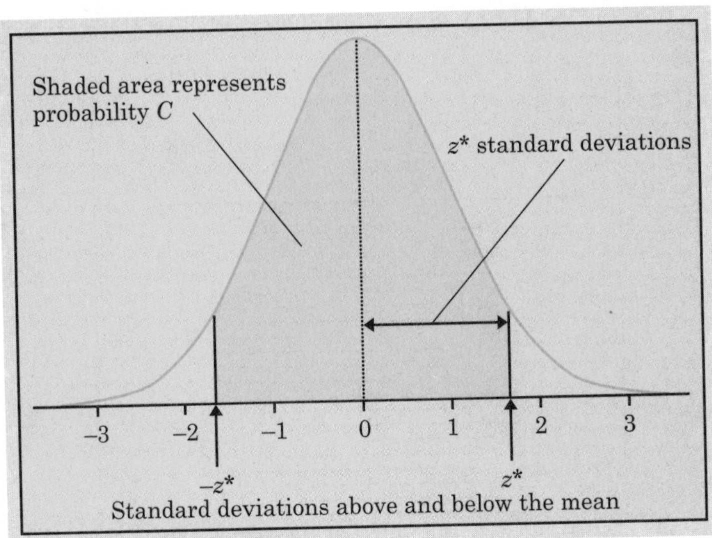

Figure 21.5 Critical values z^* of the normal distributions. In any normal distribution, there is area (probability) C under the curve between $-z^*$ and z^* standard deviations away from the mean.

Confidence intervals for a population proportion*

We used the 95 part of the 68–95–99.7 rule to get a 95% confidence interval for the population proportion. Perhaps you think that a method that works 95% of the time isn't good enough. You want to be 99% confident. For that, we need to mark off the central 99% of a normal distribution. For any probability C between 0 and 1, there is a number z^* such that any normal distribution has probability C within z^* standard deviations of the mean. Figure 21.5 shows how the probability C and the number z^* are related.

Table 21.1 gives the numbers z^* for various choices of C. For convenience, the table gives C as a confidence level in percent. The numbers z^* are called **critical values** of the normal distributions. Table 21.1 shows that any normal distribution has probability 99% within ±2.58 standard deviations of its mean. The table also shows that any normal distribution has probability 95% within ±1.96 standard deviations of its mean. The 68–95–99.7 rule uses 2 in place of the critical value $z^* = 1.96$. That is good enough for practical purposes, but the table gives the more exact value.

From Figure 21.5 we see that with probability C, the sample proportion $\hat{p}$ takes a value within z^* standard deviations of p. That is just to say that with probability C, the interval extending z^* standard deviations on either side of the observed $\hat{p}$ captures the unknown p. Using the estimated standard deviation of $\hat{p}$ produces the following recipe.

*This section is optional.

Confidence interval for a population proportion

Choose an SRS of size n from a population of individuals of which proportion p are successes. The proportion of successes in the sample is $\hat{p}$. When n is large, an approximate level C confidence interval for p is

$$\hat{p} \pm z^* \sqrt{\frac{\hat{p}(1-\hat{p})}{n}}$$

where z^* is the critical value for probability C from Table 21.1.

Example 5. A 99% confidence interval

The National AIDS Behavioral Surveys random sample of 2673 adult heterosexuals found that 170 had multiple partners in the past year. We want a 99% confidence interval for the proportion p of all heterosexuals who have multiple partners. Table 21.1 says that for 99% confidence, we must go out $z^* = 2.58$ standard deviations. Here are our calculations:

$$\hat{p} = \frac{170}{2673} = 0.0636$$

$$\begin{aligned} \hat{p} \pm z^* \sqrt{\frac{\hat{p}(1-\hat{p})}{n}} &= 0.636 \pm 2.58\sqrt{\frac{(0.0636)(0.9364)}{2673}} \\ &= 0.636 \pm (2.58)(0.0047) \\ &= 0.636 \pm 0.0121 \\ &= 0.0515 \text{ to } 0.0757 \end{aligned}$$

We are 99% confident that the true population proportion is between 5.15% and 7.57%. That is, we got this range of percents by using a method that gives a correct answer 99% of the time.

Table 21.1 Critical values of the normal distributions

Confidence level C	Critical value z^*	Confidence level C	Critical value z^*
50%	0.67	90%	1.64
60%	0.84	95%	1.96
70%	1.04	99%	2.58
80%	1.28	99.9%	3.29

Compare Example 5 with the calculation of the 95% confidence interval in Example 3. The only difference is the use of the critical value 2.58 for 99% confidence in place of 2 for 95% confidence. That makes the margin of error for 99% confidence larger and the confidence interval wider. Higher confidence isn't free—we pay for it with a wider interval. Figure 21.5 reminds us why this is true. To cover a higher percent of the area under a normal curve, we must go farther out from the center.

Statistics in Summary **Statistical inference** draws conclusions about a population on the basis of data from a sample. Because we don't have data for the entire population, our conclusions are uncertain. A **confidence interval** estimates an unknown parameter in a way that tells us how uncertain the estimate is. The interval itself says how closely we can pin down the unknown parameter. The **confidence level** is a probability that says how often in many samples the method produces an interval that does catch the parameter. We find confidence intervals starting from the **sampling distribution** of a statistic, which shows how the statistic varies in repeated sampling.

In this chapter we found one specific confidence interval, for the proportion p of "successes" in a population, based on an SRS from the population. You will find more advice on interpreting confidence intervals in Chapter 23.

CHAPTER 21 EXERCISES

21.1 A student survey. Tonya wants to estimate what proportion of the students in her dormitory like the dorm food. She interviews an SRS of 50 of the 175 students living in the dormitory. She finds that 14 think the dorm food is good.

(a) What population does Tonya want to draw conclusions about?

(b) In your own words, what is the population proportion p in this setting?

(c) What is the value of the sample proportion $\hat{p}$ from Tonya's sample?

21.2 Fire the coach? A college president says, "99% of the alumni support my firing of Coach Boggs." You contact an SRS of 200 of the college's 15,000 living alumni and find that 76 of them support firing the coach.

(a) What population does inference concern here?

(b) Explain clearly what the population proportion p is in this setting.

(c) What is the numerical value of the sample proportion $\hat{p}$?

21.3 The schools' most serious problem. The report of a sample survey of 1500 adults says, "With 95% confidence, between 27% and 33% of all American adults believe that drugs are the most serious problem facing our nation's public schools." Explain to someone who knows no statistics what the phrase "95% confidence" means in this report.

21.4 Gun violence. The Harris Poll asked a sample of 1009 adults which causes of death they thought would become more common in the future. Gun violence topped the list: 706 members of the sample thought deaths from guns would increase. Although the samples in national polls are not SRSs, they are close enough that our method gives approximately correct confidence intervals.

(a) Say in words what the population proportion p is for this poll.

(b) Find a 95% confidence interval for p.

(c) Harris announced a margin of error of plus or minus three percentage points for this poll result. How well does your work in (b) agree with this margin of error?

21.5 Polling women. A *New York Times* poll on women's issues interviewed 1025 women randomly selected from the United States, excluding Alaska and Hawaii. Of the women in the sample, 482 said they do not get enough time for themselves. Although the samples in national polls are not SRSs, they are close enough that our method gives approximately correct confidence intervals.

(a) Explain in words what the parameter p is in this setting.

(b) Use the poll results to give a 95% confidence interval for p.

(c) Write a short explanation of your findings in (b) for someone who knows no statistics.

21.6 Gun violence. In Exercise 21.4, you gave a 95% confidence interval based on a random sample of $n = 1009$ adults. How large a sample would be needed to get a margin of error half as large as the one in Exercise 21.4?

21.7 The effect of sample size. An opinion poll finds that 60% of its sample prefer balancing the federal budget to cutting taxes. Give a 95% confidence interval for the proportion of all adults who feel this way, assuming that the result $\hat{p} = 0.6$ comes from a sample of size

(a) $n = 750$

(b) $n = 1500$

(c) $n = 3000$

(d) Explain briefly what your results show about the effect of increasing the size of a sample.

21.8 Random digits. We know that the proportion of 0s among a large set of random digits is $p = 0.1$ because all 10 possible digits are equally probable. The entries in a table of random digits are a sample from the population of all random digits. To get an SRS of 200 random digits, look at the first digit in each of the 200 five-digit groups in lines 101 to 125 of Table A in the back of the book. How many of these 200 digits are 0s? Give a 95% confidence interval for the proportion of 0s in the population from which these digits are a sample. Does your interval cover the true parameter value, $p = 0.1$?

21.9 Tossing a thumbtack. If you toss a thumbtack on a hard surface, what is the probability that it will land point up? Estimate this probability p by tossing a thumbtack 100 times. The 100 tosses are an SRS of size 100 from the population of all tosses. The proportion of these 100 tosses that land point up is the sample proportion $\hat{p}$. Use the result of your tosses to give a 95% confidence interval for p. Write a brief explanation of your findings for someone who knows no statistics but wonders how often a thumbtack will land point up.

21.10 Don't forget the basics. The National AIDS Behavioral Surveys found that 170 individuals in its random sample of 2673 adult heterosexuals said they had multiple sexual partners in the past year. We used this finding to calculate confidence intervals for the proportion of all adult heterosexuals with multiple partners. This sample survey may have bias that our confidence intervals do not take into account. Why is some bias likely to be present? Does the sample proportion 6.36% probably overestimate or underestimate the true population proportion?

21.11 Count Buffon's coin. The 18th-century French naturalist Count Buffon tossed a coin 4040 times. He got 2048 heads. Give a 95% confidence interval for the probability that Buffon's coin lands heads up. Are you confident that this probability is not 1/2? Why?

21.12 Teens and their TV sets. The *New York Times* and CBS News conducted a nationwide poll of 1048 randomly selected 13- to 17-year-olds. Of these teenagers, 692 had a television in their room and 189 named Fox as their favorite television network. We can consider the sample to be an SRS.
(a) Give 95% confidence intervals for the proportion of all people in this age group who had a TV in their room at the time of the poll and the proportion who would choose Fox as their favorite network.

(b) The news article says, "In theory, in 19 cases out of 20, the poll results will differ by no more than three percentage points in either direction from what would have been obtained by seeking out all American teenagers." Explain how your results agree with this statement.

21.13 Harley motorcycles. Harley-Davidson motorcycles make up 14% of all the motorcycles registered in the United States. You plan to interview an SRS of 600 motorcycle owners.
(a) What is the sampling distribution of the proportion of your sample who own Harleys?

(b) How likely is your sample to contain 18.2% or more who own Harleys? How likely is it to contain at least 11.2% Harley owners? Use the 68–95–99.7 rule and your answer to (a).

21.14 Do you jog? Suppose that 15% of all adults jog. An opinion poll asks an SRS of 500 adults if they jog.

(a) What is the sampling distribution of the proportion $\hat{p}$ in the sample who jog?

(b) According to the 68–95–99.7 rule, what is the probability that the sample proportion who jog will be 11.8% or greater?

21.15 The quick method. The quick method of Chapter 3 uses $\hat{p} \pm 1/\sqrt{n}$ as a rough recipe for a 95% confidence interval for a population proportion. The margin of error from the quick method is a bit larger than needed. It differs most from the more accurate method of this chapter when $\hat{p}$ is close to 0 or 1. An SRS of 500 motorcycle registrations finds that 68 of the motorcycles are Harley-Davidsons. Give a 95% confidence interval for the proportion of all motorcycles that are Harleys by the quick method and then by the method of this chapter. How much larger is the quick method margin of error?

21.16 68% confidence. We used the 95 part of the 68–95–99.7 rule to give a recipe for a 95% confidence interval for a population proportion p.

(a) Use the 68 part of the rule to give a recipe for a 68% confidence interval.

(b) Explain in simple language what "68% confidence" means.

(c) Use the result of the National AIDS Behavioral Surveys (Example 3) to give a 68% confidence interval for the proportion of heterosexuals with multiple partners. How does your interval compare with the 95% interval in Example 3?

21.17 Simulating confidence intervals. In Exercise 21.16, you found the recipe for a 68% confidence interval for a population proportion p. Suppose that (unknown to anyone) 60% of the voters in her district favor the reelection of Congresswoman Caucus.

(a) How would you simulate the votes of an SRS of 25 voters?

(b) Simulate choosing 10 SRSs, using a different row in Table A for each sample. What are the 10 values of the sample proportion $\hat{p}$ who favor Caucus?

(c) Find the 68% confidence interval for p from each of your 10 samples. How many of the intervals capture the true parameter value $p = 0.6$? (Samples of size 25 are not large enough for our recipe to be very accurate, but even a small simulation illustrates how confidence intervals behave in repeated samples.)

The following exercises concern the optional section of this chapter.

21.18 Gun violence. Exercise 21.4 reports a Harris Poll in which 706 of a random sample of 1009 adults thought that deaths from gun violence would increase in the future. Give a 90% confidence interval for the proportion of all adults who feel this way. How does your interval compare with the 95% confidence interval from Exercise 21.4?

21.19 Polling women. Exercise 21.5 reports a *New York Times* poll that found that 482 in a random sample of 1025 women felt they do not get enough time for themselves. Give a 99% confidence interval for the proportion of all women who feel this way. How does your interval compare with the 95% confidence interval of Exercise 21.5?

21.20 The effect of confidence level. A random sample of 1500 adults finds that 60% favor balancing the federal budget over cutting taxes. Use this poll result and Table 21.1 to give 70%, 80%, 90%, and 99% confidence intervals for the proportion of all adults who feel this way. What do your results show about the effect of changing the confidence level?

21.21 Unhappy HMO patients. How likely are patients who file complaints with a health maintenance organization (HMO) to leave the HMO? In one recent year, 639 of the more than 400,000 members of a large New England HMO filed complaints. Fifty-four of the complainers left the HMO voluntarily. (That is, they were not forced to leave by a move or a job change.) Consider this year's complainers as an SRS of all patients who will complain in the future. Give a 90% confidence interval for the proportion of complainers who voluntarily leave the HMO.

21.22 Estimating the unemployment rate. The Bureau of Labor Statistics (BLS) uses 90% confidence in presenting unemployment results from the monthly Current Population Survey (CPS). The January 2000 survey interviewed 133,357 members of the civilian labor force. Of these, 6264 were unemployed. The CPS is not an SRS, but for the purposes of this exercise act as though the BLS took an SRS of 133,357 people. Give a 90% confidence interval for the unemployment rate (the proportion of the entire labor force that is unemployed).

21.23 Safe margin of error. The margin of error $z^* \sqrt{\hat{p}(1-\hat{p})/n}$ is 0 when $\hat{p}$ is 0 or 1 and is largest when $\hat{p}$ is 1/2. To see this, calculate $\hat{p}(1-\hat{p})$ for $\hat{p} = 0, 0.1, 0.2, \ldots, 0.9$, and 1. Plot your results vertically against the values of $\hat{p}$ horizontally. Draw a curve through the points. You have made a graph of $\hat{p}(1-\hat{p})$ Does the graph reach its highest point when $\hat{p} = 1/2$? You see that taking $\hat{p} = 1/2$ gives a margin of error that is always at least as large as needed.

Chapter 22
What Is a Test of Significance?

There goes the neighborhood?

Despite the Fair Housing Act and other laws, most black and white Americans still live in segregated neighborhoods. Whites tend to move out when a neighborhood becomes too heavily black. Have white attitudes changed over time, so that more whites are willing to stay as black families move in?

The Detroit Area Study interviewed a random sample of 1104 adults in the Detroit metropolitan area in 1976 and another random sample of 1543 adults in 1992. Detroit is the most segregated large city in the United States, so the study looked in detail at attitudes toward mixed-race neighborhoods. One question asked whites to imagine that they lived in an all-white neighborhood, illustrated by the first card in Figure 22.1. Most of the respondents did in fact live in all-white neighborhoods. Then they were shown the second card in Figure 22.1, in which 3 of the 15 houses are occupied by blacks. This is the actual proportion of blacks in the entire Detroit area. Would they try to move away from such a neighborhood?

In 1976, 24% of whites would try to leave. By 1992, this percent had dropped to 15%. It appears that changing attitudes were making "white flight" less likely. (This assumes that the respondents told the truth. It is possible that between 1976 and 1992 it just became less acceptable to give segregationist answers.) These findings come from two samples of modest size. Could it be that the difference between the two samples is just due to the luck of the draw in randomly choosing the respondents? The study authors answered that challenge by finding the probability that two random samples would differ by as much as 24% versus 15% just by chance. This probability is small, less than 0.01. So the difference is "statistically significant." That idea, and using probabilities such as that 0.01 to express it, are our topic in this chapter.

The reasoning of statistical tests

The local hot shot playground basketball player claims to make 80% of his free throws. "Show me," you say. He shoots 20 free throws and makes 8 of them. "Aha," you conclude, "if he makes 80%, he would almost never make as few as 8 of 20. So I don't believe his claim." That's the reasoning of statistical tests at the playground level: *An outcome that is very unlikely if a claim is true is good evidence that the claim is not true.*

Statistical inference uses data from a sample to draw conclusions about a population. So once we leave the playground, statistical tests deal with claims about a population. Tests ask if sample data give good evidence *against* a claim. A test says, "If we took many samples and the claim were true, we would rarely get a result like this." To get a numerical measure of how strong the sample evidence is, replace the vague term "rarely" by a probability. Here is an example of this reasoning at work.

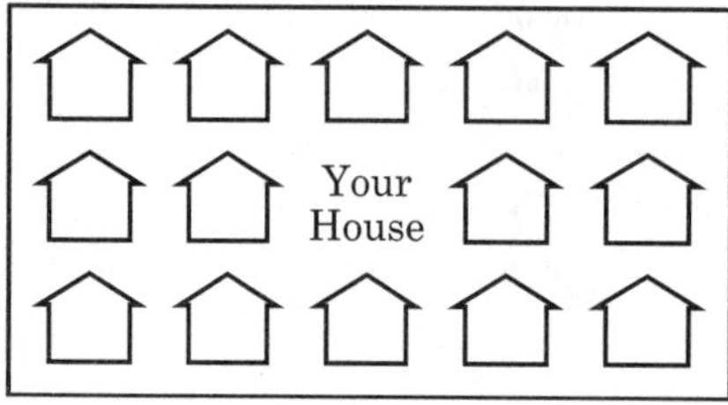

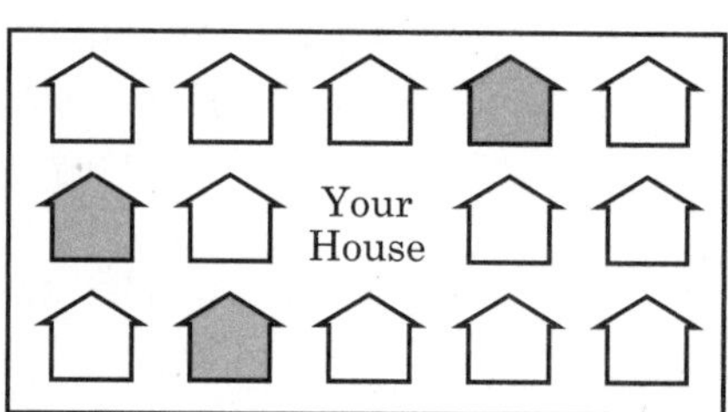

Figure 22.1 The Detroit Area Study showed white residents these cards. It asked them to imagine that their neighborhood looked like the one on the left. Now some black families move in and the neighborhood looks like the one on the right. Would the whites try to move out?

Example 1. Is the coffee fresh?

People of taste are supposed to prefer fresh-brewed coffee to the instant variety. On the other hand, perhaps many coffee-drinkers just want their caffeine fix. A skeptic claims that only half of all coffee-drinkers prefer fresh coffee. Let's do an experiment to test this claim.

Each of 50 subjects tastes two unmarked cups of coffee and says which he or she prefers. One cup in each pair contains instant coffee; the other, fresh-brewed coffee. The statistic that records the result of our experiment is the proportion $\hat{p}$ of the sample who say they like the fresh-brewed coffee better. We find that 36 of our 50 subjects choose the fresh coffee. That is,

$$\hat{p} = \frac{36}{50} = 0.72 = 72\%$$

To make a point, let's compare our outcome $\hat{p} = 0.72$ with another possible result. If only 28 of the 50 subjects like the fresh coffee better than instant coffee, the sample proportion is

$$\hat{p} = \frac{28}{50} = 0.56 = 56\%$$

Surely 72% is stronger evidence against the skeptic's claim than 56%. But how much stronger? Is even 72% in favor in a *sample* convincing evidence that a majority of the *population* prefer fresh coffee? Statistical tests answer these questions.

Here's the answer in outline form:

- **The claim.** The skeptic claims that only half of all coffee-drinkers prefer fresh-brewed coffee. That is, he claims that the population proportion p is only 0.5. *Suppose for the sake of argument that this claim is true.*
- **The sampling distribution (from page 421).** If the claim $p = 0.5$ were true and we tested many random samples of 50 coffee-drinkers, the sample proportion $\hat{p}$ would vary from sample to sample according to (approximately) the normal distribution with

$$\text{mean} = p = 0.5$$

and

$$\begin{aligned}\text{standard deviation} &= \sqrt{\frac{p(1-p)}{n}}\\ &= \sqrt{\frac{(0.5)(0.5)}{50}}\\ &= 0.0707\end{aligned}$$

Figure 22.2 displays this normal curve.

- **The data.** Place the sample proportion $\hat{p}$ on the sampling distribution. You see in Figure 22.2 that $p = 0.56$ isn't an unusual value, but that $p = 0.72$ is unusual. We would rarely get 72% of a sample of 50 coffee-drinkers preferring fresh-brewed coffee if only 50% of all coffee-drinkers felt that way. So the sample data do give evidence against the claim.

- **The probability.** We can measure the strength of the evidence against the claim by a probability. What is the probability that a sample gives a $\hat{p}$ this large or larger if the truth about the population is that $p = 0.5$? If $\hat{p} = 0.56$ this probability is the shaded area under the normal curve in Figure 22.2. This area is 0.20. Our sample actually gave $\hat{p} = 0.72$ The probability of getting a sample outcome this large is only 0.001, an area too small to see in Figure 22.2. An outcome that would occur just by chance in 20% of all samples is *not* strong evidence against the claim. But an outcome that would happen only 1 in 1000 times *is* good evidence.

Be sure you understand why this evidence is convincing. There are two possible explanations of the fact that 72% of our subjects prefer fresh to instant coffee:

(a) The skeptic is correct ($p = 0.5$) and by bad luck a very unlikely outcome occurred.

(b) In fact, the population proportion favoring fresh coffee is greater than 0.5, so the sample outcome is about what would be expected.

We cannot be certain that explanation (a) is untrue. Our taste test results *could* be due to chance alone. But the probability that such a result would occur by chance is so small (0.001) that we are quite confident that explanation (b) is right.

Hypotheses and P-values

Tests of significance refine (and perhaps hide) this basic reasoning. In most studies, we hope to show that some definite effect is present in the population. In Example 1, we suspect that a majority of coffee-drinkers prefer fresh-brewed

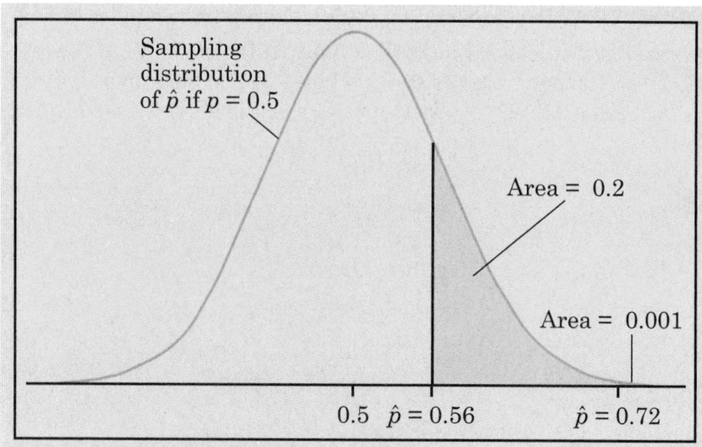

Figure 22.2 The sampling distribution of the proportion of 50 coffee-drinkers who prefer fresh-brewed coffee. This distribution would hold if the truth about all coffee-drinkers is that 50% prefer fresh coffee. The shaded area is the probability that the sample proportion is 56% or greater.

coffee. A statistical test begins by supposing for the sake of argument that the effect we seek is *not* present. We then look for evidence against this supposition and in favor of the effect we hope to find. The first step in a test of significance is to state a claim that we will try to find evidence *against*.

Null hypothesis H_0

The claim being tested in a statistical test is called the **null hypothesis**. The test is designed to assess the strength of the evidence against the null hypothesis. Usually the null hypothesis is a statement of "no effect" or "no difference."

The term "null hypothesis" is abbreviated H_0, read as "H-nought." It is a statement about the population and so must be stated in terms of a population parameter. In Example 1, the parameter is the proportion p of all coffee-drinkers who prefer fresh to instant coffee. The null hypothesis is

$$H_0: p = 0.5$$

The statement we hope or suspect is true instead of H_0 is called the **alternative hypothesis** and is abbreviated H_a. In Example 1, the alternative hypothesis is that a majority of the population favor fresh coffee. In terms of the population parameter, this is

$$H_a: p > 0.5$$

A significance tests looks for evidence against the null hypothesis and in favor of the alternative hypothesis. The evidence is strong if the outcome we observe would rarely come up if the null hypothesis is true but is more probable if the alternative hypothesis is true. For example, it would be surprising to find 36 of 50 subjects favoring fresh coffee if in fact only half of the population feel this way. How surprising? A significance test answers this question by giving a probability: the probability of getting an outcome at least as far as the actually observed outcome from what we would expect when H_0 is true. What counts as "far from what we would expect" depends on H_a as well as H_0. In the taste test, the probability we want is the probability that 36 or more of 50 subjects favor fresh coffee. If the null hypothesis $p = 0.5$ is true, this probability is very small (0.001). That's good evidence that the null hypothesis is not true.

Gotcha!

A tax examiner suspects that Ripoffs, Inc., is issuing phony checks to inflate its expenses and reduce the tax it owes. To learn the truth without examining every check, she fires up her computer. The first digits of real data follow well-known patterns that do *not* give digits 0 to 9 equal probabilities. If the check amounts don't follow this pattern, she will investigate. Down the street, a hacker is probing a company's computer files. He can't read them, because they are encrypted. But he can locate the key to the encryption anyway—it's the only long string that really *does* give equal probability to all possible characters.

P-value

The probability, computed assuming that H_0 is true, that the sample outcome would be as extreme or more extreme than the actually observed outcome is called the ***P*-value** of the test. The smaller the *P*-value is, the stronger is the evidence against H_0 provided by the data.

In practice, most statistical tests are carried out by computer software that calculates the *P*-value for us. It is usual to report the *P*-value in describing the results of studies in many fields. You should therefore understand what *P*-values say even if you don't do statistical tests yourself, just as you should understand what "95% confidence" means even if you don't calculate your own confidence intervals.

Example 2. Count Buffon's coin

The French naturalist Count Buffon (1707–1788) tossed a coin 4040 times. He got 2048 heads. The sample proportion of heads is

$$\hat{p} = \frac{2048}{4040} = 0.507$$

That's a bit more than one-half. Is this evidence that Buffon's coin was not balanced? This is a job for a significance test.

The truth about Count Buffon's coin-tossing experiment.

The hypotheses. The null hypothesis says that the coin is balanced ($p = 0.5$) We did not suspect a bias in a specific direction before we saw the data, so the alternative hypothesis is just "the coin is not balanced." The two hypotheses are

$$H_0: p = 0.5$$
$$H_a: p \neq 0.5$$

The sampling distribution. *If the null hypothesis is true,* the sample proportion of heads has approximately the normal distribution with

$$\text{mean} = p = 0.5$$

$$\begin{aligned}\text{standard deviation} &= \sqrt{\frac{p(1-p)}{n}} \\ &= \sqrt{\frac{(0.5)(0.5)}{4040}} \\ &= 0.00787\end{aligned}$$

Figure 22.3 shows this sampling distribution with Buffon's sample outcome $\hat{p} = 0.507$ marked. The picture already suggests that this is not an unlikely outcome that would give strong evidence against the claim that $p = 0.5$.

The *P*-value. How unlikely is an outcome as far from 0.5 as Buffon's $\hat{p} = 0.507$? Because the alternative hypothesis allows p to lie on either side of 0.5, values of $\hat{p}$ far from 0.5 in either direction provide evidence against H_0 and in favor of H_a. The *P*-value is therefore the probability that the observed $\hat{p}$ lies as far from 0.5 *in either direction* as the observed $\hat{p} = 0.507$. Figure 22.4 shows this probability as area under the normal curve. It is $P = 0.37$.

Conclusion. A truly balanced coin would give a result this far or farther from 0.5 in 37% of all repetitions of Buffon's trial. His result gives no reason to think that his coin was not balanced.

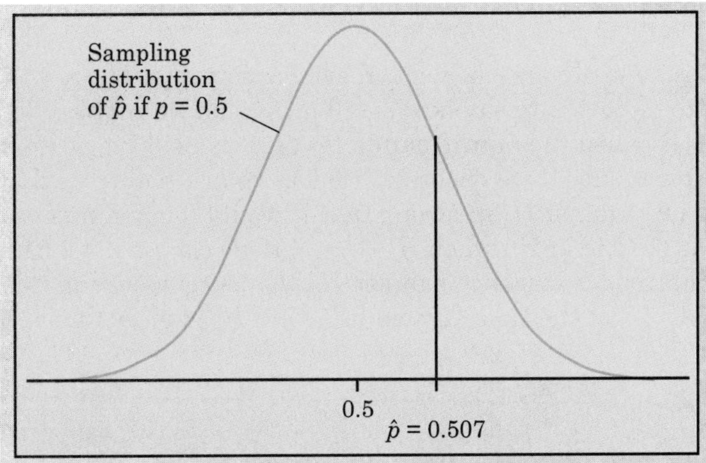

Figure 22.3 The sampling distribution of the proportion of heads in 4040 tosses of a balanced coin. Count Buffon's result, proportion 0.507 heads, is marked.

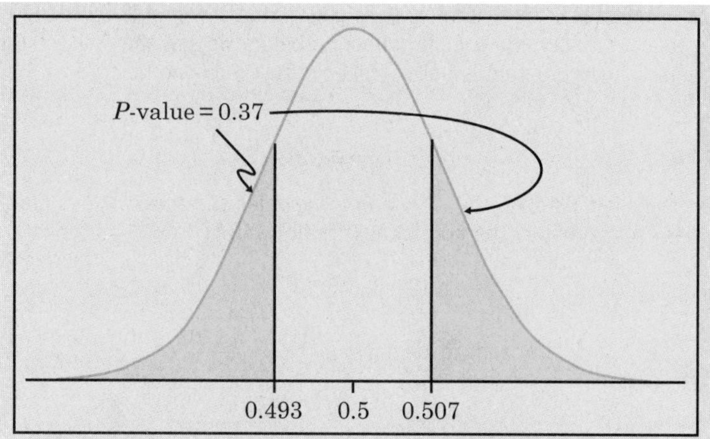

Figure 22.4 The *P*-value for testing whether Count Buffon's coin was balanced. This is the probability, calculated assuming a balanced coin, of a sample proportion as far or farther from 0.5 as Buffon's result of 0.507.

The alternative $H_a: p > 0.5$ in Example 1 is a **one-sided alternative** because the effect we seek evidence for says that the population proportion is greater than one-half. The alternative $H_a: p \neq 0.5$ in Example 2 is a **two-sided alternative** because we ask only whether or not the coin is balanced. Whether the alternative is one-sided or two-sided determines whether sample results extreme in one or in both directions count as evidence against H_0 in favor of H_a.

Statistical significance

We can decide in advance how much evidence against H_0 we will insist on. The way to do this is to say how small a *P*-value we require. The decisive value of *P* is called the **significance level**. It is usual to write it as α, the Greek letter alpha. If we choose $\alpha = 0.05$, we are requiring that the data give evidence against H_0 so strong that it would happen no more than 5% of the time (1 time in 20) when H_0 is true. If we choose $\alpha = 0.01$, we are insisting on stronger evidence against H_0, evidence so strong that it would appear only 1% of the time (1 time in 100) if H_0 is in fact true.

Statistical significance

If the *P*-value is as small or smaller than α, we say that the data are **statistically significant at level α**.

"Significant" in the statistical sense does not mean "important." It means simply "not likely to happen just by chance." We used these words in Chapter 5 (page 79). Now we have attached a number to statistical significance to say what "not likely" means. You will often see significance at level 0.01 expressed by the statement "The results were significant ($P < 0.01$)." Here P stands for the P-value.

We don't need to make use of traditional levels of significance such as 5% and 1%. The P-value is more informative, because it allows us to assess significance at any level we choose. For example, a result with $P = 0.03$ is significant at the $\alpha = 0.05$ level but not significant at the $\alpha = 0.01$ level. Nonetheless, the traditional significance levels are widely accepted guidelines for "how much evidence is enough." We might say that $P < 0.10$ indicates "some evidence" against the null hypothesis, $P < 0.05$ is "moderate evidence," and $P < 0.01$ is "strong evidence." Don't take these guidelines too literally, however. I will say more about interpreting tests in Chapter 23.

Calculating *P*-values*

Finding the P-values we gave in Examples 1 and 2 requires doing normal distribution calculations using Table B of normal percentiles. That was optional reading in Chapter 13 (pages 253–254). In practice, software does the calculation for us, but here is an example that shows how to use Table B.

Example 3. Tasting coffee

The hypotheses. In Example 1, we want to test the hypotheses

$$H_0: p = 0.5$$
$$H_a: p > 0.5$$

Here p is the proportion of the population of all coffee-drinkers who prefer fresh coffee to instant coffee.

The sampling distribution. If the null hypothesis is true, so that $p = 0.5$ we saw in Example 1 that $\hat{p}$ follows a normal distribution with mean 0.5 and standard deviation 0.0707.

The data. A sample of 50 people found that 36 preferred fresh coffee. The sample proportion is $\hat{p} = 0.72$.

The *P*-value. The alternative hypothesis is one-sided on the high side. So the P-value is the probability of getting an outcome at least as large as 0.72. Figure 22.2 displays this probability as an area under the normal sampling distribution

*This section is optional.

curve. To find any normal curve probability, move to the standard scale. The standard score for the outcome $\hat{p} = 0.72$ is

$$\text{standard score} = \frac{\text{observation} - \text{mean}}{\text{standard deviation}}$$

$$= \frac{0.72 - 0.5}{0.0707} = 3.1$$

Table B says that standard score 3.1 is the 99.9 percentile of a normal distribution. That is, the area under a normal curve to the left of 3.1 (in the standard scale) is 0.999. The area to the right is therefore 0.001, and that is our *P*-value.

Conclusion. The small *P*-value means that the data provide very strong evidence that a majority of the population prefers fresh coffee.

Exploring the Web

Confidence intervals and statistical significance appear constantly in reports of studies in many fields. Most scientific journals now have Web sites that display at least summaries of articles appearing in the journals. For example, the case study that opens Chapter 21 comes from *Circulation* (circ.ahajournals.org). The opening case in this chapter is from the *American Journal of Sociology* (www.journals.uchicago.edu/AJS/home.html).

Choose a major journal in your field of study. Use a Web search engine to find its Web site—just search on the journal's name. Look at the summaries of recent papers. If your field does studies that produce data, you will surely find phrases like "95% confidence" and "significant ($P = 0.01$)."

Statistics in Summary A confidence interval estimates an unknown parameter. A **test of significance** assesses the evidence for some claim about the value of an unknown parameter. In practice, the purpose of a statistical test is to answer the question "Could the effect we see in the sample just be an accident due to chance, or is it good evidence that the effect is really there in the population?"

Significance tests answer this question by giving the probability that a sample effect as large as the one we see in this sample would arise just by chance. This probability is the ***P*-value**. A small *P*-value says that our outcome is unlikely to happen just by chance. To set up a test, state a **null hypothesis** that says the effect you seek is *not* present in the population. The **alternative hypothesis** says that the effect *is* present. The *P*-value is the probability, calculated taking the null hypothesis to be true, of an outcome as extreme in the direction specified by the alternative hypothesis as the actually observed outcome. A sample result is **statistically significant at the 5% level** if it would occur just by chance no more than 5% of the time in repeated samples.

This chapter concerns the basic reasoning of tests and the details of tests for hypotheses about a population proportion. There is more discussion of the practical interpretation of statistical tests in Chapter 23.

CHAPTER 22 EXERCISES

22.1 Ethnocentrism. A social psychologist reports, "In our sample, ethnocentrism was significantly higher ($P < 0.05$) among church attenders than among nonattenders." Explain to someone who knows no statistics what this means.

22.2 Students' earnings. The financial aid office of a university asks a sample of students about their employment and earnings. The report says, "For academic year earnings, a significant difference ($P = 0.038$) was found between the sexes, with men earning more on the average. No difference ($P = 0.476$) was found between the earnings of black and white students." Explain both these conclusions, for the effects of sex and of race on mean earnings, in language understandable to someone who knows no statistics.

22.3 Diet and diabetes. Does eating more fiber reduce the blood cholesterol levels of patients with diabetes? A randomized clinical trial compared normal and high-fiber diets. Here is part of the researchers' conclusion: "The high-fiber diet reduced plasma total cholesterol concentrations by 6.7 percent ($P = 0.02$), triglyceride concentrations by 10.2 percent ($P = 0.02$), and very-low-density lipoprotein cholesterol concentrations by 12.5 percent ($P = 0.01$)." A doctor who knows no statistics says that a drop of 6.7% in cholesterol isn't a lot—maybe it's just an accident due to the chance assignment of patients to the two diets. Explain in simple language how "$P = 0.02$" answers this objection.

22.4 Diet and bowel cancer. It has long been thought that eating a low-fat, high-fiber diet reduces the risk of bowel cancer. A large study cast doubt on this advice. The subjects were 2079 people who had polyps removed from their bowels in the past six months. Such polyps may lead to cancer. The subjects were randomly assigned to a low-fat, high-fiber diet or to a control group in which subjects ate their usual diets. Did polyps reoccur during the next four years?

(a) Outline the design of this experiment.

(b) Surprisingly, the occurrence of new polyps "did not differ significantly between the two groups." Explain clearly what this finding means.

22.5 Pigs and prestige in ancient China. It appears that pigs in Stone Age China were not just a source of food. Owning pigs was also a display of wealth. Evidence for this comes from examining burial sites. If the skulls of sacrificed pigs tend to appear along with expensive ornaments, that suggests that the pigs, like the ornaments, signal the wealth and prestige of the person buried. A study of burials from around 3500 B.C. concluded, "there are striking differences in grave goods between burials with pig skulls and burials without them. . . . A test indicates that the two samples of total artifacts are significantly different at the 0.01 level." Explain clearly why "significantly different at the 0.01 level" gives good reason to think that there really is a systematic difference between burials that contain pig skulls and those that lack them.

22.6 Ancient Egypt. Settlements in Egypt before the time of the pharaohs are dated by measuring the presence of forms of carbon that decay over time. The first datings of settlements in the Nagada region used hair that had been excavated 60 years earlier. Now researchers have used newer methods and more recently excavated material. Do the dates differ? Here is the conclusion about one location: "There are two dates from Site KH6. Statistically, the two dates are not significantly different. They provide a weighted average corrected date of 3715 ± 90 B.C." Explain to someone interested in ancient Egypt but not interested in statistics what "not significantly different" means.

22.7 What's a gift worth? Do people value gifts from others more highly than they value the money it would take to buy the gift? We would like to think so, because we hope that "the thought counts." A survey of 209 adults asked them to list three recent gifts and then asked, "Aside from any sentimental value, if, without the giver ever knowing, you could receive an amount of money instead of the gift, what is the minimum amount of money that would make you equally happy?" It turned out that most people would need more money than the gift cost to be equally happy. The magic words "significant ($P < 0.01$)" appear in the report of this finding.

(a) The sample consisted of students and staff in a graduate program and of "members of the general public at train stations and airports in Boston and Philadelphia." The report says this sample is "not ideal." What's wrong with the sample?

(b) In simple language, what does it mean to say that the sample thought their gifts were worth "significantly more" than their actual cost?

(c) Now be more specific: what does "significant ($P < 0.01$)" mean?

22.8 Attending church. Although opinion polls have long found that about 40% of American adults say they attended religious services last week, this is almost certainly not true.

(a) Why might we expect answers to a poll to overstate true church attendance?

(b) You suspect strongly that the true percent attending church in any given week is less than 40%. You plan to watch a random sample of adults and see whether or not they go to church. What are your null and alternative hypotheses? (Be sure to say in words what the population proportion p is for your study.)

22.9 Body temperature. We have all heard that 98.6 degrees Fahrenheit (or 37 degrees Celsius) is "normal body temperature." In fact, there is evidence that most people have a slightly lower body temperature. You plan to measure the body temperature of a random sample of people very accurately. You hope to show that a majority have temperatures lower than 98.6 degrees.

(a) Say clearly what the population proportion p stands for in this setting.

(b) In terms of p, what are your null and alternative hypotheses?

22.10 Unemployment. The national unemployment rate last month was 4.3%. You think the rate may be different in your city, so you plan a sample survey that will ask the same questions as the Current Population Survey. To see if the local rate differs significantly from 4.3%, what hypotheses will you test?

22.11 First-year students. You read that 21% of all first-year college students identify themselves as politically liberal. You wonder if this percentage is different at your school, but you have no idea whether it is higher or lower. You plan a sample survey of first-year students at your school. What hypotheses will you test to see if your school differs significantly from the national result?

22.12 Do our athletes graduate? The National Collegiate Athletic Association (NCAA) requires colleges to report the graduation rates of their athletes. At one large university, 70.7% of all students who entered between 1989 and 1991 graduated within six years. Ninety-five of the 147 students who entered with athletic scholarships graduated. Consider these 95 as a sample of all athletes who will be admitted under present policies. Is there evidence that the percent of athletes who graduate is less than 70.7%?

(a) Explain in words what the parameter p is in this setting.

(b) What are the null and alternative hypotheses H_0 and H_a?

(c) What is the numerical value of the sample proportion $\hat{p}$? Of what event is the P-value the probability?

(d) The P-value is $P = 0.053$. Explain why this says there is some reason to think that graduation rates are lower among athletes than among all students.

22.13 We want to be rich. In a recent year, 73% of first-year college students responding to a national survey identified "being very well off financially" as an important personal goal. A state university finds that 132

of an SRS of 200 of its first-year students say that this goal is important. We wonder if the proportion of first-year students at this university who think being very well off is important differs from the national value, 73%.
(a) Explain in words what the parameter p is in this setting.

(b) What are the null and alternative hypotheses H_0 and H_a?

(c) What is the numerical value of the sample proportion $\hat{p}$? Of what event is the P-value the probability?

(d) The P-value is $P = 0.026$. Explain carefully why this is reasonably good evidence that H_0 is not true and that H_a is true.

22.14 Do our athletes graduate? Is the result of Exercise 22.12 statistically significant at the 10% level? At the 5% level?

22.15 We want to be rich. Is the result of Exercise 22.13 statistically significant at the 5% level? At the 1% level?

22.16 Significant at what level? Explain in plain language why a result that is significant at the 1% level must always be significant at the 5% level. If a result is significant at the 5% level, what can you say about its significance at the 1% level?

22.17 Significance means what? Asked to explain the meaning of "statistically significant at the $\alpha = 0.05$ level," a student says: "This means that the probability that the null hypothesis is true is less than 0.05." Is this explanation correct? Why or why not?

22.18 Finding a *P*-value by simulation. Is a new method of teaching reading to first graders (Method B) more effective than the method now in use (Method A)? You design a matched pairs experiment to answer this question. You form 20 pairs of first graders, with the two children in each pair carefully matched in IQ, socioeconomic status, and reading-readiness score. You assign at random one student from each pair to Method A. The other student in the pair is taught by Method B. At the end of first grade, all the children take a test of reading skill. Let p stand for the proportion of all possible matched pairs of children for which the child taught by Method B will have the higher score. Your hypotheses are

$$H_0 : p = 0.5 \quad \text{(no difference in effectiveness)}$$
$$H_a : p > 0.5 \quad \text{(Method B is more effective)}$$

The result of your experiment is that Method B gave the higher score in 12 of the 20 pairs, or $\hat{p} = 12/20 = 0.6$.
(a) If H_0 is true, the 20 pairs of students are 20 independent trials with probability 0.5 that Method B wins each trial. Explain how to use Table A to simulate these 20 trials if we assume for the sake of argument that H_0 is true.

(b) Use Table A, starting at line 105, to simulate 10 repetitions of the experiment. Estimate from your simulation the probability that Method B will do better in 12 or more of the 20 pairs when H_0 is true. (Of course, 10 repetitions are not enough to estimate the probability reliably. Once you see the idea, more repetitions are easy.)

(c) Explain why the probability you simulated in (b) is the *P*-value for your experiment. With enough patience, you could find all the *P*-values in this section by doing simulations similar to this one.

22.19 Finding a *P*-value by simulation. A classic experiment to detect extra-sensory perception (ESP) uses a shuffled deck of cards containing five suits (waves, stars, circles, squares, and crosses). As the experimenter turns over each card and concentrates on it, the subject guesses the suit of the card. A subject who lacks ESP has probability 1/5 of being right by luck on each guess. A subject who has ESP will be right more often. Julie is right in 5 of 10 tries. (Actual experiments use much longer series of guesses so that weak ESP can be spotted. No one has ever been right half the time in a long experiment!)
(a) Give H_0 and H_a for a test to see if this result is significant evidence that Julie has ESP.

(b) Explain how to simulate the experiment if we assume for the sake of argument that H_0 is true.

(c) Simulate 20 repetitions of the experiment; begin at line 121 of Table A.

(d) The actual experimental result was 5 right in 10 tries. What is the event whose probability is the *P*-value for this experimental result? Give an estimate of the *P*-value based on your simulation. How convincing was Julie's performance?

The following exercises concern the optional section on calculating P-values. To carry out a test, complete the steps (hypotheses, sampling distribution, data, picture, and calculation) illustrated in Example 3.

22.20 We want to be rich. Return to the study in Exercise 22.13, which found that 132 of 200 entering students thought "being very well off financially" was important. What do you conclude about the statistical significance of these data?

22.21 Teens and their TV sets. The *New York Times* and CBS News conducted a nationwide poll of 1048 randomly selected 13- to 17-year-olds. Of these teenagers, 692 had a television in their room. We can act as if the sample were an SRS. Is there good evidence that more than half of all teenagers have TVs in their rooms?

22.22 Side effects. An experiment on the side effects of pain relievers assigned arthritis patients to one of several over-the-counter pain medications.

Of the 440 patients who took one brand of pain reliever, 23 suffered some "adverse symptom."

(a) If 10% of all patients suffer adverse symptoms, what would be the sampling distribution of the proportion with adverse symptoms in a sample of 440 patients?

(b) Does the experiment provide strong evidence that fewer than 10% of patients who take this medication have adverse symptoms?

22.23 Do chemists have more girls? Some people think that chemists are more likely than other parents to have female children. (Perhaps chemists are exposed to something in their laboratories that affects the sex of their children.) The Washington State Department of Health lists the parents' occupations on birth certificates. Between 1980 and 1990, 555 children were born to fathers who were chemists. Of these births, 273 were girls. During this period, 48.8% of all births in Washington State were girls. Is there evidence that the proportion of girls born to chemists is higher than the state proportion?

22.24 Speeding. It often appears that most drivers on the road are driving faster than the posted speed limit. Situations differ, of course, but here is one set of data. Researchers studied the behavior of drivers on a rural interstate highway in Maryland where the speed limit was 55 miles per hour. They measured speed with an electronic device hidden in the pavement and, to eliminate large trucks, considered only vehicles less than 20 feet long. They found that 5690 out of 12,931 vehicles were exceeding the speed limit. Is this good evidence that (at least in this location) less than half of all drivers are speeding?

Chapter 23
Use and Abuse of Statistical Inference

I have perfect hindsight

Inference is subtle, so mistakes in inference are also often a bit subtle. Subtle mistakes are less important than glaring, in-your-face blunders, so don't forget that most really big statistical mistakes involve things like voluntary response samples, ignoring lurking variables, and invalid measures. Nonetheless, inference also allows room for mistakes and misunderstandings. This is particularly true of statistical significance.

Let's look at the 6700 mutual funds on sale to the investing public. Any Internet investment site worth clicking on will tell you which fund produced the highest return over the past (say) three years. As the year 2000 opened, one site claimed that the winner is the Kinetics Internet Fund. If I had bought this fund three years ago, I would have gained 112% per year. Comparing this return with the average for all funds, we find that it is significantly higher.

It should be clear that "look back and take the best" isn't a suitable foundation for a significance test. Significance tests work when we form a hypothesis such as "Kinetics Internet will have higher-than-average returns" and then wait for data. It makes no sense to look at past data, take the fund that happened to be the best out of 6700 funds, and ask if this fund was above average.

It does make sense to ask if funds that do better than average in one period tend to stay better than average. If they do, it would make sense to buy funds that have done well in the past. Professors of finance have devoted lots of computer time and lots of significance tests to this question. The answer seems to be that the worst funds do tend to stay bad (until they disappear), but that there is no statistically significant evidence for persistent good performance. In fact, Kinetics Internet Fund was in the *bottom* 25 out of 6700 mutual funds in the first half of 2000.

Using inference wisely

We have met the two major types of statistical inference: confidence intervals and significance tests. We have, however, seen only one inference method of each type, designed for inference about a population proportion p. There are libraries of both books and software filled with methods for inference about various parameters in various settings. The reasoning of confidence intervals and significance tests remains the same, but the details can seem overwhelming. The first step in using inference wisely is to understand your data and the questions you want to answer and fit the method to its setting. Here are some tips on inference, adapted to the one setting we are familiar with.

The design of the data production matters. "Where do the data come from?" remains the first question to ask in any statistical study. Any inference method is intended for use in a specific setting. For our confidence interval and test for a proportion p:

- The data must be a simple random sample (SRS) from the population of interest. When you use these methods, you are acting as if the data are an SRS. In practice, it is often not possible to actually choose an SRS from the population. Your conclusions may then be open to challenge.
- These methods are not correct for sample designs more complex than an SRS, such as stratified samples. There are other methods that fit these settings.
- There is no correct method for inference from data haphazardly collected with bias of unknown size. Fancy formulas cannot rescue badly produced data.

- Other sources of error, such as dropouts and nonresponse, are important. Remember that confidence intervals and tests use the data you give them and ignore these practical difficulties.

Know how confidence intervals behave. A confidence interval estimates the unknown value of a parameter and also tells us how uncertain the estimate is. All confidence intervals share these behaviors:

- The confidence level says how often the *method* catches the true parameter in very many uses. We never know whether this specific data set gives us an interval that contains the true parameter. All we can say is, "I got this result from a method that works 95% of the time." This data set might be one of the 5% that produce an interval that misses the parameter. If that risk bothers you, use a 99% confidence interval.
- High confidence is not free. A 99% confidence interval will be wider than a 95% confidence interval based on the same data. There is a trade-off between how closely we can pin down the parameter and how confident we are that we have caught the parameter.
- Larger samples give shorter intervals. If we want high confidence *and* a short interval, we must take a larger sample. The length of our confidence interval for p goes down in proportion to the square root of the sample size. To cut the interval in half, we must take four times as many observations. This is typical of many types of confidence intervals.

Dropping out

An experiment found that weight loss is significantly more effective than exercise for reducing high cholesterol and high blood pressure. The 170 subjects were randomly assigned to a weight-loss program, an exercise program, or a control group. Only 111 of the 170 subjects completed their assigned treatment, and the analysis used data from these 111. Did the dropouts create bias? Always ask about details of the data before trusting inference.

Know what statistical significance says. Many statistical studies hope to show that some claim is true. A clinical trial compares a new drug with a standard drug because the doctors hope that patients given the new drug will do better. A psychologist studying gender differences suspects that women will do better than men (on the average) on a test that measures social networking skills. The purpose of significance tests is to weigh the evidence that the data give in favor of such claims. That is, a test helps us know if we found what we were looking for.

To do this, we ask what would happen if the claim were *not* true. That's the null hypothesis—no difference between the two drugs, no difference between women and men. A significance test answers only one question:

"How strong is the evidence that the null hypothesis is not true?" A test answers this question by giving a *P*-value. The *P*-value tells us how unlikely our data would be if the null hypothesis were true. Data that are very unlikely are good evidence that the null hypothesis is not true. We never know whether the hypothesis is true for this specific population. All we can say is, "Data like these would occur only 5% of the time if the hypothesis were true."

This kind of indirect evidence against the null hypothesis (and for the effect we hope to find) is less straightforward than a confidence interval. I will say more about tests in the next section.

Know what your methods require. Our test and confidence interval for a proportion p require that the population be much larger than the sample. They also require that the sample itself be reasonably large so that the sampling distribution of the sample proportion $\hat{p}$ is close to normal. I have said little about the specifics of these requirements because the reasoning of inference is more important. Just as there are inference methods that fit stratified samples, there are methods that fit small samples and small populations. If you plan to use statistical inference in practice, you will need help from a statistician (or need to learn lots more statistics) to manage the details.

Most of us read about statistical studies more often than we actually work with data ourselves. Concentrate on the big issues, not on the details of whether the authors used exactly the right inference methods. Does the study ask the right questions? Where did the data come from? Do the results make sense? Does the study report confidence intervals so you can see both the estimated values of important parameters and how uncertain the estimates are? Does it report *P*-values to help convince you that findings are not just good luck?

The woes of significance tests

The purpose of a significance test is usually to give evidence for the presence of some effect in the population. The effect might be a probability of heads different from one-half for a coin or a longer mean survival time for patients given a new cancer treatment. If the effect is large, it will show up in most samples—the proportion of heads among our tosses will be far from one-half, or the patients who get the new treatment will live much longer than those in the control group. Small effects, such as a probability of heads only slightly different from one-half, will often be hidden behind the chance variation in a sample. This is as it should be: big effects are easier to detect. That is, the *P*-value will usually be small when the population truth is far from the null hypothesis.

The "woes" of testing start with the fact that a test just measures the strength of evidence against the null hypothesis. It says nothing about how big or how important the effect we seek in the population really is. For example, our hypothesis might be "This coin is balanced." We express this hypothesis in terms of the probability p of getting a head as H_0: $p = 1/2$. No real coin is exactly balanced, so we know that this hypothesis is not exactly true. If this coin has probability $p = 0.502$ of a head, we might say that for practical purposes it is balanced. A statistical test doesn't think about "practical purposes." It just asks if there is evidence that p is not exactly equal to 0.5. The focus of tests on the strength of the evidence against an exact null hypothesis is the source of much confusion in using tests.

Pay particular attention to the size of the sample when you read the result of a significance test. Here's why:

- Larger samples make tests of significance more sensitive. If we toss a coin hundreds of thousands of times, a test of H_0: $p = 0.5$ will often give a very low P-value when the truth for this coin is $p = 0.502$. The test is right—it found good evidence that p really is not exactly equal to 0.5—but it has picked up a difference so small that it is of no practical interest. **A finding can be statistically significant without being practically important.**

- On the other hand, tests of significance based on small samples are often not sensitive. If you toss a coin only 10 times, a test of H_0: $p = 0.5$ will often give a large P-value even if the truth for this coin is $p = 0.7$. Again the test is right—10 tosses are not enough to give good evidence against the null hypothesis. **Lack of significance does not mean that there is no effect, only that we do not have good evidence for an effect. Small samples often miss effects that are really present in the population.**

Whatever the truth about the population, whether $p = 0.7$ or $p = 0.502$ more observations allow us to pin down p more closely. If p is not 0.5, more observations will give more evidence of this, that is, a smaller P-value. Because significance depends strongly on the sample size as well as on the truth about the population, statistical significance tells us nothing about how large or how practically important an effect is. Large effects (like $p = 0.7$ when the null hypothesis is $p = 0.5$) often give data that are insignificant if we take only a small sample. Small effects (like $p = 0.502$) often give data that are highly significant if we take a large sample. Let's return to a favorite example to see how significance changes with sample size.

Example 1. Count Buffon's coin again

Count Buffon tossed a coin 4040 times and got 2048 heads. His sample proportion of heads was

$$\hat{p} = \frac{2048}{4040} = 0.507$$

Is the count's coin balanced? The hypotheses are

$$H_0 : p = 0.5$$
$$H_a : p \neq 0.5$$

The test of significance works by locating the sample outcome $\hat{p} = 0.507$ on the sampling distribution that describes how $\hat{p}$ would vary if the null hypothesis were true. Figure 23.1 repeats Figure 22.3. It shows that the observed $\hat{p} = 0.507$ is not surprisingly far from 0.5 and therefore is not good evidence against the hypothesis that the true p is 0.5. The *P*-value, which is 0.37, just makes this precise.

Suppose that Count Buffon got the *same result*, $\hat{p} = 0.507$, from tossing a coin 1000 times and also from tossing a coin 100,000 times. The sampling distribution of $\hat{p}$ when the null hypothesis is true always has mean 0.5, but its standard deviation gets smaller as the sample size n gets larger. Figure 23.2 displays the three sampling distributions, for $n = 1000$, and $n = 4040$, and $n = 100{,}000$ The middle curve in this figure is the same normal curve as in Figure 23.1, drawn on a scale that allows us to show the very tall and narrow curve for $n = 100{,}000$ Locating the sample outcome $\hat{p} = 0.507$ on the three curves, you see that the same outcome is more or less surprising depending on the size of the sample.

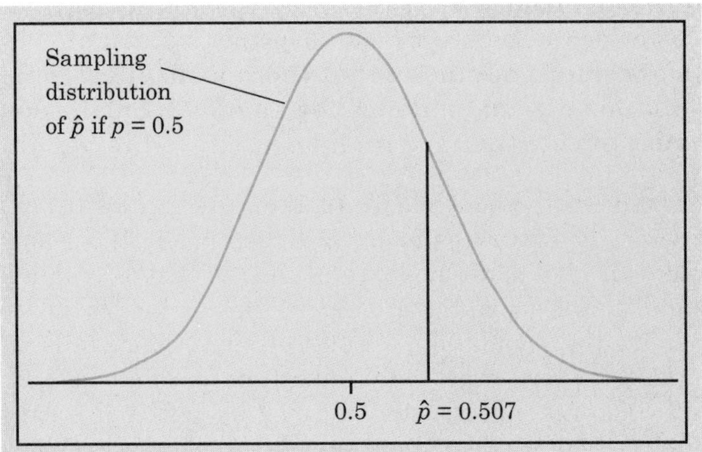

Figure 23.1 The sampling distribution of the proportion of heads in 4040 tosses of a coin if in fact the coin is balanced. Sample proportion 0.507 is not an unusual outcome.

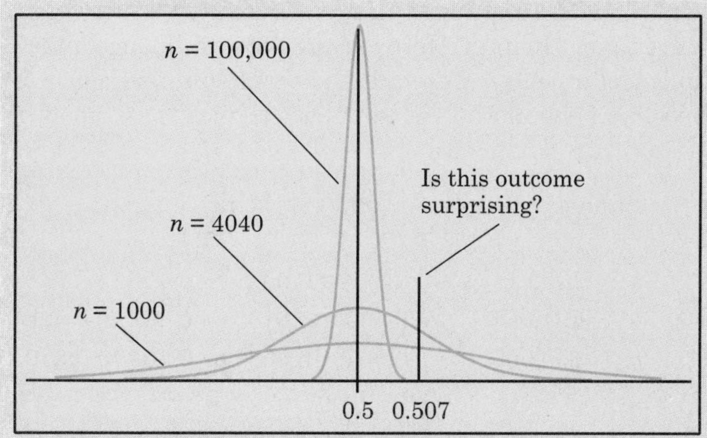

Figure 23.2 The three sampling distributions of the proportion of heads in 1000, 4040, and 100,000 tosses of a balanced coin. Sample proportion 0.507 is not unusual in 1000 or 4040 tosses but is very unusual in 100,000 tosses.

The *P*-values are $P = 0.66$ for $n = 1000$, $P = 0.37$ for $n = 4040$, and $P = 0.000009$ for $n = 100{,}000$. Imagine tossing a balanced coin 1000 times repeatedly. You will get a proportion of heads at least as far from one-half as Buffon's 0.507 in about two-thirds of your repetitions. If you toss a balanced coin 100,000 times, however, you will almost never (9 times in a million repeats) get an outcome this unbalanced.

The outcome $\hat{p} = 0.507$ is not evidence against the hypothesis that the coin is balanced if it comes up in 1000 tosses or in 4040 tosses. It is completely convincing evidence if it comes up in 100,000 tosses.

Beware the naked *P*-value

The *P*-value of a significance test depends strongly on the size of the sample, as well as on the truth about the population.

It is bad practice to report a *P*-value without also giving the sample size and a statistic or statistics that describe the sample outcome.

The advantages of confidence intervals

Example 1 suggests that we not rely on significance alone in understanding a statistical study. Just knowing that the sample proportion was $\hat{p} = 0.507$ helps a lot. You can decide whether this deviation from one-half is large enough to interest you. Of course, $\hat{p} = 0.507$ isn't the exact truth about the coin, just the chance result of the count's tosses. So a confidence interval, whose width shows how closely we can pin down the truth about the coin, is even more helpful. Here are the 95% confidence intervals for the true probability of a head p, based on the three sample sizes in Example 1. You can check that the method of Chapter 21 gives these answers.

Number of tosses	95% confidence interval
$n = 1000$	0.507 ± 0.031, or 0.476 to 0.538
$n = 4040$	0.507 ± 0.015, or 0.492 to 0.522
$n = 100{,}000$	0.507 ± 0.003, or 0.504 to 0.510

The confidence intervals make clear what we know (with 95% confidence) about the true p. The intervals for 1000 and 4040 tosses include 0.5, so we are not confident the coin is unbalanced. For 100,000 tosses, however, we are confident that the true p lies between 0.504 and 0.510. In particular, we are confident that it is not 0.5.

Give a confidence interval

Confidence intervals are more informative than tests because they actually estimate a population parameter. They are also easier to interpret. It is good practice to give confidence intervals whenever possible.

Significance at the 5% level isn't magical

The purpose of a test of significance is to describe the degree of evidence provided by the sample against the null hypothesis. The P-value does this. But how small a P-value is convincing evidence against the null hypothesis? This depends mainly on two circumstances:

- *How plausible is H_0?* If H_0 represents an assumption that the people you must convince have believed for years, strong evidence (small P) will be needed to persuade them.
- *What are the consequences of rejecting H_0?* If rejecting H_0 in favor of H_a means making an expensive changeover from one type of product packaging to another, you need strong evidence that the new packaging will boost sales.

These criteria are a bit subjective. Different people will often insist on different levels of significance. Giving the P-value allows each of us to decide individually if the evidence is sufficiently strong.

Statistical Controversies

Should Significance Tests Be Banned?

Research studies in many fields rely on tests of significance. Often habit leads to over-reliance. Robert Rosenthal of Harvard, an eminent psychologist well known for statistical work, says, "Many of us were trained that we're not supposed to look too carefully at the data. You come up with a hypothesis, decide on a statistical test, do the test, and if your results are significant at .05, you've supported your hypothesis. If not, you stick it all in a drawer and never look at your data."

That should shock you. "Always plot your data" has been one of our mottoes, along with "Always ask where the data come from." Psychologists generally think carefully about how their data are produced. How can it be that many psychologists hardly glance at the data? The tyranny of significance tests and of 0.05 as the magical sign that a result is important, say some psychologists. In particular, custom dictates that results should be significant at the 5% level in order to be published, so researchers fall into the bad habits that Rosenthal describes. Significant at 5%, good. Not significant at 5%, failure. The limitations of tests are so severe, the risks of misinterpretation so high, and bad habits so ingrained, say these critics, that significance tests should be banned from professional journals in psychology.

In response, the American Psychological Association appointed a Task Force on Statistical Inference. Robert Rosenthal was one of the cochairs of this group. The Task Force did not want to ban tests. Its report was in fact a summary of good statistical practice. Define your population clearly. Describe your data production and prefer randomized methods whenever possible. Describe your variables and how they were measured. Give your sample size and explain how you decided on the sample size. If there were dropouts or other practical problems, mention them. "As soon as you have collected your data, before you compute *any* statistics, *look at your data.*" Ask whether the results of computations make sense to you. Recognize that "inferring causality from nonrandomized designs is a risky enterprise."

There is more, but here is the punch line about statistical tests: "It is hard to imagine a situation in which a dichotomous accept-reject decision is better than reporting an actual p value or, better still, a confidence interval. . . . Always provide some effect-size estimate when reporting a p value." Ban tests? "Although this might eliminate some abuses, the committee thought there were enough counterexamples to justify forbearance."

Users of statistics have often emphasized standard levels of significance such as 10%, 5%, and 1%. For example, courts have tended to accept 5% as a standard in discrimination cases. This emphasis reflects the time when tables of critical values rather than computer software dominated statistical practice. The 5% level ($\alpha = 0.05$) is particularly common. **There is no sharp border between "significant" and "insignificant," only increasingly strong evidence as the *P*-value decreases.** There is no practical distinction between the *P*-values 0.049 and 0.051. It makes no sense to treat $P \leq 0.05$ as a universal rule for what is significant.

Beware of searching for significance

Statistical significance ought to mean that you have found an effect that you were looking for. The reasoning behind statistical significance works well if you decide what effect you are seeking, design a study to search for it, and use a test of significance to weigh the evidence you get. In other settings, significance may have little meaning. We saw one example at the beginning of this chapter: it makes no sense to look at 6700 mutual funds, find the one that happened to have the highest return, then ask if this fund "significantly" outperformed the average. Here is another example.

Example 2. Predicting success of trainees

You want to learn what distinguishes managerial trainees who eventually become executives from those who, after expensive training, don't succeed and leave the company. You have abundant data on past trainees—data on their personalities and goals, their college preparation and performance, even their family backgrounds and their hobbies. Statistical software makes it easy to perform dozens of significance tests on these dozens of variables to see which ones best predict later success. Aha! You find that future executives are significantly more likely than washouts to have an urban or suburban upbringing and an undergraduate degree in a technical field.

Before you base future recruiting on these findings, recall that results significant at the 5% level occur 5 times in 100 in the long run even when H_0 is true. When you make dozens of tests at the 5% level, you expect a few of them to be significant by chance alone. Running one test and reaching the $\alpha = 0.05$ level is reasonably good evidence that you have found something. Running several dozen tests and reaching that level once or twice is not.

In the mutual fund example, we looked for the best, then tested it as if we had not first sought it out. In Example 2, we just tested everything and took the most significant. Both bad practices confuse the roles of exploratory analysis of data and formal statistical inference.

Searching data for suggestive patterns is certainly legitimate. Exploratory data analysis is an important part of statistics. But the reasoning of formal inference does not apply when your search for a striking effect in the data is successful. The remedy is clear. Once you have a hypothesis, design a study to search specifically for the effect you now think is there. If the result of this study is statistically significant, you have real evidence.

Exploring the Web

The report of the American Psychological Association's Task Force on Statistical Inference is an excellent brief introduction to wise use of inference. The report appeared in the journal *American Psychologist* in 1999. You can find it on the Web in the list of "Selected Articles" from this journal at www.apa.org/journals/amp.html.

Statistics in Summary Statistical inference is less widely applicable than exploratory analysis of data. Any inference method requires the right setting, in particular the right design for a random sample or randomized experiment. Understanding the meaning of confidence levels and statistical significance helps avoid improper conclusions. Increasing the number of observations has a straightforward effect on confidence intervals—the interval gets shorter for the same level of confidence. More observations usually drive down the *P*-value of a test when the truth about the population stays the same, making tests harder to interpret than confidence intervals. A finding with a small *P*-value may not be practically interesting if the sample is large, and an important truth about the population may fail to be significant if the sample is small. Avoid depending on fixed significance levels such as 5% to make decisions.

CHAPTER 23 EXERCISES

23.1 A television poll. A television news program conducts a call-in poll about a proposed city ban on handgun ownership. Of the 2372 calls, 1921 oppose the ban. The station, following recommended practice, makes a confidence statement: "81% of the Channel 13 Pulse Poll sample opposed the ban. We can be 95% confident that the true proportion of citizens opposing a handgun ban is within 1.6% of the sample result." The confidence interval calculation is correct, but the conclusion is not justified. Why?

23.2 Ages of presidents. Joe is writing a report on the backgrounds of American presidents. He looks up the ages of all 43 presidents when they entered office. Because Joe took a statistics course, he uses these 43 numbers to get a 95% confidence interval for the mean age of all men who have been president. This makes no sense. Why not?

23.3 Who will win? A poll taken shortly before an election finds that 52% of the voters favor candidate Shrub over candidate Snort. The poll has a margin of sampling error of plus or minus three percentage points at 95% confidence. The poll press release says the election is too close to call. Why?

23.4 How far do rich parents take us? How much education children get is strongly associated with the wealth and social status of their parents. In social science jargon, this is "socioeconomic status," or SES. But the SES of parents has little influence on whether children who have graduated from college go on to yet more education. One study looked at whether college graduates took the graduate admissions tests for business, law, and other graduate programs. The effects of the parents' SES on taking the LSAT test for law school were "both statistically insignificant and small."
(a) What does "statistically insignificant" mean?
(b) Why is it important that the effects were small in size as well as insignificant?

23.5 Searching for ESP. A researcher looking for evidence of extrasensory perception (ESP) tests 500 subjects. Four of these subjects do significantly better ($P < 0.01$) than random guessing.
(a) Is it proper to conclude that these four people have ESP? Explain your answer.
(b) What should the researcher now do to test whether any of these four subjects have ESP?

23.6 Comparing package designs. A company compares two package designs for a laundry detergent by placing bottles with both designs on the shelves of several markets. Checkout scanner data on more than 5000 bottles bought show that more shoppers bought Design A than Design B. The difference is statistically significant ($P = 0.02$). Can we conclude that consumers strongly prefer Design A? Explain your answer.

23.7 Color blindness in Africa. An anthropologist suspects that color blindness is less common in societies that live by hunting and gathering than in settled agricultural societies. He tests a number of adults in two populations in Africa, one of each type. The proportion of color-blind people is significantly lower ($P < 0.05$) in the hunter-gatherer population. What additional information would you want to help you decide whether you accept the claim about color blindness?

23.8 Blood types in Southeast Asia. One way to assess whether two human groups should be considered separate populations is to compare their distributions of blood types. An anthropologist finds significantly different ($P = 0.01$) proportions of the main human blood types (A, B, AB, O) in different tribes in central Malaysia. What other information would you want before you agree that these tribes are separate populations?

23.9 Why we seek significance. Asked why statistical significance appears so often in research reports, a student says, "Because saying that results are significant tells us that they cannot easily be explained by chance variation alone." Do you think that this statement is essentially correct? Explain your answer.

23.10 What is significance good for? Which of the following questions does a test of significance answer?
(a) Is the sample or experiment properly designed?

(b) Is the observed effect due to chance?

(c) Is the observed effect important?

23.11 What distinguishes schizophrenics? Psychologists once measured 77 variables on a sample of schizophrenic people and a sample of people who were not schizophrenic. They compared the two samples using 77 separate significance tests. Two of these tests were significant at the 5% level. Suppose that there is in fact no difference on any of the 77 variables between people who are and people who are not schizophrenic in the adult population. That is, all 77 null hypotheses are true.
(a) What is the probability that one specific test shows a difference significant at the 5% level?

(b) Why is it not surprising that 2 of the 77 tests were significant at the 5% level?

23.12 Why are larger samples better? Statisticians prefer large samples. Describe briefly the effect of increasing the size of a sample (or the number of subjects in an experiment) on each of the following:
(a) The margin of error of a 95% confidence interval.

(b) The *P*-value of a test, when H_0 is false and all facts about the population remain unchanged as n increases.

23.13 Is this convincing? You are planning to test a new vaccine for a virus that now has no vaccine. Since the disease is usually not serious, you will expose 100 volunteers to the virus. After some time, you will record whether or not each volunteer has been infected.
(a) Explain how you would use these 100 volunteers in a designed experiment to test the vaccine. Include all important details of designing the experiment (but don't actually do any random allocation).

(b) You hope to show that the vaccine is more effective than a placebo. State H_0 and H_a. (Notice that this test compares *two* population proportions.)

(c) The experiment gave a *P*-value of 0.25. Explain carefully what this means.

(d) Your fellow researchers do not consider this evidence strong enough to recommend regular use of the vaccine. Do you agree?

The following exercises require carrying out the methods described in the optional sections of Chapters 21 and 22.

23.14 Do our athletes graduate? Return to the study in Exercise 22.12 (page 447), which found that 95 of 147 athletes admitted to a large university graduated within six years. The proportion of athletes who graduated was significantly lower ($P = 0.053$) than the 70.7% graduation rate for all students. It may be more informative to give a 95% confidence interval for the graduation rate of athletes. Do this.

23.15 We want to be rich. Return to the study in Exercise 22.13 (page 447), which found that 132 of 200 entering students at a state university thought "being very well-off financially" was important. This differed significantly ($P = 0.026$) from the 73% of all first-year students in the country who feel this way. It may be more informative to give a 95% confidence interval for the proportion of this university's entering students who consider being rich an important goal. Do this.

23.16 Is it significant? Over several years and many thousands of students, 89% of the high school students in a large city have passed the competency test that is one of the requirements for a diploma. Now reformers claim that a new mathematics curriculum will increase the percent who pass. A random sample of 1000 students follow the new curriculum. The school board wants to see an improvement that is statistically significant at the 5% level before it will adopt the new program for all students. If p is the proportion of all students who would pass the exam if they followed the new curriculum, we must test

$$H_0: p = 0.89$$
$$H_a: p > 0.89$$

(a) Suppose that 906 of the 1000 students in the sample pass the test. Show that this is *not* significant at the 5% level. (Follow the method of Example 3 in Chapter 22.)

(b) Suppose that 907 of the 1000 students pass. Show that this *is* significant at the 5% level.

(c) Is there a practical difference between 906 successes in 1000 tries and 907 successes? What can you conclude about the importance of significance at the 5% level?

23.17 We like confidence intervals. The previous exercise compared significance tests about the proportion p of all students who would pass a competency test, based on data showing that either 906 or 907 of an SRS of 1000 students passed. Give the 95% confidence interval for p in both cases. The intervals make clear how uncertain we are about the true value of p and how little difference there is between the two sample outcomes.

Chapter 24
Two-Way Tables and the Chi-Square Test*

Female college professors

Purdue, where I teach, is a Big Ten university that emphasizes engineering, scientific, and technical fields. In the 1998–1999 academic year, Purdue had 1621 professors, of whom 335 were women. That's just over 20%, or one out of every five professors. These numbers don't tell us much about the place of women on the faculty. As usual, we must look at relationships among several variables, not just at gender alone. For example, female faculty are more common in the humanities than in agriculture.

Let's look at the relationship between gender and a variable particularly important to faculty members, academic rank. Professors typically start as assistant professors, are promoted to associate professor (and gain tenure then), and finally reach the rank of full professor. Universities tend to be run by full professors. Here is a *two-way table* that breaks down Purdue's 1621 faculty members by both gender and academic rank:

	Female	Male	Total
Assistant professors	126	213	339
Associate professors	149	411	560
Professors	60	662	722
Total	335	1286	1621

*This more advanced chapter is optional.

The table makes the place of women on the faculty much clearer. The number of men goes up as we climb the ladder of ranks, and the number of women goes down. Rates speak more clearly than counts, so let's calculate some percentages. More than 37% of the assistant professors are women, but women make up only about 27% of the associate professors and only about 8% of the full professors. Women are strikingly underrepresented in the highest rank.

The table reports the facts but does not explain them. It typically takes about 6 years as an assistant professor before promotion to associate professor and still more time to become a full professor. It may be that many women joined the faculty only in the past decade and are still in the lower ranks because they are young. Or, more seriously, it may be that women have a harder time gaining promotion. More data are needed to say which explanation fits.

Two-way tables

The rank and gender of college faculty are both *categorical variables*. That is, they place individuals into categories but do not have numerical values that allow us to describe relationships by scatterplots, correlation, or regression lines. To display relationships between two categorical variables, use a **two-way table** like the table of rank and gender of Purdue faculty. Rank is the **row variable** because each row in the table describes faculty in one rank. Gender is the **column variable** because each column describes one gender. The entries in the table are the counts of faculty in each rank-by-gender class. Although both rank and gender are categorical variables, rank has a natural order from lowest to highest. The order of the rows in the table reflects the order of the categories.

How can we best grasp the information contained in this table? First, *look at the distribution of each variable separately*. The distribution of a categorical variable says how often each outcome occurred. The "Total" column at the right of the table contains the totals for each of the rows. These row totals give the distribution of rank for all faculty, men and women combined. The "Total" row at the bottom of the table gives the distribution of gender for faculty, all ranks combined. It is often clearer to present these distributions using percents. We might report the distribution of gender as

$$\text{percent female} = \frac{335}{1621} = 0.207 = 20.7\%$$

$$\text{percent male} = \frac{1286}{1621} = 0.793 = 79.3\%$$

The two-way table contains more information than the two distributions of rank alone and gender alone. The nature of the relationship between rank and gender cannot be deduced from the separate distributions but requires the full table. **To describe relationships among categorical variables, calculate appropriate percents from the counts given.**

Example 1. Rank and gender on the faculty

Because there are only two genders, we can see the relationship between gender and academic rank by comparing the percents of women in the three ranks:

Assistant professors	Associate professors	Full professors
$\frac{126}{339} = 37.2\%$	$\frac{149}{560} = 26.6\%$	$\frac{60}{722} = 8.3\%$

The percentages make it clear that women are less common in the higher ranks. That's the nature of the association between gender and rank.

In working with two-way tables, you must calculate lots of percents. Here's a tip to help decide what fraction gives the percent you want. Ask, "What group represents the total that I want a percent of?" The count for that group is the denominator of the fraction that leads to the percent. In Example 1, we wanted the percent *of each rank* who are women, so the counts in the ranks form the denominators.

Simpson's paradox

As is the case with quantitative variables, the effects of lurking variables can change or even reverse relationships between two categorical variables. Let's continue the theme of gender and higher education in looking at an example. The numbers here are artificial for simplicity, but they illustrate a phenomenon that often appears in real data.

Example 2. Discrimination in admissions?

A university offers only two degree programs, one in electrical engineering and one in English. Admission to these programs is competitive, and the women's caucus suspects discrimination against women in the admissions process. The caucus obtains the following data from the university, a two-way table of all applicants by gender and admission decision:

	Male	Female
Admit	35	20
Deny	45	40
Total	80	60

These data do show an association between the gender of applicants and their success in obtaining admission. To describe this association more precisely, we compute some percents from the data.

$$\text{percent of male applicants admitted} = \frac{35}{80} = 44\%$$

$$\text{percent of female applicants admitted} = \frac{20}{60} = 33\%$$

Aha! Almost half of the males but only one-third of the females who applied were admitted.

The university replies that although the observed association is correct, it is not due to discrimination. In its defense, the university produces a **three-way table** that classifies applicants by sex, admission decision, and the program to which they applied. We present a three-way table as several two-way tables side by side, one for each value of the third variable. In this case there are two two-way tables, one for each program:

	Engineering		English	
	Male	Female	Male	Female
Admit	30	10	5	10
Deny	30	10	15	30
Total	60	20	20	40

Check that these entries add to the entries in the two-way table. The university has simply broken down that table by department. We now see that engineering admitted exactly half of all applicants, both male and female, and that English admitted one-fourth of both males and females. There is *no association* between sex and admission decision in either program.

How can no association in either program produce strong association when the two are combined? Look at the data: English is hard to get into, and mainly females apply to that program. Electrical engineering is easier to get into and attracts mainly male applicants. English had 40 female and 20 male applicants, while engineering had 60 male and only 20 female applicants. The original two-way table, which did not take account of the difference between programs, was misleading. This is an example of *Simpson's paradox*.

Simpson's paradox

An association or comparison that holds for all of several groups can disappear or even reverse direction when the data are combined to form a single group.

Simpson's paradox is just an extreme form of the fact that observed associations can be misleading when there are lurking variables. Remember the caution from Chapter 15: *Beware the lurking variable.*

The Inventor of Simpson's Paradox at work.

"No, I am **NOT** Homer Simpson."

Example 3. Discrimination in mortgage lending?

Studies of applications for home mortgage loans from banks show a strong racial pattern: banks reject a higher percentage of black applicants than of white applicants. One lawsuit in the Washington, D.C., area contends that a bank rejected 17.5% of blacks but only 3.3% of whites.

The bank replies that lurking variables explain the difference in rejection rates. Blacks have (on the average) lower incomes, poorer credit records, and less secure jobs than whites. Unlike race, these are legitimate reasons to turn down a mortgage application. It is because these lurking variables are confounded with race, the bank says, that it rejects a higher percentage of black applicants. It is even possible, thinking of Simpson's paradox, that the bank accepts a *higher* percentage of black applicants than of white applicants if we look at people with the same income and credit record.

Who is right? Both sides will hire statisticians to examine the effects of the lurking variables. The court will eventually decide.

Inference for a two-way table

We often gather data and arrange them in a two-way table to see if two categorical variables are related to each other. The sample data are easy to investigate: turn them into percentages and look for an association between the row and column variables. Is the association in the sample evidence of an association between these variables in the entire population? Or could the sample association easily arise just from the luck of random sampling? This is a question for a significance test.

Example 4. Treating cocaine addiction

Cocaine addicts need the drug to feel pleasure. Perhaps giving them a medication that fights depression will help them stay off cocaine. A three-year study compared an antidepressant called desipramine with lithium (a standard treatment for cocaine addiction) and a placebo. The subjects were 72 chronic users of cocaine who wanted to break their drug habit. Twenty-four of the subjects were randomly assigned to each treatment. Here are the counts and percentages of the subjects who succeeded in staying off cocaine during the study:

Group	Treatment	Subjects	Successes	Percent
1	Desipramine	24	14	58.3%
2	Lithium	24	6	25.0%
3	Placebo	24	4	16.7%

The sample proportions of subjects who stayed off cocaine are quite different. In particular, desipramine was much more successful than lithium or a placebo. The bar graph in Figure 24.1 compares the results visually. Are these data good evidence that there is a relationship between treatment and outcome in the population of all cocaine addicts?

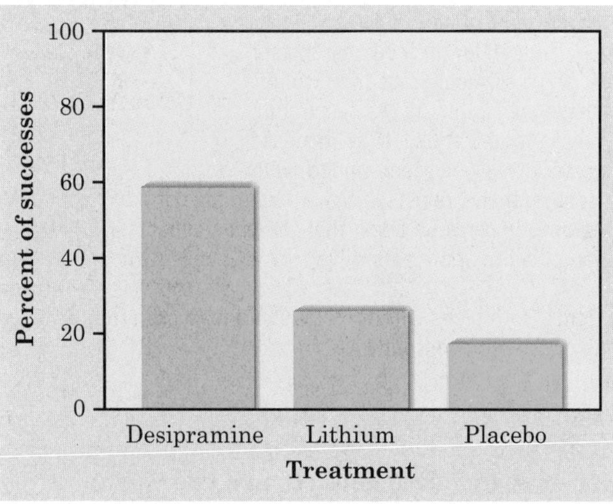

Figure 24.1 Bar graph comparing the success rates of three treatments for cocaine addiction.

The test that answers this question starts with a two-way table. Here's the table for the data of Example 4:

	Success	Failure	Total
Desipramine	14	10	24
Lithium	6	18	24
Placebo	4	20	24
Total	24	48	72

Our null hypothesis, as usual, says that the treatments have no effect. That is, addicts do equally well on any of the three treatments. The differences in the sample are just the play of chance. Our null hypothesis is

H_0: There is no association between treatment and success in the population of all cocaine addicts.

Expressing this hypothesis in terms of population parameters can be a bit messy, so we will be content with the verbal statement. The alternative hypothesis just says, "Yes, there is some association between the treatment an addict receives and whether or not he succeeds in staying off cocaine." The alternative doesn't specify the nature of the relationship. It doesn't say, for example, "Addicts who take desipramine are more likely to succeed than addicts given lithium or a placebo."

To test H_0, we compare the observed counts in a two-way table with the *expected counts*, the counts we would expect—except for random variation—if H_0 were true. If the observed counts are far from the expected counts, that is evidence against H_0. We can guess the expected counts for the cocaine study. In all, 24 of the 72 subjects succeeded. That's overall success rate one-third, because 24/72 is one-third. If the null hypothesis is true, there is no difference among the treatments. So we expect one-third of the subjects in each group to succeed. There were 24 subjects in each group, so we expect 8 successes and 16 failures in each group. If the treatment groups differ in size, the expected counts will differ also, even though we still expect the same proportion in each group to succeed. Fortunately, there is a rule that makes it easy to find expected counts. Here it is.

More tests

There are also tests for hypotheses more specific than "no relationship." Place people in classes by social status, wait ten years, then classify the same people again. The row and column variables are the classes at the two times. We might test the hypothesis that there has been no change in the overall distribution of social status. Or we might ask if moves up in status are balanced by matching moves down. There are statistical tests for these and other hypotheses.

Expected counts

The **expected count** in any cell of a two-way table when H_0 is true is

$$\text{expected count} = \frac{\text{row total} \times \text{column total}}{\text{table total}}$$

Try it. For example, the expected count of successes in the desipramine group is

$$\begin{aligned}\text{expected count} &= \frac{\text{row 1 total} \times \text{column 1 total}}{\text{table total}} \\ &= \frac{(24)(24)}{72} = 8\end{aligned}$$

If the null hypothesis of no treatment differences is true, we expect 8 of the 24 desipramine subjects to succeed. That's just what we guessed.

The chi-square test

To see if the data give evidence against the null hypothesis of "no relationship," compare the counts in the two-way table with the counts we would expect if there really were no relationship. If the observed counts are far from the expected counts, that's the evidence we were seeking. The test uses a statistic that measures how far apart the observed and expected counts are.

Chi-square statistic

The **chi-square statistic** is a measure of how far the observed counts in a two-way table are from the expected counts. The formula for the statistic is

$$X^2 = \sum \frac{(\text{observed count} - \text{expected count})^2}{\text{expected count}}$$

The symbol Σ means "sum over all cells in the table."

The chi-square statistic is a sum of terms, one for each cell in the table. In the cocaine example, 14 of the desipramine group succeeded. The expected count for this cell is 8. So the term in the chi-square statistic from this cell is

$$\begin{aligned}\frac{(\text{observed count} - \text{expected count})^2}{\text{expected count}} &= \frac{(14 - 8)^2}{8} \\ &= \frac{36}{8} = 4.5\end{aligned}$$

Example 5. The cocaine study

Here are the observed and expected counts for the cocaine study side by side:

	Observed		Expected	
	Success	Failure	Success	Failure
Desipramine	14	10	8	16
Lithium	6	18	8	16
Placebo	4	20	8	16

We can now find the chi-square statistic, adding 6 terms for the 6 cells in the two-way table:

$$\begin{aligned} X^2 &= \frac{(14-8)^2}{8} + \frac{(10-16)^2}{16} + \frac{(6-8)^2}{8} \\ &\quad + \frac{(18-16)^2}{16} + \frac{(4-8)^2}{8} + \frac{(20-16)^2}{16} \\ &= 4.50 + 2.25 + 0.50 + 0.25 + 2.00 + 1.00 = 10.50 \end{aligned}$$

Because X^2 measures how far the observed counts are from what would be expected if H_0 were true, large values are evidence against H_0. Is $X^2 = 10.5$ a large value? You know the drill: compare the observed value 10.5 against the *sampling distribution* that shows how X^2 would vary if the null hypothesis were true. This sampling distribution is *not* a normal distribution. It is a right-skewed distribution that allows only positive values because X^2 can never be negative. Moreover, the sampling distribution is different for two-way tables of different sizes. Here are the facts.

The chi-square distributions

The sampling distribution of the chi-square statistic X^2 when the null hypothesis of no association is true is called a **chi-square distribution**.

The chi-square distributions are a family of distributions that take only positive values and are skewed to the right. A specific chi-square distribution is specified by giving its **degrees of freedom**.

The chi-square test for a two-way table with r rows and c columns uses critical values from the chi-square distribution with $(r-1)(c-1)$ degrees of freedom.

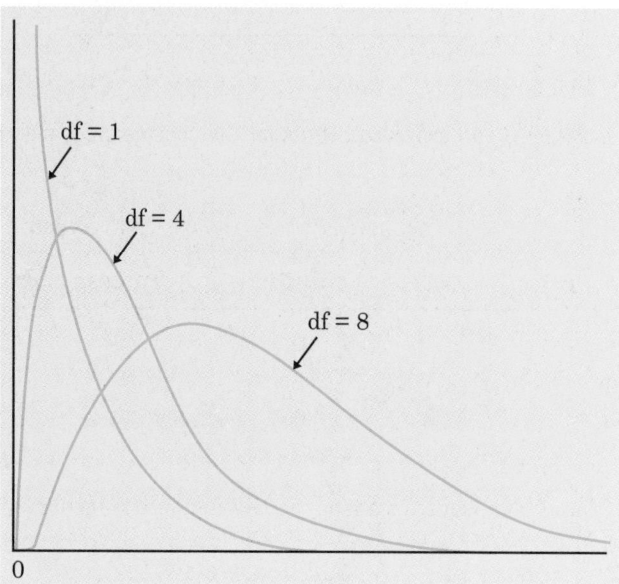

Figure 24.2 The density curves for three members of the chi-square family of distributions. The sampling distributions of chi-square statistics belong to this family.

Figure 24.2 shows the density curves for three members of the chi-square family of distributions. As the degrees of freedom (df) increase, the density curves become less skewed and larger values become more probable. We can't find *P*-values as areas under a chi-square curve by hand, though software can do it for us. Table 24.1 is a shortcut. It shows how large the chi-square statistic X^2 must be in order to be significant at various levels. This isn't as good as an actual *P*-value, but it is often good enough. Each number of degrees of freedom has a separate row in the table. We see, for example, that a chi-square statistic with 3 degrees of freedom is significant at the 5% level if it is greater than 7.81 and is significant at the 1% level if it is greater than 11.34.

Example 6. The cocaine study, conclusion

We have seen that desipramine produced markedly more successes and fewer failures than lithium or a placebo. Comparing observed and expected counts gave the chi-square statistic $X^2 = 10.5$. The last step is to assess significance.

The two-way table of 3 treatments by 2 outcomes for the cocaine study has 3 rows and 2 columns. That is, $r = 3$ and $c = 2$ The chi-square statistic therefore has degrees of freedom

$$(r - 1)(c - 1) = (3 - 1)(2 - 1) = (2)(1) = 2$$

Table 24.1 To be significant at level α, a chi-square statistic must be larger than the table entry for α

	Significance level α						
df	0.25	0.20	0.15	0.10	0.05	0.01	0.001
1	1.32	1.64	2.07	2.71	3.84	6.63	10.83
2	2.77	3.22	3.79	4.61	5.99	9.21	13.82
3	4.11	4.64	5.32	6.25	7.81	11.34	16.27
4	5.39	5.99	6.74	7.78	9.49	13.28	18.47
5	6.63	7.29	8.12	9.24	11.07	15.09	20.51
6	7.84	8.56	9.45	10.64	12.59	16.81	22.46
7	9.04	9.80	10.75	12.02	14.07	18.48	24.32
8	10.22	11.03	12.03	13.36	15.51	20.09	26.12
9	11.39	12.24	13.29	14.68	16.92	21.67	27.88

Look in the df = 2 row of Table 24.1. We see that $X^2 = 10.5$ is larger than the critical value 9.21 required for significance at the $\alpha = 0.01$ level but smaller than the critical value 13.82 for $\alpha = 0.001$. The cocaine study shows a significant relationship ($P < 0.01$) between treatment and success.

The significance test only says we have strong evidence of *some* association between treatment and success. We must look at the two-way table to see the nature of the relationship: desipramine works better than the other treatments.

Using the chi-square test

Like our test for a population proportion, the chi-square uses some approximations that become more accurate as we take more observations. Here is a rough rule for when it is safe to use this test.

Cell counts required for the chi-square test

You can safely use the chi-square test when no more than 20% of the expected counts are less than 5 and all individual expected counts are 1 or greater.

The cocaine study easily passes this test: all the expected cell counts are either 8 or 16. Here is a concluding example that outlines the examination of a two-way table.

Example 7. Do angry people have more heart disease?

People who get angry easily tend to have more heart disease. That's the conclusion of a study that followed a random sample of 12,986 people from three locations for about four years. All subjects were free of heart disease at the beginning of the study. The subjects took the Spielberger Trait Anger Scale, which measures how prone a person is to sudden anger. Here are data for the 8474 people in the sample who had normal blood pressure. CHD stands for "coronary heart disease." This includes people who had heart attacks and those who needed medical treatment for heart disease.

	Anger score		
	Low	Moderate	High
Sample size	3110	4731	633
CHD count	53	110	27
CHD percent	1.7%	2.3%	4.3%

There is a clear trend: as the anger score increases, so does the percent who suffer heart disease. Is this relationship between anger and heart disease statistically significant?

The first step is to write the data as a two-way table by adding the counts of subjects who did not suffer from heart disease. We also add the row and column totals, which we need to find the expected counts.

	Low anger	Moderate anger	High anger	Total
CHD	53	110	27	190
No CHD	3057	4621	606	8284
Total	3110	4731	633	8474

We can now follow the steps for a significance test, familiar from Chapter 22.

The hypotheses. The chi-square method tests these hypotheses:

H_0: no relationship between anger and CHD

H_a: some relationship between anger and CHD

The sampling distribution. We will see that all the expected cell counts are larger than 5, so we can safely apply the chi-square test. The two-way table of anger versus CHD has 2 rows and 3 columns. We will use critical values from the chi-square distribution with degrees of freedom df = (2 – 1)(3 – 1) = 2.

The data. First find the expected cell counts. For example, the expected count of high-anger people with CHD is

$$\text{expected count} = \frac{\text{row 1 total} \times \text{column 3 total}}{\text{table total}} = \frac{(190)(633)}{8474} = 14.19$$

Here is the complete table of observed and expected counts side by side:

	Observed			Expected		
	Low	Moderate	High	Low	Moderate	High
CHD	53	110	27	69.73	106.08	14.19
No CHD	3057	4621	606	3040.27	4624.92	618.81

Looking at these counts, we see that the high-anger group has more CHD than expected and the low-anger group has less CHD than expected. This is consistent with what the percents in Example 7 show. The chi-square statistic is

$$\begin{aligned} X^2 &= \frac{(53-69.73)^2}{69.73} + \frac{(110-106.08)^2}{106.08} + \frac{(27-14.19)^2}{14.19} \\ &+ \frac{(3057-3040.27)^2}{3040.27} + \frac{(4621-4624.92)^2}{4624.92} + \frac{(606-618.81)^2}{618.81} \\ &= 4.014 + 0.145 + 11.557 + 0.092 + 0.003 + 0.265 = 16.077 \end{aligned}$$

In practice, statistical software can do all this arithmetic for you. Look at the 6 terms that we sum to get X^2. Most of the total comes from just one cell: high-anger people have more CHD than expected.

Significance? Look at the df = 2 line of Table 24.1. The observed chi-square $X^2 = 16.077$ is larger than the critical value 13.82 for $\alpha = 0.001$. We have highly significant evidence ($P < 0.001$) that anger and heart disease are related. Statistical software can give the actual P-value. It is $P = 0.0003$

Can we conclude that proneness to anger *causes* heart disease? This is an observational study, not an experiment. It isn't surprising that some lurking variables are confounded with anger. For example, people prone to anger are more likely than others to be men who drink and smoke. The study report used advanced statistics to adjust for many differences among the three anger groups. The adjustments raised the P-value from $P = 0.0003$ to $P = 0.002$ because the lurking variables explain some of the heart disease. This is still good evidence for a relationship. Because the study started with a random sample of people who had no CHD and followed them forward in time, and because many lurking variables were measured and accounted for, it does give some evidence for causation. The next step might be an experiment that shows anger-prone people how to change. Will this reduce their risk of heart disease?

Statistics in Summary Categorical variables group individuals into classes. To display the relationship between two categorical variables, make a **two-way table** of counts in the classes. We describe the nature of an association between categorical variables by comparing selected percentages. As always, lurking variables can make an observed association misleading. In some cases, an association that holds for every level of a lurking variable disappears or changes direction when we lump all levels together. This is **Simpson's paradox.**

The **chi-square test** tells us whether an observed association in a two-way table is statistically significant. The **chi-square statistic** compares the counts in the table with the counts we would expect if there were no association between the row and column variables. The sampling distribution is not normal. It is a new distribution, the **chi-square distribution**.

CHAPTER 24 EXERCISES

24.1 Extracurricular activities and grades. North Carolina State University studied student performance in a course required by its chemical engineering major. One question of interest is the relationship between time spent in extracurricular activities and whether a student earned a C or better in the course. Here are the data for the 119 students who answered a question about extracurricular activities:

	Extracurricular activities (hours per week)		
	<2	2 to 12	>12
C or better	11	68	3
D or F	9	23	5

Calculate percents that describe the nature of the relationship between time spent on extracurricular activities and performance in the course. Give a brief summary in words.

24.2 Smoking by students and their parents. How are the smoking habits of students related to their parents' smoking? Here is a two-way table from a survey of students in eight Arizona high schools:

	Student smokes	Student does not smoke
Both parents smoke	400	1380
One parent smokes	416	1823
Neither parent smokes	188	1168

Write a brief answer to the question posed, including a comparison of selected percents.

24.3 Python eggs. How is the hatching of water python eggs influenced by the temperature of the snake's nest? Researchers assigned newly laid eggs to one of three temperatures: hot, neutral, or cold. Hot duplicates the extra warmth provided by the mother python, and cold duplicates the absence of the mother. Here are the data on the number of eggs and the number that hatched:

	Eggs	Hatched
Cold	27	16
Neutral	56	38
Hot	104	75

(a) Make a two-way table of temperature by outcome (hatched or not).

(b) Calculate the percent of eggs in each group that hatched. The researchers anticipated that eggs would not hatch in cold water. Do the data support that anticipation?

24.4 Firearm deaths. Firearms are second to motor vehicles as a cause of nondisease deaths in the United States. Here are counts from a study of all firearm-related deaths in Milwaukee, Wisconsin, between 1990 and 1994. We want to compare the types of firearms used in homicides and in suicides. We suspect that long guns (shotguns and rifles) will more often be used in suicides because many people keep them at home for hunting. Make a careful comparison of homicides and suicides, with a bar graph. What do you find about long guns versus handguns?

	Handgun	Shotgun	Rifle	Unknown	Total
Homicides	468	28	15	13	524
Suicides	124	22	24	5	175

24.5 Who earns academic degrees? How do women and men compare in the pursuit of academic degrees? The table below presents counts (in thousands) from the *Statistical Abstract* of degrees earned in 1996 categorized by the level of the degree and the gender of the recipient.

	Bachelor's	Master's	Professional	Doctorate
Female	642	227	32	18
Male	522	179	45	27
Total	1165	406	77	45

(a) How many people earned bachelor's degrees? Why do you think the "Total" entry for bachelor's degrees is not quite equal to the sum for women and men?

(b) What percent of each level of degree is earned by women? Write a brief description of what the data show about the relationship between gender and degree level.

24.6 Majors for men and women in business. A study of the career plans of young women and men sent questionnaires to all 722 members of the senior class in the College of Business Administration at the University of Illinois. One question asked which major within the business program the student had chosen. Here are the data from the students who responded:

	Female	Male
Accounting	68	56
Administration	91	40
Economics	5	6
Finance	61	59

Describe the differences between the distributions of majors for women and men with percents, with a graph, and in words.

24.7 Totals aren't enough. Here are the row and column totals for a two-way table with two rows and two columns:

a	b	50
c	d	50
60	40	100

Find *two different* sets of counts a, b, c, and d for the body of the table that give these same totals. This shows that the relationship between two variables cannot be obtained from the two individual distributions of the variables.

24.8 Airline flight delays. Here are the numbers of flights on time and delayed for two airlines at five airports in one month. Overall on-time percentages for each airline are often reported in the news. The airport that flights serve is a lurking variable that can make such reports misleading.

	Alaska Airlines		America West	
	On time	Delayed	On time	Delayed
Los Angeles	497	62	694	117
Phoenix	221	12	4840	415
San Diego	212	20	383	65
San Francisco	503	102	320	129
Seattle	1841	305	201	61

(a) What percent of all Alaska Airlines flights were delayed? What percent of all America West flights were delayed? These are the numbers usually reported.

(b) Now find the percent of delayed flights for Alaska Airlines at each of the five airports. Do the same for America West.

(c) America West does worse at *every one* of the five airports, yet does better overall. That sounds impossible. Explain carefully, referring to the data, how this can happen. (The weather in Phoenix and Seattle lies behind this example of Simpson's paradox.)

24.9 Race and the death penalty. Whether a convicted murderer gets the death penalty seems to be influenced by the race of the victim. Here are data on 326 cases in which the defendant was convicted of murder:

White Defendant	White victim	Black victim
Death	19	0
Not	132	9

Black Defendant	White victim	Black victim
Death	11	6
Not	52	97

(a) Use these data to make a two-way table of defendant's race (white or black) versus death penalty (yes or no).

(b) Show that Simpson's paradox holds: a higher percent of white defendants are sentenced to death overall, but for both black and white victims a higher percent of black defendants are sentenced to death.

(c) Use the data to explain why the paradox holds in language that a judge could understand.

24.10 Majors for men and women in business. Exercise 24.6 gives the responses to questionnaires sent to all 722 members of a business school graduating class.

(a) Two of the observed cell counts are small. Do these data satisfy our guidelines for safe use of the chi-square test?

(b) Is there a statistically significant relationship between the gender and major of business students?

(c) What percent of the students did not respond to the questionnaire? The nonresponse weakens conclusions drawn from these data.

24.11 Extracurricular activities and grades. In Exercise 24.1, you described the relationship between extracurricular activities and success in a required course. Is the observed association between these variables statistically significant? To find out, proceed as follows.

(a) Add the row and column totals to the two-way table in Exercise 24.1 and find the expected cell counts. Which observed counts differ most from the expected counts?

(b) Find the chi-square statistic. Which cells contribute most to this statistic?

(c) What are the degrees of freedom? Use Table 24.1 to say how significant the chi-square test is. Write a brief conclusion for your study.

24.12 Smoking by students and their parents. In Exercise 24.2, you saw that there is an association between smoking by parents and smoking by their children. Children are more likely to smoke as their parents smoke more. We want to know whether this association is statistically significant.

(a) State the hypotheses for the chi-square test. What do you think the population is?

(b) Find the expected cell counts. Write a sentence that explains in simple language what "expected counts" are.

(c) Find the chi-square statistic and its degrees of freedom. What is your conclusion about significance?

24.13 Python eggs. Exercise 24.3 presents data on the hatching of python eggs in water at three different temperatures. Does temperature have a significant effect on hatching? Write a clear summary of your work and your conclusion.

24.14 Stress and heart attacks. You read a newspaper article that describes a study of whether stress management can help reduce heart attacks. The 107 subjects all had reduced blood flow to the heart and so were at risk of a heart attack. They were assigned at random to three groups. The article goes on to say:

> *One group took a four-month stress management program, another underwent a four-month exercise program and the third received usual heart care from their personal physicians.*
>
> *In the next three years, only three of the 33 people in the stress management group suffered "cardiac events," defined as a fatal or non-fatal heart attack or a surgical procedure such as a bypass or angioplasty. In the same period, seven of the 34 people in the exercise group and 12 out of the 40 patients in usual care suffered such events.*

(a) Use the information in the news article to make a two-way table that describes the study results.

(b) What are the success rates of the three treatments in avoiding cardiac events?

(c) Find the expected cell counts under the null hypothesis that there is no difference among the treatments. Verify that the expected counts meet our guideline for use of the chi-square test.

(d) Is there a significant difference among the success rates for the three treatments?

24.15 Standards for child care. Do unregulated providers of child care in their homes follow different health and safety practices in different cities? A study looked at people who regularly provided care for someone else's children in poor areas of three cities. The numbers who required medical releases from parents to allow medical care in an emergency were 42 of 73 providers in Newark, N.J., 29 of 101 in Camden, N.J., and 48 of 107 in South Chicago, Ill.

(a) Use the chi-square test to see if there are significant differences among the proportions of child-care providers who require medical releases in the three cities. What do you conclude?

(b) How should the data be produced in order for your test to be valid? (In fact, the samples came in part from asking parents who were subjects in another study who provided their child care. The author of the study wisely did not use a statistical test. He wrote: "Application of conventional statistical procedures appropriate for random samples may produce biased and misleading results.")

Chapter 25 Inference about a Population Mean*

Is 98.6 degrees normal?

We have all heard that 98.6 degrees Fahrenheit (that's 37.0 degrees Celsius) is "normal body temperature" as measured by a thermometer stuck under the tongue. Of course, your actual body temperature varies during the day. Body temperature is usually highest around 6 A.M. and lowest around 4 to 6 P.M. Body temperature also varies among people and is a bit higher in children. So that 98.6 degrees really refers to a *mean* temperature. Presumably the claim is that if we measured all healthy adults at many times during the day, the mean temperature would be 98.6 degrees.

Is this claim correct? Some enterprising doctors measured 148 healthy adults (aged 18 to 40 years) four times a day for three days. They claimed that the traditional value is wrong and that the correct mean is 98.2 degrees rather than 98.6 degrees. The old value really is old—it goes back to 1861. We might ask if new data, from just 148 subjects but with more accurate thermometers, justifies giving a new value. That's a question of inference about the mean of a population. A confidence interval can answer the question "What can we say about the true value of the mean?" A significance test can answer the question "Is the true mean different from 98.6 degrees?"

---

*This more advanced chapter requires the optional material in Chapter 21.

A computer will do the inference if we give it the data. We should also ask some questions that the computer won't ask for us. Why are we interested in the mean—don't we really want to know how high or low body temperature must be to signal that something is wrong? Is 98.2 versus 98.6 too small a difference to be medically important? Was the 1861 result rounded to the nearest degree (37 degrees Celsius), so that translating it to 98.6 degrees Fahrenheit claims more accuracy than is justified? These are good questions. They remind us even more than the new study that "98.6 degrees is normal" is too simple a claim to be much help to doctors or to patients.

Although the reasoning of confidence intervals and significance tests is ever the same, specific recipes vary greatly. The form of an inference procedure depends first on the parameter you want information about—a population proportion or mean or median or whatever. The second influence is the design of the sample or experiment. Estimating a population proportion from a stratified sample requires a different recipe than if the data come from an SRS. This chapter presents a confidence interval and a significance test for inference about a population mean when the data are an SRS from the population.

The sampling distribution of a sample mean

What is the mean number of hours your college's first-year students study each week? What was their mean grade point average in high school? We often want to estimate the mean of a population. To distinguish the population mean (a parameter) from the sample mean $\bar{x}$, we write the population mean as μ, the Greek letter mu. We use the mean $\bar{x}$ of an SRS to estimate the unknown mean μ of the population.

Like the sample proportion $\hat{p}$, the sample mean $\bar{x}$ from a large SRS has a sampling distribution that is close to normal. Because the sample mean of an SRS is an unbiased estimator of μ, the sampling distribution of $\bar{x}$ has μ as its mean. The standard deviation of $\bar{x}$ depends on the standard deviation of the population, which is usually written as σ (the Greek letter sigma). By mathematics we can discover the following facts.

Sampling distribution of a sample mean

Choose an SRS of size n from a population in which individuals have mean μ and standard deviation σ. Let $\bar{x}$ be the mean of the sample. Then:

- The sampling distribution of $\bar{x}$ is **approximately normal** when the sample size n is large.
- The **mean** of the sampling distribution is equal to μ.
- The **standard deviation** of the sampling distribution is $\sigma/\sqrt{n}$.

It isn't surprising that the values that $\bar{x}$ takes in many samples are centered at the true mean μ of the population. That's the lack of bias in random sampling once again. The other two facts about the sampling distribution make precise two very important properties of the sample mean $\bar{x}$:

- The mean of a number of observations is less variable than individual observations.
- The distribution of a mean of a number of observations is more normal than the distribution of individual observations.

Figure 25.1 illustrates the first of these properties. It compares the distribution of a single observation with the distribution of the mean $\bar{x}$ of 10 observations. Both have the same center, but the distribution of $\bar{x}$ is less

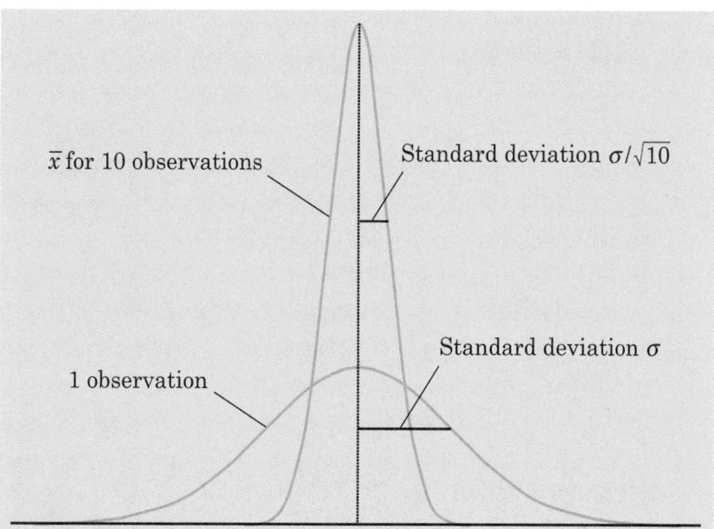

Figure 25.1 The sampling distribution of the sample mean $\bar{x}$ of 10 observations compared with the distribution of individual observations.

spread out. In Figure 25.1, the distribution of individual observations is normal. If that is true, then the sampling distribution of $\bar{x}$ is exactly normal for any size sample, not just approximately normal for large samples. A remarkable statistical fact, called the **central limit theorem**, says that as we take more and more observations at random from *any* population, the distribution of the mean of these observations eventually gets close to a normal distribution. (There are some technical qualifications to this big fact, but in practice we can ignore them.) The central limit theorem lies behind the use of normal sampling distributions for sample means.

Example 1. The central limit theorem in action

Figure 25.2 shows the central limit theorem in action. The top left density curve describes individual observations from a population. It is strongly right-skewed. Distributions like this describe the time it takes to repair a household appliance, for example. Most repairs are quickly done, but some are lengthy.

The other three density curves in Figure 25.2 show the sampling distributions of the sample means of 2, 10, and 25 observations from this population. As the sample size n increases, the shape becomes more normal. The mean remains fixed and the standard deviation decreases, following the pattern $\sigma/\sqrt{n}$. The distribution for 10 observations is still somewhat skewed to the right but already resembles a normal curve. The density curve for $n = 25$ is yet more normal. The contrast between the shapes of the population distribution and of the distribution of the mean of 10 or 25 observations is striking.

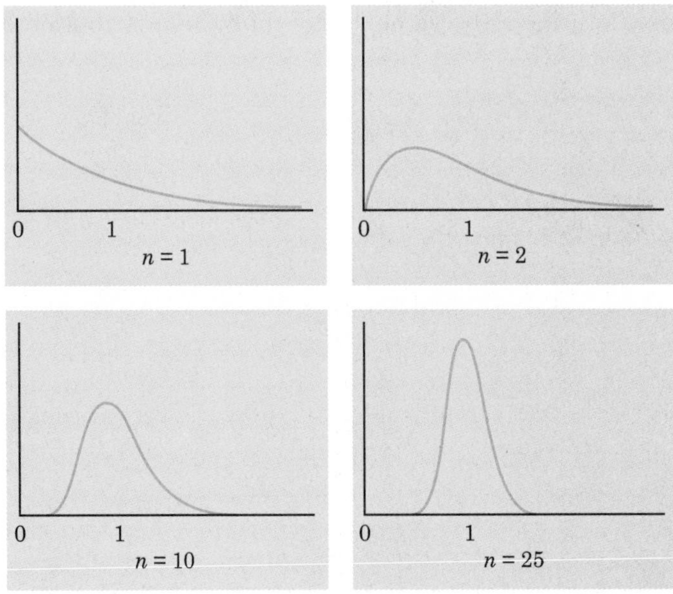

Figure 25.2 The distribution of a sample $\bar{x}$ becomes more normal as the size of the sample increases. The distribution of individual observations ($n = 1$) is far from normal. The distribution of means 2, 10, and finally 25 observations move closer to the normal shape.

Confidence intervals for a population mean

The standard deviation of $\bar{x}$ depends on both the sample size n and the standard deviation σ of individuals in the population. We know n but not σ. When n is large, the sample standard deviation s is close to σ and can be used to estimate it, just as we use the sample mean $\bar{x}$ to estimate the population mean μ. The estimated standard deviation of $\bar{x}$ is therefore $s/\sqrt{n}$. Now we can find confidence intervals for μ following the same reasoning that led us to confidence intervals for a proportion p in Chapter 21. The big idea is that to cover the central area C under a normal curve, we must go out a distance z^* on either side of the mean. Look again at Figure 21.5 (page 428) to see how C and z^* are related.

Confidence interval for a population mean

Choose an SRS of size n from a large population of individuals having mean μ. The mean of the sample observations is $\bar{x}$. When n is large, an approximate level C confidence interval for μ is

$$\bar{x} \pm z^* \frac{s}{\sqrt{n}}$$

where z^* is the critical value for confidence level C from Table 21.1 (page 429).

The cautions we noted in estimating p apply here as well. The recipe is valid only when an SRS is drawn and the sample size n is reasonably large. The margin of error again decreases only at a rate proportional to $\sqrt{n}$ as the sample size n increases. One additional caution: Remember that $\bar{x}$ and s are strongly influenced by outliers. Inference using $\bar{x}$ and s is suspect when outliers are present. Always look at your data.

Example 2. NAEP quantitative scores

The National Assessment of Educational Progress (NAEP) includes a short test of quantitative skills, covering mainly basic arithmetic and the ability to apply it to realistic problems. Scores on the test range from 0 to 500. For example, a person who scores 233 can add the amounts of two checks appearing on a bank deposit slip; someone scoring 325 can determine the price of a meal from a menu; a person scoring 375 can transform a price in cents per ounce into dollars per pound.

In a recent year, 840 men 21 to 25 years of age were in the NAEP sample. Their mean quantitative score was $\bar{x} = 272$ and the standard deviation of their scores was $s = 59$. These 840 men are a simple random sample from the population of all young men. On the basis of this sample, what can we say about the mean score μ in the population of all 9.5 million young men of these ages?

The 95% confidence interval for μ uses the critical value $z^* = 1.96$ from Table 21.1. The interval is

$$\bar{x} \pm z^* \frac{s}{\sqrt{n}} = 272 \pm 1.96 \frac{59}{\sqrt{840}}$$
$$= 272 \pm (1.96)(2.036) = 272 \pm 4.0$$

We are 95% confident that the mean for all young men lies between 268 and 276.

Catching cheaters

Lots of students take a long multiple-choice exam. Can the computer that scores the exam also screen for papers that are suspiciously similar? Clever people have created measures that take into account not just identical answers but the popularity of those answers and the total score on the similar papers. The measure has close to a normal distribution, and the computer flags pairs of papers with a measure outside ±4 standard deviations as significant.

Tests for a population mean

As with confidence intervals, the reasoning that leads to significance tests for hypotheses about a population mean μ follows the reasoning that led to tests about a population proportion p. The big idea is to use the sampling distribution that the sample mean $\bar{x}$ would have if the null hypothesis were true. Locate the $\bar{x}$ from your data on this distribution and see if it is unlikely. A value of $\bar{x}$ that would rarely appear if H_0 were true is evidence that H_0 is not true. The four steps are also similar to those in tests for a proportion. Here are two examples, the first one-sided and the second two-sided.

Example 3. Can you balance your checkbook?

In a discussion of the education level of the American workforce, a pessimist says, "The average young person can't even balance a checkbook." The NAEP survey says that a score of 275 or higher on its quantitative test reflects the skill needed to balance a checkbook. The NAEP random sample of 840 young men had mean score $\bar{x} = 272$, a bit below the checkbook-balancing level. Is this sample result good evidence that the mean for *all* young men is less than 275? The standard deviation of the scores in the sample was $s = 59$.

The hypotheses. The pessimist's claim is that the mean NAEP score is less than 275. That's our alternative hypothesis, the statement we seek evidence *for*. The hypotheses are

$$H_0 : \mu = 275$$
$$H_a : \mu < 275$$

The sampling distribution. *If the null hypothesis is true,* the sample mean $\bar{x}$ has approximately the normal distribution with mean $\mu = 275$ and standard deviation

$$\frac{s}{\sqrt{n}} = \frac{59}{\sqrt{840}} = 2.036$$

We once again use the sample standard deviation s in place of the unknown population standard deviation σ.

The data. The NAEP sample gave $\bar{x} = 272$.The standard score for this outcome is

$$\text{standard score} = \frac{\text{observation} - \text{mean}}{\text{standard deviation}}$$

$$= \frac{272 - 275}{2.036} = -1.47$$

That is, the sample result is about 1.47 standard deviations below the mean we would expect if on the average a young man had just enough skill to balance a checkbook.

The *P*-value. Figure 25.3 locates the sample outcome −1.47 (in the standard scale) on the normal curve that represents the sampling distribution if H_0 is true. This curve has mean 0 and standard deviation 1 because we are using the standard scale. The *P*-value for our one-sided test is the shaded area to the left of −1.47. To use Table B, round the standard score to −1.5. Table B says that −1.5 is the 6.68 percentile, so the area to its left is 0.0668. That is our *P*-value. (This result is approximate because we rounded the standard score to use Table B. Software gives $P = 0.071$.)

Conclusion. A *P*-value of about $P = 0.07$ suggests that the mean score for all young men is below the checkbook-balancing level, but it is not convincing evidence.

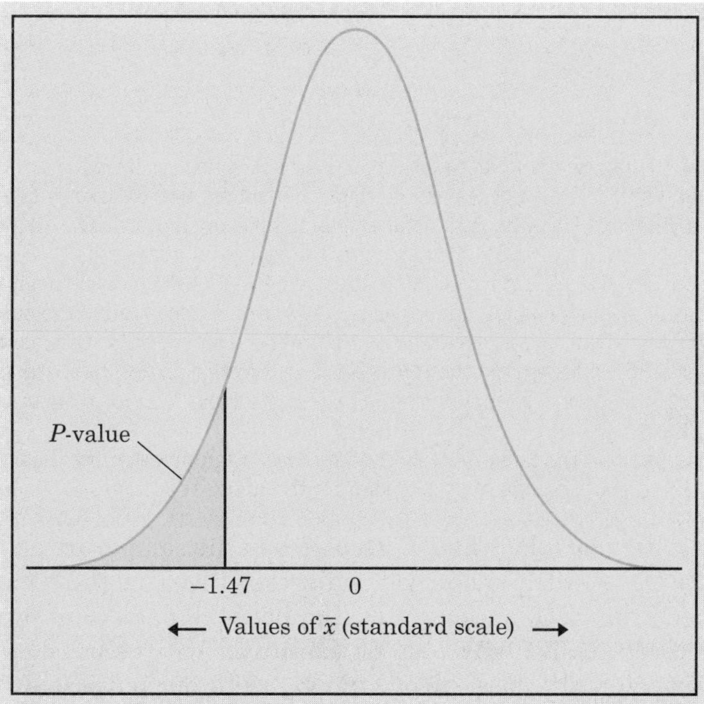

Figure 25.3 The *P*-value for a one-sided test when the standard score for the sample mean is −1.47.

Example 4. Executives' blood pressures

The National Center for Health Statistics reports that the mean systolic blood pressure for males 35 to 44 years of age is 128. The medical director of a large company looks at the medical records of 72 executives in this age group and finds that the mean systolic blood pressure in this sample is $\overline{x} = 126.1$ and that the standard deviation is $s = 15.2$ Is this evidence that the company's executives have a different mean blood pressure from the general population?

The hypotheses. The null hypothesis is "no difference" from the national mean. The alternative is two-sided, because the medical director did not have a particular direction in mind before examining the data. So the hypotheses about the unknown mean μ of the executive population are

$$H_0 : \mu = 128$$
$$H_a : \mu \neq 128$$

The sampling distribution. *If the null hypothesis is true*, the sample mean $\overline{x}$ has approximately the normal distribution with mean $\mu = 128$ and standard deviation

$$\frac{s}{\sqrt{n}} = \frac{15.2}{\sqrt{72}} = 1.79$$

The data. The sample mean is $\overline{x} = 126.1$ The standard score for this outcome is

$$\text{standard score} = \frac{\text{observation} - \text{mean}}{\text{standard deviation}}$$
$$= \frac{126.1 - 128}{1.79} = -1.06$$

We know that an outcome just more than 1 standard deviation away from the mean of a normal distribution is not very surprising. The last step is to make this formal.

The *P*-value. Figure 25.4 locates the sample outcome –1.06 (in the standard scale) on the normal curve that represents the sampling distribution if H_0 is true. The two-sided *P*-value is the probability of an outcome at least this far out *in either direction*. This is the shaded area under the curve. To use Table B, round the standard score to –1.1. This is the 13.57 percentile of a normal distribution. So the area to the left of –1.1 is 0.1357. The area to the left of –1.1 and to the right of 1.1 is double this, or about 0.27. This is our approximate *P*-value. (The exact *P*-value, from software, is $P = 0.289$.)

Conclusion. The large *P*-value gives no reason to think that the mean blood pressure of the executive population differs from the national average.

The test assumes that the 72 executives in the sample are an SRS from the population of all middle-aged male executives in the company. We should check this assumption by asking how the data were produced. If medical records are available only for executives with recent medical problems, for example, the data are of little value for our purpose. It turns out that all executives are given a free annual medical exam, and that the medical director selected 72 exam results at random.

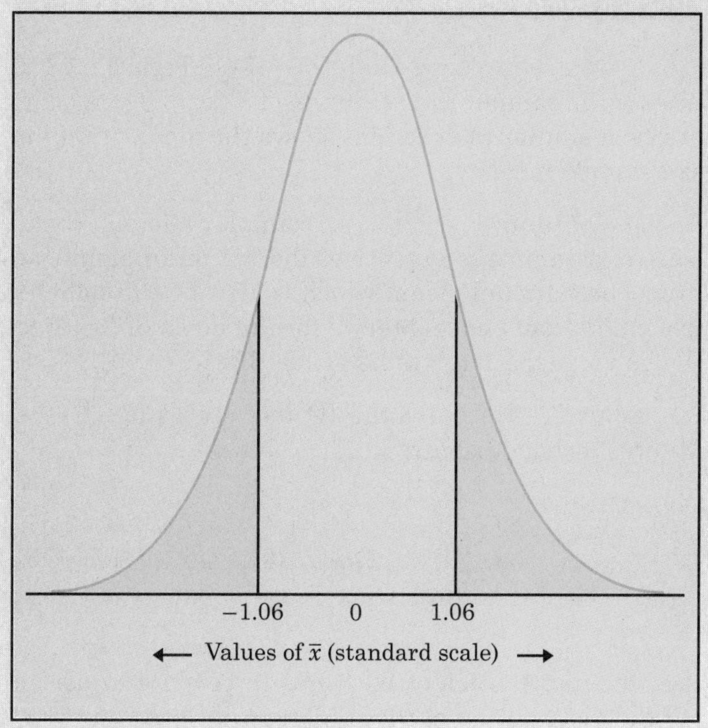

Figure 25.4 The *P*-value for a two-sided test when the standard score for the sample mean is −1.06.

The data in Example 4 do *not* establish that the mean blood pressure μ for this company's executives is 128. We sought evidence that μ differed from 128 and failed to find convincing evidence. That is all we can say. No doubt the mean blood pressure of the entire executive population is not exactly equal to 128. A large enough sample would give evidence of the difference, even if it is very small.

Statistics in Summary We estimate a population mean μ using the sample mean $\bar{x}$ of an SRS from the population. Confidence intervals and significance tests for μ are based on the **sampling distribution** of $\bar{x}$. When the sample size n is large, the **central limit theorem** says that this distribution is approximately normal. Although the details of the methods differ, inference about μ is quite similar to inference about a population proportion p because both are based on normal sampling distributions.

CHAPTER 25 EXERCISES

25.1 The idea of a sampling distribution. Figure 21.1 (page 422) shows the idea of the sampling distribution of a sample proportion $\hat{p}$ in picture form. Draw a similar picture that shows the idea of the sampling distribution of a sample mean $\bar{x}$.

25.2 Executives' blood pressure. Example 4 found that the mean blood pressure of a sample of executives did not differ significantly at the 10% level from the national mean, which is 128. Use the data in that example to give a 90% confidence interval for the mean of the executive population. Why do you expect 128 to be inside this interval?

25.3 IQ test scores. Here are the IQ test scores of 31 seventh-grade girls in a Midwest school district:

114	100	104	89	102	91	114	114	103	105	
108	130	120	132	111	128	118	119	86	72	
111	103	74	112	107	103	98	96	112	112	93

(a) We expect the distribution of IQ scores to be close to normal. Make a histogram of the distribution of these 31 scores. Does your plot show outliers, clear skewness, or other nonnormal features? Using a calculator, find the mean and standard deviation of these scores.

(b) Treat the 31 girls as an SRS of all middle-school girls in the school district. Give a 95% confidence interval for the mean score in the population.

(c) In fact, the scores are those of all seventh-grade girls in one of the several schools in the district. Explain carefully why we cannot trust the confidence interval from (b).

25.4 Confidence level and margin of error. The NAEP test (Example 2) was also given to a sample of 1077 women of ages 21 to 25 years. Their mean quantitative score was 275 and the standard deviation was 58.
(a) Give a 95% confidence interval for the mean score μ in the population of all young women.

(b) Give the 90% and 99% confidence intervals for μ.

(c) What are the margins of error for 90%, 95%, and 99% confidence? How does increasing the confidence level affect the margin of error of a confidence interval?

25.5 IQ test scores. The mean IQ for the entire population in any age group is supposed to be 100. Treat the IQ scores in Exercise 25.3 as if they were an SRS from all middle-school girls in this district. Do the scores provide good evidence that the mean IQ of this population is not 100?

25.6 Averages versus individuals. Scores on the American College Testing (ACT) college admissions examination vary normally with mean $\mu = 18$ and standard deviation $\sigma = 6$. The range of reported scores is 1 to 36.

(a) What range contains the middle 95% of all individual scores?

(b) If the ACT scores of 25 randomly selected students are averaged, what range contains the middle 95% of the averages $\bar{x}$?

25.7 Student attitudes. The Survey of Study Habits and Attitudes (SSHA) is a psychological test that measures students' study habits and attitude toward school. Scores range from 0 to 200. The mean score for U.S. college students is about 115, and the standard deviation is about 30. A teacher suspects that older students have better attitudes toward school. She gives the SSHA to 25 students who are at least 30 years old. Assume that scores in the population of older students are normally distributed with standard deviation $\sigma = 6$. The teacher wants to test the hypotheses

$$H_0: \mu = 115$$
$$H_a: \mu > 115$$

(a) What is the sampling distribution of the mean score $\bar{x}$ of a sample of 25 older students if the null hypothesis is true? Sketch the density curve of this distribution. (Hint: Sketch a normal curve first, then mark the axis using what you know about locating μ and σ on a normal curve.)

(b) Suppose that the sample data give $\bar{x} = 118.6$. Mark this point on the axis of your sketch. In fact, the outcome was $\bar{x} = 125.7$. Mark this point on your sketch. Using your sketch, explain in simple language why one outcome is good evidence that the mean score of all older students is greater than 115 and why the other outcome is not.

(c) Shade the area under the curve that is the *P*-value for the sample result $\bar{x} = 118.6$.

25.8 Blood pressure. A randomized comparative experiment studied the effect of diet on blood pressure. Researchers divided 54 healthy white males at random into two groups. One group received a calcium supplement; the other, a placebo. At the beginning of the study, the researchers measured many variables on the subjects. The paper reporting the study gives $\bar{x} = 114.9$ and $s = 9.3$ for the seated systolic blood pressure of the 27 members of the placebo group.

(a) Give a 95% confidence interval for the mean blood pressure of the population from which the subjects were recruited.

(b) The recipe you used in part (a) requires an important assumption about the 27 men who provided the data. What is this assumption?

25.9 Mice in a maze. Experiments on learning in animals sometimes measure how long it takes mice to find their way through a maze. The mean time is 18 seconds for one particular maze. A researcher thinks that a loud noise will cause the mice to complete the maze faster. She measures how long each of several mice takes with a noise as stimulus. What are the null hypothesis H_0 and alternative hypothesis H_a?

25.10 Response time. Last year, your company's service technicians took an average of 2.6 hours to respond to trouble calls from business customers who had purchased service contracts. Do this year's data show a significantly different average response time? What null and alternative hypotheses should you test to answer this question?

25.11 Testing a random number generator. My statistical software has a "random number generator" that is supposed to produce numbers scattered at random between 0 to 1. If this is true, the numbers generated come from a population with $\mu = 0.5$. A command to generate 100 random numbers gives outcomes with mean $\bar{x} = 0.532$ and $s = 0.316$. Is this good evidence that the mean of all numbers produced by this software is not 0.5?

25.12 Will they charge more? A bank wonders whether omitting the annual credit card fee for customers who charge at least $2400 in a year will increase the amount charged on its credit cards. The bank makes this offer to an SRS of 200 of its credit card customers. It then compares how much these customers charge this year with the amount that they charged last year. The mean increase in the sample is $332, and the standard deviation is $108. Is there significant evidence at the 1% level that the mean amount charged increases under the no-fee offer? State H_0 and H_a and carry out a test.

25.13 Testing a random number generator. Give a 90% confidence interval for the mean of all numbers produced by the software in Exercise 25.11.

25.14 Will they charge more? Exercise 25.12 reports a bank's trial of eliminating credit card fees for clients who charge at least $2400. Give a 99% confidence interval for the mean amount charges would have increased if this benefit had been extended to all such customers.

25.15 A sampling distribution. Exercises 25.11 and 25.13 concern the mean of the random numbers generated by a computer program. The mean is supposed to be 0.5 because the numbers are supposed to be spread at random between 0 and 1. I asked the software to generate samples of 100 random numbers repeatedly. Here are the sample means $\bar{x}$ for 50 samples of size 100:

0.532	0.450	0.481	0.508	0.510	0.530	0.499	0.461	0.543	0.490
0.497	0.552	0.473	0.425	0.449	0.507	0.472	0.438	0.527	0.536
0.492	0.484	0.498	0.536	0.492	0.483	0.529	0.490	0.548	0.439
0.473	0.516	0.534	0.540	0.525	0.540	0.464	0.507	0.483	0.436
0.497	0.493	0.458	0.527	0.458	0.510	0.498	0.480	0.479	0.499

The sampling distribution of $\bar{x}$ is the distribution of the means from all possible samples. We actually have the means from 50 samples. Make a histogram of these 50 observations. Does the distribution appear to be roughly normal, as the central limit theorem says will happen for large enough samples?

25.16 Will they charge more? In Exercises 25.12 and 25.14, you carried out the calculations for a test and confidence interval based on a bank's experiment in changing the rules for its credit cards. You ought to ask some questions about this study.
(a) The distribution of the amount charged is skewed to the right, but outliers are prevented by the credit limit that the bank enforces on each card. Why can we use a test and confidence interval based on a normal sampling distribution for the sample mean $\bar{x}$?

(b) The bank's experiment was not comparative. The increase in amount charged over last year may be explained by lurking variables rather than by the rule change. What are some plausible reasons why charges might go up? Outline the design of a comparative randomized experiment to answer the bank's question.

25.17 A sampling distribution, continued. Exercise 25.15 presents 50 sample means $\bar{x}$ from 50 random samples of size 100. Using a calculator, find the mean and standard deviation of these 50 values. Then answer these questions.
(a) The mean of the population from which the 50 samples were drawn is $\mu = 0.5$ if the random number generator is accurate. What do you expect the mean of the distribution of $\bar{x}$'s from all possible samples to be? Is the mean of these 50 samples close to this value?

(b) The standard deviation of the distribution of $\bar{x}$ from samples of size $n = 100$ is supposed to be $\sigma/10$, where σ is the standard deviation of individuals in the population. Use this fact with the standard deviation you calculated for the 50 $\bar{x}$'s to estimate σ.

Part IV Review

Statistical inference draws conclusions about a population on the basis of sample data and uses probability to indicate how reliable the conclusions are. A confidence interval estimates an unknown parameter. A significance test shows how strong the evidence is for some claim about a parameter. Chapters 21 and 22 present the reasoning of confidence intervals and tests and give optional details for inference about a population proportion p.

The probabilities in both confidence intervals and tests tell us what would happen if we used the formula for the interval or test very many times. A confidence level is the probability that the formula for a confidence interval actually produces an interval that contains the unknown parameter. A 95% confidence interval gives a correct result 95% of the time when we use it repeatedly. Figure IV.1 illustrates the reasoning using the approximate 95% confidence interval for a population proportion p.

A P-value is the probability that the test would produce a result at least as extreme as the observed result if the null hypothesis really were true. Figure IV.2 illustrates the reasoning, placing the sample proportion $\hat{p}$ from our one sample on the normal curve that shows how $\hat{p}$ would vary in all possible samples *if the null hypothesis were true.* A P-value tells us how surprising the observed outcome is. Very surprising outcomes (small P-values) are good evidence that the null hypothesis is not true.

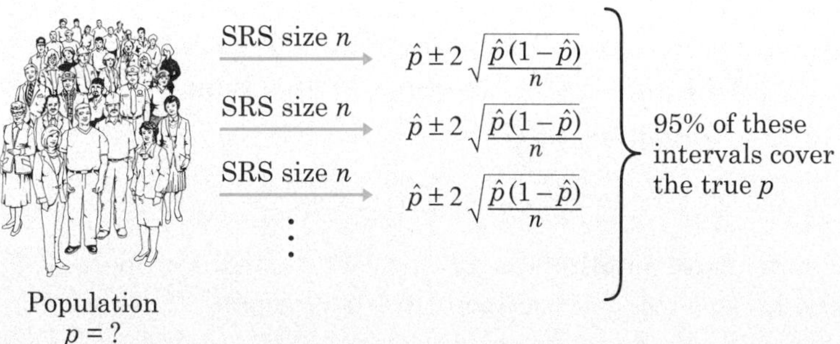

Figure IV.1

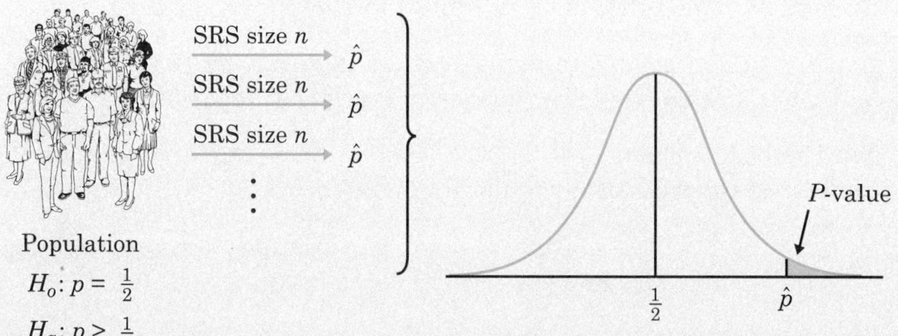

Figure IV.2

To detect sound and unsound uses of inference, you must know the basic reasoning and also be aware of some fine points and pitfalls. Chapter 23 will help. The optional Chapters 24 and 25 discuss a few more specific inference procedures. Chapter 24 deals with two-way tables, both for description and for inference. The descriptive part of this chapter completes Part II's discussion of relationships among variables by describing relationships among categorical variables. Chapter 25 concerns inference about the mean of a population.

PART IV SUMMARY

Here are the most important skills you should have after reading Chapters 21 to 25. Asterisks mark skills that appear in optional sections of the text.

A. SAMPLING DISTRIBUTIONS

1. Explain the idea of a sampling distribution. See Figure 21.1 (page 422).

2. Use the normal sampling distribution of a sample proportion $\hat{p}$ and the 68–95–99.7 rule to find probabilities involving $\hat{p}$.

3. *Use the normal sampling distribution of a sample mean $\bar{x}$ to find probabilities involving $\bar{x}$.

B. CONFIDENCE INTERVALS

1. Explain the idea of a confidence interval. See Figure IV.1.

2. Explain in nontechnical language what is meant by "95% confidence" and other statements of confidence in statistical reports.

3. Use the basic formula $\hat{p} \pm 2\sqrt{\hat{p}(1-\hat{p})/n}$ to obtain an approximate 95% confidence interval for a population proportion p.

4. Understand how the margin of error of a confidence interval changes with the sample size and the level of confidence.

5. Detect major mistakes in applying inference, such as improper data production, selecting the best of many outcomes, ignoring high nonresponse, and ignoring outliers.

6. *Use the detailed formula $\hat{p} \pm z^*\sqrt{\hat{p}(1-\hat{p})/n}$ and critical values z^* for normal distributions to obtain confidence intervals for a population proportion p.

7. *Use the formula $\bar{x} \pm z^* s/\sqrt{n}$ to obtain confidence intervals for a population mean μ.

C. SIGNIFICANCE TESTS

1. Explain the idea of a significance test. See Figure IV.2.

2. State the null and alternative hypotheses in a testing situation when the parameter in question is a population proportion p.

3. Explain in nontechnical language the meaning of the P-value when you are given the numerical value of P for a test.

4. Explain the meaning of "statistically significant at the 5% level" and other statements of significance. Explain why significance at a specific level such as 5% is less informative than a P-value.

5. Recognize that significance testing does not measure the size or importance of an effect.

6. Recognize and explain the effect of small and large samples on the significance of an outcome.

7. *Use Table B of percentiles of normal distributions to find the *P*-value for a test about a proportion p.

8. *Carry out one-sided and two-sided tests about a mean μ using the sample mean $\bar{x}$ and Table B.

D. *TWO-WAY TABLES

1. Arrange data on two categorical variables for a number of individuals into a two-way table of counts.

2. Use percents to describe the relationship between any two categorical variables starting from the counts in a two-way table.

3. Explain what null hypothesis the chi-square statistic tests in a specific two-way table.

4. Calculate expected cell counts, the chi-square statistic, and its degrees of freedom from a two-way table.

5. Use Table 24.1 for chi-square distributions to assess significance. Interpret the test result in the setting of a specific two-way table.

PART IV REVIEW EXERCISES

Review exercises are short and straightforward exercises that help you solidify the basic ideas and skills in each part of this book.

IV.1 Big events. An SRS of 489 adults found that 347 chose World War II from a list of events as the most important event of the 20th century. Give a 95% confidence interval for the proportion of all adults who think World War II is the century's most important event.

IV.2 Drinking problems. An SRS of 1039 adults found that 374 said that drinking had been a problem in their families. Give a 95% confidence interval for the proportion of all adults who have had a drinking-related problem in their families.

IV.3 Big events. Exercise IV.1 concerns an SRS of 489 adults. Suppose that (unknown to the pollsters) exactly 70% of all adults would choose World War II as the most important event of the 20th century. Imagine that we take very many SRSs of size 489 from this population and record the percent in each sample who choose World War II. Where would the middle 95% of all values of this percent lie?

IV.4 Drinking problems. Exercise IV.2 concerns an SRS of 1039 adults. Suppose that in the population of all adults, exactly 35% would say that drinking had been a problem in their families. Imagine that we take a very large number of SRSs of size 1039. For each sample, record the proportion $\hat{p}$ of the sample who have had a drinking problem in their families.
(a) What is the sampling distribution that describes the values $\hat{p}$ would take in our samples?

(b) Use this distribution and the 68–95–99.7 rule to find the approximate percent of all samples in which more than 36.5% of the respondents have had a drinking problem in their families.

IV.5 Drinking too much. An SRS of 684 adults who drink alcohol found that 164 agreed that they "sometimes drink more alcoholic beverages than you should." Give a 95% confidence interval for the proportion of all drinkers who sometimes drink more than they should.

IV.6 Roulette. A roulette wheel has 18 red slots among its 38 slots. You observe many spins and record the number of times that red occurs. Now you want to use these data to test whether the probability p of a red has the value that is correct for a fair roulette wheel. State the hypotheses H_0 and H_a that you will test.

IV.7 Why not? Table 11.1 (page 199) records the percent of residents aged 65 or older in each of the 50 states. You can check that this percent is 14% or higher in 11 of the states. So the sample proportion of states with at least 14% of elderly residents is $\hat{p} = 11/50 = 0.22$. Explain why it does *not* make sense to go on to obtain a 95% confidence interval for the population proportion p.

IV.8 Helping welfare mothers. A study compares two groups of mothers with young children who were on welfare two years ago. One group attended a voluntary training program that was offered free of charge at a local vocational school and was advertised in the local news media. The other group did not choose to attend the training program. The study finds a significant difference ($P < 0.01$) between the proportions of the mothers in the two groups who are still on welfare. The difference is not only significant but quite large. The report says that with 95% confidence the percent of the nonattending group still on welfare is 21% ± 4% higher than that of the group who attended the program. You are on the staff of a member of Congress who is interested in the plight of welfare mothers, and who asks you about the report.
(a) Explain in simple language what "a significant difference ($P < 0.01$)" means.

(b) Explain clearly and briefly what "95% confidence" means.

(c) This study is not good evidence that requiring job training of all welfare mothers would greatly reduce the percent who remain on welfare. Explain this to the member of Congress.

IV.9 Beating the system. Some doctors think that health plan rules restrict their ability to treat their patients effectively, so they bend the rules to help patients get reimbursed by their health plans. Here's a sentence from a study on this topic: "Physicians who agree with the statement 'Today it is necessary to game the system to provide high-quality care' reported manipulating reimbursement systems more often than those who did not agree with the statement (64.3% vs 35.7%; $P < 0.01$).

(a) Explain to a doctor what ($P < 0.01$) means in the context of this specific study.

(b) A result that is statistically significant can still be too small to be of practical interest. How do you know this is not true here?

IV.10 Alternative medicine. A nationwide random survey of 1500 adults asked about attitudes toward "alternative medicine" such as acupuncture, massage therapy, and herbal therapy. Among the respondents, 660 said they would use alternative medicine if traditional medicine was not producing the results they wanted.

(a) Give a 95% confidence interval for the proportion of all adults who would use alternative medicine.

(b) Write a short paragraph for a news report based on the survey results.

IV.11 When shall we call you? As you might guess, telephone sample surveys get better response rates during the evening than during the weekday daytime. One study called 2304 randomly chosen telephone numbers on weekday mornings. Of these, 1313 calls were answered and only 207 resulted in interviews. Of 2454 calls on weekday evenings, 1840 were answered and 712 interviews resulted. Give two 95% confidence intervals, for the proportions of all calls that are answered on weekday mornings and on weekday evenings. Are you confident that the proportion is higher in the evening?

IV.12 Alternative medicine. Does the survey of Exercise IV.10 provide good evidence that more than 1/3 of all adults would use alternative medicine if traditional medicine did not produce the results they wanted?

(a) State the hypotheses to be tested.

(b) If your null hypothesis is true, what is the sampling distribution of the sample proportion $\hat{p}$? Sketch this distribution.

(c) Mark the actual value of $\hat{p}$ on the curve. Does it appear surprising enough to give good evidence against the null hypothesis?

IV.13 When shall we call you? Suppose we knew that 57% of all calls made by sample surveys on weekday mornings are answered. We make 2454 calls to randomly chosen numbers during weekday evenings. Of these, 1840 are answered. Is this good evidence that the proportion of answered calls is higher in the evening?

(a) State the hypotheses to be tested.

(b) If your null hypothesis is true, what is the sampling distribution of the sample proportion $\hat{p}$? Sketch this distribution.

(c) Mark the actual value of $\hat{p}$ on the curve. Does it appear surprising enough to give good evidence against the null hypothesis?

IV.14 Not significant. The study cited in Exercise IV.9 looked at the factors that may affect whether doctors bend medical plan rules. Perhaps doctors who fear being prosecuted will bend the rules less often. The study report said, "Notably, greater worry about prosecution for fraud did not affect physicians' use of these tactics ($P = .34$)." Explain why the result $P = .34$ supports the conclusion that worry about prosecution did not affect behavior.

IV.15 Going to church. Opinion polls show that about 40% of Americans say they attended religious services in the last week. This result has stayed stable for decades. Studies of what people actually *do*, as opposed to what they *say* they do, suggest that actual church attendance is much lower. One study calculated 95% confidence intervals based on what a sample of Catholics said and then based on a sample of actual behavior. In Chicago, for example, the 95% confidence interval from the opinion poll said that between 45.7% and 51.3% of Catholics attend mass weekly. The 95% confidence interval from actual counts said that between 25.7% and 28.9% attended mass weekly.

(a) Why might we expect opinion polls on church attendance to be biased in the direction of overestimating true attendance?

(b) The poll in Chicago found that 48.5% of Catholics claimed to attend mass weekly. Why don't we just say "48.5% of all Catholics in Chicago claim to attend mass" instead of giving the interval 45.7% to 51.3%?

(c) The two results, from reported and observed behavior, are quite different. What does it mean to say that we are "95% confident" in each of the two intervals given?

The following exercises are based on the optional sections of Chapters 21 and 22.

IV.16 Big events. An SRS of 489 adults found that 347 chose World War II from a list of events as the most important event of the 20th century. Give 90% and 99% confidence intervals for the proportion of all adults who think World War II is the century's most important event. Explain briefly what important fact about confidence intervals is illustrated by comparing these two intervals and the 95% confidence interval from Exercise IV.1.

IV.17 Drinking problems. An SRS of 1039 adults found that 374 said that drinking had been a problem in their families. Is this good evidence that drinking has been a problem in the families of more than 1/3 of American adults? Show the five steps of the test (hypotheses, sampling distribution, data, *P*-value, conclusion) clearly.

IV.18 Drinking too much. An SRS of 684 adults who drink alcohol found that 164 agreed that they "sometimes drink more alcoholic beverages than you should." Give a 90% confidence interval for the proportion of all drinkers who sometimes drink more than they should. In what way is a 90% confidence interval less appealing than a 95% confidence interval? In what way is a 90% interval more appealing?

IV.19 A poll of voters. You are the polling consultant to a member of Congress. An SRS of 500 registered voters finds that 28% name "environmental problems" as the most important issue facing the nation. Give a 90% confidence interval for the proportion of all voters who hold this opinion. Then explain carefully to the member of Congress what your conclusion tells us about voters' opinions.

IV.20 Alternative medicine. Carry out the significance test called for in Exercise IV.12 in all detail. Show the five steps of the test (hypotheses, sampling distribution, data, *P*-value, conclusion) clearly.

IV.21 When shall we call you? Carry out the significance test called for in Exercise IV.13 in all detail. Show the five steps of the test (hypotheses, sampling distribution, data, *P*-value, conclusion) clearly.

The following exercises are based on the optional Chapters 24 and 25.

IV.22 CEO pay. A study of 104 corporations found that the pay of their chief executive officers had increased an average of $\bar{x} = 6.9\%$ per year in real terms. The standard deviation of the percent increases was $s = 17.4\%$

(a) The 104 individual percent increases have a right-skewed distribution. Explain why the central limit theorem says we can nonetheless act as if the mean increase has a normal distribution.

(b) Give a 95% confidence interval for the mean percent increase in pay for all corporate CEOs.

(c) What must we know about the 104 corporations studied to justify the inference you did in (b)?

IV.23 Water quality. An environmentalist group collects a liter of water from each of 45 random locations along a stream and measures the amount of dissolved oxygen in each specimen. The mean is 4.62 milligrams (mg) and the standard deviation is 0.92 mg. Is this strong evidence that the stream has a mean oxygen content of less than 5 mg per liter?

IV.24 Pleasant smells. Do pleasant odors help work go faster? Twenty-one subjects worked a paper-and-pencil maze wearing a mask that was either unscented or carried the smell of flowers. Each subject worked the maze three times with each mask, in random order. (This is a matched pairs design.) Here are the differences in their average times (in seconds), unscented minus scented. If the floral smell speeds work, the difference will be positive because the time with the scent will be lower.

−7.37	−3.14	4.10	−4.40	19.47	−10.80	−0.87
8.70	2.94	−17.24	14.30	−24.57	16.17	−7.84
8.60	−10.77	24.97	−4.47	11.90	−6.26	6.67

(a) We hope to show that work is faster on the average with the scented mask. State null and alternative hypotheses in terms of the mean difference in times μ for the population of all adults.
(b) Using a calculator, find the mean and standard deviation of the 21 observations. Did the subjects work faster with the scented mask? Is the mean improvement big enough to be important?
(c) Make a stemplot of the data (round to the nearest whole second). Are there outliers or other problems that might hinder inference?
(d) Test the hypotheses you stated in (a). Is the improvement statistically significant?

IV.25 Sharks. Great white sharks are big and hungry. Here are the lengths in feet of 44 great whites:

18.7	12.3	18.6	16.4	15.7	18.3	14.6	15.8	14.9	17.6	12.1
16.4	16.7	17.8	16.2	12.6	17.8	13.8	12.2	15.2	14.7	12.4
13.2	15.8	14.3	16.6	9.4	18.2	13.2	13.6	15.3	16.1	13.5
19.1	16.2	22.8	16.8	13.6	13.2	15.7	19.7	18.7	13.2	16.8

(a) Make a stemplot with feet as the stems and 10ths of feet as the leaves. There are 2 outliers, one in each direction. These won't change $\overline{x}$ much but will pull up the standard deviation s.
(b) Give a 90% confidence interval for the mean length of all great white sharks. (The interval may be too wide due to the influence of the outliers on s.)
(c) What do we need to know about these sharks in order to interpret your result in (b)?

IV.26 Pleasant smells. Return to the data in Exercise IV.24. Give a 95% confidence interval for the mean improvement in time to solve a maze when wearing a mask with a floral scent. Are you confident that the scent does improve mean working time?

IV.27 Sharks. Return to the data in Exercise IV.25. Is there good evidence that the mean length of sharks in the population these sharks represent is greater than 15 feet?

IV.28 Simpson's paradox. If we compare average NAEP quantitative scores, we find that eighth-grade students in Nebraska do better than eighth-grade students in New Jersey. But if we look only at white students, New Jersey does better. If we look only at minority students, New Jersey again does better. That's Simpson's paradox: the comparison reverses when we lump all students together. Explain carefully why this makes sense, using the fact that a much higher percent of Nebraska eighth graders are white.

IV.29 Unhappy HMO patients. A study of complaints by HMO members compared those who filed complaints about medical treatment and those who filed nonmedical complaints with an SRS of members who did not complain that year. Here are the data on the total number in each group and the number who voluntarily left the HMO:

	No complaint	Medical complaint	Nonmedical complaint
Total	743	199	440
Left	22	26	28

(a) Find the percent of each group who left.

(b) Make a two-way table of complaint status by left or not.

(c) Find the expected counts and check that you can safely use the chi-square test.

(d) The chi-square statistic for this table is $X^2 = 31.765$. What null and alternative hypotheses does this statistic test? What are its degrees of freedom? How significant is it? What do you conclude about the relationship between complaints and leaving the HMO?

IV.30 Treating ulcers. Gastric freezing was once a recommended treatment for stomach ulcers. Use of gastric freezing stopped after experiments showed it had no effect. One randomized comparative experiment found that 28 of the 82 gastric freezing patients improved, while 30 of the 78 patients in the placebo group improved.

(a) Outline the design of this experiment.

(b) Make a two-way table of treatment versus outcome (subject improved or not). Is there a significant relationship between treatment and outcome?

(c) Write a brief summary that includes the test result and also percents that compare the success of the two treatments.

IV.31 When shall we call you? In Exercise IV.11, we learned of a study that dialed telephone numbers at random during two periods of the day. Of 2304 numbers called on weekday mornings, 1313 answered. Of 2454 calls on weekday evenings, 1840 were answered.

(a) Make a two-way table of time of day versus answered or not. What percent of calls were answered in each time period?

(b) It should be obvious that there is a highly significant relationship between time of day and answering. Why?

(c) Nonetheless, carry out the chi-square test. What do you conclude?

PART IV PROJECTS

Projects are longer exercises that require gathering information or producing data and emphasize writing a short essay to describe your work. Many are suitable for teams of students.

Project 1. Reporting a medical study. Many of the major articles in medical journals concern statistically designed studies and report the results of inference, usually either *P*-values or 95% confidence intervals. You can find summaries of current articles on the Web sites of the *Journal of the American Medical Association* (jama.ama-assn.org) and the *New England Journal of Medicine* (www.nejm.org). A full copy may require paying a fee or visiting the library. Choose an article that describes a medical experiment on a topic that is understandable to those of us who lack medical training—anger and heart attacks and fiber in the diet to reduce cholesterol are two examples used in Chapters 21 and 22. Write a two-paragraph news article explaining the results of the study.

Then write a brief discussion of how you decided what to put in the news article and what to leave out. For example, if you omitted details of statistical significance or of confidence intervals, explain why. What did you say about the design of the study, and why? News writers must regularly make decisions like these.

Project 2. Use and abuse of inference. Few accounts of really complex statistical methods are readable without extensive training. One that is, and that is also an excellent essay on the abuse of statistical inference, is "The Real Error of Cyril Burt," a chapter in Stephen Jay Gould's *The Mismeasure of Man* (W. W. Norton, 1981). We met Cyril Burt under suspicious circumstances in Exercise 9.20 on page 158. Gould's long chapter shows Burt and others engaged in discovering dubious patterns by complex statistics. Read it, and write a brief explanation of why "factor analysis" failed to give a firm picture of the structure of mental ability.

Project 3. Roll your own statistical study. Collect your own data on two categorical variables whose relationship seems interesting. A simple example is the gender of a student and his or her political party preference. A more elaborate example is the year in school of a college undergraduate and his or her plans following graduation (immediate employment, further study, take some time off, . . .). I won't insist on a proper SRS.

Collect your data and make a two-way table. Do an analysis that includes comparing percents to describe the relationship between your two variables and using the chi-square statistic to assess its significance. Write a description of your study and its findings. Was your sample so small that lack of significance may not be surprising?

Project 4. Car colors? I have heard that more white cars are sold than any other color. (This no doubt requires allowing "white" to include various shades of "pearl" and "cream.") What percent of the cars driven by students at your school are white? Answer this question by collecting data and giving a confidence interval for the proportion of white cars. You might collect data by questioning a sample of students or by looking at cars in student parking areas. In your discussion, explain how you attempted to get data that are close to an SRS of student cars.

Notes and Data Sources

TO THE TEACHER

Pages xiii–xiv For more thoughts on statistics as a liberal discipline, see my presidential address to the American Statistical Association: David S. Moore, "Statistics among the liberal arts," *Journal of the American Statistical Association*, 93 (1998), pp. 1253–1259.

Page xiii The quotations are from a summary of the committee's report that was unanimously endorsed by the Board of Directors of the American Statistical Association. The full report is George Cobb, "Teaching statistics," in L. A. Steen (ed.), *Heeding the Call for Change: Suggestions for Curricular Action*, Mathematical Association of America, 1990, pp. 3–43.

PRELUDE

Page xix The following later passages treat examples in the Prelude: Power lines and childhood leukemia: Example 3 in Chapter 1. Ann Landers: Example 2 in Chapter 2. Gastric freezing to treat ulcers: Example 2 in Chapter 5. Casinos and crime: Chapter 20, page 392. Education and income: Chapter 12, page 216 and Example 3; Chapter 23, Exercise 4. Normal body temperature: Chapter 25, page 485. Measuring unemployment: Chapter 8, Examples 3 and 10.

Page xx Tim Hammonds is quoted as saying: "The public has become used to conflicting opinions. . . Many have come to feel that for every Ph.D., there is an equal and opposite Ph.D." *Chance*, 7, No. 2 (1994), p. 9.

Page xxii A. C. Nielsen, Jr., "Statistics in marketing," in *Making Statistics More Effective in Schools of Business*, Graduate School of Business, University of Chicago, 1986.

Page xxii The information about mammograms comes from H. C. Sox, "Editorial: benefit and harm associated with screening for breast cancer," *New England Journal of Medicine*, 338 (1998), p. 1145.

Page xxiii On suicide, see Howard I. Kushner, *Self Destruction in the Promised Land*, Rutgers University Press, 1989.

PART I

Page 1 Music sales data are collected electronically from store registers by SoundScan. Various results appear on many Web sites. I found the 1999 year-end results cited at backstreet.net.

CHAPTER 1

Page 3 Data on voting from "Only four in ten Americans 'always vote,'" Gallup poll press release by Lydia Saad, November 2, 1999. Read online at www.gallup.com/poll/releases.

Page 5 Example 1 is suggested by Maxine Pfannkuch and Chris J. Wild, "Statistical thinking and statistical practice: themes gleaned from professional statisticians," unpublished manuscript, 1998.

Page 7 Example 3: M. S. Linet et al., "Residential exposure to magnetic fields and acute lymphoblastic leukemia in children," *New England Journal of Medicine*, 337 (1997), pp. 1–7.

Page 11 The estimates of the census undercount come from Howard Hogan, "The 1990 post-enumeration survey: operations and results," *Journal of the American Statistical Association*, 88 (1993), pp. 1047–1060.

Page 11 The Web site for the American Community Survey is www.census.gov/acs/www.

CHAPTER 2

Page 19 "Acadian ambulance officials, workers flood call-in poll," *Baton Rouge Advocate*, January 22, 1999.

Page 25 Example 4: "Majority of smokers want to quit, consider themselves addicted," Gallup Poll press release by Mark Gillespie, November 18, 1999. Read online at www.gallup.com/poll/releases.

Page 27 Exercise 2.6: D. Horvitz in his contribution to "Pseudo-opinion polls: SLOP or useful data?" *Chance*, 8, No. 2 (1995), pp. 16–25.

CHAPTER 3

Pages 30–31 The Gallup Organization, "Gambling in America–1999: a comparison of adults and teenagers." Topline and Trends detailed report; read online at www.gallup.com/poll.

Page 36 Information about restrictions on releasing election forecasts from Richard Morin, "Crackdown on pollsters," *Washington Post*, January 19, 1998.

Page 44 Exercise 3.7: Warren McIsaac and Vivek Goel, "Is access to physician services in Ontario equitable?" Institute for Clinical Evaluative Sciences in Ontario, October 18, 1993.

Page 46 Exercise 3.15: *New York Times*, August 21, 1989.

Page 46 Exercise 3.16: see www.louisharris.com. This Web site is also the source for Exercise 3.19.

Page 48 Exercise 3.28: based on a report at the Gallup Web site, www.gallup.com.

CHAPTER 4

Pages 49–50 Gregory Flemming and Kimberly Parker, "Race and reluctant respondents: possible consequences of non-response for pre-election surveys," Pew Research Center for the People and the Press, 1997, found at www.people-press.org.

Page 52 For more detail on nonsampling errors, along with references, see P. E. Converse and M. W. Traugott, "Assessing the accuracy of polls and surveys," *Science*, 234 (1986), pp. 1094–1098.

Page 53 For more detail on the limits of memory in surveys, see N. M. Bradburn, L. J. Rips, and S. K. Shevell, "Answering autobiographical questions: the impact of memory and inference on surveys," *Science*, 236 (1987), pp. 157–161.

Page 54 Example 4: The nonresponse rate for the CPS comes from "Technical notes to household survey data published in Employment and Earnings," found on the Bureau of Labor Statistics Web site at http://stats.bls.gov/cpshome.htm. The General Social Survey reports its response rate on its Web site, www.norc.uchicago.edu/gss/homepage.htm. The claim that "pollsters say response rates have fallen as low as 20 percent in some recent polls" is made in Don Van Natta, Jr., "Polling's 'dirty little secret': no response," *New York Times*, November 11, 1999.

Page 55 Example 5: The responses on welfare are from a *New York Times*/CBS News Poll reported in the *New York Times*, July 5, 1992. Those for Scotland are from "All set for independence?" *Economist*, September 12, 1998.

Page 56 Example 6: D. Goleman, "Pollsters enlist psychologists in quest for unbiased results," *New York Times*, September 7, 1993.

Page 56 The quotation on weighting is from Adam Clymer and Janet Elder, "Poll finds greater confidence in Democrats," *New York Times*, November 10, 1999.

Page 58 Example 7: The most recent account of the design of the CPS is Bureau of Labor Statistics, *Design and Methodology*, Current Population Survey Technical Paper 63, March 2000. Available in print or online at www.bls.census.gov/cps/tp/tp63.htm. The account in Example 7 omits many complications, such as the need to separately sample "group quarters" like college dormitories.

Page 63 Exercise 4.1: The quotation is typical of Gallup polls, from the Gallup Web site, www.gallup.com.

Page 63 Exercise 4.4: P. H. Lewis, "Technology" column, *New York Times*, May 29, 1995.

Page 64 Exercise 4.6: Giuliana Coccia, "An overview of non-response in Italian telephone surveys," *Proceedings of the 99th Session of the International Statistical Institute, 1993*, Book 3, pp. 271–272.

Page 64 Exercise 4.7: Richard Morin, "It depends on what your definition of 'do' is," *Washington Post*, December 21, 1998.

Page 65 Exercise 4.12: John Simons, "For risk takers, system is no longer sacred," *Wall Street Journal*, March 11, 1999.

Pages 65–66 Exercise 4.13: Adam Clymer and Janet Elder, "Poll finds greater confidence in Democrats," *New York Times*, November 10, 1999.

Page 66 Exercise 4.15: D. Goleman, "Pollsters enlist psychologists in quest for unbiased results," *New York Times*, September 7, 1993.

Page 68 Exercise 4.23: from the online "Supplementary Material" for G. Gaskell et al., "Worlds apart? The reception of genetically modified foods in Europe and the U.S.," *Science*, 285 (1999), pp. 384–387.

CHAPTER 5

Page 71 Allan H. Schulman and Randi L. Sims, "Learning in an online format versus an in-class format: an experimental study," *T.H.E. Journal*, June 1999, pp. 54–56. L. L. Miao, "Gastric freezing: an example of the evaluation of medical therapy by randomized clinical trials," in J. P. Bunker, B. A. Barnes, and F. Mosteller (eds.), *Costs, Risks and Benefits of Surgery*, Oxford University Press, 1977, pp. 198–211. This is also the source for Example 2. Details of the Carolina Abecedarian Project, including references to published work, can be found online at www.fpg.unc.edu/overview/abc/abc-ov.htm.

Page 75 Example 3: Samuel Charache et al., "Effects of hydroxyurea on the frequency of painful crises in sickle cell anemia," *New England Journal of Medicine*, 332 (1995), pp. 1317–1322.

Page 78 H. Sacks, T. C. Chalmers, and H. Smith, Jr., "Randomized versus historical controls for clinical trials," *American Journal of Medicine*, 72 (1982), pp. 233–240.

Page 80 Example 5: Marilyn Ellis, "Attending church found factor in longer life," *USA Today*, August 9, 1999.

Page 81 Example 6: Dr. Daniel B. Mark, in Associated Press, "Age, not bias, may explain differences in treatment," *New York Times*, April 26, 1994. Dr. Mark was commenting on Daniel B. Mark et al., "Absence of sex bias in the referral of patients for cardiac catheterization," *New England Journal of Medicine*, 330 (1994), pp. 1101–1106. See the correspondence from D. Douglas Miller and Leslee Shaw, "Sex bias in the care of patients with cardiovascular disease," *New England Journal of Medicine*, 331 (1994), p. 883, for comments on an opposed study.

Page 84 Exercise 5.10: G. Kolata, "New study finds vitamins are not cancer preventers," *New York Times*, July 21, 1994. Look in the *Journal of the American Medical Association*/ of the same date for the details.

Page 85 Exercise 5.12: letter to the editor by Stan Metzenberg, *Science*, 286 (1999), p. 2083.

Page 85 Exercise 5.14 is based on Christopher Anderson, "Measuring what works in health care," *Science*, 263 (1994), pp. 1080–1082.

Page 87 Exercise 5.23: L. E. Moses and F. Mosteller, "Safety of anesthetics," in J. M. Tanur et al. (eds.), *Statistics: A Guide to the Unknown*, 3rd edition, Wadsworth, 1989, pp. 15–24.

CHAPTER 6

Page 89 Martin Enserink, "Fickle mice highlight test problems," *Science*, 284 (1999), pp. 1599–1600. There is a full report of the study in the same issue.

Pages 90–91 Example 1: The fact about rats comes from E. Street and M. B. Carroll, "Preliminary evaluation of a new food product," in J. M. Tanur et al. (eds.), *Statistics: A Guide to the Unknown*, 3rd edition, Wadsworth, 1989, pp. 161–169.

Page 91 Example 2: The placebo effect examples are from Sandra Blakeslee, "Placebos prove so powerful even experts are surprised," *New York Times*, October 13, 1998. The "three-quarters" estimate is cited by Martin Enserink, "Can the placebo be the cure?" *Science*, 284 (1999), pp. 238–240. An extended treatment is Anne Harrington (ed.), *The Placebo Effect: An Interdisciplinary Exploration*, Harvard University Press, 1997.

Page 93 The flu trial quotation is from Kristin L. Nichol et al., "Effectiveness of live, attenuated intranasal influenza virus vaccine in healthy, working adults," *Journal of the American Medical Association*, 282 (1999), pp. 137–144.

Page 94 Example 3: "Cancer clinical trials: barriers to African American participation," *Closing the Gap*, newsletter of the Office of Minority Health, December 1997–January 1998.

Pages 94–95 Example 4: Michael H. Davidson et al., "Weight control and risk factor reduction in obese subjects treated for 2 years with orlistat: a randomized controlled trial," *Journal of the American Medical Association*, 281 (1999), pp. 235–242.

Page 98 Example 8: This is a simpler version of Arno J. Rethans, John L. Swasy, and Lawrence J. Marks, "Effects of television commercial repetition, receiver knowledge, and commercial length: a test of the two-factor model," *Journal of Marketing Research*, 23 (February 1986), pp. 50–61.

Page 99 Example 9: "Advertising: the cola war," *Newsweek*, August 30, 1976, p. 67.

Page 102 Exercise 6.3: Edward P. Sloan et al., "Diaspirin cross-linked hemoglobin (DCLHb) in the treatment of severe traumatic hemorrhagic shock," *Journal of the American Medical Association*, 282 (1999), 1857–1864.

Page 104 Exercise 6.10: W. E. Nelson, R. C. Henderson, L. C. Almekinders, R. A. DeMasi, and T. N. Taft, "An evaluation of pre- and postoperative nonsteroidal antiinflammatory drugs in patients undergoing knee arthroscopy," *Journal of Sports Medicine*, 21 (1994), pp. 510–516.

Page 107 Exercise 6.22: Mary O. Mundinger et al., "Primary care outcomes in patients treated by nurse practitioners or physicians," *Journal of the American Medical Association*, 238 (2000), pp. 59–68.

CHAPTER 7

Page 108 The quotation is from Thomas B. Freeman et al., "Use of placebo surgery in controlled trials of a cellular-based therapy for Parkinson's disease," *New England Journal of Medicine*, 341 (1999), pp. 988–992. Freeman supports the Parkinson's disease trial. The opposition is represented by Ruth Macklin, "The ethical problems with sham surgery in clinical research," *New England Journal of Medicine*, 341 (1999), pp. 992–996.

Page 109 Example 1: John C. Bailar III, "The real threats to the integrity of science," *The Chronicle of Higher Education*, April 21, 1995, pp. B1–B2.

Page 111 Example 2: The difficulties of interpreting guidelines for informed consent and for the work of institutional review boards in medical research are a main theme of Beverly Woodward, "Challenges to human subject protections in U.S. medical research," *Journal of the American Medical Association*, 282 (1999), pp. 1947–1952. The references in this paper point to other discussions.

Page 114 Example 4: quotation from the *Report of the Tuskegee Syphilis Study Legacy Committee*, May 20, 1996. A detailed history is James H. Jones, *Bad Blood: The Tuskegee Syphilis Experiment*, Free Press, 1993.

Page 114 Dr. Hennekens's words are from an interview in the Annenberg/Corporation for Public Broadcasting video series *Against All Odds: Inside Statistics*.

Page 115 Quotations from Gina Kolata and Kurt Eichenwald, "Business thrives on unproven care leaving science behind," *New York Times*, October 3, 1999. Background and details about the first clinical trials appear in a National Cancer Institute press release "Questions and answers: high-dose chemotherapy with bone marrow or stem cell transplants for breast cancer," April 15, 1999. That one of the studies reported there involved falsified data is reported by Denise Grady, "Breast cancer researcher admits falsifying data," *New York Times*, February 5, 2000.

Page 117 Example 7: R. D. Middlemist, E. S. Knowles, and C. F. Matter, "Personal space invasions in the lavatory: suggestive evidence for arousal," *Journal of Personality and Social Psychology*, 33 (1976), pp. 541–546.

Page 122 Exercise 7.13: Dr. C. Warren Olanow, chief of neurology, Mt. Sinai School of Medicine, quoted in Margaret Talbot, "The placebo prescription," *New York Times Magazine*, January 8, 2000, pp. 34–39, 44, 58–60.

Page 122 Exercise 7.14: For extensive background, see Jon Cohen, "AIDS trials ethics questioned," *Science*, 276 (1997), pp. 520–523. Search the archives at www.sciencemag.org for recent episodes in the continuing controversies of Exercises 7.14, 7.15, and 7.16.

Page 125 Exercise 7.24: news item in *Science*, 192 (1976), p. 1086.

CHAPTER 8

Page 126 Janny Scott, "Working hard, more or less: studies of leisure time point both up and down," *New York Times*, July 10, 1999. I learned of this article at the *Chance News* Web site, www.dartmouth.edu/~chance/chance_news.

Page 127 Example 1: "Trial and error," *Economist*, October 31, 1998, pp. 87–88.

Page 132 Example 7: quotation from a FairTest press release dated August 31, 1999, and appearing on the organization's Web page, www.fairtest.org. Most other information in this example and the Statistical Controversies feature comes from chapter 2 of the National Science Foundation report *Women, Minorities, and Persons with Disabilities in Science and Engineering: 1998*, NSF99-338, 1999. The table reports r^2-values, calculated from correlations given on the College Board Web site, www.collegeboard.org.

Page 136 Example 9: The deviations of NIST time from BIPM time are from the Web site of the NIST Time and Frequency Division, www.boulder.nist.gov/timefreq.

CHAPTER 9

Pages 146–147 The "missing vans" case is based on news articles in the *New York Times*, April 18, 1992, and September 10, 1992.

Page 147 Example 1: R. J. Newan, "Snow job on the slopes," *US News and World Report*, December 17, 1994, pp. 62–65.

Page 148 Example 2: Robert Niles of the *Los Angeles Times*, at nilesonline.com.

Page 148 Example 3: "Colleges inflate SATs and graduation rates in popular guidebooks," *Wall Street Journal*, April 5, 1995.

Page 149 Example 4: Robyn Meredith, "Oops, Cadillac says, Lincoln won after all," *New York Times*, May 6, 1999.

Page 149 Example 5: B. Yuncker, "The strange case of the painted mice," *Saturday Review / World*, November 30, 1974, p. 53.

Page 150 Example 6: letter by J. L. Hoffman, *Science*, 255 (1992), p. 665. The original article appeared on December 6, 1991.

Page 150 Example 7: E. Marshall, "San Diego's tough stand on research fraud," *Science*, 234 (1986), pp. 534–535.

Page 151 Example 8: both items are from *Chance News*, www.dartmouth.edu/~chance/chance_news.

Page 151 Darryl Nester spotted the ad featured in Example 9.

Pages 151–152 Example 10: *Science*, 192 (1976), p. 1081.

Page 153 Example 12: quotation from the American Heart Association statement "Cardiovascular disease in women," found at www.americanheart.org.

Pages 153–154 Example 13: Edwin S. Rubenstein, "Inequality," *Forbes*, November 1, 1999, pp. 158–160, and Bureau of the Census, *Money Income in the United States, 1998.*

Page 155 Exercise 9.1: *Providence Journal* (Rhode Island), December 24, 1999. I found it in *Chance News* 9.02.

Page 155 Exercise 9.4: *Fine Gardening*, September/October 1989, p. 76.

Page 155 Exercise 9.6: *Condé Nast Traveler* magazine, June 1992.

Page 156 Exercise 9.7: *Science*, 189 (1975), p. 373.

Page 156 Exercise 9.8: *Lafayette (Ind.) Journal and Courier*, October 23, 1988.

Page 156 Exercise 9.9: letter by L. Jarvik in the *New York Times*, May 4, 1993. The editorial, "Muggings in the Kitchen," appeared on April 23, 1993.

Page 156 Exercise 9.10: *New York Times*, April 21, 1986.

Page 157 Exercise 9.15: *Purdue Exponent*, September 21, 1977.

Pages 157–158 *Organic Gardening*, July 1983 for Exercise 9.17 and March 1983 for Exercise 9.18.

PART I REVIEW

Page 165 Exercises I.10 to I.12: Gallup Polls described on the Gallup Web site, www.gallup.com. The quotation in Exercise I.10 is from David W. Moore, "Americans support teaching creationism as well as evolution in public schools," Gallup press release dated August 30, 1999.

Page 166 Exercises I.16 to I.19: Kristin L. Nichol et al., "Effectiveness of live, attenuated intranasal influenza virus vaccine in healthy, working adults," *Journal of the American Medical Association*, 282 (1999), pp. 137–144.

PART II

CHAPTER 10

Page 180 Example 5: The prices in Figure 10.7 combine the Consumer Price Index component for unleaded regular gasoline (an index number) with the Energy Information Administration report of the actual price in January 1999. This information is available online.

Page 191 Exercise 10.16: F. Norris, "Market Watch" column, *New York Times*, July 15, 1992.

Page 192 Exercise 10.18: Bureau of Economic Analysis, *Survey of Current Business*, available online at www.bea.doc.gov.

Page 193 Exercise 10.21: *1997 Statistical Yearbook of the Immigration and Naturalization Service*, U.S. Department of Justice, 1999.

Page 194 Exercise 10.25: Centers for Disease Control and Prevention, *National Vital Statistics Reports*, June 30, 1999.

CHAPTER 11

Page 197 Figures 11.1 and 11.2: 1999–2000 tuition and fees from the College Board's *Annual Survey of Colleges, 1999–2000.*

Page 204 Figure 11.5: C. B. Williams, *Style and Vocabulary: Numerical Studies*, Griffin, 1970.

Page 207 Figure 11.7 is based on an episode in the Annenberg/Corporation for Public Broadcasting telecourse *Against All Odds: Inside Statistics.*

Page 208 Figure 11.9: data from the government's Survey of Earned Doctorates, in the online supplement to Jeffery Mervis, "Wanted: a better way to boost numbers of minority PhDs," *Science*, 281 (1998), pp. 1268–1270.

Page 209 Figure 11.10: based on J. K. Ford, "Diversification: how many stocks will suffice?" *American Association of Individual Investors Journal*, January 1990, pp. 14–16.

Page 209 Table 11.2: Environmental Protection Agency, *Model Year 2000 Fuel Economy Guide*, 1999.

Page 210 Table 11.3: Ali H. Mokdad et al., "The spread of the obesity epidemic in the United States, 1991–1998," *Journal of the American Medical Association*, 282 (1999), pp. 1519–1522.

Page 212 Exercise 11.13: *Consumer Reports*, June 1986, pp. 366–367. A more recent study of hot dogs appears in *Consumer Reports*, July 1993, pp. 415–419. The newer data cover few brands of poultry hot dogs and take calorie counts mainly from the package labels, resulting in suspiciously round numbers.

Page 214 Table 11.6: data from the National Oceanic and Atmospheric Administration Web site, www.ncdc.noaa.gov.

CHAPTER 12

Pages 216–217 Data on income and education from the March 1999 CPS supplement, downloaded from the Census Bureau Web site using the government's FERRET system. These are raw survey data, including many people with no income or negative income. To simplify comparisons, I report only four education classes.

Pages 224–225 The data for Figures 12.4 and 12.5 come from U.S. Bureau of the Census, *Money Income in the United States, 1998*, 1999, Tables B-2 and B-3. You can find a detailed argument from the political right in John H. Hinderaker and Scott W. Johnson, "The truth about income inequality," Center of the American Experiment, 1995 (www.amexp.org/ publications/). They cite the study of tax returns. The study of children is Peter Gottschalk and Sheldon Danziger, "Income mobility and exits from poverty of American children, 1970–1992," Boston University Department of Economics Working Paper 430, 1999. Figure 12.6 uses data from U.S. Bureau of the Census, *Poverty in the United States, 1998*, 1999.

Page 231 The salaries in Table 12.1 are estimates from several Web sites. Most players have complex multiyear contracts.

Pages 235 and 238 Exercises 12.9 and 12.21: data from *Consumer Reports*, June 1986, pp. 366–367.

Page 236 Exercise 12.13: data from the 1999 *Statistical Abstract of the United States.*

Pages 236 and 237 The raw data behind Exercises 12.15 and 12.19 come from College Board Online, www.collegeboard.org.

Page 238 Exercise 12.20: Environmental Protection Agency, *Model Year 2000 Fuel Economy Guide*, 1999.

Page 240 Exercise 12.27: part of the larger set of data in Table 15.2. See the source note for Exercise 15.17.

Page 241 Exercise 12.30: quotation from the *New York Times*, May 31, 1989.

CHAPTER 13

Page 257 The IQ scores in Figure 13.13 were collected by Darlene Gordon, Purdue University School of Education.

Page 258 Exercise 13.10: Stephen Jay Gould, "Entropic homogeneity isn't why no one hits .400 anymore," *Discover*, August 1986, pp. 60–66.

Pages 260–261 Exercise 13.22: Ulric Neisser, "Rising scores on intelligence tests," *American Scientist*, September-October 1997, online edition.

CHAPTER 14

Pages 263 and 264 The data for Figures 14.1 and 14.2, for 1998, come from the College Board Web site, www.collegeboard.org.

Pages 265–266 Example 1: Data from the World Bank's *1999 World Development Indicators*. Life expectancy is estimated for 1997 and GDP per capita (purchasing power parity basis) for 1998.

Page 267 Example 2: M. A. Houck et al., "Allometric scaling in the earliest fossil bird, *Archaeopteryx lithographica*," *Science*, 247 (1990), pp. 195–198. The authors conclude from a variety of evidence that all specimens represent the same species.

Page 280 Exercise 14.22: quotation from the *T. Rowe Price Report*, winter 1997, p. 4.

Page 281 Exercise 14.23: *Philadelphia City Paper*, May 23–29, 1997. Because the sodas served vary in size, I have converted soda prices to the price of a 16-ounce soda at each price per ounce.

Pages 281–282 Exercise 14.25: W. L. Colville and D. P. McGill, "Effect of rate and method of planting on several plant characters and yield of irrigated corn," *Agronomy Journal*, 54 (1962), pp. 235–238.

CHAPTER 15

Page 284 The quotation is from the Galateia Corporation Web site, at www.voicenet.com/~mitochon.

Page 293 Example 7: Laura L. Calderon et al., "Risk factors for obesity in Mexican-American girls: dietary factors, anthropometric factors, physical activity, and hours of television viewing," *Journal of the American Dietetic Association*, 96 (1996), pp. 1177–1179.

Page 295 Studies mentioned in the Statistical Controversies feature are John R. Lott Jr., *More Guns, Less Crime: Understanding Crime and Gun Control Laws*, University of Chicago Press, 1998; Andrés Villaveces et al., "Effect of a ban on carrying firearms on homicide rates in 2 Colombian cities," *Journal of the American Medical Association*, 283 (2000), pp. 1205–1209; see also the editorial by Lawrence W. Sherman in the same issue. Lawrence W. Sherman, James W. Shaw, and Dennis P. Rogan, "The Kansas City gun experiment," National Institute of Justice, 1995.

Pages 299–300 Exercise 15.7: data from M. H. Criqui, University of California, San Diego, reported in the *New York Times*, December 28, 1994.

Pages 300–301 Exercise 15.8: data estimated from a graph in G. D. Martinsen, E. M. Driebe, and T. G. Whitham, "Indirect interactions mediated by changing plant chemistry: beaver browsing benefits beetles," *Ecology*, 79 (1998), pp. 192–200.

Page 302 Exercise 15.13: W. M. Lewis and M. C. Grant, "Acid precipitation in the western United States," *Science*, 207 (1980), pp. 176–177.

Page 305 Exercise 15.24: David E. Bloom and David Canning, "The health and wealth of nations," *Science*, 287 (2000), pp. 1207–1208.

Pages 305–306 Exercise 15.25: quotation from a Gannett News Service article appearing in the *Lafayette (Ind.) Journal and Courier*, April 23, 1994.

Page 306 Marigene Arnold of Kalamazoo College contributed the orchestra ad in Exercise 15.29.

CHAPTER 16

Pages 308–309 The quotations are from Bureau of Labor Statistics, "Updated response to the recommendations of the advisory commission to study the Consumer Price Index," June 1998. The entire December 1996 issue of the BLS *Monthly Labor Review* is devoted to the major revision of the CPI that became effective in January of 1998. The effect of changes in the CPI is discussed by Kenneth J. Stewart and Stephen B. Reed, "CPI research series using current methods, 1978–98," *Monthly Labor Review*, 122 (1999), pp. 29–38. All of these are available on the BLS Web site.

Page 318 The government statistics survey is reported in "The good statistics guide," *Economist*, September 13, 1993, p. 65.

Page 326 Exercise 16.25: Gordon M. Fisher, "Is there such a thing as an absolute poverty line over time? Evidence from the United States, Britain, Canada, and Australia on the income elasticity of the poverty line," U.S. Census Bureau Poverty Measurement Working Papers, 1995.

Page 326 Exercise 16.26: G. J. Borjas, "The internationalization of the U.S. labor market and the wage structure," *Federal Reserve Bank of New York Economic Policy Review*, 1, No. 1 (1995), pp. 3–8. The quote appears on p. 3. This entire issue is devoted to articles seeking to explain stagnant earnings and the income gap. The consensus: we don't know.

PART II REVIEW

Page 335 The data plotted in Figure II.3 come from G. A. Sacher and E. F. Staffelt, "Relation of gestation time to brain weight for placental mammals: implications for the theory of vertebrate growth," *American Naturalist*, 108 (1974), pp. 593–613. I found them in Fred L. Ramsey and Daniel W. Schafer, *The Statistical Sleuth: A Course in Methods of Data Analysis*, Duxbury, 1997, p. 228.

Page 337 Exercise II.24: Antoni Basinski, "Almost never on Sunday: implications of the patterns of admission and discharge for common conditions," Institute for Clinical Evaluative Sciences in Ontario, October 18, 1993.

Page 339 Exercise II.28: official final election results from the Web site of the Federal Election Commission, www.fec.gov.

Page 340 Exercise II.29: correlations for 36 monthly returns ending in December 1999, reported in the *Fidelity Insight* newsletter for January 2000.

PART III

CHAPTER 17

Page 345 Richard Feynman's comments on the probability of a shuttle failure can be found in his appendix to the Presidential Commission report, "Personal Observations on the Reliability of the Shuttle by R. P. Feynman." See the Kennedy Space Center Web site, at www.ksc.nasa.gov/shuttle/missions/51-l/docs/rogers-commission/Appendix-F.txt.

Page 349 More historical detail can be found in the opening chapters of F. N. David, *Games, Gods and Gambling*, Charles Griffin and Co., 1962. The historical information given here comes from this excellent and entertaining book.

Page 350 For a discussion and amusing examples, see A. E. Watkins, "The law of averages," *Chance*, 8, No. 2 (1995), pp. 28–32.

Page 351 Example 5: R. Vallone and A. Tversky, "The hot hand in basketball: on the misperception of random sequences," *Cognitive Psychology*, 17 (1985), pp. 295–314.

Page 352 Example 7: for Woburn, see S. W. Lagakos, B. J. Wessen, and M. Zelen, "An analysis of contaminated well water and health effects in Woburn, Massachusetts," *Journal of the American Statistical Association*, 81 (1986), pp. 583–596, and the following discussion. For Randolph, see R. Day, J. H. Ware, D. Wartenberg, and M. Zelen, "An investigation of a reported cancer cluster in Randolph, Ma.," Harvard School of Public Health Technical Report, June 27, 1988.

Pages 353–356 The presentation of personal probability and long-term proportion as distinct ideas is influenced by psychological research that appears to show that people judge single-case questions differently from distributional or relative-frequency questions. At least some of the biases found in the classic work of Tversky and Kahneman on our perception of chance seem to disappear when it is made clear to subjects which interpretation is intended. This is a complex area and *SCC* is a simple book, but I judged it somewhat behind the times to put too much emphasis on the Tversky-Kahneman findings. See Gerd Gigerenzer, "How to make cognitive illusions disappear: beyond heuristics and biases," in Wolfgang Stroebe and Miles Hewstone (eds.), *European Review of Social Psychology*, Volume 2, Wiley, 1991, pp. 83–115.

Page 355 Example 9: estimated probabilities from R. D'Agostino, Jr., and R. Wilson, "Asbestos: the hazard, the risk, and public policy," in K. R. Foster, D. E. Bernstein, and P. W. Huber (eds.), *Phantom Risk: Scientific Inference and the Law*, MIT Press, 1994, pp. 183–210. See also the similar conclusions in B. T. Mossman et al., "Asbestos: scientific developments and implications for public policy," *Science*, 247 (1990), pp. 294–301.

Page 356 The quotation is from R. J. Zeckhauser and W. K. Viscusi, "Risk within reason," *Science*, 248 (1990), pp. 559–564.

Page 359 Exercise 17.13: see T. Hill, "Random-number guessing and the first digit phenomenon," *Psychological Reports*, 62 (1988), pp. 967–971.

CHAPTER 19

Pages 376–377 The introductory case study is based in part on Nico M. van Dijk et al., "Designing the Westerscheldetunnel toll plaza using a combination of queueing and simulation," in P. A. Farrington et al. (eds.), *Proceedings of the 1999 Winter Simulation Conference*, INFORMS, 1999, pp. 1272–1279.

Page 388 Exercise 19.13: F. N. David, *Games, Gods and Gambling*, Charles Griffin and Co., 1962.

Pages 388–389 Exercise 19.14: stochastic beetles are well known in the folklore of simulation, if not in entomology. They are said to be the invention of Arthur Engle of the School Mathematics Study Group.

CHAPTER 20

Pages 392–393 Andrew Pollack, "In the gaming industry, the house can have bad luck, too," *New York Times*, July 25, 1999, reports that Mirage Resorts issued a quarterly profit warning due in part to bad luck at baccarat.

Page 399 For the Statistical Controversies feature, see the Web sites in the Exploring the Web box and "Gambling on the future," *Economist*, June 26, 1999. The comments on the poor are based on the NGISC staff report on lotteries, at www.ngisc.gov/research.

Page 401 Exercise 20.6: obituary by Karen Freeman, *New York Times*, June 6, 1996.

Pages 401–402 Exercise 20.7: based on A. Tversky and D. Kahneman, "Extensional versus intuitive reasoning: the conjunction fallacy in probability judgment," *Psychological Review*, 90 (1983), pp. 293–315.

PART III REVIEW

Page 411 Exercise III.13: from a poll commissioned in 2000 by the National Center for Public Policy and Higher Education, found at www.highereducation.org.

PART IV

CHAPTER 21

Pages 419–420 Janice E. Williams et al., "Anger proneness predicts coronary heart disease risk," *Circulation*, 101 (2000), pp. 2034–2039.

Page 420 Example 1: Joseph H. Catania et al., "Prevalence of AIDS-related risk factors and condom use in the United States," *Science*, 258 (1992), pp. 1101–1106.

Page 427 For the state of the art in confidence intervals for p, see Alan Agresti and Brent Coull, "Approximate is better than 'exact' for interval estimation of binomial proportions," *American Statistician*, 52 (1998), pp. 119–126. The authors note that the accuracy of our confidence interval for p can be greatly improved by simply "adding 2 successes and 2 failures." That is, replace $\hat{p}$ by (count of successes + 2)/(n + 4). Texts on sample surveys give confidence intervals that take into account the fact that the population has finite size and also give intervals for sample designs more complex than an SRS.

Page 432 Exercise 21.12: Laurie Goodstein and Marjorie Connelly, "Teenage poll finds support for tradition," *New York Times*, April 30, 1998.

Page 434 Exercise 21.21: Sara J. Solnick and David Hemenway, "Complaints and disenrollment at a health maintenance organization," *Journal of Consumer Affairs*, 26 (1992), pp. 90–103.

CHAPTER 22

Pages 435–436 This is a somewhat simplified account of part of the study described by Reynolds Farley et al., "Stereotypes and segregation: neighborhoods in the Detroit area," *American Journal of Sociology*, 100 (1994), pp. 750–780. Figure 22.1 is based on part of a figure on p. 754 of this article.

Page 445 Exercise 22.3: Manisha Chandalia et al., "Beneficial effects of high dietary fiber intake in patients with type 2 diabetes mellitus," *New England Journal of Medicine*, 342 (2000), pp. 1392–1398.

Page 445 Exercise 22.4: Arthur Schatzkin et al., "Lack of effect of a low-fat, high-fiber diet on the recurrence of colorectal adenomas," *New England Journal of Medicine*, 342 (2000), pp. 1149–1155.

Pages 445–446 Exercise 22.5: Seung-Ok Kim, "Burials, pigs, and political prestige in neolithic China," *Current Anthropology*, 35 (1994), pp. 119–141.

Page 446 Exercise 22.6: Fekri A. Hassan, "Radiocarbon chronology of predynastic Nagada settlements, upper Egypt," *Current Anthropology*, 25 (1984), pp. 681–683.

Page 446 Exercise 22.7: Sara J. Solnick and David Hemenway, "The deadweight loss of Christmas: comment," *American Economic Review*, 86 (1996), pp. 1299–1305.

Page 449 Exercise 22.21: Laurie Goodstein and Marjorie Connelly, "Teenage poll finds support for tradition," *New York Times*, April 30, 1998.

Page 450 Exercise 22.23: Eric Ossiander, letter to the editor, *Science*, 257 (1992), p. 1461.

Page 450 Exercise 22.24: N. Teed, K. L. Adrian, and R. Knoblouch, "The duration of speed reductions attributable to radar detectors," *Accident Analysis and Prevention*, 25 (1991), pp. 131–137. This is one of the Electronic Encyclopedia of Statistical Examples and Exercises (EESEE) case studies.

CHAPTER 23

Pages 451–452 I looked at the list of top funds at the *Smart Money* site, www.smartmoney.com, in early May 2000. Here's part of the abstract of a typical recent study, Mark Carhart, "On persistence in mutual fund performance," *Journal of Finance*, 52 (1997), pp. 57–82: "Using a sample free of survivor bias, I demonstrate that common factors in stock returns and investment expenses almost completely explain persistence in equity mutual funds' mean and risk-adjusted returns. . . The only significant persistence not explained is concentrated in strong underperformance by the worst-return mutual funds. The results do not support the existence of skilled or informed mutual fund portfolio managers."

Page 459 Robert Rosenthal is quoted in B. Azar, "APA statistics task force prepares to release recommendations for public comment," *APA Monitor Online*, 30 (May 1999), at www.apa.org/monitor. The task force report, Leland Wilkinson et al., "Statistical methods in psychology journals: guidelines and explanations," *American Psychologist*, 54 (August 1999), offers a summary of the elements of good statistical practice.

Page 462 Exercise 23.4: Ross M. Stolzenberg, "Educational continuation by college graduates," *American Journal of Sociology*, 99 (1994), pp. 1042–1077.

CHAPTER 24

Pages 467–468 Example 2: an example of real data similar in spirit is P. J. Bickel and J. W. O'Connell, "Is there a sex bias in graduate admissions?" *Science*, 187 (1975), pp. 398–404.

Page 469 Example 3: data for Nationsbank, from S. A. Holmes, "All a matter of perspective," *New York Times*, October 11, 1995.

Page 470 Example 4: D. M. Barnes, "Breaking the cycle of addiction," *Science*, 241 (1988), pp. 1029–1030.

Page 475 There are many computer studies of the accuracy of chi-square critical values for X^2. For a brief discussion and some references, see Section 3.2.5 of David S. Moore, "Tests of chi-squared type," in Ralph B. D'Agostino and Michael A. Stephens (eds.), *Goodness-of-Fit Techniques*, Marcel Dekker, 1986, pp. 63–95.

Page 476 Example 7: Janice E. Williams et al., "Anger proneness predicts coronary heart disease risk," *Circulation*, 101 (2000), pp. 2034–2039.

Pages 478–479 Exercise 24.1: Richard M. Felder et al., "Who gets it and who doesn't: a study of student performance in an introductory chemical engineering course," *1992 ASEE Annual Conference Proceedings*, American Society for Engineering Education, 1992, pp. 1516–1519.

Page 479 Exercise 24.2: S. V. Zagona (ed.), *Studies and Issues in Smoking Behavior*, University of Arizona Press, 1967, pp. 157–180.

Page 479 Exercise 24.3: R. Shine, T. R. L. Madsen, M. J. Elphick, and P. S. Harlow, "The influence of nest temperatures and maternal brooding on hatchling phenotypes in water pythons," *Ecology*, 78 (1997), pp. 1713–1721.

Pages 479–480 Exercise 24.4: S. W. Hargarten et al., "Characteristics of firearms involved in fatalities," *Journal of the American Medical Association*, 275 (1996), pp. 42–45.

Page 480 Exercise 24.5: *Digest of Education Statistics 1997*, accessed on the National Center for Education Statistics Web site, http://www.ed.gov/NCES.

Page 480 Exercise 24.6: Francine D. Blau and Marianne A. Ferber, "Career plans and expectations of young women and men," *Journal of Human Resources*, 26 (1991), pp. 581–607.

Page 481 Exercise 24.8: From reports submitted by airlines to the Department of Transportation, found in A. Barnett, "How numbers can trick you," *Technology Review*, October 1994, pp. 38–45.

Pages 481–482 Exercise 24.9: M. Radelet, "Racial characteristics and imposition of the death penalty," *American Sociological Review*, 46 (1981), pp. 918–927.

Page 483 Exercise 24.14: Brenda C. Coleman, "Study: heart attack risk cut 74% by stress management," Associated Press dispatch appearing in the *Lafayette (Ind.) Journal and Courier*, October 20, 1997.

Pages 483–484 Exercise 24.15: David M. Blau, "The child care labor market," *Journal of Human Resources,* 27 (1992), pp. 9–39.

CHAPTER 25

Pages 485–486 The study is P. A. Mackowiak, S. S. Wasserman, and M. M. Levine, "A critical appraisal of 98.6 degrees F, the upper limit of normal body temperature, and other legacies of Carl Reinhold August Wunderlich," *Journal of the American Medical Association*, 268 (1992), pp. 1578–1580. I owe the reference, and data based on plots in the original article, to Allen L. Shoemaker, "What's normal? Temperature, gender, and heart rate," *Journal of Statistics Education*, 1996. (This electronic journal is found at www.amstat.org/publications/jse.)

Pages 489–490 Example 2: Information about the NAEP test is from Francisco L. Rivera-Batiz, "Quantitative literacy and the likelihood of employment among young adults," *Journal of Human Resources*, 27 (1992), pp. 313–328.

Page 494 Exercise 25.3: data provided by Darlene Gordon, Purdue University.

PART IV REVIEW

Pages 501–502 Exercises IV.1, IV.2, and IV.5 cite results from Gallup Polls, found at www.gallup.com.

Pages 503 and 504 Exercises IV.9 and IV.14: Matthew K. Wynia et al., "Physician manipulation of reimbursement rules for patients," *Journal of the American Medical Association*, 283 (2000), pp. 1858–1865.

Page 503 Exercise IV.10: Jane E. Brody, "Alternative medicine makes inroads," *New York Times*, April 28, 1998.

Pages 503–504 Exercise IV.13: Michael F. Weeks, Richard A. Kulka, and Stephanie A. Pierson, "Optimal call scheduling for a telephone survey," *Public Opinion Quarterly*, 51 (1987), pp. 540–549.

Page 504 Exercise IV.15: C. Kirk Hadaway, Penny Long Marler, and Mark Chaves, "What the polls don't show: a closer look at U.S. church attendance," *American Sociological Review*, 58 (1993), pp. 741–752.

Page 505 Exercise IV.22: Charles W. L. Hill and Phillip Phan, "CEO tenure as a determinant of CEO pay," *Academy of Management Journal*, 34 (1991), pp. 707–717.

Page 506 Exercise IV.24: A. R. Hirsch and L. H. Johnston, "Odors and learning," *Journal of Neurological and Orthopedic Medicine and Surgery*, 17 (1996), pp. 119–126. I found the data in a case study on the Electronic Encyclopedia of Statistical Examples and Exercises (EESEE), W. H. Freeman, 2000.

Page 506 The shark data in Exercise IV.25 were provided by Chris Olsen, who found the information in scuba diving magazines.

Page 507 I found the example in Exercise IV.28 in Howard Wainer's "Visual revelations" column, *Chance*, 12, No. 2 (1999), pp. 43–44.

Page 507 Exercise IV.29: Sara J. Solnick and David Hemenway, "Complaints and disenrollment at a health maintenance organization," *Journal of Consumer Affairs*, 26 (1992), pp. 90–103.

Page 507 Exercise IV.30: L. L. Miao, "Gastric freezing: an example of the evaluation of medical therapy by randomized clinical trials," in J. P. Bunker, B. A. Barnes, and F. Mosteller (eds.), *Costs, Risks and Benefits of Surgery*, Oxford University Press, 1977, pp. 198–211.

Solutions to Selected Exercises

Note to students: These answers should be considered guides or checkpoints, not complete solutions. Although explanations and details of computations are not given here, they should be part of a complete answer to an exercise.

Chapter 1

1.1 **(a)** Cars. **(b)** Make/model, vehicle type, transmission type, number of cylinders, city MPG, and highway MPG. The last three are numerical.

1.3 E.g., whether or not a household uses its recycling bin.

1.5 Approval of the president's job performance; adult U.S. residents; the 1210 adults interviewed.

1.7 **(a)** Adult residents of the United States. **(b)** All pieces of hardwood in the lot. **(c)** All U.S. households.

1.9 A sample survey.

1.11 **(a)** Doctors (rather than researchers) chose the treatment for each patient. **(b)** Each patient should be randomly assigned to a treatment.

1.13 **(a)** No; subjects are assigned to groups based on fitness level. **(b)** Population: e.g., "business executives," or "men in leadership positions," or "adults." Variables: fitness level and leadership ability.

1.15 **(a)** Sample survey. **(b)** Experiment. **(c)** Observational study.

Chapter 2

2.1 Voluntary response samples might not represent the population well.

2.3 72% is too high.

2.5 Possible answers: **(a)** A call-in poll. **(b)** Interviewing students as they enter the student center.

2.7 Number from 01 to 28 and choose 04, 10, 17, 19, 12, 13.

2.9 **(a)** Use labels 001 to 440. **(b)** 381, 262, 183, 322, 341, 185, 414, 273, 190, 325, 330, 029, 079, 078, 118, 209, 354, 239, 421, 426, 435, 437, 193, 099, 224.

2.11 **(a)** Label from 0001 to 3478. **(b)** 2940, 0769, 1481, 2975, 1315.

2.13 **(a)** False. **(b)** True. **(c)** False.

2.15 "Phone-answerers" and "non–phone-answerers" might have different characteristics; both groups should be represented.

2.17 **(a)** Random selection seems reasonable. **(b)** Random selection is not used here. **(c)** Random selection seems best.

Chapter 3

3.1 Statistic.

3.3 Statistic, parameter.

3.5 **(a)** Choose 19, 22, 39, 50, 34; $x = 0.6$. **(b)** 0.4, 0.6, 0.4, 0, 0.6, 0.2, 0.8, 0.8, 0.6. **(d)** Four samples have $\hat{p} = 0.6$.

3.7 **(a)** Population: Ontario residents; sample: the 61,239 people interviewed. **(b)** Such a large sample should represent both men and women fairly well.

3.9 57% would not be surprising, but 37% would be.

3.11 **(a)** High bias, high variability. **(b)** Low bias, low variability. **(c)** Low bias, high variability. **(d)** High bias, low variability.

3.13 The margin of error is half as big with the larger sample.

3.15 **(a)** 44% to 50%. **(b)** Sample results can differ from population values. **(c)** The procedure used gives correct results 95% of the time.

3.17 Smaller sample size means larger margin of error.

3.19 **(a)** About 706 (or between 702 and 711). **(b)** The results are based on a method that is usually within 3 percentage points of the true value.

3.21 0.031.

3.23 **(a)** 0.030 (about 3%). **(b)** We are 95% confident that between 57% and 63% of all adults believe that there is a hell.

3.25 Quadruple the sample size, to $n = 4036$.

3.27 Larger.

3.29 The histogram does appear to be centered at 20%; the samples vary from 1 (4%) to 9 (36%) unemployed students. Of the 50 samples, 32 (64%) had 4, 5, or 6 unemployed students.

Chapter 4

4.1 E.g., undercoverage or nonresponse.

4.3 **(a)** Sampling error. **(b)** Nonsampling error. **(c)** Sampling error.

4.5 This is (random) sampling error, which is accounted for in the margin of error.

4.7 "Fight the charges" and "continue to serve" might evoke different emotional responses.

4.9 The specific programs listed in the second wording pull respondents away from a tax cut.

4.13 Survey carried out by *New York Times*/CBS News. The sample was randomly chosen from the population of adult U.S. residents, and included 1162 people, contacted by phone between Nov. 4 and Nov. 7 (1999). Neither the response rate nor the exact questions asked are given.

4.15 **(a)** The final result is most accurate, since anonymity should promote honesty.

4.17 Each student has a 1/10 chance of being interviewed, but each sample must contain exactly three over-21 students and two under-21 students.

4.19 **(a)** Assign labels 0001, . . . , 2000 for men and 001, . . . , 500 for women, then use the table twice. The first five women: 138, 159, 052, 087, 359. The first five men: 1369, 0815, 0727, 1025, 1868. **(b)** Men: 0.1; women: 0.4.

4.21 The margin of error depends on the sample size, not the population size.

4.23 **(a)** The sample was picked by a procedure involving several steps, perhaps similar to the Current Population Survey sample. **(b)** The strata are the 17 countries. **(c)** At least one of the stages in choosing the sample involved random selection.

4.25 Choose houses 35, 75, 115, 155, and 195.

4.27 **(a)** Assign labels 0001 through 3500 and use random digits. **(b)** Randomly select one of the first 14 students, then take the students who are 14, 28, 42, . . . places down the list. **(c)** Choose SRSs of size 200 and 50 from each group.

4.29 Mall interviews do not choose randomly from any (useful) population; they only represent people who shop at a mall.

Chapter 5

5.1 **(a)** Explanatory: treatment method; response: survival times. **(b)** Women (or their doctors) chose which treatment to use. **(c)** Doctors may recommend a treatment based in part on how advanced the case is.

5.3 The mothers' prior attitudes toward their babies—which probably influenced their choice of feeding method—is the lurking variable; the prior attitude is confounded with the effect of the nursing.

5.5 **(a)** Subjects: physicians; explanatory variable: medication (aspirin or placebo); response variable: health (heart attacks or not).

5.7 We can never know how much of the change in attitudes was due to the explanatory variable (reading propaganda) and how much to the historical events of that time.

5.9 Students registering for a course should be randomly assigned to a classroom or online version of the course. Scores on a standardized test can then be compared.

5.11 If this year is somehow different from last year (e.g., warmer), we cannot compare electricity consumption. Such differences would confound the effects of the treatments.

5.13 **(a)** Pairs of pieces of package liner. **(b)** Explanatory variable: jaw temperature (250°F, 275°F, 300° F, 325° F). **(c)** Response variable: peeling force.

5.15 Choose two digits at a time; Group 1 is 16, 04, 19, 07, 10; Group 2 is 13, 15, 05, 09, 08; Group 3 is 18, 03, 01, 06, 11. The others are in Group 4.

5.17 The difference in blood pressures was so great that it was unlikely to have occurred by chance (if calcium is not effective).

5.19 **(a)** Measure blood pressure for all subjects, then randomly select half to get a calcium supplement, with the other half getting a placebo. After some period of time, observe the change in blood pressure. **(b)** Assign labels 01 to 40 and choose 18, 20, 26, 35, 39, 16, 04, 21, 19, 37, 29, 07, 34, 22, 10, 25, 13, 38, 15, 05.

5.21 The effects of factors other than the nonphysical treatment have been eliminated or accounted for, so the differences in improvement observed between the subjects can be attributed to the differences in treatments.

5.23 The nature and seriousness of the illness and the patient's overall physical condition may influence the choice of anesthetic and also influence the death rate. We should consider the type of surgery and the age, sex, and condition of the patient.

Chapter 6

6.1 Some subjects were given placebos, which allow those doing the study to observe the improvement or benefits arising from participating in the experiment. "Double-blind" means that neither subjects nor those who work with them know who is getting which treatment; this method prevents the researchers' expectations from affecting the way in which the subjects' conditions are diagnosed.

6.3 The patients did not know which treatment they were getting, but those treating the patients knew. This is necessary for the safety of the patients.

6.5 E.g., do rat tumors arising from high exposures for a relatively short period indicate that humans would get tumors from low doses for a longer period?

6.7 **(a)** The placebo effect. **(b)** Use a three-treatment completely randomized design. **(c)** No. **(d)** Since the patient assesses the effectiveness of the treatment, it does not need to be.

6.9 Placebos can provide genuine pain relief.

6.11 **(a)** Individuals: chicks; response variable: weight gain. **(b)** Two explanatory variables (corn type and % protein); nine treatments. The diagram should be a 3×3 table. 90 chicks are needed. **(c)** *Note*: This diagram is quite large.

6.13 For each person, flip a coin to decide which hand to use first. Record the difference in hand strength for each person.

6.15 **(a)** Make a diagram similar to Figure 5.2. **(b)** Have each subject do the task twice, once under each temperature condition, randomly choosing which temperature comes first. Compute the difference in each subject's performances at the two temperatures.

6.17 **(a)** Ordered by increasing weight, the five blocks are Williams, Deng, Hernandez, Moses; Santiago, Kendall, Mann, Smith; Brunk, Obrach, Rodriguez, Loren; Jackson, Stall, Brown, Cruz; Birnbaum, Tran, Nevesky, Wilansky. **(b)** The simplest method is to number from 1 to 4 within each block, then assign each member of block 1 to a different weight-loss treatment, then assign block 2, and so on.

6.19 E.g., all letters have typed addresses and standard envelope size and are all mailed at varying times on (say) Tuesday at the same post office to eliminate some variation. Days in transit is the response variable. Mail some letters with ZIP codes and some without. Possible lurking variables: destination; day of the week the letter is mailed, and so on.

Chapter 7

7.1 Opinions may vary: **(a)** seems to qualify as minimal risk, while **(c)** seems to be excessive.

7.9 This offers anonymity, since names are never revealed.

7.17 **(a)** The subjects should be told what kinds of questions the survey will ask and about how much time it will take. **(b)** For example, respondents may wish to contact the organization if they feel they have been treated unfairly by the interviewer. **(c)** Respondents should not know the poll's sponsor (responses might be affected), but sponsors should be revealed when results are published.

Chapter 8

8.1 A count does not allow for the growing size of the labor force.

8.3 Compare death rates (deaths in buses and in cars divided by the number of passengers in each).

8.5 Texas (8.48 executions per million) is highest; Florida (3.01) is the lowest of these seven.

8.9 **(a)** Cancer deaths increase as the population becomes older. **(b)** Cancer death rates rise as the health of the population improves and fewer people die of other causes. Also, perhaps the environment is becoming more hazardous. **(c)** Survival times could increase if the disease is being discovered earlier; that is, if diagnosis (not treatment) becomes more effective.

8.13 **(a)** The average should be close to 0 if length estimates are unbiased (neither too high nor too low). This is a reasonable expectation with many pieces of string. **(b)** Perfect reliability means identical guesses for repeated measurements on the same piece of string.

8.15 **(a)** Bias. **(b)** E.g., randomly assign workers to a training program, rather than letting them choose whether or not to participate.

8.17 Local agencies may deliberately under-report crime to the FBI, or may simply be careless in keeping records. Survey respondents may remember dates inaccurately, lie, or not understand what constitutes a crime.

8.19 This method omits the "fraction of a beat" between the last beat and the end of the fixed time period. It will always result in an answer that is a multiple of 10, and the measurement could easily vary by 10 even if pulse rate is constant.

8.21 We need to know class size.

8.23 Compare service call rates: 22% for Brand A, and 40% for Brand B.

Chapter 9

9.1 Friday through Sunday is 42.8% of the week.

9.3 Both Anacin and Bufferin are brands of aspirin. Bufferin also contains ingredients to prevent upset stomach, and is likely specified for that reason.

9.5 E.g., over 20 years, 150,000 suicides means an average of 20 suicides per day, which would be hard to ignore.

9.7 57% of 20 equals 11.4 studies, and 42% would be 8.4 studies.

9.9 **(a)** It is 2.1 million beatings, not 21 million. **(b)** The survey counts cases, not incidents: a woman in an abusive relationship (one case) likely is beaten more than once a year (several incidents).

9.11 3.41%.

9.13 This would mean no bags were lost.

9.15 Percents of distinct groups cannot be added.

9.17 1300 separate coast-to-coast highways, which is implausibly high.

9.19 More than twice as many unmarried men as women is far too high.

9.21 From 1989 to 1997, a 0.64% increase. From 1993 to 1997, the increase is 8.57%.

9.23 Deaths are almost certain to be reported, so the listed number of deaths is quite accurate; the increase in injuries from 1970 to 1990 is likely due to increased reporting, not to an increase in actual injuries.

Part I Review

I.3 It is a convenience sample.

I.5 20, 11, 07, 24, 17.

I.7 Sampling error; no.

I.9 Only the error in I.8 would be reduced.

I.11 **(a)** Adults who have used the Internet. **(b)** Margin of error: about 4.1%.

I.13 Each student has a 1/10 chance of being interviewed, but each sample must contain exactly three over-21 students, and two under-21 students. This is a stratified random sample.

I.15 Randomly select 10 subjects to try each method and compare success rates.

I.21 Reliability means similar results in repeated measurements. Reliability can be improved by taking many measurements and averaging them.

I.23 **(a)** This is an observational study, because no treatments were assigned. **(b)** Unlikely to happen by chance. **(c)** For example, some nondrinkers might avoid drinking because of other health concerns.

I.25 **(a)** 14.8% drop.

I.27 Implausible: you are much more likely to hear of a woman over 40 getting married than killed by a terrorist. Information from *Statistical Abstract* could be used.

Chapter 10

10.1 A pie chart could be used.

10.3 **(a)** 43,148,000. **(c)** A pie chart could be used.

10.5 This is a pictogram and therefore is misleadingly scaled.

10.7 **(a)** The price fluctuates regularly, rising and falling over the course of each year, as the supply of oranges varies. **(b)** Overall, prices rise gradually over the decade.

10.9 **(a)** This is a pictogram and therefore is misleadingly scaled.

10.11 Adjust the proportions and the maximum and minimum values on the vertical axis.

10.13 **(a)** A downward trend. **(b)** An upward trend. **(c)** No clear trend.

10.15 This is likely just seasonal variation (holiday sales).

10.17 The cycle is about 11 years. There might be a trend, in that the peaks appear to be higher in the middle and end of the century.

10.19 This is a correct display.

10.21 Use a line graph.

10.23 **(a)** Use a line graph. **(b)** Without exception, average prices increased. **(c)** Fastest: 1978–80; slowest: either 1972–74 or 1986–87.

10.25 **(a)** 46%, 13%, 11%, 4%, 4%, and 22% (20,340 deaths from other causes). **(b)** Use a bar graph or pie chart.

10.27 **(b)** Winning times initially declined (with some fluctuation), but have remained relatively constant since the mid-1980s.

Chapter 11

11.1 Roughly symmetric, centered at or about noon, spread from 6:30 A.M. to 5:30 P.M.; no outliers.

11.3 Strongly right-skewed, trailing off rapidly from the peak at 0 through 4, spread from 0 to 55, with few universities awarding more than 20 doctorates to minorities.

11.5 A stemplot would have more information (too many digits) than could easily be absorbed.

11.7 Use a histogram or a stemplot. The distribution is roughly symmetric, perhaps slightly left-skewed, spread from 12.7% to 22.9%, centered around 17% or 18%.

11.9 **(a)** Strongly right-skewed (many short words, a few quite long words). **(b)** Shakespeare uses more short words (especially 3 and 4 letters) and fewer very long words.

11.11 It will likely be roughly symmetric, but perhaps slightly right-skewed.

11.13 The distribution is irregular in shape. There are two distinct groups, plus a low outlier (the veal brand).

11.15 Roughly symmetric, centered at 46 (a "typical" year); 60 is not an outlier.

11.17 Use a histogram or a stemplot. Distribution is right-skewed, with peak between 10 and 12.

Chapter 12

12.1 Half the households make more, half make less.

12.3 **(a)** The mean is greater. **(b)** Incomes might have been exaggerated.

12.5 **(a)** Distribution is roughly symmetric. **(b)** $\bar{x}$ =13.818% and M = 13.9%.

12.7 Distribution is right-skewed, so the mean ($3.04 million) is greater than the median ($2.57 million).

12.9 Five-number summary: 107, 139, 153, 179, 195. The distribution is irregular; there are two groups with a gap between them, plus a low outlier.

12.11 **(a)** 1, 29, 58, 87, 115. **(b)** Five-number summary: 0 to 4, 0 to 4, 5 to 9, 10 to 14, 50 to 54. The top quarter is about 10 or more.

12.13 The distribution is strongly right-skewed with at least two high outliers; the five-number summary is 0.6, 1.6, 5.15, 17.3, 123.7.

12.15 There are two peaks, so choosing a single "center" is not appropriate.

12.17 **(a)** Mean. **(b)** Median.

12.19 The distributions are quite similar. SATM scores are generally slightly higher.

12.21 Poultry hot dogs are lower in calories than are meat and beef hot dogs.

12.23 Both have mean 3; **(a)** is more spread out and has s = 2.19, while **(b)** has s = 1.41.

12.25 Cars: $\bar{x}$ = 26.6 and s = 2.70 mpg. SUVs: $\bar{x}$ = 19.5 and s = 2.47 mpg.

12.27 The means and standard deviations are both 7.50 and 2.03, respectively. Set A is left-skewed, while B has a high outlier.

12.29 Both measures of spread will rise.

Chapter 13

13.3 **(a)** The curve is a square. **(b)** 0.5. **(c)** 40%.

13.5 84%.

13.7 **(a)** 234 to 298 days. **(b)** Shorter than 234 days.

13.9 **(a)** 327 to 345 days. **(b)** 16%.

13.11 **(a)** Sarah: $z = 1$; her mother: $z = 1.2$. **(b)** Sarah's mother; Sarah.

13.13 **(a)** 2.5%. **(b)** 65 to 75 inches. **(c)** 16%.

13.15 No; it has two peaks because of the two distinct subgroups (men and women).

13.17 About 2.5%.

13.19 About 8%.

13.21 About 38%.

13.23 About the 76th percentile.

13.25 About $\pm$ 0.7.

13.27 About 127 or more.

Chapter 14

14.1 **(a)** –1 to 1. **(b)** Any positive number.

14.3 **(a)** About 103 and 0.5. **(b)** A, B: low GPA, moderate IQ. C: low IQ, moderate GPA.

14.5 Clearly positive but not near 1.

14.7 Figure 14.9; it shows a stronger linear relationship.

14.9 **(a)** Time is explanatory. **(b)** Negative; lower time requires greater exertion. **(c)** Linear; moderate.

14.11 $r = 1$.

14.13 **(a)** $r = -0.746$. **(b)** It would not change.

14.15 The relationship is not linear.

14.17 **(a)** Years. **(b)** Seconds. **(c)** No units. **(d)** Years.

14.19 **(a)** Gender is not quantitative. **(b)** r cannot exceed 1. **(c)** r has no units.

14.21 **(a)** Negative. **(b)** Negative. **(c)** Positive. **(d)** Small.

14.23 The relationship, if any, is weakly positive. The points for the Mets and the Cardinals might be considered outliers.

14.25 **(a)** Planting rate. **(c)** Curved; neither positive nor negative. Yield falls off if the plants are too crowded.

Chapter 15

15.1 Inactive girls are more likely to be obese. 3.2%.

15.3 40.2%; 59.8%.

15.5 GPA goes up 0.101 points for every one-point increase in IQ; 8.055.

15.7 **(b)** Moderately strong negative association, linear or slightly curved.

15.9 1 liter: 237.63; 8 liters: 76.84 deaths per 100,000.

15.11 E.g., diets and genetic (ethnic) background vary between countries.

15.13 **(a)** Negative; pH decreases (and acidity increases) over time. **(b)** Beginning: 5.425; end: 4.635. **(c)** $b = -0.0053$; pH decreased by 0.0053 per week on the average.

15.15 **(a)** Weight $y = 100 + 40x$ g; slope 40 g/week. **(c)** 4260 g is about 9.4 pounds.

15.17 **(a)** When $x = 5$, $y = 5.5$; when $x = 10$, $y = 8$. **(b)** Only Set A.

15.19 **(b)** The population cannot continue to decline at the same rate as the plot shows.

15.21 Number of firefighters and amount of damage are common responses to the seriousness of the fire.

15.23 Seriousness of the illness affects both choice of hospital and length of stay.

15.25 Stronger students are more likely to choose such math courses; weaker students may avoid them.

15.27 Lack of parental supervision or parental interest in school performance.

15.29 Students with music experience may have other advantages (wealthier parents, better school systems, etc.).

15.31 **(b)** Time $y = 43.10–0.0574x$, so the predicted time is 34.38 minutes. **(c)** The results depend on which variable is viewed as explanatory.

Chapter 16

16.1 89.2, 100, 81.3.

16.3 **(a)** A 7.9 point (8.9%) decrease. **(b)** An 18.7 point (18.7%) decrease.

16.5 **(a)** The New York CPI is greater than the Los Angeles CPI. **(b)** We do not know how prices compared in the base period.

16.7 The costs are \$66.45 (1985) and \$94.55; the GPI is 142.3.

16.9 About \$38,300.

16.11 Assuming the CPI continues to increase, the answer should be less than \$16.

16.13 \$32,000 in 1940 is worth about \$55,100 in 1950; his real income increased by about 80%.

16.15 $5900 in 1976 is about $17,300 in 1999; the cost of Harvard increased faster.

16.17 The adjusted Purdue figures are $1158, $1307, $1376, $1453, $1490, $1551, $1696, $1823, $1889, $1977. The real cost of Purdue tuition has increased about 71% from 1981 to 1999.

16.19 A $3852 increase is 11% of $35,033; a $30,324 increase is 30% of $101,875.

16.21 The weight reflects the difference in cost and the proportions of people buying versus renting.

16.23 Seasonal variations might affect those who receive those payments.

16.25 The poverty level is going up, even after adjusting for inflation.

16.27 The more we sample, the more we know.

Part II Review

II.1 Somewhat right-skewed, with no outliers.

II.3 8.2%, 9.3%, 11.3%, 14.3%, 16.7%.

II.5 11.98%. Without Mississippi: 11.79%.

II.7 **(a)** 500. **(b)** 68%.

II.9 **(a)** Centimeters. **(b)** Centimeters. **(c)** Centimeters. **(d)** No units.

II.11 Dolphin: 180 kg, 1600 g. Hippo: 1400 kg, 600 g.

II.13 **(a)** Decrease. **(b)** Decrease.

II.15 About 800 g.

II.17 Close to –1.

II.19 –56.1 grams; prediction outside the range of the available data is risky.

II.21 About $43,990.

II.23 Not a good investment: in 1983 dollars, an ounce of gold was worth only $176 in 1999.

II.25 **(a)** Roughly symmetric; the two highest and one lowest time might be considered outliers.

II.27 The mean is higher.

II.29 **(a)** Fidelity Technology Fund. **(b)** No.

Chapter 17

17.1 Long trials of this experiment suggest about 40% heads.

17.3 The proportion from Table A is 0.105.

17.5 Obviously, results will vary with the type of thumbtack used.

17.7 **(a)** 0. **(b)** 1. **(c)** 0.01. **(d)** 0.6.

17.9 **(b)** A personal probability might take into account specific information about your driving habits. **(c)** Most people believe that they are better-than-average drivers.

17.15 If two people talk at length, they will eventually discover *something* in common.

17.17 0, 0.47, 0.497, 0.4997.

17.19 The "law of averages" is no more reliable for forecasting the weather (in the short run) than it is for other predictions.

17.23 51, 510, 5100, and 51,000 heads; 1, 10, 100, and 1000 heads away from half the number of tosses.

Chapter 18

18.1 0.54.

18.3 **(a)** 0.65. **(b)** 0.38. **(c)** 0.62.

18.5 In Models 1, 3, and 4, the probabilities do not sum to 1; Model 4 has probabilities greater than 1. Model 2 is legitimate.

18.7 Each possible value (1, 2, 3, 4) has probability 1/4.

18.9 Possible totals: 2 through 8; probabilities 1/16, 2/16, 3/16, 4/16, 3/16, 2/16, 1/16.

18.11 **(a)** 0.1. **(b)** 0.3. **(c)** 0.5; 0.4.

18.13 **(a)** 0.438 to 0.502. **(b)** 16%.

18.15 0.62%.

18.17 **(a)** About 50%. **(b)** About 68%. **(c)** About 32%.

18.19 **(a)** 69.4. **(b)** Answers will vary. **(c)** Answers will vary.

Chapter 19

19.1 **(a)** 0–4 for Democrats, 5–9 for Republicans. **(b)** 0–5 for Democrats, 6–9 for Republicans. **(c)** 0–3 for Democrats, 4–7 for Republicans, 8–9 undecided. **(d)** 00–52 for Democrats, 53–99 for Republicans.

19.3 **(a)** 4 chose Democrats, 6 chose Republicans. **(b)** 3 Democrats, 7 Republicans. **(c)** 2 Democrats, 4 Republicans, 4 undecided. **(d)** 6 Democrats, 4 Republicans.

19.5 **(a)** 0.2. **(b)** 0 or 1 is A, 2–4 is B, 5–7 is C, 8 or 9 is D/F.

19.7 Results vary with starting point in Table A.

19.9 **(a)** With 0–4 meaning a made free throw, he hits 8 or more three times (if 0–4 is a miss, seven times). **(b)** Six made in a row; eight missed (or vice versa).

19.11 **(a)** 0 or 1 is a pass, 2–9 a failure. **(b)** 0.5. **(c)** No: probability of passing likely increases on each trial.

19.13 **(a)** 0 is narrow flat side, 1–4 is broad concave side, 5–8 is broad convex side, and 9 is narrow hollow side. **(b)** Results vary with the starting line from Table A.

19.15 **(a)** 0–8 means System A works. **(b)** 0–7 means System B works. **(c)** Results vary with the starting line from Table A.

19.17 00–24 means a passenger does not show up. Results vary with the starting line from Table A.

19.19 0–5 means the van is available.

19.21 **(a)** BBB, BBG, BGB, GBB, GGB, GBG, BGG, GGG. The first has probability 0.132651; the next three 0.127449; the next three 0.122451; and the last 0.117649. (In practice, these should be rounded to 2 or 3 decimal places.) **(b)** 0.867349.

Chapter 20

20.1 $0.60.

20.3 About $0.9474.

20.5 About $0.9474 (the same as in 20.3).

20.7 **(a)** The first sequence. **(b)** The second sequence looks "more random."

20.9 **(a)** $0.75. **(b)** $0.72.

20.11 **(a)** 1.3 female offspring. **(b)** Every generation has (on the average) 1.3 times as many females as the previous generation.

20.13 Use 0–1 for 0 female offspring, 2–4 for 1 female, and 5–9 for 2. Results vary with starting point in Table A.

20.15 2.55 people.

20.17 00–48 is a girl, 49–99 is a boy. Results vary with starting point in Table A.

20.19 0 or 1 is a correct guess, and 2–9 is an incorrect guess. Results vary with starting point in Table A.

20.21 $np = 0.5$; simulation gives 0.502.

Part III Review

III.1 **(b)** If all are equally likely, this would be 30%.

III.3 **(a)** 0.1. **(b)** Use 0–3 for type O, 4–6 for A, 7 and 8 for B, and 9 for AB.

III.5 Results vary with the starting line in Table A.

III.7 3.5.

III.9 **(a)** 0.03. **(b)** Use 00–91 for one pair or worse and 92–99 for two pairs or better.

III.11 **(a)** All probabilities are between 0 and 1 and add to 1. **(b)** 0.41. **(c)** 0.38.

III.13 **(a)** 0.27. **(b)** 0.57.

III.15 **(a)** 68%. **(b)** 95%.

III.17 0.70%

III.19 Correct; each face value has probability 1/13.

III.21 Michigan, Ohio State, and Purdue: 0.1556 (rounded); Illinois, Indiana, and Wisconsin: 0.0778 (also rounded).

Chapter 21

21.1 **(a)** The 175 residents of Tonya's dorm. **(b)** The proportion who like the food. **(c)** $\hat{p} = 0.28$.

21.3 The method that produced this interval gives correct results 95% of the time.

21.5 **(a)** Proportion of all women who say they do not get enough time for themselves. **(b)** 0.439 to 0.501.

21.7 **(a)** 0.564 to 0.636. **(b)** 0.575 to 0.625. **(c)** 0.582 to 0.618. **(d)** The interval gets narrower as sample size increases.

21.9 Results vary with different kinds of thumbtacks.

21.11 0.491 to 0.523; 0.5 is in this interval, so the coin could be fair.

21.13 **(a)** Normal with mean 0.14 and standard deviation 0.0142. **(b)** Above 18.2% is unlikely (0.15%); above 11.2% is likely (97.5%).

21.15 Quick method: 0.0447; new method: 0.307.

21.17 **(a)** Use 0–5 for those who favor Caucus. **(b)** Results vary with the starting line in Table A.

21.19 0.43 to 0.51; this interval is wider.

21.21 0.0665 to 0.1026.

21.23 When $\hat{p} = 0.5$, $\hat{p}(1-\hat{p}) = 0.25$; all other values are smaller.

Chapter 22

22.1 If church attenders were no more ethnocentric than nonattenders, then sample results like the one observed would rarely occur.

22.3 Such a drop has only a 2% chance of happening by accident.

22.5 The differences observed would occur less than 1% of the time if pig skulls had no special meaning.

22.7 **(a)** It is a convenience sample. **(b)** This use of "significant" means "quite a bit." **(c)** Results like those observed would occur less than 1% of the time if people did not truly want more money.

22.9 **(a)** The proportion of all people with body temperature below 98.6°. **(b)** $H_0: p = 0.5$; $H_a: p > 0.5$.

22.11 $H_0: p = 0.21$; $H_a: p \neq 0.21$.

22.13 **(a)** The proportion of all students at this university who want to be well off. **(b)** $H_0: p = 0.73$; $H_a: p \neq 0.73$. **(c)** 0.66; the event "$\hat{p} < 0.66$ or $\hat{p} > 0.79$." **(d)** Values of $\hat{p}$ like 0.66 would occur only 3.7% of the time if H_0 were true.

22.15 Significant at 5% but not 1%.

22.17 No; this statement means that if H_0 is true, we have observed outcomes that occur less than 5% of the time.

22.19 **(a)** $H_0: p = 0.2$; $H_a: p > 0.2$. **(b)** Use 0 or 1 for a correct guess and 2–9 for incorrect guesses. **(c)** 0 correct once, 1 correct 4 times, 2 correct 9 times, 3 correct 5 times, 6 correct 1 time. **(d)** The event "$\hat{p} > 0.5$." We estimate $P = 0.05$.

22.21 Standard score –2.23; not quite significant at $\alpha = 0.025$.

22.23 **(a)** Normal with mean 0.10 and standard deviation 0.0143. **(b)** Standard score –3.34; significant at $\alpha = 0.0005$.

22.25 Standard score –13.6; very good evidence that less than half are speeding.

Chapter 23

23.1 Our confidence interval method can only be applied to an SRS.

23.3 Our confidence interval contains some values of p that give the election to Snort, and the true estimation error could be even larger.

23.5 **(a)** In a sample of size 500, we expect to see about 5 with $P < 0.01$. **(b)** Test these four again.

23.7 Were these random samples? How big were the samples?

23.9 It is essentially correct.

23.11 **(a)** 5%. **(b)** Some will show a significant difference by chance.

23.13 **(a)** Choose 50 to get the vaccine; compare infection rates. **(b)** $H_0: p_1 = p_2$; $H_a: p_1 > p_2$. **(c)** Differences like those observed would occur 25% of the time by chance. **(d)** Yes.

23.15 0.594 to 0.726.

23.17 906 passing: 0.888 to 0.924. 907 passing: 0.889 to 0.925.

Chapter 24

24.1 In each group, the percent with C or better is 55%, 74.7%, 37.5%. Some (but not too much) time spent in extracurricular activities seems to be beneficial.

24.3 **(a)** Hatched: 16, 38, 75; did not hatch: 11, 18, 29. **(b)** Cold: 59.3%; Neutral: 67.9%; Hot: 72.1%. Cold water did not prevent hatching but made it less likely.

24.5 **(a)** 1,693,000; table entries are rounded to nearest thousand. **(b)** 55.1%, 55.9%, 41.6%, 40.0%. Women earn a majority of bachelor's and master's degrees but smaller percentages of professional and doctoral degrees.

24.7 Start by setting a equal to any number from 10 to 50.

24.9 **(a)** White defendant: 19 yes, 141 no. Black defendant: 17 yes, 149 no. **(b)** Overall death penalty: 11.9% of white defendants, 10.2% of black defendants. For white victims, 12.6% and 17.5%; for black victims, 0% and 5.8%. **(c)** The death penalty was more likely when the victim was white (14%) rather than black (5.4%). White defendants killed whites 94.3% of the time, but were less likely to get the death penalty than blacks who killed whites.

24.11 **(a)** Row totals: 82, 37. Column totals: 20, 91, 8. Grand total: 119. Expected counts: 13.8, 62.7, 5.5; 6.2, 28.3, 2.5. Middle column counts differ the most. **(b)** 6.93; the largest contribution is from the lower right, with slightly less from the middle column. **(c)** df = 2; significant at $\alpha = 0.05$ but not 0.01.

24.13 $X^2 = 1.703$, df = 2, not significant.

24.15 **(a)** $X^2 = 14.863$; $P < 0.001$. The differences are significant. **(b)** The data should come from independent SRSs of the (unregulated) child-care providers in each city.

Chapter 25

25.3 **(a)** Fairly normal, but with two low outliers $\bar{x} = 105.84$ and $s = 14.27$. **(b)** 100.8 to 110.9. **(c)** Confidence interval methods assume that the numbers represent an SRS.

25.5 Standard score –2.28; significant at $\alpha = 0.025$.

25.7 **(a)** Normal with mean 115 and standard deviation 6. **(b)** 118.6 is fairly close to the middle of the curve and would not be too surprising if H_0 were true. Meanwhile, 125.7 lies out toward the high tail of the curve and would rarely occur when $\mu = 115$.

25.9 H_0: $\mu = 18$; H_a: $\mu < 18$ seconds.

25.11 Standard score 1.01; not significant.

25.13 0.480 to 0.584.

25.15 *Very* roughly normal.

25.17 **(a)** $\bar{x} = 0.496$, which is close to 0.5. **(b)** We estimate $\sigma = 0.3251$.

Part IV Review

IV.1 0.669 to 0.750.

IV.3 About 66% to 74%.

IV.5 0.208 to 0.272.

IV.7 You have information about all states, not just a sample.

IV.9 **(a)** If there were no difference between the two groups of doctors, results like these would be very rare. **(b)** The two proportions are given for comparison.

IV.11 Mornings: 0.550 to 0.590. Evenings: 0.733 to 0.767. The evening proportion is quite a bit higher.

IV.13 **(a)** H_0: $p = 0.57$; H_a: $p > 0.57$. **(b)** Normal with mean 0.57 and standard deviation 0.01. **(c)** Yes.

IV.15 **(a)** People are either reluctant to admit that they don't attend regularly, or they believe they are more regular than they truly are. **(b)** Sample results vary from population truth. **(c)** Both intervals are based on methods that work 95% of the time.

IV.17 H_0: $p = 1/3$; H_a: $p > 1/3$; standard score 1.82; significant at $\alpha = 0.05$ but not 0.025.

IV.19 0.247 to 0.313.

IV.21 H_0: $p = 0.57$; H_a: $p > 0.57$; standard score 17.99; significant for any reasonable choice of α.

IV.23 Standard score –2.77; significant at $\alpha = 0.005$.

IV.25 **(b)** $\bar{x} = 15.59$ ft and $s = 2.550$ ft; 15.0 to 16.2 ft. **(c)** What population are we examining: Full-grown sharks? Male sharks?

IV.27 Standard score 1.53; significant at $\alpha = 0.10$ but not 0.05.

IV.29 **(a)** 2.96%, 13.07%, 6.36%. **(b)** The rows are 721, 173, 412; 22, 26, 28. **(c)** Expected counts: 702.14, 188.06, 415.80; 40.86, 10.94, 24.20. All expected counts are greater than 5. **(d)** H_0: there is no relationship between a member's complaining and leaving the HMO; H_a: there is some relationship. df = 2; this is very significant.

IV.31 **(a)** Rows: 1313, 1840; 991, 614. Morning: 57.0%; evening: 75.0%. **(b)** We have very large samples with very different proportions. **(c)** $X^2 = 172$ with df = 1; very significant.

Table A Random digits

Line								
101	19223	95034	05756	28713	96409	12531	42544	82853
102	73676	47150	99400	01927	27754	42648	82425	36290
103	45467	71709	77558	00095	32863	29485	82226	90056
104	52711	38889	93074	60227	40011	85848	48767	52573
105	95592	94007	69971	91481	60779	53791	17297	59335
106	68417	35013	15529	72765	85089	57067	50211	47487
107	82739	57890	20807	47511	81676	55300	94383	14893
108	60940	72024	17868	24943	61790	90656	87964	18883
109	36009	19365	15412	39638	85453	46816	83485	41979
110	38448	48789	18338	24697	39364	42006	76688	08708
111	81486	69487	60513	09297	00412	71238	27649	39950
112	59636	88804	04634	71197	19352	73089	84898	45785
113	62568	70206	40325	03699	71080	22553	11486	11776
114	45149	32992	75730	66280	03819	56202	02938	70915
115	61041	77684	94322	24709	73698	14526	31893	32592
116	14459	26056	31424	80371	65103	62253	50490	61181
117	38167	98532	62183	70632	23417	26185	41448	75532
118	73190	32533	04470	29669	84407	90785	65956	86382
119	95857	07118	87664	92099	58806	66979	98624	84826
120	35476	55972	39421	65850	04266	35435	43742	11937
121	71487	09984	29077	14863	61683	47052	62224	51025
122	13873	81598	95052	90908	73592	75186	87136	95761
123	54580	81507	27102	56027	55892	33063	41842	81868
124	71035	09001	43367	49497	72719	96758	27611	91596
125	96746	12149	37823	71868	18442	35119	62103	39244
126	96927	19931	36809	74192	77567	88741	48409	41903
127	43909	99477	25330	64359	40085	16925	85117	36071
128	15689	14227	06565	14374	13352	49367	81982	87209
129	36759	58984	68288	22913	18638	54303	00795	08727
130	69051	64817	87174	09517	84534	06489	87201	97245
131	05007	16632	81194	14873	04197	85576	45195	96565
132	68732	55259	84292	08796	43165	93739	31685	97150
133	45740	41807	65561	33302	07051	93623	18132	09547

Table A Random digits (continued)

Line								
134	27816	78416	18329	21337	35213	37741	04312	68508
135	66925	55658	39100	78458	11206	19876	87151	31260
136	08421	44753	77377	28744	75592	08563	79140	92454
137	53645	66812	61421	47836	12609	15373	98481	14592
138	66831	68908	40772	21558	47781	33586	79177	06928
139	55588	99404	70708	41098	43563	56934	48394	51719
140	12975	13258	13048	45144	72321	81940	00360	02428
141	96767	35964	23822	96012	94591	65194	50842	53372
142	72829	50232	97892	63408	77919	44575	24870	04178
143	88565	42628	17797	49376	61762	16953	88604	12724
144	62964	88145	83083	69453	46109	59505	69680	00900
145	19687	12633	57857	95806	09931	02150	43163	58636
146	37609	59057	66967	83401	60705	02384	90597	93600
147	54973	86278	88737	74351	47500	84552	19909	67181
148	00694	05977	19664	65441	20903	62371	22725	53340
149	71546	05233	53946	68743	72460	27601	45403	88692
150	07511	88915	41267	16853	84569	79367	32337	03316

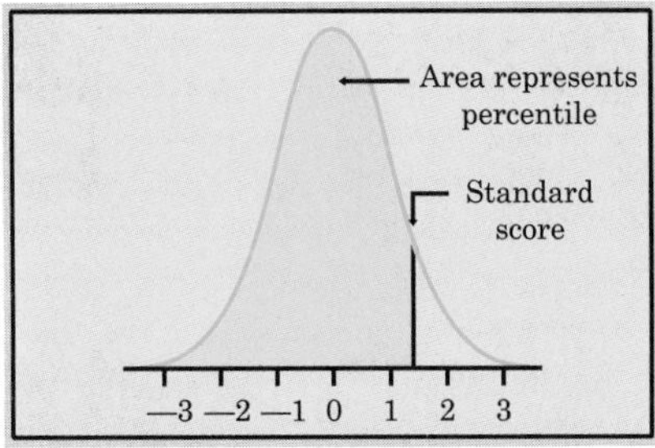

Table B Percentiles of the normal distributions

Standard score	Percentile	Standard score	Percentile	Standard score	Percentile
–3.4	0.03	–1.1	13.57	1.2	88.49
–3.3	0.05	–1.0	15.87	1.3	90.32
–3.2	0.07	–0.9	18.41	1.4	91.92
–3.1	0.10	–0.8	21.19	1.5	93.32
–3.0	0.13	–0.7	24.20	1.6	94.52
–2.9	0.19	–0.6	27.42	1.7	95.54
–2.8	0.26	–0.5	30.85	1.8	96.41
–2.7	0.35	–0.4	34.46	1.9	97.13
–2.6	0.47	–0.3	38.21	2.0	97.73
–2.5	0.62	–0.2	42.07	2.1	98.21
–2.4	0.82	–0.1	46.02	2.2	98.61
–2.3	1.07	0.0	50.00	2.3	98.93
–2.2	1.39	0.1	53.98	2.4	99.18
–2.1	1.79	0.2	57.93	2.5	99.38
–2.0	2.27	0.3	61.79	2.6	99.53
–1.9	2.87	0.4	65.54	2.7	99.65
–1.8	3.59	0.5	69.15	2.8	99.74
–1.7	4.46	0.6	72.58	2.9	99.81
–1.6	5.48	0.7	75.80	3.0	99.87
–1.5	6.68	0.8	78.81	3.1	99.90
–1.4	8.08	0.9	81.59	3.2	99.93
–1.3	9.68	1.0	84.13	3.3	99.95
–1.2	11.51	1.1	86.43	3.4	99.97

Index of Tables

Subject Index